Mathematical Programming and Game Theory for Decision Making

Statistical Science and Interdisciplinary Research

Series Editor: Sankar K. Pal *(Indian Statistical Institute)*

Description:
In conjunction with the Platinum Jubilee celebrations of the Indian Statistical Institute, a series of books will be produced to cover various topics, such as Statistics and Mathematics, Computer Science, Machine Intelligence, Econometrics, other Physical Sciences, and Social and Natural Sciences. This series of edited volumes in the mentioned disciplines culminate mostly out of significant events — conferences, workshops and lectures — held at the ten branches and centers of ISI to commemorate the long history of the institute.

Vol. 1 Mathematical Programming and Game Theory for Decision Making
edited by S. K. Neogy, R. B. Bapat, A. K. Das & T. Parthasarathy (Indian Statistical Institute, India)

Platinum Jubilee Series

Statistical Science and
Interdisciplinary Research — Vol. 1

Mathematical Programming and Game Theory for Decision Making

Editors

S. K. Neogy

R. B. Bapat

A. K. Das

T. Parthasarathy

Indian Statistical Institute, India

Series Editor: ***Sankar K. Pal***

World Scientific

NEW JERSEY • LONDON • SINGAPORE • BEIJING • SHANGHAI • HONG KONG • TAIPEI • CHENNAI

Published by

World Scientific Publishing Co. Pte. Ltd.
5 Toh Tuck Link, Singapore 596224
USA office: 27 Warren Street, Suite 401-402, Hackensack, NJ 07601
UK office: 57 Shelton Street, Covent Garden, London WC2H 9HE

British Library Cataloguing-in-Publication Data
A catalogue record for this book is available from the British Library.

MATHEMATICAL PROGRAMMING AND GAME THEORY FOR DECISION MAKING
Statistical Science and Interdisciplinary Research — Vol. 1

ISBN-13 978-981-281-321-3
ISBN-10 981-281-321-7

Printed in Singapore by World Scientific Printers

Foreword

The Indian Statistical Institute (ISI) was established on 17th December, 1931 by a great visionary Professor Prasanta Chandra Mahalanobis to promote research in the theory and applications of statistics as a new scientific discipline in India. In 1959, Pandit Jawaharlal Nehru, the then Prime Minister of India introduced the ISI Act in the parliament and designated it as an *Institution of National Importance* because of its remarkable achievements in statistical work as well as its contribution to economic planning.

Today, the Indian Statistical Institute occupies a prestigious position in the academic firmament. It has been a haven for bright and talented academics working in a number of disciplines. Its research faculty has made India proud in the arenas of Statistics, Mathematics, Economics, Computer Science, among others. Over seventy five years, it has grown into a massive banyan tree, like the institute emblem. The Institute now serves the nation as a unified and monolithic organization from different places, namely Kolkata, the Head Quarter, Delhi and Bangalore, two centers, a network of six SQC-OR Units located at Mumbai, Pune, Baroda, Hyderabad, Chennai and Coimbatore, and a branch (field station) at Giridih.

The platinum jubilee celebrations of ISI have been launched by Honorable Prime Minister Dr. Manmohan Singh on December 24, 2006, and the Government of India has declared 29th June as the "Statistics Day" to commemorate the birthday of Professor Mahalanobis nationally.

Professor Mahalanobis was a great believer in interdisciplinary research, because he thought that this will promote the development of not only statistics, but also the other natural and social sciences. To promote interdisciplinary research, major strides were made in the areas of computer science, statistical quality control, economics, biological and social sciences, physical and earth sciences.

The Institute's motto of 'unity in diversity' has been the guiding principle of all its activities since its inception. It highlights the unifying role of statistics in relation to various scientific activities.

In tune with this hallowed tradition, a comprehensive academic programme, involving Nobel Laureates, Fellows of the Royal Society, and other dignitaries has been implemented throughout the Platinum Jubilee year, highlighting the emerging areas of ongoing frontline research in its various scientific divisions, centres, and outlying units. It includes international and national-level seminars, symposia, conferences and workshops, as well as series of special lectures. As an outcome of these events, the Institute is bringing out a series of comprehensive volumes in different subjects under the title *Statistical Science and Interdisciplinary Research*, published by World Scientific.

The present volume titled *Mathematical Programming and Game Theory for Decision Making* is the first one in the series. It has twenty five chapters, written by eminent scientists including a Nobel Laureate, from different parts of the world, dealing with the application of the theory and methods of mathematical programming to problems in statistics, finance, electrical networks and game theory. I believe, the state of the art studies presented in this book will be very useful to readers.

Thanks to the contributors for their excellent research contributions and to volume editors Dr. S. K. Neogy, Prof. R. B. Bapat, Dr. A. K. Das and Prof. T. Parthasarathy for their sincere effort in bringing out the volume nicely in time. The active role of the Platinum Jubilee Core Committee is appreciated. Thanks are also due to World Scientific for their initiative in publishing the series and being a part of the Platinum Jubilee endeavor of the Institute.

December 2007 S. K. Pal
Kolkata Series Editor and Director, ISI

Preface

This volume is dedicated to the presentation and discussion of state of the art studies in Mathematical Programming and Game Theory for decision making problem in the form of twenty five papers. It is a peer reviewed volume under the Platinum Jubilee Volume Series of Indian Statistical Institute. The topics of this volume include the application of the theory and methods of mathematical programming to problems in statistics, finance, electrical networks and game theory. Mathematical programming comprises a variety of paradigms (theoretical frameworks) tailored to different kinds of problems and it is extremely useful to problems in strategic decision making. Support vector machines, bilevel programming, neural network models, cooperative games, non-cooperative games and stochastic games appear in this volume. It is hoped that the research articles of this volume will significantly aid in the dissemination of research efforts in these areas. In this volume some pioneers of the field, as well as some prominent younger researchers have contributed articles which are briefly mentioned below.

Mathematical programming has long been recognized as a vital modelling approach to solve optimization problems. In Chapter 1, Lyn C Thomas presents a review on some of the applications of mathematical programming in finance which includes prominent and well documented applications in long-term financial planning and portfolio problems. This includes asset-liability management for pension plans and insurance companies, integrated risk management for intermediaries, and long-term planning for individuals. In this chapter, it is discussed how one can use linear programming to estimate the term structure of interest rates for the prices of bonds.

Even though several anti-cycling pivot selection rules exist for the simplex method for a general linear program (LP), none is known to avoid stalling (an exponential sequence of degenerate pivots). Santosh N. Kabadi and Abraham P. Punnen discuss an anti-stalling pivot rule for linear programs with totally unimodular coefficient matrix in Chapter 2. For an LP with m constraints and totally unimodular coefficient matrix, pivot selection rule presented in this chapter guarantees that the simplex method performs at most m consecutive degenerate pivots or declares that the current solution is optimal.

Katta G. Murty developed a new interior point method for linear programming, based on a new centering strategy that moves any interior feasible solution x^0 to the center of the intersection of the feasible region with the objective hyperplane through x^0, before beginning the descent moves. Using this centering strategy, that method obtains an optimum solution for an LP by a very efficient descent method that uses no matrix inversions. In Chapter 3, he extended this method into a descent method for solving quadratic programs (QP). Compared to other existing methods for QP, the new method is able to handle it with minimal matrix inversion computations.

Chapter 4 by Richard Caron and Tim Traynor is about the analysis of sets of constraints, with no explicit assumptions. The relationship between the minimal representation problem and a certain set covering problem of Boneh is explored. This provides a framework that shows the connection between minimal representations, irreducible infeasible systems, minimal infeasibility sets, as well as other attributes of the preprocessing of mathematical programs.

Most research on algorithms for combinatorial optimization uses the costs of the elements in the ground set for making decisions about the solutions that the algorithms would output. For traveling salesman problems, this implies that algorithms generally use arc lengths to decide on whether an arc is included in a partial solution or not. In Chapter 5, Diptesh Ghosh, Boris Goldengorin, Gregory Gutin and Gerold Jäger study the effect of using element tolerances for making these decisions and several greedy algorithms for it based on tolerances are proposed for traveling salesman problem.

In Chapter 6, T. S. Arthanari studies the membership problem for the pedigree polytope. In this chapter, it is shown that a necessary condition for membership in the pedigree polytope is the existence of a multicommodity flow with value equal to unity in a layered network.

Many real life scheduling problems involve the use of a graph coloring problem where the vertices of a graph $G(V, E)$ are colored such that the coloured graph satisfies certain desired properties. Nirmala Achuthan, N. R. Achuthan and R. Collinson discuss one such graph coloring problem in Chapter 7. The $k - defective$ *chromatic number* $\chi_k(G)$ of a graph G is the least positive integer m for which G is (m, k) – *colorable.* In this chapter, exact algorithms based on partial enumeration methods to determine the one defective chromatic number $\chi_1(G)$, of a graph G are developed.

The vertical block matrix arises naturally in the literature of stochastic games where the states are represented by the columns and actions in each state are represented by rows in a particular block. S. K. Neogy, A. K. Das and P. Das present some results related to complementarty problem involving vertical block matrices in Chapter 8. A neural network algorithm for solving a vertical linear complementarity problem is also discussed.

In Chapter 9, Reshma Khemchandani, Jayadeva and Suresh Chandra present a fuzzy extension to twin support vector machines for binary data classification. The approach can be used to obtain an improved classification when one has an estimate of the fuzziness of samples in either class.

Except for constrained least squares, seldom is linear regression by least squares presented as an optimization problem whereas regression by minimum sum of absolute errors (MSAE) regression is always framed as a linear optimization problem. However, most students of statistics are unfamiliar with methods of mathematical programming. Given the dearth of treatment to regression by MSAE in textbooks, literature reviews and updates to MSAE regression such as Chapter 10 by Subhash C. Narula and John F. Wellington becomes an important learning resources to the student, researcher, and practitioner.

Consider a stochastic securities market model with a finite state space and a finite number of trading dates. In Chapter 11, Stephen A. Clark and Cidambi Srinivasan discuss how arbitrage price theory is modified by a no short-selling constraint. The principle of No Arbitrage is characterized by the existence of an equivalent supermartingale measure. In this chapter, it is shown that the Law of One Price holds for marketed claims if and only if there exists an equivalent martingale measure. Given that the Law of One Price prevails, then a contingent claim has a unique fundamental value if and only if it is the difference of two marketed claims. The main tool for arbitrage analysis in this essay is finite-dimensional LP duality theory.

In Chapter 12, H. Narayanan discusses about solving min cost flow problems approximately by transforming them to network analysis problems. This Chapter provides a relook at commonly used algorithms in computational linear algebra by associating an electrical network with the linear equations.

In Chapter 13, A. K. Bardhan and Udayan Chanda present optimal control policies of quality level and price for the introduction of a new product with two competing technology generations in a dynamic environment and also proposes a new model in this regard. The proposed model in this chapter is a combination of diffusion models and the cost function, which is capable of estimating the future profit trends.

Katta G. Murty presents a simple and easy method to implement nonparametric technique to forecast the demand distribution based on statistical learning, and ordering policies in Chapter 14. An application of this nonparametric forecasting method to portfolio management is also presented.

Chapter 15 by S. Dempe, J. Dutta and B. S. Mordukhovich is devoted to an application of advanced tools of modern variational analysis and generalized differentiation to problems of optimistic bilevel programming. Some new necessary optimality conditions are derived for two major classes of bilevel programs: those with partially convex and with fully convex lower-level problems.

Chapter 16 contains a summary of the talk by R. J. Aumann, Nobel Laureate, which contains a discussion on Game Engineering.

In Chapter 17, Pradeep Dubey and Rahul Garg consider a communications network in which users transmit beneficial information to each other at a cost. Conditions under which the induced cooperative game is supermodular (convex) is presented. This analysis is in a lattice-theoretic framework, which is at once simple and able to encompass a wide variety of seemingly disparate models.

Magnus Hennlock define a robust feedback Nash equilibrium in Chapter 18 and solve analytically in a differential climate model with N regions based on an approach of IPCC 2001 scientific report for calculating radiative forcing due to anthropogenic CO_2 emissions. In addition, uncertainty is introduced by perturbing the climate change dynamics such that future radiative forcing and global mean temperature will have unknown outcomes and probability distributions. There are n asymmetric investors, each investing in a portfolio containing N regional capital stocks used in production that generates CO_2 emissions. In each region there is one policy maker, acting as a regional social planner, that chooses regionally optimal abatement policies. Dynamic maximin decision criteria are applied for the policy makers in a robust feedback Nash equilibrium for N policy makers' abatement strategies and n investors' investment strategies.

In Chapter 19, Haruo Imai and Katsuhiko Yonezaki consider a multi-person bargaining problem where players interests are correlated. This chapter investigates the limit outcomes of the stationary subgame perfect equilibrium outcomes of the sequential bargaining game with a coalition under two different bargaining protocols and correlation of interests are found within each coalition. Here limit means the case where the interval between the two consecutive offers vanishes. The result shows that an endogenous delegation occurs in each coalition to its toughest member. The outcome exhibits a sharp distinction that under the fixed order rule, the size of coalition does not matter, while under the predetermined proposer rule, it matters.

Chapter 20 by Dawidson Razafimahatolotra investigates stability properties of effectivity functions. The Bargaining Set in effectivity function generalizes the concept of cycles and connects it with the well known stability notion of bargaining sets. The first part devotes to the study of relations between cycles and implement a class of effectivity functions for which theses cycles are equivalent. Part two of this Chapter is devoted to analyze the stability of the bargaining sets and give relations between them. Bargaining sets considered are by Zhou, the Mass-Colell and the Aumann Davis Maschler's bargaining sets.

In Chapter 21, Agnieszka Wiszniewska-Matyszkiel considers a game modelling a market consisting of two firms with market power and a continuum of consumers. A specific feature of a market for toys is considered with each firm producing two kinds of distinguishable goods. The problem of finding a Nash equilibrium implies firms' optimal advertising and production plans over time, where the aggregate of demands of consumers may depend on firms' past decisions. Equilibria at this market may have strange properties, like oscillatory production and advertising strategies.

R. B. Bapat introduces two classes of games in Chapter 22 and shows that they are balanced. In regression games, the observations in a regression model are controlled by players, and the worth of a coalition is inversely proportional to the variance of the estimate of the regression parameter. In connectivity games the players control the edges of a graph and the worth of a coalition is directly proportional to the degree of connectivity of the subgraph formed by the corresponding edges.

Chapter 23 by Somdeb Lahiri presents the concept of the induced combinatorial auction of a nonnegative TU game and shows that the existence of market equilibrium of the induced combinatorial auction implies the existence of a possibly different market equilibrium as well, which corresponds very naturally to an outcome in the matching core of the TU game. In this Chapter, it is shown that the matching core of the nonnegative TU game is non-empty if and only if the induced combinatorial auction has a market equilibrium.

Arrow formulated an important conceptual framework enabling one to discuss various collective decision making problems in an axiomatic fashion. There is, nevertheless, no topological structure given in Arrow's social choice framework to make it possible to discuss continuity of social welfare functions. In the turn of 1980s Chichilnisky had a systematic framework to discuss continuity of certain type of social welfare functions. In Chapter 24 by Kari Saukkonen, it is explained what continuity of a social welfare function is for Chichilnisky. It is then pointed out that there are difficulties, if this viewpoint is extended to cover continuity of Arrovian social welfare function, because of too specific assumption about the topological structure and dimension of the state sets. The discussion suggests that Chichilnisky's framework is not of much help in formulating appropriate topological foundations for the Arrovian social choice theory conceptualizing, for example, the workings of capitalistic democracy.

Finally in Chapter 25, S. K. Neogy, A. K. Das, S. Sinha and A. Gupta consider a mixture class of zero-sum stochastic game in which the set of states are partitioned into sets S_1, S_2 and S_3 so that the law of motion is controlled by Player I alone when the game is played in S_1, Player II alone when the game is played in S_2 and in S_3 the reward and transition probabilities are additive. It is proved that the game with SC/AR-AT mixture has the ordered field property by showing that the problem of solving the value vector v_s^β and optimal stationary strategies $f^\beta(s)$ for Player I and $g^\beta(s)$ for Player II for such a mixture type of game can be formulated as a complementarity problem. This gives an alternative proof of the ordered field property that holds for such a mixture type of game.

The 25 refereed articles contained in this volume are selected from 43 papers presented in International Symposium on Mathematical Programming for Decision Making: Theory and Applications which was organized as a part of the Platinum Jubilee Celebrations of the Indian Statistical Institute during January 10-11, 2007 at Indian Statistical Institute, Delhi Centre. The symposium was inaugurated by Professor Robert J. Aumann who delivered the inaugural talk on Game Engineering. The welcome address was delivered by Professor S. K. Pal, Director, Indian Statistical Institute. This symposium provided a forum for national and international academicians, researchers and practitioners to exchange ideas and approaches, to present research findings and state-of-the-art solutions, to discuss new developments in the theory and applications of mathematical programming to the problems in business and industries. A session titled S. R. Mohan Memorial Session was arranged to recall the memory of our colleague Professor S. R. Mohan (who passed away in October 2005) and his contribution in the area of Mathematical Programming and Game Theory. In fact, some of the papers are dedicated to the memory of Professor S. R. Mohan. It is the hope of the editors that the majority of the papers will simulate questions and possible solutions that are of interest to researchers of these areas.

S. K. Neogy, R. B. Bapat, A. K. Das and T. Parthasarathy
(Editors)

Acknowledgments

The editors are thankful to the following referees who have helped in reviewing the articles for this volume.

- Peter Zörnig, Univ. of Braslia, Braslia, Brazil.
- Wayne Goddard, Clemson University, Clemson SC 29634, USA.
- Roberto Cellini, Università di Catania, Italy.
- L. Lambertini, Università di Bologna, Italy.
- Anirban Kar, University of Warwick, UK.
- S. Chandra, Indian Institute of Technology, Delhi.
- Engelbert Josef Dockner, University of Vienna, Austria.
- Adedeji Badiru, University of Central Florida, USA.
- Arnon Boneh, Tel-Aviv University, Tel-Aviv, Israel.
- John W. Chinneck, Carleton University, Canada.
- Sidartha Gordon, University of Montreal, Canada.
- Jose Carlos Dias, Instituto Superior de Contabilidade e Administra de Coimbra, Portugal.
- J. Liu, Cornell University, Ithaca, NY 14853, USA.
- Hong Xia, Wuhan University, China.
- Janez Zerovnik, Institute of Mathematics, Physics and Mechanics, Jadranska 19, 1111, Ljubljana, Slovenia.
- J. Y. Guo, Nankai University, China.
- Maite Mármol, Universitat de Barcelona, Spain.
- Martin Jacobsen, University of Copenhegan, Denmark.
- Gordon E. Willmot, University of Waterloo, Canada.
- Amitava Bhattacharya, Univ. of Illinois at Chicago, USA.
- Le Dung Muu, Institute of Mathematics, Vietnam.
- Janez Brest, University of Maribor, Slovenia.
- Z. Wan, Wuhan University, China.

- T. S. Arthanari, University of Auckland, New Zealand.
- E. Somanathan, Indian Statistical Institute, Delhi Centre.
- A. Sen, Indian Statistical Institute, Delhi Centre.
- P. Roy Chaudhury, Indian Statistical Institute, Delhi Centre.
- G. S. R. Murthy, Indian Statistical Institute, Hyderabad.
- D. Mishra, Indian Statistical Institute, Delhi Centre.
- Joaquim J. Judice, Universidade de Coimbra, Departamento de Matemática, Portugal.
- C. A. Murty, Indian Statistical Institute, Kolkata.
- Ana Maria Faustino, Universidade do Porto, Portugal.

We are grateful to our colleagues and many other members of the staff at Indian Statistical Institute for preparation of this Platinum Jubilee volume. Finally, we thank World Scientific for their cooperation at all stages in publishing this volume.

S. K. Neogy, R. B. Bapat, A. K. Das and T. Parthasarathy
(Editors)

Contents

Chapter 1

Mathematical Programming and its Applications in Finance

Lyn C Thomas
Quantitative Financial Risk management Centre
School of Mathematics
University of Southampton
Southampton, UK
e-mail: L.Thomas@soton.ac.uk

Abstract

This article reviews some of the applications of mathematical programming in finance. Of course mathematical programming has long been recognised as a vital modelling approach to solve optimization problems in finance. Markowitz's Nobel Prize winning work on portfolio optimization showed how important a technique it is. Other prominent and well documented applications in long-term financial planning and portfolio problems include asset-liability management for pension plans and insurance companies, integrated risk management for intermediaries, and long-term planning for individuals. Nowadays there is an emphasis on the interaction between optimization and simulation techniques in these problems

There are though many uses of mathematical programming in finance which are not purely about optimizing the return on a portfolio and we will also discuss these applications. For example we discuss how one can use linear programming to estimate the term structure of interest rates for the prices of bonds. In the personal sector finance, where the lending is far greater than the higher profile corporate sector, the use of linear programming as a way of developing credit scorecards is proving extremely valuable.

Key Words: Mathematical programming, optimization problems in finance, portfolio optimization, credit scorecards, linear programming, asset-liability Models

1.1 Introduction

Mathematical programming was one of the key tools used in the earliest modelling in finance, namely Markowitz Nobel prize winning work [Markowitz (1952), (1959)] on optimising portfolios of shares or other financial instruments. This lead to a quadratic programming formulation, which has subsequently been extended in many ways. A related problem but from a different area of finance is the asset- liability problem faced by many insurance companies. Despite a long tradition of statistical and actuarial models of the liabilities involved in insurance and variants of portfolio optimisation problems to determine how to hold the assets, it is only in the last decade that these two complementary sides to an insurance company have been put together in one model. This leads to very large scale stochastic programming problems. These are the high profile applications of mathematical programming in finance and continue to be heavily researched not least because of the size of the programmes needed to solve real applications.

There continue though to be new applications of linear programming which are perhaps less well known but equally important to those specific areas of finance. One of these is the way of calculating the yield curve - the markets forecast of what the future of interest rates will be, which are implicit in the prices of bonds.

Another example occurs in consumer credit risk. This area of finance does not receive any of the research attention that corporate lending, equity models and the pricing of equity derivatives has received in the last twenty years. Yet the lending to consumers in most developed countries is much higher than the lending to companies (30% more in the US than the total of business lending). The tool used to assess the risk of lending to customers is to develop a credit scorecard and linear programming has some real advantages in developing such scorecards.

So in this review we briefly outline these four applications and the types of mathematical programming models that can solve them

1.2 Portfolio Optimization

The literature on financial optimization models dates back to the ground breaking application of Markowitz on optimizing a portfolio of financial products by concentrating on the mean return and taking the variance of the return as a measure of the risk. In this basic model one is interested in investing in a single period, there is an initial portfolio available and one of the assets is risk free, i.e. cash.

Assume there are N traded assets labelled $i, i = 1, 2, ..N$ and R_i is the random variable of the return in that one period on asset i, where $Exp(R_i) = r_i$ and the covariance of the returns on the assets is given by the matrix Σ. Assume that w_i^0 is the initial holding of asset i where $w_i^0 \geq 0$. Let x_i be the amount of asset i traded, where positive values means more of the asset is purchased and negative values means some of the asset is sold. The returns and risk (variance) of the resulting portfolio is

$$E[R^T(w^0 + x)] = r^T(w^0 + x); V[R^T(w^0 + x)] = (w^0 + x)^T\Sigma(w^0 + x)$$

Hence if one wants a portfolio with at least an expected return of t but with minimum variance one needs to solve the quadratic programming problem

$$\begin{aligned} \text{Minimise} \quad & (w^0 + x)^T\Sigma(w^0 + x) \\ \text{subject to} \quad & r^T(w^0 + x) \geq t \\ & (1, 1, 1, \ldots)^T x = 0 \end{aligned}$$

where the first constraint ensures the return is at least t (in reality it will always be exactly t) and the second constraint means the trading is self financing. The problem is then solved for different values of t to get a risk return trade-off and the investor chooses the outcome where his utility as a function of risk and return is maximised.

Although this model was fundamental in understanding the portfolio investment problem, it is of limited use in practice because it does not model all the aspects of the real situation. Some of these - limits on short selling, and the need for diversification - can be dealt with by adding appropriate constraints. Short selling is when one sells an asset at the start of the period, which one does not own. At the end of the period the asset has to be bought and passed on to the original buyer. As it stands there is no limit on how much of this can de done but one could put a limit on how much of this can be done by introducing the constraint

$$w_i^0 + x_i \geq -s_i.$$

One can also limit the amounts invested in an asset in three different ways as follows

Limit the amount invested in each asset
$w_i^0 + x_i \leq b_i$
Limit the relative amount invested in each asset
$w_i^0 + x_i \leq \gamma_i \sum_j (w_j^0 + x_j)$
Limit the relative amount invested in a group of assets J
$\sum_{j \in J} (w_j^0 + x_j) \leq \alpha \sum_j (w_j^0 + x_j)$

The original model also ignored the transaction costs involved in trading. If these can be considered to be piece wise linear in the level of the transaction then they can be added to the return constraint without losing linearity. Alternatively as [Mulvey (1993)] suggests one can mimic transaction costs by putting upper limits on the transaction for classes which have high such costs.

There are two real drawbacks to this formulation. The first is that variance is not always the way investors want to measure the risk. In particular it penalises returns which are well above the mean in the same way as those that are below the mean. So other risk measures have been suggested. [Mansini, Ogryczak and Speranza (2003)] reviewed the different measures that could still lead to linear programming formulations. Following [Sharpe (1973)] there have been a number of attempts to linearize the portfolio optimization problem. However if a portfolio is to take advantage of diversification then no risk measure can just be a linear function of the x. The way around it has been to assume there are a number of different scenarios $S_k - k = 1, \ldots, K$ with specific values of the return for each asset in each scenario and a probability p_k of that scenario occurring. In this way one can model other risk measures such as the mean semideviation, which looks at the expected shortfall below the mean value, i.e. for any trading policy $\mathbf{x}$., if $r(\mathbf{x})$ is the mean return and $R(k, \mathbf{x})$ is the actual return under scenario k, the mean semideviation is $sd(x) = E_k[\max\{r(\mathbf{x}) - R(k, \mathbf{x}), 0\}]$.

One can translate that into a convex piecewise linear function of the variables $\mathbf{x}$ by defining the following optimisation problem

$$sd(\mathbf{x}) = \min \sum_{k=1}^{k} d_k p_k$$

$$\text{subject to } d_k \geq r(\mathbf{x}) - R(k, \mathbf{x}), \;\; d_k \geq 0$$

In a similar way one can use other shortfall and stochastic dominance measures of risk such as mean below target deviation, minimise the maximum semideviation and the Gini mean difference which corresponds to the mean worst return. However for other risk measures this may not be possible.

The second issue that brings the standard models into question is that it is a one period model which may be inappropriate for investment problems with long time horizons. This would lead one to using stochastic linear programming models and their application in finance is surveyed in the paper by [Yu, Ji and Wang (2003)]. Thus we would need to extend the models to stochastic programming ones. Instead of doing this, we will consider in the next section the finance problem which has been most modelled as a stochastic programme in the last decade, namely the asset liability problem.

1.3 Asset-liability Models

Asset-Liability management looks at the problem of how to construct a portfolio of securities that will cover the cost of a set of liabilities, which are themselves varying as they depend on external economic conditions. This is exactly the problem that insurance companies have to face. For over a century they have had models which allow them to assess the costs of their liabilities. Fifty years ago the advent of the models in the previous section allowed them to optimise their portfolio of assets. Thus it is surprising that it is only in the last decade or so that they have sought to combine the two sides of their business into one model.

The time scales (many years) involved in such asset-liability problems and the need to allow for the possibility of rebalancing the portfolio at future times in response to new information means one is driven to model these problems as stochastic programming ones. We outline a formulation related to that suggested in [Bradley and Crane (1972)], [Klaassen (1998)] and in the review of [Sodhi (2005)], though other models have been used with considerable success by a number of U.S. and European insurance companies.

One of the problems in these models is what to do about the requirements at the end of the time horizon. It is reasonable to assume that the company will wish to continue trading thereafter but if one wants to minimise the initial cost of the asset portfolio one needs to cover the liabilities one is drawn to trying to make this residue as close to zero as possible. Alternatively one assumes the initial position is given and tries to maximise the surplus at the end of the time horizon provided all the liabilities have been met. That is the approach we will take here

Consider a T time horizon. Let $s(t)$ be a scenario which ends at time t and gives rise to two new scenarios $s^{(t+1)}$ with equal probability, which share the same history as $s(t)$ until time t. Thus the probabilities of all scenarios at time t are $2^{-(t-1)}$.

The variables and constraints in the resultant stochastic programme are as follows

Decision variables:

$x_{i,s(t)}$ - amount of asset i bought in period t in scenario $s(t)$
$y_{i,s(t)}$ - amount of asset i sold in period in scenario $s(t)$
$l_{s(t)}$ - amount lent at current short rate in period t in scenario $s(t)$
$b_{s(t)}$ - amount borrowed at current short rate in period t in scenario $s(t)$
$x_{i,s(t)}$ - amount of asset i bought in period t in scenario $s(t)$

Costs and Profits:

$c_{i,s(t)}$ - cash flow (dividents etc) from asset i in period t in scenario $s(t)$
$p_{i,s(t)}$ - price (ex-divident) from asset i in period t in scenario $s(t)$
$\nu_{s(t)}$ - present value of a cash flow of 1 in period t in scenario $s(t)$
$\eta_{s(t)}$ - one period interest rate in period t in scenario $s(t)$
$L_{s(t)}$ - liability due in period t in scenario $s(t)$
α_i - transaction cost as proportion of value of trade in asset i.
$h_{i,0}, l_0, b_0$ are initial asset holdings, lendings and borrowings

Model:

$$\text{Maximise } \; 2^{-(T-1)} \sum_{s(T)} \nu_{s(T)} \left(\sum_i p_{i,s(T)} h_{i,s(T)} + l_{sT)} - b_{s(T)} \right)$$

$$\sum_i c_{i,s(t)} h_{i,s(t-1)} + l_{s(t-1)}(1+\eta_{s(t-1)}) + b_{s(t)} + \sum_i (1-\alpha_i) p_{i,s(t)} y_{i,s(t)}$$

$$-\sum_i (1+\alpha_i) p_{i,s(t)} x_{i,s(t)} - l_{s(t)} - b_{s(t-1)}(1+\gamma_{s(t-1)}) = L_{s(t)} \;\; \forall s(t), t = 1, 2, \ldots, T$$

$$h_{i,s(t)} = h_{i,s(t-1)} + x_{i,s(t)} - y_{i,s(t)} \;\; \forall i, s(t), t = 1, 2, \ldots, T$$

$$h_{i,s(t)}, x_{i,s(t)}, y_{i,s(t)}, l_{s(t)}, b_{s(t)} \forall i, s(t), t = 1, 2, \ldots, T$$

One can see from this formulation how critical is the choice of scenarios to represent the uncertainties throughout the whole period of the problem. The scenarios describe the asset prices and also the term structures for the interest rates. Thus scenario generation becomes crucial to building useful models. There are three approaches that are commonly used to do this -i) bootstrapping using historical data ii) modelling the economy and asset returns with vector autoregressive models and iii) using simulations based on multivariate normal distributions of the values at risk from different classes, where the parameters in the normal distribution are obtained using time series analysis.

The other real difficulty is that the size of the scenario tree can make computation almost impossible. This has stimulated even further the work in stochastic programming on how to solve approximately such large problems. Obviously one way is not to have too many stages and so amalgamate together many of the periods towards the end of the time horizon into much larger time periods. However the real advantages come from using aggregation to combine nodes of the tree where appropriate and/or using decomposition approaches such as Benders decomposition, and the more recent interior point methods which can exploit the problem structure. These together with parallel processing of the computation and using object oriented parallel solvers mean that one can solve problems with 1,000,000,000 decision variables [Gondzio and Grothey (2006)].

1.4 Yield Curves

In financial markets the price of bonds can be used to estimate what interest rates will do in the future. This is because bond pricing models really model the current term structure of interest spot rates using both risk free (Treasury) and risky (corporate) securities. The spot rate can be extracted from the prices of zero coupon bonds which would repay only on maturity. However there are very few zero coupon bonds in the market and it is thus necessary to extract the spot interest rates from bonds, both Treasury and corporate, which pay coupon payments throughout their duration as well as making a final repayments.

The standard methods of stripping coupons from bonds are bootstrapping [Fabozzi (1998)] or linear regression [Carleton and Cooper (1976)]. If for each period there is one and only one coupon bond that matures, these techniques generate a unique set of spot interest rates over the period. However if there are periods where no bonds mature or other periods when several bonds mature at the same tine, then there is not a unique solution to the spot rates and in some cases these approaches giver rise to rates with unacceptable features. For example the rates might suggest that receiving one unit later in time is worth more than receiving it earlier in time, which would imply there were negative interest rates between the two times. One could also get results where the price for a high risk zero-coupon bond is higher than for a lower risk zero coupon bond maturing at the same time, which defies logical explanation.

To remedy the mispricing caused by bootstrapping, [Allen, Thomas and Zheng (2000)] suggested using linear programming to strip out the coupons of risk-free and risky bonds in such a way that there are no such difficulties. This approach will produce the same spot interest rates as the bootstrapping technique if there is one and only one coupon bond maturing in each time period.

Suppose there are only risk free bonds, labelled $i, i = 1, \ldots, N_0$ in the market, and bond i has a current price of P_i and $c_i(t)$ is its cash flow at time t. Then one can estimate the pure discounted bond prices $v_0(t)$ of risk free zero-coupon bonds paying 1 at a set of agreed times $t = 0, 1, \ldots, T$ by solving the following linear programming problem

$$\text{Minimize} \sum_{i=1}^{N_0} (a_i + b_i)$$

$$\text{subject to } P_i + a_i = \sum_{i=1}^{T} c_i(t) v_0(t) + b_i$$

$$v_0(t) \geq (1 + m(t)) v_0(t+1)$$

$$a_i, b_i \geq 0$$

$$\text{for } i = 1, \ldots N_0; \text{ and } t = 0, 1, \ldots, T-1$$

where $m(t)$ is the minimum expected forward interest from t to $t+1$.

The first constraint seeks to match the present value P_i to the discounted cash flows $c_i(t)$ and a_i and b_i are the mispricing errors. a_i is positive and $b_i = 0$ if the price is "too low" and the other way around if the price is "too high". The second constraint ensures there is no mispricing with respect to maturity (if $m(t) = 0$, one has the constraint that bonds of longer maturity should be priced at or below those with shorter maturity).

If one has calculated the values $v_0(t)$ for $t = 0, 1, 2, \ldots T$ one can transform these values into the spot interest rates $i(0, t)$ over the same period by taking

$$v_0(t) = \frac{1}{(1 + i(0,t))^t}$$

One can also use the price of risky bonds not only to determine the term structures of interest rates when applied to bonds of that risk class but also to help determine the term structure of interest rates of all classes including the risk free ones. Suppose bonds are rated according to their riskiness with 1 being the highest quality and M the lowest quality, with 0 remaining the grade ascribed to risk free bonds. Suppose there are N bonds observable in the market . Bond i has current price P_i, maturity date T_i, cashflow $c_i(t)$ for $t = 1, 2, \ldots, T_i$ and credit rating $d(i)$. Suppose for the class of bonds with credit rating $j, j = 0, 1, 2, \ldots, M$ the price of a bond stripped of its coupon paying 1 at t is $v_j(t)$ for $t = 1, 2, \ldots, T$ then we can calculate the best fit for these values from the bond prices given by solving the following Linear Programme

$$\text{Minimize} \sum_{i=1}^{N} (a_i + b_i)$$

$$\text{subject to } P_i + a_i = \sum_{t=1}^{T} c_i(t) v_{d(i)}(t) + b_i$$

$$v_0(t) \geq (1 + m(t)) v_0(t+1)$$

$$v_j(t+1) - v_{j+1}(t+1) \geq v_j(t+1) - v_{j+1}(t+1)$$

$$a_i, b_i \geq 0$$

$$\text{for } i = 1, \ldots N; j = 0, 1, \ldots, M-1, \text{ and } t = 0, 1, \ldots, T-1$$

$m(t)$ is again the minimum expected risk free forward interest rate from t to $t+1$ at time $t = 0$. The third constraint guarantees both that the price of a longer maturity bond is cheaper than that of a shorter maturity bond and that the price of a less risky zero coupon bond is higher than that of a riskier rated one of the same maturity.

Finally note that one could introduce the liquidity of the market into the optimization of the bond price and hence the term structure by recognizing that the issue amounts of different bonds will be quite different. Bonds which have a large amount issued are likely to be more liquid and hence their prices more accurately reflect the market's view than those where far less in value was issued. What is important is the issue value of the bond and if this is w_i for bond i, one can reflect the relative likelihood of the bonds being accurately prices by changing the objective function is the Linear programme above to

$$\text{Minimize} \sum_{i=1}^{N} w_i(a_i + b_i)$$

Whether we use liquidity or not, these linear programmes allow one to calculate the spot price interest rates $i(j,t)$ $j = 0, 1, ..M, t =, 1, T$ for risk free and risky bonds using

$$v_j(t) = \frac{1}{(1 + i(j,t))^t}.$$

1.5 Credit Scorecards

Most financial mathematics courses and text books concentrate exclusively on interest rate models, equities, bonds, their derivatives and corporate lending. However in most first world countries lending to consumers far exceeds lending to companies and yet that area of finance is hardly ever mentioned. Yet at the start of the twenty first century consumer credit is the driving force behind the economies of most of the leading industrial countries. Without it, the phenomenal growth in home ownership and consumer spending of the last fifty tears would not have occurred.

In 2004 the total debt owed by consumers in the US was \$10.3 trillion (\$10,300,000,000,000) of which \$7.5 was on mortgages and \$2.2 trillion on consumer credit (personal bank loans, credit cards, overdrafts, motor and retail loans). This is now 30% more than the \$7.8 trillion owed by all US industry and almost double the \$5.5 trillion of corporate borrowing (the rest being borrowing by small and medium sized companies and agricultural organisations). Figure 1 shows the growth in this borrowing since the 1960s and emphasises how consumer credit has been growing faster than corporate borrowing for most of that period.

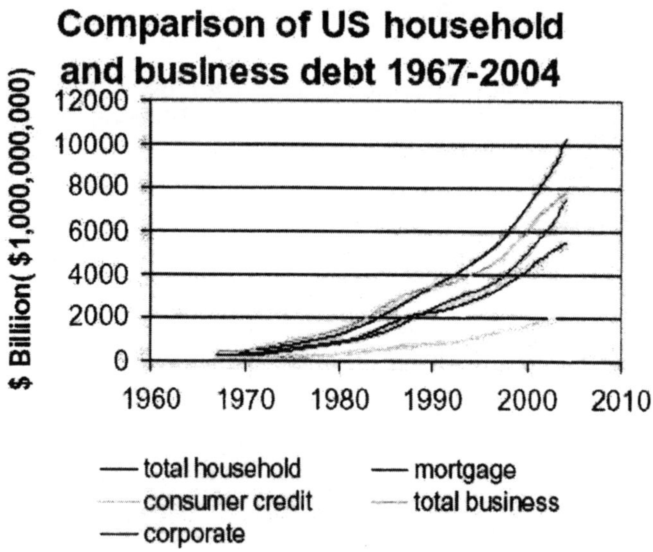

Figure 1: US household and business debt

This growth would not have been possible without credit scoring, the development of automatic risk assessment systems which assess the probability that new and existing customers will default on their loans within a fixed future time period (usually 12 months). The approach to building such credit scorecards is essentially one of classification. A sample of previous customers is taken and each classified as a defaulter or a non-defaulter according to their subsequent performance. The idea then is to identify which combination of application and/or performance attributes of consumers best separate the two groups. This idea of such statistical classification began with [Fisher (1936)] work on discriminant analysis which can be reinterpreted by saying that it is essentially a regression which tries to estimate p, the probability of non-default, as a linear function of the attributes of the consumer $x_1, x_2, \ldots, x_m$ by

$$p = w_0 + w_1 x_1 + w_2 x_2 + \ldots + w_m x_m$$

One of the succesful modifications used in credit scoring is that the x_i are not the orginal characteristics like age, but coarse classified variants of them. So one will split age into a number of age bands and the x_i are then either binary indicator variables of whether consumers are in that band, or weights of evidence transformations so each band is ranked according to the ratio of defaulters to non-defaulters in that band. This is one way of dealing with the non linearity of the relationship between default risk and age.

Of course the probability of non-defaulting in the above equation for those in the sample will be either 0 or 1. When one has used this sample to determine the regression equation one has an equation where the right hand side could take any value from $-\infty$ to $+\infty$ but the left hand side is a probability and so for any new applicants should only take values between 0 and 1. It would be better if the left hand side was a function of p which also could take a wider range of values. One such function is the log of the probability odds. This leads to the logistic regression approach where one matches the log of the probability odds by a linear combination of the consumer attributes, i.e.,

$$\log(p/(1-p)) = s = w_0 + w_1x_1 + w_2x_2 + \quad + w_mx_m$$

The right hand side of the equation is considered the credit score of the individual and ranks the consumers according to their chance of defaulting. One then chooses some cut-off score c, and give loans or credit cards to those applicants with scores above c and refuses it to those with scores below c. For existing customers, the scores are used to determine what changes in credit limit should be allowed and whether one should offer other products to that consumer.

Linear programming can also be used as a classification approach and also ends up with a linear scorecard. [Mangasarian (1965)] was the first to recognise that linear programming could be used for discrimination, but it was the papers by [Freed and Glover (1981a,b)] that sparked off the interest.

Suppose one has a sample of n_G goods and n_B bads and a set of m predictive variables from the application form answers so borrower i has predictive variable values $(x_{i1}, x_{i2}, \ldots, x_{im})$. One seeks to develop a linear scorecard where all the goods will have a value above the cut-off score c and all the bads have a score below the cut-off score. This cannot happen in all cases so we introduce variables a_i which allow for the possible errors - all of which are positive or zero. If we seek to find the weights $(w_1, w_2, \ldots, w_m)$ that minimise the sum of the absolute values of these errors we end up with the following linear programme

Minimise $a_1 + a_2 + \ldots + a_{n_G+n_B}$

subject to

$w_1 x_{i1} + w_2 x_{i2} + \ldots + w_m x_{im} \geq c - a_i,\ 1 \geq i \geq n_G$

$w_1 x_{i1} + w_2 x_{i2} + \ldots + w_m x_{im} \leq c + a_i,\ n_G + 1 \leq i \leq n_G + n_B$

$a_i \geq 0 \quad 1 \leq i \leq n_G + n_B$

In essence this approach is minimising the errors using the l_1 norm, while linear regression minimises the errors using the l_2 norm. One could also use linear programming to minimise the l_∞ norm, i.e., minimise the maximum error, by changing a_i to a in each constraint.

Linear programming is used by several organisations in building their scorecards because it allows one to build the best scorecard with any particular bias. For example, a lender might want to target consumers who are under 25 more than those who are over 25. In the linear programming formulation this can be easily done. For example if x_{25-} is the binary indicator variable that someone is under 25 and x_{25+} is the indicator variable for being over 25, then one requires that the corresponding weights satisfy. $w_{25-} > w_{25+}$. In this way one can construct the scorecard which best classifies the two groups but also has the required bias in it, something which is much harder to do in the standard regression approaches.

Bibliography

Allen D. E., Thomas L. C. and Zheng H. (2000). Stripping Coupons with Linear Programming, *Journal of Fixed Income* **10**, pp. 80–87.

Bradley p. and Crane D. B.(1972). A dynamic model for bond portfolio management, *Management Science* **19**, pp. 139–151.

Carleton W. and Cooper I. (1976), Estimation and uses of the Term Structure of Interest Rates, *Journal of Finance* **31**, pp. 1067–1083.

Fabozzi F. (1998). Valuation of Fixed Income Securities and Derivatives, *Frank J. Fabozzi Associates*, (New Hope, PA).

Fisher R. A. (1936). The use of multiple measurements in taxonomic problems, *Annals of Eugenics* **7**, pp. 179–188.

Freed N. and Glover F. (1981a). A linear programming approach to the discriminant problem, *Decision Sciences* **12**, 68-74.

Freed N. and Glover F. (1981b). Simple but powerful goal programming formulations for the discriminant problem, *European J. Operational Research* **7**, pp. 44–60.

Gondzio J. and Grothey A. (2006). Solving nonlinear Financial planning problems with 109 decision variables on massively parallel architecture, in *Computational Finance and its Application II*, ed by Constantino M., Brebbia C.A., WIT Transaction on Modelling and Simulation **43**, (WIT Press).

Klaassen P., (1998). Financial asset-pricing theory and stochastic programming models for asset-liability management: A synthesis, *Management Science* **44**, pp. 31–48.

Mangasarian O. L. (1965). Linear and nonlinear separation of patterns by linear programming, *Operations Research* **13**, pp. 444–452.

Mansini R., Ogryczak W. and Speranza M. G. (2003). LP Solvable Models for Portfolio Optimization: a classification and computational comparison, *IMA J. Mathematics in Management* **14**, pp. 187–220.

Mulvey J. M. (1993). Incorporating transaction costs in models for asset allocation, in *Financial Optimization(ed. Zenios S.A.)*, pp. 243–259, (Cambridge University Press, Cambridge).

Sharpe W. F. (1973). A linear programming approximation for the general portfolio analysis problem, *J. Fin. Quant Analysis* **6**, pp. 1263–1275.

Sodhi M. S. (2005), LP Modeling for Asset Liability management: A survey of choices and simplifications, *Operations Research* **53**, pp. 181–196.

Thomas L. C, Edelman, D. B. and Crook, J. N, (2002). Credit Scoring and its Applications, (*SIAM*, Philadelphia).

Yu L-Y, Ji X-D, Wang S_Y, (2003). Stochastic programming Models in Financial optimization: a survey, *Advanced Modelling and Optimization* **5**, pp. 1–26.

Chapter 2

Anti-stalling Pivot Rule for Linear Programs with Totally Unimodular Coefficient Matrix

Santosh N. Kabadi
Faculty of Business Administration, University of New Brunswick, Fredericton, New Brunswick, Canada E3B 5A3
e.mail: kabadi@unb.ca
Abraham P. Punnen
Department of Mathematics, Simon Fraser University Surrey, Central City, 250-13450 102nd AV, Surrey, British Columbia, V3T 0A3, Canada
e.mail: apunnen@sfu.ca

Abstract

Although several anti-cycling pivot selection rules exist for the simplex method for a general linear program (LP), none is known to avoid stalling (an exponential sequence of degenerate pivots). In this paper we develop a pivot selection rule that prevents stalling when the coefficient matrix of the LP is totally unimodular. For an LP with m constraints and totally unimodular coefficient matrix, our pivot selection rule guarantees that the simplex method performs at most m consecutive degenerate pivots or declares that the current solution is optimal. This extends a corresponding result available for minimum cost flows.

Key Words: Linear programming, totally unimodular coefficient matrix, anti-cycling pivot selection rule, simplex method

2.1 Introduction

Simplex method for linear programming [Dantzig (1963); Murty (1983)] is perhaps the most popular algorithm for solving linear programming problems. For the simplex method for a general linear program, no pivot selection rule is known for which the algorithm has a polynomial time worst case complexity and it is one of the outstanding open problems related to linear programs. However, rules generating polynomial sequence of pivots are known for specially structured problems [Akgul (1993); Goldfarb and Hag (1991); Goldfarb, Hao and Kai (1990); Hung M.S (1983); Sokkalingam, Sharma and Ahuja (1997)]. When the entering and leaving variables are not selected carefully, the simplex algorithm could even get into cycling in presence of degeneracy [Dantzig (1963); Kotiah and Steinberg (1978); Lee (1997); Marshall and Suurballe (1969)]. Several pivot selection rules are available in literature to avoid cycling [Avis and Chavtal (1978); Bland (1977); Clausen (1987); Magnanti and Orlin (1988); Pan (1988); Wolfe (1963); Zhang S. (1991)] that guarantees finite convergence of the algorithm.

Another phenomenon closely related to cycling is called *stalling* - an exponential sequence of consecutive degenerate pivots. No pivot selection rules discussed in literature is known to avoid stalling in simplex method for a general linear program. However, for specially structured linear programs, pivot selection rules are available that prevent stalling [Ahuja, Orlin, Sharma and Sokkalingam (2002); Akgul (1993); Goldfarb and Hag (1991); Goldfarb, Hao and Kai (1990); Hung M.S (1983); Sokkalingam, Sharma and Ahuja (1997)]. Recently, Ahuja, Orlin, Sharma and Sokalingam [Ahuja, Orlin, Sharma and Sokkalingam (2002)] proposed a pivot selection rule that guarantees the number consecutive degenerate pivots for the network simplex method to be $O(n)$ where n is the number of nodes in the network. We consider a generalization of this case where the coefficient matrix is totally unimodular and present a pivot selection rule that performs at the most m consecutive degenerate pivots or declares that the current solution is optimal. Here m is the number of constraints.

It may be noted that the result in Ahuja et al [Ahuja, Orlin, Sharma and Sokkalingam (2002)] is designed specifically for network matrices and does not apply for any other totally unimodular matrix that is not a network matrix. For example, when the coefficient matrix is transpose of a node-arc incidence matrix or if it is a mixture of network and transpose of network matrices. Unlike the method in Ahuja et al [Ahuja, Orlin, Sharma and Sokkalingam (2002)], we do not require use of strongly feasible bases. This is crucial since this concept, as used in [Ahuja, Orlin, Sharma and Sokkalingam (2002)], is available only for the case when coefficient matrix is a node-arc matrix.

2.2 Pivot Selection Rule

Consider the linear programming problem (LPP)

$$\text{LPP:} \qquad \begin{aligned} &\text{Minimize} \quad cx \\ &\text{Subject to} \\ &\qquad\qquad Ax = b \\ &\qquad\qquad x \geq 0. \end{aligned}$$

where A is an $m \times n$ matrix with full row-rank, $b = (b_1, b_2, \ldots, b_m)^T$ is an m-vector, $c = (c_1, c_2, \ldots, c_n)$ is an n-vector and $x = (x_1, x_2, \ldots, x_n)^T$ is an n-vector.

We assume that the matrix A is totally unimodular (i.e. any square sub-matrix of A has determinant ± 1 or 0). The jth column of A is denoted by $A_{.j}$. Let B be any feasible basis matrix. Let the ith column of B be the B_i^{th} column of A, (i.e. $B_{.i} = A_{.B_i}$). Then $\{x_{B_1}, \ldots, x_{B_m}\}$ is the corresponding set of basic variables and the corresponding basic feasible solution (BFS) is represented by $(x^B, 0)$, where $x^B = B^{-1}b = (x^B_{B_1}, x^B_{B_2}, \ldots, x^B_{B_m})$ gives the values of basic variables and the value of each non-basic variable is zero. If $x^B_{B_i} > 0$ for all i, then $(x^B, 0)$ is called a *non-degenerate BFS*. If $x^B_{B_i} = 0$ for some i then $(x^B, 0)$ is called a *degenerate BFS*. Let $c^B = (c_{B_1}, c_{B_2}, \ldots, c_{B_m})$ be the cost vector associated with the basic variables. For each variable x_j, its *reduced cost*, denoted by $\bar{c}_j$ is defined as $\bar{c}_j = c_j - c^B B^{-1} A_{.j}$. Note that for each basic variable x_j, $\bar{c}_j = 0$. The following theorem summarizes the optimality criterion in simplex method.

Theorem 2.1. *[Dantzig (1963); Murty (1983)] If the reduced cost $\bar{c}_j \geq 0$ for all j, then the BFS $(x^B, 0)$ is optimal.*

The converse of the above theorem is true for non-degenerate BFS. A degenerate BFS could be represented by a large number of basis matrices. If a degenerate BFS is optimal, there exists an associated basis B such that the corresponding reduced cost $\bar{c}_j \geq 0$ for all j.

Let B^0 be a basis at any iteration of the simplex method. Without loss of generality we assume that $B^0 = [A_{.1}, A_{.2}, \ldots, A_{.m}]$ and we denote $(x^{B^0}, 0)$ by $x^0 = (x^0_1, x^0_2, \ldots, x^0_n)$. Let $N^0 = \{m+1, m+2, \ldots, n\}$, the set of non-basic variables. Let $\bar{A} = (B^0)^{-1} A$.

Assume that B^0 is a degenerate basis. Without loss of generality assume that $x_i^0 = 0$ for $i = 1, 2, \ldots, k$ and $x_i^0 > 0$ for $i = k+1, k+2, \ldots, m$ for some $1 \leq k < m$. Of course, $x_i^0 = 0$ for $i = m+1, m+2, \ldots n$. For any basis B let $S(B) = \{j : \bar{c}_j < 0\}$. In the pivot selection rule, the *minimum ratio* [Dantzig (1963); Murty (1983)] is strictly positive for some $j \in S(B^0)$, then we perform a non-degenerate pivot using such a j as the entering column. So, we assume that the minimum ratio is zero for each $j \in S(B^0)$ as the choice of entering column. This implies that for each $j \in S(B^0)$, $\bar{A}_{ij} > 0$ for some $i \in \{1, 2, \ldots, k\}$.

The following theorem is well-known [Dantzig (1963); Murty (1983)].

Theorem 2.2. *The BFS x^0 is not optimal if and only if there exists $y = (y_1, y_2, \ldots, y_n)$ such that $Ay = 0$, $cy < 0$ and $y_i \geq 0$ for all $i \in \{1, 2, \ldots, k\} \cup \{m+1, m+2, \ldots, n\}$.*

The proof of the above theorem follows from the fact that y is an improving feasible direction from x^0 if and only if it satisfies the conditions of the theorem.

Suppose there exists an n-vector y satisfying the conditions of Theorem 2.2. Choose such a vector y^0 with a minimal set of non-zero coefficients. This can be achieved using standard reduction techniques. Since A is totally unimodular we assume that $y_i^0 \in \{0, 1, -1\}$ $\forall$ i. Let $Q_1 = \{i : i \in \{1, 2, \ldots, k\}; y_i^0 \neq 0\}$ and $Q_2 = \{i : i \in \{m+1, m+2, \ldots, n\}; y_i^0 \neq 0\}$. For convenience, let us assume without loss of generality that $Q_1 = \{1, 2, \ldots, r\}$ and $Q_2 = \{m+1, m+2, \ldots, m+t\}$ for some $1 \leq r \leq k$ and $1 \leq t \leq n-m$. It follows from Theorem 2.2 that $y_j^0 = 1$ for all $j \in Q_1 \cup Q_2$ and $y_j^0 = 0$ for all $j \in \{r+1, r+2, \ldots, k, m+t+1, m+t+2, \ldots, n\}$. Since $cy^0 < 0$, it follows that $\bar{c}y^0 < 0$, and hence, $S(B^0) \cap Q_2 \neq \emptyset$. Choose any $f \in S(B^0) \cap Q_2$. Recall that $\bar{A} = (B^0)^{-1}A$.

Case 1:$\bar{A}_{if} > 0$ *for some* $i \in \{r+1, r+2, \ldots, k\}$: In this case perform a degenerate pivot on $\bar{A}_{if}$. Let B' be the new basis obtained. The same vector y^0 satisfies the conditions of Theorem 2.2 with respect to the new basis B'. Note that $y_j^0 > 0$ for t indices not in the basis B^0. We pivot in the column f and $y_f^0 > 0$ by choice of f. The leaving column belongs to $\{r+1, r+2, \ldots, k\}$ and $y_j^0 = 0$ $\forall$ $j \in \{r+1, r+2, \ldots, k\}$. Thus $|\{j : y_j^0 > 0$ and j non-basic in $B'\}| = t-1$. Now repeat the process with B' and y^0.

Case 2: $\bar{A}_{if} \leq 0$ *for all* $i \in \{r+1, r+2, \ldots, k\}$ *but* $\bar{A}_{if} > 0$ *for some* $i \in \{1, 2, \ldots, r\}$*:* Without loss of generality assume $\bar{A}_{1f} > 0$. Consider the following n-vector z:

$$z_j = \begin{cases} -\bar{A}_{jf}, & \text{for } j = 1, 2, \ldots, m; \\ 1, & \text{for } j = f; \\ 0, & \text{otherwise.} \end{cases} \tag{2.1}$$

Note that $z_f = 1, z_1 = -1$, $z_i \geq 0$ for $i \in \{r+1, r+2, \ldots, k\}$ and $cz = \bar{c}_f < 0$. Now consider the vector $z + y^0$. Clearly $A(z + y^0) = 0$. Further, $z_1 + y_1^0 = 0$, $z_f + y_f^0 = 2$, $z_i + y_i^0 \geq 0$ for all $i \in \{1, 2, \ldots, k\}$, $z_i + y_i^0 = 1$ for all $i \in \{m+1, m+2, \ldots, m+t\} \setminus \{f\}$, $z_i + y_i^0 = 0$ for all $i \in \{m+t+1, \ldots, n\}$ and $c(z + y^0) < 0$. Since the matrix A is totally unimodular, we can write $z + y^0$ as $z + y^0 = h^1 + h^2 + \cdots + h^d$, where for each $i \in \{1, 2, \ldots, n\}$ and $j = 1, 2, \ldots, d$, (i) $h_i^j \in \{0, \pm 1\}$; (ii) $h_i^j = 1$ implies $z_i + y_i^0 > 0$; and (iii)$h_i^j = -1$ implies $z_i + y_i^0 < 0$. In addition, obviously, $h_f^1 + h_f^2 + \cdots + h_f^d = 2$ and $ch^j \leq c(z + y^0)$ for at least one $j = \{1, 2, \ldots, d\}$. Without loss of generality assume that $ch^1 \leq c(z + y^0)$. Then the vector h^1 satisfies the conditions of Theorem 2.2 with respect to basis B^0 and $|\{i : h_i^1 > 0$ and i is non-basic in $B^0\}| < t$. Repeat the process with basis B^0 and with vector $y^1 = h^1$.

Thus, in each iteration, the value of t, (the number of non-zero indices of the current vector y^i that are non-basic in the current feasible basis B^j) goes down by at least 1. Hence, using our pivot rule, the number of consecutive degenerate pivots can be at most m. We thus have our main theorem.

Theorem 2.3. *When the coefficient matrix A of the LP is totally unimodular, there exists a pivot rule that limits the number of consecutive degenerate pivots to at most m.*

Bibliography

Ahuja R. K., Orlin J. B., Sharma P., and Sokkalingam P. T (2002). A network simplex algorithm with O(n) consecutive degenerate pivots. *Operations Research Letters* **30**, pp. 141–148.

Akgul M. (1993). A geneuinely polynomial primal simplex algorithm for the assignment problem, *Discrete Applied Mathematics* **45**, pp. 93–115.

Avis D. and Chavtal V. (1978). Notes on Blands rule, *Mathematical Programming Study* **8**, pp. 24–34.

Bland R. G. (1977). New finite pivoting rules for the simplex method, *Matheamtics of Operations Research* **2**, pp. 103–107.

Clausen J. (1987). A note on Edmonds-Fukuda pivoting rule for simplex method, *European Journal of Operationsl Research* **29**, pp. 378–383.

Cunningham W. H. (1976). A network simplex method. *Mathematical Programming* **11**, pp.105–116.

Dantzig G. B. (1963). *Linear programming and Extensions*, (Princeton University Press, Princeton, NJ).

Goldfarb, D and Hag J. (1991). On strongly polynomial variants of the network simplex algorithm for the maximum flow problem. *Operations Research Letters* **10**, pp. 383–387.

Goldfarb D., Hao J. and Kai S-R. (1990). Efficient shortest path simplex algorithms, *Operations Research* **38**, pp. 624–628.

Hung M.S (1983). A polynomial simplex method for the assignment problem, *Operations Research* **31**, pp. 595–600.

Klee V and Minty G. J.(1972). How good is simplex algorithm, in O. Shisha (ed) *Inequalities III*, pp. 159–175, (Academic Press, New York).

Kotiah T. C. T, and Steinberg D. I. (1978). On the possibility of cycling with the simplex method, *Operations Research* **26**, pp. 374–376.

Lee J. (1997). Hoffman.s circle untangled, *Siam Review* **39**, pp. 98–105.

Magnanti T. L. and Orlin J. B.(1988). Parametric linear programming and anticycling pivoting rules, *Mathematical Programming* **41**, pp. 317-325.

Marshall K. T. and Suurballe J. W.(1969). A note on cycling in the simplex method, *Naval Res. Logist. Quart.* **16**, pp. 121-137.

Murty K. G (1983). *Linear Programming*, (John Wiley & Sons, New York).

Orlin J.B (1997). A polynomial time primal network simplex algorithm for minimum cost flows, *Mathematical Programming: Series A*, **78**, pp. 109–129.

Pan P.-Q (1988). Practical finite pivoting rules for the simplex method, *OR Spektrum* **12**, pp. 219–225.

Sokkalingam P. T, Sharma P. and Ahuja R. K. (1997). A new pivot selection rule for the network simplex algorithm, *Mathematical Programming* **78**, pp. 149–158.

Wolfe P. (1963). A technique for resolving degeneracy in linear programming, *Journal of the Society for Industrial and Applied Mathematics* **11**, pp. 205–211.

Zhang S.(1991). On anticycling pivoting rules for the simplex method, *Operations Research Letters* **10**, pp. 189–191.

Chapter 3

A New Practically Efficient Interior Point Method for Convex Quadratic Programming

Katta G. Murty
Department of Industrial and Operations Engineering
University of Michigan
Ann Arbor, MI 48109-2117, USA
e-mail: murty@umich.edu

Abstract

Murty developed a new interior point method for linear programming (LP), based on a new centering strategy that moves any interior feasible solution x^0 to the center of the intersection of the feasible region with the objective hyperplane through x^0, before beginning the descent moves. Using this centering strategy, that method obtains an optimum solution for an LP by a very efficient descent method that uses no matrix inversions. Here we extend that method into a descent method for solving quadratic programs (QP). The advantages of this method are: (i) all the constraints in the problem never appear together in any matrix inversion operations performed in the algorithm, (ii) each iteration in the algorithm consists of essentially three steps, one step requires no matrix inversions, a second step requires solving a system of linear equations involving a small subset of constraints, a third step involves matrix operations involving only the coefficient matrix of the objective function. So, compared to other existing methods for QP, the new method is able to handle it with minimal matrix inversion computations.

Key Words: Convex quadratic programming, interior point method, centering strategy, descent method

3.1 Introduction

We consider the quadratic program (QP)

$$\begin{array}{ll}\text{Minimize} & Q(x) = cx + (1/2)x^T Dx \\ \text{subject to} & Ax \geq b\end{array} \tag{1}$$

where the objective coefficient matrix D is a symmetric matrix of order n, the constraint coefficient matrix A is of order $m \times n$, and b, c are column and row vectors of appropriate orders [Cottle, Pang, and Stone (1992)], [Murty (1988)], [Ye (1997)]. Let K denote the set of feasible solutions. For simplicity we assume that K is bounded. We also assume that an interior point x^0 of K (i.e., a point satisfying $Ax^0 > b$) is available.

In this paper we assume that D is positive definite, i.e., that $Q(x)$ is strictly convex. Strategies for relaxing this assumption are discussed briefly in Section 3.7.

Let $K^0 = \{x : Ax > b\}$, it is the interior of K. We assume that the row vectors of A, denoted by $A_{i.}$ for $i = 1$ to m, are normalized so that their Euclidean norm $||A_{i.}|| = 1$ for all i. For each $x \in K^0$, we define $\delta(x) = \min\{A_{i.}x - b_i : i = 1 \text{ to } m\}$, $\delta(x)$ is the radius of the largest ball that can be inscribed within K with its center at x.

In [Murty (2006)], in the iteration when x^0 is the current interior feasible solution, the centering step has the aim of finding an $x \in K^0$ on the objective plane through x^0, that maximizes $\delta(x)$ so as to get the largest ball inscribed in K with center at an interior feasible solution that has the same objective value as x^0. In our problem here, the set of all points with the same objective value as x^0 is a nonlinear surface and not a hyperplane; so we will not constrain the center to have the same objective value as x^0 in the centering step here, but will allow only moves that keep the objective value the same or decrease it while increasing $\delta(x)$.

3.2 The Centering Strategy

When x^0 is the current interior feasible solution for (1), the problem of finding the largest inscribed sphere inside K with center at a point where the objective value $Q(x)$ is $\leq Q(x^0)$, is the following constrained max-min problem:

$$\begin{array}{ll}\text{Maximize} & \delta \\ \text{subject to} & \delta - A_{i.}x \leq -b_i, \quad i = 1, \ldots, m \\ & Q(x) \leq Q(x^0)\end{array} \tag{2}$$

If $(\bar{x}, \bar{\delta})$ is an optimum solution of this problem, then $\bar{\delta} = \delta(\bar{x})$, and the ball $B(\bar{x}, \bar{\delta})$ with $\bar{x}$ as center, and $\bar{\delta}$ as radius, is a largest inscribed sphere required. This problem (2) is itself a quadratic program. This type of model may have to be solved several times before we get a solution for our original QP (1), and for implementing our algorithm an exact solution of (2) is not essential, so solving (2) exactly will be counterproductive. Using the special max-min structure of (2), we now develop an efficient procedure for getting an approximate solution to (2), similar to the one developed in [Murty (2006)] for the corresponding centering problem in the algorithm discussed there for LP.

Procedure for Getting an Approximate Solution for (2)

Since our goal is to increase the minimum distance of x from the facetal hyperplanes of K, an approximate solution of (2) can be obtained through line searches in directions perpendicular to the facetal hyperplanes of K. So, in this procedure, for finding the new center $x \in K^0 \cap \{x : Q(x) \leq Q(x^0)\}$, we only consider moves in directions among $\Gamma = \{A_{i.}^T, -A_{i.}^T : i = 1, ..., m\}$ which are descent directions for $Q(x)$ at the current point.

So, this procedure consists of a series of moves beginning with x^0, generating a sequence of points $x^r \in K^0 \cap \{x : Q(x) \leq Q(x^0)\}$, $r = 1, 2, \ldots$. When at x^r look for a **profitable direction to move** at x^r, which is a direction $p \in \Gamma = \{A_{i.}^T, -A_{i.}^T : i = 1, ..., m\}$ satisfying:

(i): $\nabla Q(x^r)p < 0$, and

(ii): $\delta(x^r + \alpha p)$ increases as α changes from 0 to positive values.

For any $x \in K^0$ define $T(x) = \{i : 1 \leq i \leq m$, and i ties for the minimum in $\delta(x) = \text{minimum}\{A_{i.}x - b_i : i = 1, ..., m\}\}$. $T(x)$ is known as the **index set of touching constraints** at x, because it is the index set of facetal hyperplanes of K which are tangents to the ball $B(x, \delta(x))$ if each constraint in (1) defines a facetal hyperplane for K. In [Murty (2006)], it has been shown that a direction p satisfies condition (ii) above at x^r iff all the entries in $\{A_{t.}p : t \in T(x^r)\}$ are of the same sign. So, for any given direction p, both (i), (ii) can be checked easily to determine if p is a profitable direction to move at x^r.

If a profitable direction $p \in \Gamma$ to move at x^r has been found, the step length α to move at x^r in the direction p to get the next point in the sequence $x^{r+1} = x^r + \alpha p$ is defined to be: $\alpha = \text{minimum}\{\beta_1, \beta_2\}$ where

$\beta_1 =$ the value of β that minimizes $Q(x^r + \beta p)$ over $\beta \geq 0$. Finding β_1 therefore requires minimizing a quadratic function in the single variable β, which can be solved easily.

$\beta_2 =$ the value of β that maximizes $\delta(x^r + \beta p)$ over $\beta \geq 0$. In [Murty (2006)] it has been shown that this can be found by solving the following 2-variable linear program in which the variables are θ, β.

$$\begin{array}{ll} \text{Maximize} & \theta \\ \text{subject to} & \theta - \beta A_{i.}p \leq A_{i.}x^r - b_i, \; i = 1, \ldots, m \\ & \theta, \beta \geq 0 \end{array}$$

which can be found with at most $O(m)$ effort. [Murty (2006)] discusses how to solve this efficiently.

Once β_1, β_2 are determined, let $\alpha = \text{minimum}\{\beta_1, \beta_2\}$, take the next point in the sequence to be $x^{r+1} = x^r + \alpha p$, and continue the procedure in the same way with x^{r+1}.

The procedure continues as long as profitable directions $p \in \Gamma$ to move at the current point can be found.

When there are several profitable directions to move at the current point in this procedure, efficient selection criteria to choose the best among them can be developed. In fact, additional directions can be included in Γ to improve the quality of the approximation obtained. When there are no profitable directions to move at the current point, or when improvement in the value of the radius of the inscribed ball becomes smaller than some selected tolerance, take the current point in the sequence as the center selected by this procedure.

As can be seen, the procedure used in this centering strategy does not need any matrix inversion, and only solves a series of 2-variable LPs, and single variable quadratic function minimization problems, which can be solved very efficiently. Hence this centering strategy can be expected to be efficient.

What is the Purpose of Maximizing the Radius of the Inscribed Ball in this Centering Step?

Our goal is to find an optimum solution to the original quadratic program (1). Then, why are we focussing on the seemingly unrelated problem of maximizing the radius of the inscribed ball in this centering step? The reason is the following.

Let $B(\bar{x}, \bar{\delta})$, the ball with center $\bar{x}$ and radius $\bar{\delta}$ be the ball constructed in this centering step. Then in this iteration the algorithm uses the direction $\hat{x} - \bar{x}$ as a descent direction for a line search step to minimize $Q(x)$ over $\{\bar{x}+\lambda(\hat{x}-\bar{x}) : \lambda \geq 0$, and λ such that $\bar{x}+\lambda(\hat{x}-\bar{x}) \in K\}\}$, where $\hat{x}$ is a point that minimizes $Q(x)$ over the ball $B(\bar{x}, \bar{\delta})$. There are efficient polynomial time algorithms for computing $\hat{x}$, but its computation is perhaps the most expensive computational operation in this algorithm. Maximizing $\bar{\delta}$, the radius of the ball found in this centering step, helps to reduce the number of times this expensive step has to be used in this algorithm.

3.3 Descent Step Using a Descent Direction

Let $B(\bar{x}, \bar{\delta}) = \{x : (x - \bar{x})^T(x - \bar{x}) \leq \bar{\delta}^2\}$ be the ball with center $\bar{x}$, and radius $\bar{\delta}$, obtained in the centering step. In this step we solve the problem

$$\begin{array}{ll} \text{Minimize} & Q(x) = cx + (1/2)x^T Dx \\ \text{subject to} & (x - \bar{x})^T(x - \bar{x}) \leq \bar{\delta}^2 \end{array} \tag{3}$$

This is the problem of minimizing a quadratic function inside a ball for which efficient polynomial time algorithms exist. Associating the Lagrange multiplier $\lambda \in R^1$ with the constraint, the KKT optimality conditions for this problem are

$$\begin{aligned} c^T + Dx + 2\lambda(x - \bar{x}) &= 0 \\ \lambda \geq 0,\ \bar{\delta}^2 - (x - \bar{x})^T(x - \bar{x}) &\geq 0 \\ \lambda(\bar{\delta}^2 - (x - \bar{x})^T(x - \bar{x})) &= 0 \end{aligned}$$

Since $\lambda \in R^1$, this problem can be solved efficiently (in polynomial time) using the KKT conditions, see [Conn, Gould and Toint (2000)], [Ye (1997)] for complete details of this algorithm. The algorithm becomes simpler when D is positive definite or semidefinite, but even if D is not positive semidefinite, it can be solved efficiently using the KKT conditions.

Let $\hat{x}$ be the optimum solution computed for (3). If $\hat{x}$ is an interior point of $B(\bar{x}, \bar{\delta})$, or if it is a boundary point of both $B(\bar{x}, \bar{\delta})$ and K, or if $\nabla Q(\hat{x}) = 0$; then $\hat{x}$ is an optimum solution of (1), terminate.

Otherwise, using $\hat{x} - \bar{x}$ as the descent direction for $Q(x)$ at $\bar{x}$, do a line search to minimize $Q(x)$ on the line segment $\{\bar{x} + \lambda(\hat{x} - \bar{x}) : \lambda \geq 0$, and λ such that $\bar{x} + \lambda(\hat{x} - \bar{x}) \in K\}$. Let λ_1 be the optimum step length for this line search. If $\bar{x} + \lambda_1(\hat{x} - \bar{x})$ is an interior point of K; then terminate if $\nabla Q(x) = 0$ at this point, otherwise define this point as the output of this step.

If however, $\bar{x} + \lambda_1(\hat{x} - \bar{x})$ is a boundary point of K, let $I = \{i : i\text{-th}$ constraint in (1) is satisfied as an equation by $\bar{x}+\lambda_1(\hat{x}-\bar{x})\}$. If the following system in Lagrange multipliers $\pi_I = (\pi_i : i \in I)$

$$\begin{array}{l} c + (\bar{x} + \lambda_1(\hat{x} - \bar{x}))^T D - \sum_{i \in I} \pi_i A_{i.} = 0 \\ \pi_i \geq 0, \text{ for all } i \in I \end{array} \tag{4}$$

has a feasible solution, then $\bar{x} + \lambda_1(\hat{x} - \bar{x})$ is an optimum solution of (1), terminate. However, it may not be productive to check if system (4) is feasible every time this step ends up at this stage. If this operation of checking the feasibility of (4) is not carried out, or if (4) turns out to be infeasible, then take the output of this step as $\bar{x} + (\lambda_1 - \epsilon)(\hat{x} - \bar{x})$ where ϵ is some preselected positive tolerance for the current point to be an interior point of K.

3.4 Descent Step Using the Touching Constraints

We will first provide the motivation for this step. Assume that the centering step is carried out exactly, and suppose $B(\bar{x}, \bar{\delta}) = \{x : (x-\bar{x})^T(x-\bar{x}) \leq \bar{\delta}^2\}$ is the ball with center $\bar{x}$ and radius $\bar{\delta}$ obtained in the centering step in this iteration. $T(\bar{x}) = \{i : A_{i.}\bar{x} = b_i + \bar{\delta}\}$ is the index set of **touching constraints** in this iteration, this is the index set of facetal hyperplanes of K that are touching the ball $B(\bar{x}, \bar{\delta})$ and hence are tangent hyperplanes for it. Actually $T(\bar{x})$ is the index set of linear constraints in (2) that are active at its optimum solution, all other linear constraints in (2) are inactive at its optimum solution; and the same thing is also true for the problem obtained by replacing x^0 in (2) by $\bar{x}$. So, $(\bar{x}, \bar{\delta})$ is an optimum solution for (2) when x^0 there is replaced by $\bar{x}$, i.e., for

$$\begin{array}{ll} \text{Maximize} & \delta \\ \text{subject to} & \delta - A_{i.}x \leq -b_i, \ i = 1, ..., m \\ & Q(x) \leq Q(\bar{x}) \end{array} \tag{5}$$

It often happens the the index set of touching constraints for the ball obtained from an optimum solution of (5) with $Q(\bar{x})$ replaced by $Q(\bar{x}) - \gamma$ remains the same as $T(\bar{x})$, for a range of values of γ, say $0 \leq \gamma \leq \gamma_1$. In this range $0 \leq \gamma \leq \gamma_1$, let $\delta(\gamma)$ denote the optimum radius of the ball, and $x(\gamma)$ the center. Beginning with $\delta(0) = \bar{\delta}$, clearly, $\delta(\gamma)$ decreases as γ increases to γ_1. From these facts we see that in the range $\delta(0) \geq \delta(\gamma) \geq \delta(\gamma_1)$, $x(\gamma)$ is the optimum solution of

$$\begin{array}{ll} \text{Minimize} & Q(x) \\ \text{subject to} & A_{i.}x = b_i + \delta(\gamma), \ i \in T(\bar{x}) \end{array} \tag{6}$$

Replacing the parameter $\delta(\gamma)$ by the symbol s, an optimum solution for (6) can be obtained by solving

$$\begin{aligned} c^T + Dx - \textstyle\sum_{i \in T(\bar{x})} \pi_i A_{i.} = 0 \\ A_{i.}x = b_i + s, \ i \in T(\bar{x}) \end{aligned} \tag{7}$$

where $\pi_{T(\bar{x})} = (\pi_i : i \in T(\bar{x}))$ is the vector of lagrange multipliers for (6). If $(x(s), \pi_{T(\bar{x}}(s))$ is a solution of (7) as a function of the parameter s, then $x(s)$ defines a straight line in R^n in terms of the parameter s. The above argument shows that by carrying out a line search step on this straight line, we can decrease the value of $Q(x)$ to reach $Q(x(\gamma_1))$; and any further decrease in the value of $Q(x)$ below this will lead to an optimal touching constraint index set for the ball different from $T(\bar{x})$.

Even when (2) is solved approximately, we may improve the objective value by carrying out this work with the ball obtained. That is what this step does.

Denoting the ball obtained in the centering step by the same symbol $B(\bar{x}, \bar{\delta}) = \{x : (x - \bar{x})^T(x - \bar{x}) \leq \bar{\delta}^2\}$, denote the touching constraint index set by the same symbol as above $T(\bar{x}) = \{i : A_{i.}\bar{x} = b_i + \bar{\delta}\}$. With this $T(\bar{x})$, get the solution $(x(s), \pi_{T(\bar{x})})$ for system (7). Then do a line search to minimize $Q(x)$ over the line segment $\{x(s) : s$ such that $x(s) \in K\}$. Suppose $s = s_1$ gives the optimum $x(s)$ in this line search step.

If $x(s_1)$ is an interior point of K; then terminate if $\nabla Q(x) = 0$ at this point, otherwise define this point as the output of this step.

If however, $x(s_1)$ is a boundary point of K, let $I = \{i : i\text{-th constraint in}$ (1) is satisfied as an equation by $x(s_1)\}$. If the following system in Lagrange multipliers $\pi_I = (\pi_i : i \in I)$

$$\begin{aligned} c + x(s_1)^T D - \textstyle\sum_{i \in I} \pi_i A_{i.} = 0 \\ \pi_i \geq 0 \text{ for all } i \in I \end{aligned} \tag{8}$$

has a feasible solution, then $x(s_1)$ is an optimum solution of (1), terminate. However, it may not be productive to check if system (8) is feasible every time this step ends up at this stage. If this operation of checking the feasibility of (8) is not carried out, or if (8) turns out to be infeasible, then take the output of this step as a point on the line segment $\{x(s) : s \in R^1\}$ close to $x(s_1)$ but in the interior of K.

3.5 The Algorithm

The algorithm consists of repititions of the following iteration beginning with an initial interior point of K. We will now describe the general iteration. In each iteration, Steps 2.1 and 2.2 are parallel steps, both of which begin with the ball obtained in the centering step in the iteration.

A General Iteration

Let x^0 be the current interior feasible solution.

1. Centering Strategy: Apply the centering strategy described in Section 3.2 beginning with the current interior feasble solution. Let $B(\bar{x}, \bar{\delta})$ denote the ball obtained with center $\bar{x}$ and radius $\bar{\delta}$. Let $T(\bar{x}) = \{i : A_{i.}\bar{x} = b_i + \bar{\delta}\}$ is the index set of touching constraints for this ball.

2.1. Descent Step Using a Descent Direction: Apply this strategy described in Section 3.3 beginning with the ball $B(\bar{x}, \bar{\delta})$. If termination did not occur in this step, let x^1 denote the interior feasible solution of (1) which is the output point in this step.

2.2. Descent Step Using the Touching Constraints: Apply this strategy described in Section 3.4 beginning with the ball $B(\bar{x}, \bar{\delta})$. If termination did not occur in this step, let x^2 denote the interior feasible solution of (1) which is the output point in this step.

3. Move to Next Iteration: Define the new current interior feasible solution as the point among x^1, x^2 obtained in Steps 2.1, 2.2, which gives the smallest value for $Q(x)$. With it, go to the next iteration.

3.6 Convergence Results

In this section we discuss convergence results on the algorithm under the assumption that the centering problem is solved to optimality in every iteration.

Theorem 1: Cosider the following version of (2) with $Q(x^0)$ replaced by a parameter t.

$$\begin{array}{rl} & \delta[t] = \text{Maximum value of } \delta \\ \text{subject to} & \delta - A_{i.}x \leq -b_i, \; i = 1, \ldots, m \\ & Q(x) \leq t \end{array} \tag{9}$$

$\delta[t]$ is a concave function of t in the interval of values of t for which the above problem has a feasible solution.

Proof. Let $t_{\min}, t_{\max}$ be the minimum and maximum values of $Q(x)$ over $x \in K$. Let t_1, t_2 be any pair of values in the interval $[t_{\min}, t_{\max}]$; and suppose $(x^1, \delta^1), (x^2, \delta^2)$ are optimum solutions of (9) when $t = t_1, t_2$ respectively. Let $0 < \alpha < 1$.

Since $Q(x)$ is convex, we have $Q(\alpha x^1 + (1-\alpha)x^2) \leq \alpha Q(x^1) + (1-\alpha)Q(x^2) \leq \alpha t_1 + (1-\alpha)t_2$. From this we verify that $(\alpha x^1 + (1-\alpha)x^2, \alpha\delta^1 + (1-\alpha)\delta^2)$ is feasible to (9) when $t = \alpha t_1 + (1-\alpha)t_2$, but it may not be an optimum solution of (9).

Therefore the optimum objective value in (9) when $t = \alpha t_1 + (1-\alpha)t_2$, $\delta[\alpha t_1 + (1-\alpha)t_2] \geq \alpha\delta^1 + (1-\alpha)\delta^2 = \alpha\delta[t_1] + (1-\alpha)\delta[t_2]$. This shows that $\delta[t]$ satisfies Jensen's inequality required for being concave. □

Let $P(t)$ denote the set of feasible solutions of (9). Clearly, for $t_1 < t_2$, we have $P(t_1) \subset P(t_2)$. So, $\delta[t]$ decreases monotonically as t decreases; and since it is concave its slope decreases as t increases.

Theorem 2: The index set of touching constraints for the ball obtained in the centering step changes after each iteration in the algorithm.

Proof. This follows since the output point in each iteration in the algorithm, is selected as the best among the outputs in Steps 2.1, 2.2 in that iteration. □

Theorem 3: The algorithm terminates after at most $2m$ iterations.

Proof. Select an index between 1 to m, say i_1. As t is decreasing to $t_{\min}$, suppose the index i_1 is in the touching constraint index set for the ball obtained from (9) when $t = t_1$, and drops out of this set when t decreases from t_1. This implies that the system of constraints

$$\begin{aligned} &\delta - A_{i_1.}x = -b_{i_1}, \\ &\delta - A_{i.}x \leq -b_i, \ i \neq i_1 \\ &Q(x) \leq t \end{aligned} \tag{10}$$

is feasible when $t = t_1$, and infeasible when t is slightly smaller than t_1. From convexity of $Q(x)$ we know that the set of values of t for which (10) is feasible is an interval. These facts imply that (10) is infeasible for all $t < t_1$, i.e., as t decreases below t_1, the index i_1 can never be in the touching constraint index set. So, once an index drops out of the touchnig constraint index set in the algorithm, it can never enter it in a subsequent iteration. Since the touching constraint index set changes in every iteration, these facts prove the theorem. □

Even if the centering step is carried out approximately, these results indicate that if it is carried to good accuracy, the algorithm will have superior performance.

3.7 The Case When the Matrix D is Not Positive Definite

Relaxing the positive definiteness assumption on the matrix D leads to a vast number of applications for the model (1). For example, an important model with many applications is the following 0−1 mixed integer programming (MIP) model:

$$\begin{array}{ll} \text{Minimize} & cx \\ \text{subject to} & Ax \geq b \\ & x \geq 0 \\ & x_j = \ 0 \text{ or } 1 \text{ for each }, j \in J \end{array} \tag{11}$$

where J is the subscript set for variables which are required to be binary. Solving this problem is equivalent to finding the global minimum in the quadratic program

$$\begin{array}{ll} \text{Minimize} & cx + M\sum_{j\in J} x_j(1-x_j) \\ \text{subject to} & Ax \geq b \\ & x_j \geq 0, \text{ for } j \notin J \\ & 0 \leq x_j \leq 1 \text{ for each } j \in J \end{array} \tag{12}$$

where M is a large positive penalty coefficient; which is in the form (1) with D negative semidefinite. Unlike the model (1) when D is positive definite, (12) may have many local minima, and we need to find the global minimum for (12).

Some of the steps in this algorithm can still be carried out. The approximate centering procedure can be carried out. Also, Steps 2.1 can be carried out exactly. For Step 2.2, the system of equations (6) may typically have a unique solution. Even when (6) has many feasible solutions, a solution to (7) may not even be a local minimum for (6), in fact it may be a local maximum for (6). So, the value of including Step 2.2 in the algorithm is not clear in this case. Also, many of the proofs in Section 3.6 based on convexity will not be valid in this nonconvex case.

However, since the ball minimization problems in Step 2.1 can be solved exactly, there is reason to hope that by adjusting the value of the penalty cost coefficient M during the algorithm, the algorithm can be made to lead to a good local minimum, and thereby offer a good heuristic approach. For this general case, these and other issues need to be pursued.

Bibliography

Conn A. R., Gould N. I. M and Toint P. L.(2000). *Trust-Region Methods*, (MPS-SIAM Series on Optimization).

Cottle R. W., Pang J. S. and Stone R. E. (1992). *The Linear Complementarity Problem*, (Academic Press, Boston, MA).

Murty K. G. (1988). *Linear Complementarity, Linear and Nonlinear Programming*, (Helderman Verlag, Berlin, Germany, 1988) (can be accessed on the web from Murty's web page at: http:// www-personal.umich.edu/ murty/).

Murty K. G. (2006). A new practically efficient interior point method for LP. *Algorithmic Operations Research* **1**, pp. 3–19.

Ye Y. (1997). *Interior Point Algorithms, Theory and Analysis*, (Wiley-Interscience, New York).

Chapter 4

A General Framework for the Analysis of Sets of Constraints

Richard Caron
Department of Mathematics and Statistics
University of Windsor
Windsor, Ontario, N9B3P4, Canada
e-mail: rcaron@uwindsor.ca
Tim Traynor
Department of Mathematics and Statistics
University of Windsor
Windsor, Ontario, N9B3P4, Canada
e-mail: tt@uwindsor.ca

Abstract

This paper is about the analysis of sets of constraints, with no further assumptions. We explore the relationship between the minimal representation problem and a certain set covering problem of Boneh. This provides a framework that shows the connection between minimal representations, irreducible infeasible systems, minimal infeasibility sets, as well as other attributes of the preprocessing of mathematical programs. The framework facilitates the development of preprocessing algorithms for a variety of mathematical programs. As some such algorithms require random sampling, we present results to identify those sets of constraints for which all information can be sampled with nonzero probability.

Key Words: Optimization, preprocessing, redundancy, irreducible infeasible systems, set covering, minimal infeasibility sets

4.1 Introduction

We consider an indexed family $\{A_i, B_i\}$ of partitions of an abstract space X. We think of A_i as the set of points satisfied by the ith constraint of an optimization problem, and $B_i = A_i^c$, the set of those that violate it. For example we could be given a family of constraint functions g_i on X and $A_i = \{x \in X : g_i(x) \leq 0\}$ and $B_i = \{x \in X : g_i(x) > 0\}$. In general, the function $g_i = \delta_i$, defined by

$$\delta_i(x) = \begin{cases} 0, & x \in A_i \\ 1, & x \in B_i; \end{cases} \tag{4.1}$$

namely, the indicator (or characteristic) function of B_i, could serve as such a constraint function. Often, in applications, the A_i will be closed sets.

The **feasible set** for the family $\{A_i, B_i\}_{i \in I}$ is given by $Z(I) = \bigcap_{i \in I} A_i$. This may, or may not, be empty. If it is empty, we will say that the family is **infeasible**; otherwise, **feasible**. In either case, we are interested in subsets J of I such that $Z(J) = Z(I)$. In this situation, we call J a **reduction** of I and the family $\{A_i, B_i\}_{i \in J}$, a reduction of $\{A_i, B_i\}_{i \in I}$. The family $\{A_i, B_i\}_{i \in J}$ is **irreducible** if $J \subseteq I$ and there is no proper reduction J' of J.

In the feasible case ($Z(I) \neq \emptyset$), the search for such subsets J is equivalent to the detection of redundancy, one aspect of preprocessing. For linear programs, the importance of preprocessing has been established, for example, by [Karwan *et al.* (1983)], [Andersen and Andersen (1995)], [Bixby (1994)], and [Lustig *et al.* (1994)]. Descriptions of deterministic methods to detect redundancy can be found in [Brearly *et al.* (1975)], [Tomlin and Welch (1986)], [Karwan *et al.* (1983)], [Greenberg (1996)], and [Caron *et al.* (1989)]. Probabilistic methods are described in [Boneh (1983)], [Berbee *et al.* (1987)] and [Caron *et al.* (1990)].

In the case of infeasible ($Z(I) = \emptyset$) linear programs, the search for reductions J is equivalent to the search for irreducible infeasible systems (IIS's). The paper by [Chinneck and Dravnieks (1991)] describes the powerful IIS algorithms that are available in many professional linear programming codes. Related to the problem of finding an IIS is that of finding a minimal infeasibility set (MIS), that is, a set J of minimum cardinality such that $Z(I \setminus J) \neq \emptyset$. [Chakravarti (1994)] showed that finding an MIS is NP-hard.

According to [Caron (2001)], very little attention has been paid to nonlinear constraint sets. Some exceptions are in [Obuchowska and Caron

(1995)], [Jibrin (2002)], and [Obuchowska (2000)], for quadratic, positive semidefinite, and convex programming problems, respectively. The growing importance of more general mathematical programs to very large scale engineering design problems, among others, together with strong evidence of the importance of presolve and infeasibility analysis for linear programmes, indicates the need for this deficiency to be corrected. This paper is such a contribution.

In 1984, [Boneh (1984)] introduced and exploited an equivalence between the minimal representation problem and a set covering problem to develop a tool for detecting and removing redundant constraints. His implementation involved a random sampling of points x in X each of which created a row $\delta(x)$ in a set covering matrix. With the introduction of an objective function this becomes a set covering problem. Boneh showed that while any feasible solution gives a reduction, an optimal solution produces an irreducible reduction. As the set covering problem is NP-hard, polynomial time heuristics were suggested to find near-optimal reductions.

This paper develops Boneh's equivalence further. Our initial contribution, in Section 4.2, is a new presentation of the concepts, making, we believe, the correspondence between sets of constraints and set covering problems, and the proof of key results, more transparent and shorter. It also becomes clear that the equivalence holds regardless of feasibility, yielding the first theoretical framework to address minimal representations, IIS's and MIS's without a priori knowledge of feasibility.

4.2 The Set Covering Formulation

Consider the feasible set $Z(I)$. Since, for each i, $A_i = B_i^c$, the complement of $Z(I)$, the set of **infeasible points**, is

$$Z(I)^c = \bigcup_{i \in I} B_i.$$

Thus, $\{B_i : i \in I\}$ is a cover of $Z(I)^c$ and a reduction J of I defining $Z(I)$ amounts to a reduction of the cover, in that $\{B_i : i \in J\}$ also covers $Z(I)^c$:

$$\bigcup_{i \in J} B_i \supset Z(I)^c. \tag{4.2}$$

This inclusion characterizing reduction tells us that each infeasible point is in some B_i, $i \in J$ and thus violates some constraint in the reduction J. Applied to irreducible reductions, this is the content of "The Main Theorem" of [Boneh (1984)]. In informal speech, when an irreducible reduction

J is found, the constraints indexed by J are called **necessary** or **non-redundant**, and the others **redundant**. An irreducible reduction is not unique, but:

Theorem 1. *[Boneh (1984)] For each infeasible point x, some constraint violated by x must be necessary in each irreducible set of constraints.*

We emphasize that, as noted, this is actually true for every *reduction*, even if it is not irreducible, since in practice it is usually not possible to obtain a truly irreducible one.

Corollary 2. *If x violates only one constraint, that constraint is necessary in each reduction.*

Let's gather the indicator functions δ_i defined above, into one "binary word valued" indicator function $\delta = \delta_I = (\delta_i)_{i \in I}$, mapping X to $\{0,1\}^I$. Its sets of constancy, in other words, the equivalence classes

$$[x] = \delta^{-1}(\delta(x)) = \{y : \delta(y) = \delta(x)\},$$

partition X. The resulting partition $\mathcal{P} = \mathcal{P}_I$ is the coarsest partition finer than each of the $\{A_i, B_i\}$ and

$$[x] = \left(\bigcap_{i:\delta_i(x)=0} A_i \right) \bigcap \left(\bigcap_{i:\delta_i(x)=1} B_i \right).$$

Since each class in $\mathcal{P}$ is determined by an element $\delta(x) \in \{0,1\}^I$, an upper bound on the cardinality of $\mathcal{P}$ is $2^{|I|}$. At times we will refer to $\delta(x)$ as the word or observation associated with the point x. Since $Z(I) = \bigcap_{i \in I} A_i = \delta^{-1}(0)$, the zero set of δ, it is also one of the classes in $\mathcal{P}$.

We extend these notions to subfamilies $\{A_i, B_i\}_{i \in J}$ of the original family $\{A_i, B_i\}_{i \in I}$. Thus, δ_J is the function on X with values in $\{0,1\}^J$ whose i^{th} component is δ_i, for each $i \in J$; in other words, δ_J is the composition of δ with the projection of $\{0,1\}^I$ onto $\{0,1\}^J$. We see that the partition $\mathcal{P}_J$ induced by δ_J is coarser than that of δ_I.

Theorem 3. *Let $y = \mathbf{1}_J \in \{0,1\}^I$. Then, J is a reduction of I if and only if $\delta(x) \cdot y \geq 1$ for all x with $\delta(x) \neq 0$.*

Proof. We have done all the work in setting up the notation. In terms of the indicator functions, the inclusion (4.2) says J is a reduction of I if and only if $\delta_I(x) \neq 0$ implies $\delta_J(x) \neq 0$ and this latter holds if and only if $\delta(x) \cdot \mathbf{1}_J \geq \mathbf{1}$. □

Thus, one can find an irreducible reduction of I by solving the set covering problem

$$\begin{array}{ll} \text{minimize} & |y| = \sum_{i \in I} y_i \\ \text{subject to} & \delta(x) \cdot y \geq 1, \text{ for all } x \text{ with } \delta(x) \neq 0. \end{array}$$

It is convenient to let E be the set of all possible words $\delta(x)$ other than 0, and think of it as a matrix whose rows are indexed by the infeasible equivalence classes (elements of the partition $\mathcal{P}$). Then this becomes a standard set-covering (SC) problem

$$\begin{array}{l} \text{minimize } \ |y| = 1^\top y \\ \text{subject to } Ey \geq \mathbf{1}, \ y \text{ binary}, \end{array} \tag{4.3}$$

where $\mathbf{1}$ is a vector of ones of appropriate dimension.

In the case of linear programs, the corresponding SC problem can be solved in linear time. (This can be achieved by carefully applying Corollary 2.) This is not the case for more general problems. Fortunately, since any *feasible* y in the SC problem (4.3) corresponds to a reduction, one needn't actually find an *optimal* solution to obtain a benefit, and heuristics, such as the greedy algorithm in [Chvatal (1979)], can produce excellent results.

The partition $\mathcal{P}$, represented by the complete matrix E, provides a common framework for the concepts of minimal representation, irreducible infeasible system, and minimal infeasibility set. Suppose that we have a optimal solution to the set covering problem with corresponding irreducible reduction J. If the family is feasible, J provides a minimal representation. If the family is infeasible, J provides an irreducible infeasible system and the word with smallest row sum indicates a minimal infeasibility set. Thus, the concepts need no longer be treated separately.

Concerning the matrix E, we notice that:

(1) If columns k and ℓ in E are identical, then constraints k and ℓ are duplicate.
(2) If columns k and ℓ in E are complementary, then constraints k and ℓ are opposite, i.e., $A_k = B_\ell$.
(3) If column k is a column of zeros, then constraint k is everywhere satisfied.
(4) If column k is a column of ones, then constraint k is everywhere violated.

This next observation was suggested by A. Boneh in private conversation with the first author, and appeared in the master's major paper [Krishnamurthy (2001)] supervised by the authors.

Lemma 4. *If E contains $\binom{|I|}{m}$ rows with row sum m, then any reduction J of I has at least $(|I| - m + 1)$ elements.*

Proof. Note that $\binom{|I|}{m}$ is the number of possible rows of row sum m. If J is a subset of I with fewer than $|I| - m + 1$ elements, its complement in I contains a set K of m elements, which provides a row $e = \mathbf{1}_K^\top$ of E with $e \cdot \mathbf{1}_J = 0$, so that $E1_J \geq \mathbf{1}$ is not satisfied. □

The same argument can be applied to subsets I_0 of I. Thus, if I_0 has k elements and there are $\binom{k}{m}$ rows with exactly m non-zero entries in I_0, then $k+m-1$ of the constraints in I_0 are necessary. In practice, it may be difficult to use this version, because it would require searching through too many submatrices. If $m = 1$, it would be easy for we could simply take I_0 to be the set of all i corresponding to row sum 1, but this is already covered by Corollary 2. If m is 2, then this version would say that all but 1 of the members of I_0 are necessary.

Reducing the Set Covering Matrix. By a **reduction of the set covering matrix** E, we mean a subset F of E such that, for "column" binary words y, $Fy \geq \mathbf{1}$ implies $Ey \geq \mathbf{1}$. Clearly, the set covering problem obtained from the original by replacing E by F has the same feasible solutions, hence the same optimal solutions. The **bitwise partial ordering** on $\{0,1\}^I$, $e \leq f$, is given by $e \leq f$ if $e_i \leq f_i$, for all $i \in I$.

Lemma 5. *For $F \subseteq E \subseteq \{0,1\}^I$, F is a reduction of E (as a set covering matrix) iff for every $e \in E$, there exists $f \in F$, with $f \leq e$.*

Proof. If the condition holds, then for each $e \in E$, we can choose f with $e \geq f$, and then for each y, $ey \geq fy \geq 1$.

Conversely, suppose F is a reduction of E, but the condition is not satisfied; say, $e \in E$, but there is no $f \in F$ with $e \geq f$. Then, for each $f \in F$, we may choose $j = j_f$ with $1 = f_j > e_j = 0$. Let $J = \{j_f : f \in F\}$ and $y = \mathbf{1}_J$. Then $Fy \geq \mathbf{1}$, but $ey = 0$, a contradiction. □

Thus, E is irreducible (that is, has no proper reduction) if and only if no two elements are comparable. This is not to say that, if E corresponds to the family of constraints $\{A_i : i \in I\}$, the latter is irreducible.

4.3 Random Sampling

One way of collecting the elements of E is by sampling points $x \in X$ and calculating the corresponding $\delta(x)$.

Boneh's example. In [Boneh (1984)] the author presented a seemingly straightforward example to demonstrate his SC method. In a 1985 private communication, McDonald and Caron pointed out that a failure to sample on classes (members of the partion $\mathcal{P}$) of measure zero caused rows of E to be overlooked and led to incorrect conclusions. The 1999 Master's thesis by Feng [Feng (1999)], co-supervised by the authors, presented the first theoretical results aimed at the identification of a class of problems for which all classes can be sampled with nonzero probability. In the present paper, we provide a refined theorem and proof.

The **support** of a Borel measure is the complement of largest open set of zero measure. (See for example [Rao (2004)].) For a probability distribution on the Borel sets of a metric subspace of $\mathbf{R}^n$, this amounts to the smallest closed set of probability 1 (equivalently the set of points, each of whose neighbourhoods have positive probability — see [Chung (1974)], page 31.) We say the distribution P is **supported on** X if the support of P is X. Thus, if the distribution P is supported on X and $a \in X$, then every neighbourhood of a will intersect X in a set of positive probability. For example, X could be an interval (box) of $\mathbf{R}^n$ with non-empty interior and the distribution could be uniform on X or (the restriction to X of) a multivariate normal distribution. More generally, if X is a metric subspace of $\mathbf{R}^n$ and P has a continuous density f, with $(f > 0)$ dense in X, then P is supported on X.

Theorem 6. *Suppose that each A_i is given by $(g_i \leq 0) = \{x \in X : g_i(x) \leq 0\}$ where the g_i are continuous functions. For each $J \subseteq I$, put $g_J(x) = \max_{j \in J} g_j(x)$. If 0 is not a local minimum of any g_J, then each non-zero value of δ will be sampled with non-zero probability under any distribution supported on X.*

Proof. For a given $x \in X$, let J be the set of indices j with $g_j(x) = 0$. Then the equivalence class $[x]$ is $Z(J) \cap U(J^c)$, where $Z(J) = \bigcap_{j \in J} A_j = (g_J \leq 0)$ and $U(J^c) = \bigcap_{j \in J^c} B_j$, is an open set. If 0 is not a local minimizer, then the open set $(g_J < 0)$ contains a point a of $[x]$. Thus, the open set $(g_J < 0) \cap U(J^c)$ is a neighbourhood of a, hence intersects X in a set of positive probability. $\square$

In Boneh's example, mentioned above, the hypotheses of Theorem 6 are not satisfied, since there is a local minimum of 0 for some g_J. The next result gives conditions under which the hypotheses of the theorem are satisfied.

A constraint ($g_i \leq 0$) is said to be an **implicit equality** if there is a subset J of I with $Z(J) \neq \emptyset$ such that $g_i = 0$ on $Z(J)$. (The definition here is modified from that in [Obuchowska and Caron (1995)] to take into account the possibility of an infeasible family of constraints. The original concept was introduced by [Telgen (1983)].)

Corollary 7. *If the constraint functions are convex and there are no implicit equalities, all non-zero values of δ are chosen with positive probability under any distribution supported on X.*

Proof. In case all the functions g_j are convex, so are the g_J. The existence of a local minimum 0 would give a global minimum 0 and hence, $g_J = 0$ on $Z(J)$, which means the constraints g_j, $j \in J$ induce implicit equalities: on $Z(J)$, all g_i are 0. Thus, if there are no implicit equalities, 0 is not a local minimum for any g_J, and the theorem applies. □

We illustrate these ideas with families of (non-linear) convex quadratic constraints in 2 variables. Here X is an interval $[0, 10] \times [0, 10]$, the A_i are the intersections with X of elliptical regions with non-empty interior. Points are selected uniformly in X and the corresponding observations $\delta(x)$ are calculated. The distinct values of $\delta(x)$ are put into a set E and treated as the set covering matrix, although some rows may be missing. (Since the constraint functions are strictly convex, there can be no implicit equalities; hence, according to Corollary 7, each region will be sampled with positive probability.) Figure 1 shows an infeasible family of constraints, a plot of 1000 random points, the corresponding matrix E, and beside it an irreducible reduction, from which we see that irreducible reductions of the original problem are given by $\{2, 7\}$ and $\{3, 7\}$. Since the family is infeasible, these are Irreducible Infeasiblity Sets. The figure itself indicates that these results are probably correct. Chvatal's algorithm applied to this E yielded the reduction $\{2, 7\}$. Figure 2 shows a feasible family, its corresponding E and its (unique) corresponding irreducible reduction, which consists of only words with a single bit 1. This determines the minimal representation $\{1, 2, 3, 5, 6\}$. Note, by the way, that constraint 7 here turned out to be the entire interval X, so was always satisfied. This is reflected in the column of 0's in the matrix E.

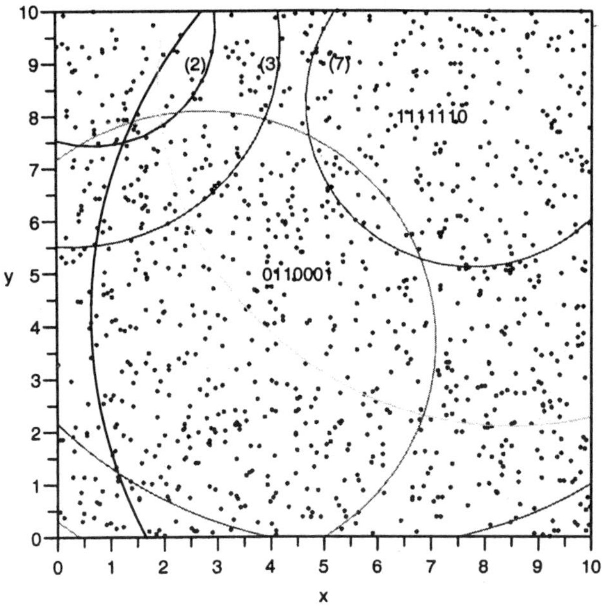

E	reduction of E
0101001	
0111000	
1111101	
1110101	
1110001	
1001011	
1111001	
1001001	
0110000	
0100001	
1110111	
0110001	
0111001	
1100001	
1100011	
1110011	
0001001	0110000
1000001	0100001
1000011	0001001
0001011	1000001

Figure 1: Infeasible family: IIS given by $\{2,7\}$ or $\{3,7\}$.

Hit-and-Run variations. In [Boneh and Golan (1979)] a "hit-and-run" algorithm was introduced for the identification of redundancy and feasible region boundedness. This led to the development of a family of variations of the method ([Boneh (1983); Smith (1984); Telgen (1979); Berbee *et al.* (1987); Bélisle *et al.* (1998, 1993)]). Consider an absolutely continuous distribution π on an open set G of $\mathbf{R}^n$, with an almost-everywhere continuous Lebesgue density f, positive on G. Let ν be a distribution on the unit sphere of $\mathbf{R}^n$. Define a discrete time Markov chain by taking as transition kernel $P(x, B)$ the result of first choosing a direction s according to the distribution ν and then choosing a point according to the distribution π conditioned[1] on the line $L(x,s)$ through x in the direction s. In [Bélisle *et al.* (1993)] it is shown that if the support of ν spans $\mathbf{R}^n$ and if connected components ν-communicate (in particular if G is connected), then the chain

[1]Since $L = L(x,s)$ has measure 0, this requires interpretation. One takes $f\mathbf{1}_L$ (normalized) as density with respect to 1 dimensional measure on L.

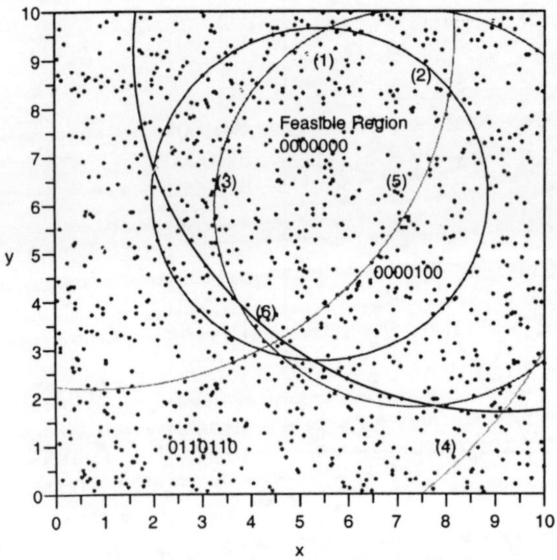

E	reduction of E
0110110	
0100100	
0010000	
0000100	
1110010	
0100000	
0110010	
1010000	
1100000	
1000000	
0100110	
1110000	
1110110	
0010010	
0000110	
0111100	1000000
0111110	0100000
1110100	0010000
0110100	0000100
0000010	0000010

Figure 2: Feasible family: constraints 1, 2, 3, 5, 6 are necessary.

converges in total variation to π. In particular, if the support of ν spans $\mathbf{R}^n$ and $P(x, B)$ comes from choosing a direction s according to ν, then a point from the uniform distibution on the intersection of $L(x, s)$ with a bounded open connected set G, specifically,

$$P(x,B) = \int \frac{\boldsymbol{\lambda}^1(B \cap L(x,G))}{\boldsymbol{\lambda}^1(G \cap L(x,s))}\,\nu(ds),$$

then the Markov chain converges in total variation to uniform distribution on G. Here $\boldsymbol{\lambda}^1$ denotes 1 dimensional Lebesgue (actually Hausdorff) measure in $\mathbf{R}^n$.

In application, if the space X is a subset of $\mathbf{R}^n$ whose interior is connected and whose boundary has measure 0, the analogous results would hold for X. Note that one only needs the support of ν to span $\mathbf{R}^n$, so one can simply use the uniform distribution on the n positive coordinate directions. This is the CD version of the Hit-and-run methods.

Once one decides to generate points along straight lines, $L(x,s)$, one can introduce modifications that have further advantages. For example, instead of just looking at the observation $\delta(x)$, one can collect many - in special cases all possible - observations along that line and easily retain an equivalent irreducible set of them. To illustrate, suppose X is convex and the A_j are of the form $(g_j \leq 0)$ with the g_j strictly convex, so that the $A_j \cap L(x,s)$ are completely determined by the solutions of the form $x+\sigma s$ to $g_j = 0$. Say these solutions are $a_i = x + \sigma_i s$, with $\sigma_i \leq \sigma_{i+1}$, for $1 \leq i < N$, $g_{j_i}(a_i) = 0$. Each index j_i will appear at most twice because of the convexity and as the parameter σ crosses σ_i, the j_i^{th} bit of $\delta(x+\sigma)$ will change from 1 to 0 or from 0 to 1. Thus, we can determine all the possible observations along that line.

This sets us up to use the following result, which enables one to select an equivalent irreducible set from the set of all observations collected along the line.

Theorem 8. *Let E be a set $\{e_1, \ldots, e_N\}$ of binary words in $\{0,1\}^J$. Suppose*

(1) for each $i < N$, either $e_i < e_{i+1}$ or $e_i > e_{i+1}$ and
(2) for each j there do not exist $i < k < \ell$ with $e_{ij} = 0$, $e_{kj} = 1$, and $e_{\ell j} = 0$.

Let $E' = \{e_i : i \in I'\}$ be the set of local minima of E in the partial ordering $\leq$. Thus, if $1 < i < N$, $i \in I'$ if and only if $e_{i-1} > e_i < e_{i+1}$, with the obvious modification for the cases $i = 1$ and $i = N$. Then, E' is a reduction of E for the set covering problem, and E' consists of incomparable words.

Condition (1) here is satisfied if there exists a unique j with $e_{i+1,j} \neq e_{ij}$.

Proof. To prove no two elements of E' are comparable, let $i, k \in I'$, with $i < k$. By the local minimality, $e_i < e_{i+1}$ and $e_{k-1} > e_k$. Choose j so that $e_{ij} < e_{i+1,j}$ and j' so that $e_{k-1,j'} > e_{k,j'}$. By condition (2) $e_{ij} < e_{kj}$ and $e_{ij'} > e_{kj'}$. Thus, neither $e_i \geq e_k$, nor $e_i \leq e_k$.

To show E' is a reduction of E, suppose $e_\ell \in E \setminus E'$. Then, either $e_\ell > e_{\ell+1}$ or $e_\ell > e_{\ell-1}$. In the first case, let i be the largest index $\geq \ell+1$ with $e_{i-1} > e_i$. Then, $e_\ell > e_i$ and $e_i \in E'$. The case $e_\ell > e_{\ell-1}$ is similar.□

In conclusion, we would like to remind the reader, that although our illustrations here emphasized convex constraints, the framework is completely general: the sets A_i, B_i need not have any special geometric or topological properties.

Acknowledgement

This work has been supported by NSERC, the Natural Sciences and Engineering Research Council of Canada.

Bibliography

Andersen, E. and Andersen, K. (1995). Presolving in linear programming, *Mathematical Programming* **71**, pp. 221–245.

Bélisle, C., Boneh, A. and Caron, R. (1998). Convergence properties of hit-and-run samples, *Stochastic Models* **14**, pp. 767–800.

Bélisle, C., Romeijn, H. and Smith, R. (1993). Hit-and-run algorithms for generating multivariate distributions, *Mathematics of Operations Research* **18**, pp. 255–266.

Berbee, H., Boender, C., Rinnooy Kan, A. H., Romeijn, H., Scheffer, C., Smith, R. L. and Telgen, J. (1987). Hit-and-run algorithms for the identification of nonredundant linear equalities, *Mathematical Programming* **37**, pp. 184–207.

Bixby, R. (1994). Progress in linear programming, *ORSA Journal on Computing* **6**, pp. 15–22.

Boneh, A. (1983). PREDUCE - a probabilistic algorithm for identifying redundancy by a random feasible point generator (RFPG), in [Karwan *et al.* (1983)], pp. 108–134.

Boneh, A. (1984). Identification of redundancy by a set-covering equivalence, in J. Brans (ed.), *Operational Research '84* (Elsevier Science Publishers B.V.(North Holland), Amsterdam), pp. 407–422.

Boneh, A. and Golan, A. (1979). Constraints redundancy and feasible region boundedness by random feasible point generator (RFPG), Presented at the EURO III Conference, Amsterdam, April 9-11(22pp).

Brearly, A., Mitra, G. and Williams, H. (1975). Analysis of mathematical programming problems prior to applying the SIMPLEX method, *Mathematical Programming* **8**, pp. 54–83.

Caron, R. (2001). Redundancy in nonlinear programs, in C. Floudas and P. Pardalos (eds.), *Encyclopedia of Optimization* (Kluwer Academic Publishers), pp. 12–15.

Caron, R. J., Hlynka, M. and McDonald, J. F. (1990). On the best case performance of hit and run methods for detecting necessary constraints, *Mathematical Programming* **54**, pp. 233–249.

Caron, R. J., McDonald, J. F. and Ponic, C. M. (1989). A degenerate extreme point strategy for the classification of linear inequalities as redundant or necessary, *Journal of Optimization Theory and Applications* **62**, pp. 225–237.

Chakravarti, N. (1994). Some results concerning post-infeasibility analysis, *European Journal of Operational Research* **73**, pp. 139–143.

Chinneck, J. and Dravnieks, E. (1991). Locating minimal infeasible constraint sets in linear programs, *ORSA Journal on Computing* **3**, pp. 157–167.

Chung, K. L. (1974). *A course in probability theory*, 2nd edn. (Academic Press [A subsidiary of Harcourt Brace Jovanovich, Publishers], New York-London), probability and Mathematical Statistics, Vol. 21.

Chvatal, V. (1979). A greedy heuristic for the set-covering problem, *Mathematics of Operations Research* **4**, pp. 233–235.

Feng, J. (1999). *Redundancy in Nonlinear Systems: a Set Covering Approach*, Master's thesis, University of Windsor.

Greenberg, H. (1996). Consistency, redundancy and implied equalities in linear systsems, *Annals of Mathematics and Artificial Intelligence* **17**, pp. 37–83.

Jibrin, S. (2002). Detecting redundancy in optimization problems over intersection of ellipsoids, *J. Interdiscip. Math.* **5**, 2, pp. 183–194.

Karwan, M. H., Lotfi, V., Telgen, J. and Zionts, S. (eds.) (1983). *Redundancy in Mathematical Programming* (Springer-Verlag, Berlin).

Krishnamurthy, A. (2001). *Classification of constraints by set covering*, Master's thesis, University of Windsor.

Lustig, I., Marsten, R. and Shanno, D. (1994). Interior point methods for linear programming: Computational state of the art, *ORSA Journal on Computing* **6**, pp. 1–14.

Obuchowska, W. (2000). Minimal representation of convex region defined by analytic constraints, *Journal of Mathematical Analysis and its Applications* **246**, pp. 100–121.

Obuchowska, W. T. and Caron, R. J. (1995). Minimal representation of quadratically constrained convex feasible regions, *Mathematical Programming* **68**, pp. 169–186.

Rao, M. M. (2004). *Measure theory and integration*, *Monographs and Textbooks in Pure and Applied Mathematics*, Vol. 265, 2nd edn. (Marcel Dekker Inc., New York), ISBN 0-8247-5401-8.

Smith, R. L. (1984). Efficient monte carlo procedures for generating points uniformly distributed over bounded regions, *Operations Research* **32**, pp. 1296–1308.

Telgen, J. (1979). The CD algorithm, Private communication.

Telgen, J. (1983). Identifying redundant constraints and implicit equalities in systems of linear constraints, *Management Science* **29**, pp. 1209–1222.

Tomlin, J. A. and Welch, J. F. (1986). Finding duplicate rows in a linear programming model, *Operations Research Letters* **5**, pp. 7–11.

Chapter 5

Tolerance-based Algorithms for the Traveling Salesman Problem

Diptesh Ghosh
P&QM Area, IIM Ahmedabad, India.
e-mail: diptesh@iimahd.ernet.in
Boris Goldengorin
Faculty of Economic Sciences, University of Groningen, The Netherlands.
Gregory Gutin
Department of Computer Science, Royal Holloway University of London, Egham, Surrey TW20 OEX, UK and *Department of Computer Science, University of Haifa, Israel.*
Gerold Jäger
Computer Science Institute, University of Halle-Wittenberg, Germany.

Abstract

Most research on algorithms for combinatorial optimization use the costs of the elements in the ground set for making decisions about the solutions that the algorithms would output. For traveling salesman problems, this implies that algorithms generally use arc lengths to decide on whether an arc is included in a partial solution or not. In this paper we study the effect of using element tolerances for making these decisions. We choose the traveling salesman problem as a model combinatorial optimization problem and propose several greedy algorithms for it based on tolerances. We report extensive computational experiments on benchmark instances that clearly demonstrate that our tolerance-based algorithms outperform their weight-based counterpart. This indicates that the potential for using tolerance-based algorithms for various optimization problems is high and motivates further investigation of the approach.

Key Words: Traveling salesman problems, greedy algorithms, arc tolerances

5.1 Introduction

In this paper we propose several algorithms for the traveling salesman problem (TSP). In a TSP instance of size n, we are given a weighted complete digraph $D = (V, A, C)$ where V is the set of n vertices, A the set of arcs between vertices in V, and $C = [c(i,j)]$ is the $n \times n$-matrix of non-negative arc weights, and we are required to find a Hamilton cycle (called a *tour*) such that the sum of the weights of the arcs in the tour is as small as possible. A TSP instance is called a *symmetric TSP* (STSP) instance if for each pair of vertices i and j, $c(i,j) = c(j,i)$; and an *asymmetric TSP* (ATSP) instance otherwise. Also, a TSP instance is defined by the weight matrix C.

Most algorithms for solving the TSP make use of the arc weights to decide whether or not to include an arc in the solution that they finally output. For example, the weight-based greedy algorithm and its variations are popular heuristics to produce initial tours for local search and other improvement heuristics (see, e.g., [Gamboa *et al.* (2006)]). However, as pointed out in [Goldengorin *et al.* (2004)] and [Turkensteen (2005)], arc *tolerances* are better indicators than arc weights for generating good tours.

An arc tolerance (see e.g., [Goldengorin *et al.* (2006)], [Goldengorin and Sierksma (2003)], [Libura (1991)]) is the maximum amount by which the weight of the arc that is in (not in) an optimal tour can be increased (respectively, decreased) while keeping other arc weights unchanged for the tour to remain optimal. Among currently known algorithms for the TSP, only Helsgaun's version of the Lin-Kernighan heuristic for the STSP (see [Helsgaun (2000)]) explicitly applies tolerances in algorithm design. Implicit applications of tolerances in algorithm design are found in the Vogel's method for the Transportation Problem and in the MAX-REGRET heuristic for solving the Three-Index Assignment Problem (see [Balas and Saltzman (1991)]).

To the best of our knowledge the concept of tolerances has not been applied to the design of greedy algorithms for the TSP prior to this paper. Our aim is to motivate research on the use of tolerances for decision making in fast TSP heuristics. The algorithms that we propose may therefore not be the best of breed, but they demonstrate the superiority of tolerance-based algorithms over their arc-weights counterparts. Our results thus indicate a high potential of tolerance-based algorithms for various optimization problems and motivate further investigation of the approach.

In the next section, we develop concepts that will help us to describe the algorithms that we introduce for the TSP in Section 5.3. Our tolerance-based greedy algorithms are described in Section 5.3. We report computational experience with our greedy algorithms in Section 5.4. We conclude the paper in Section 5.5 with a summary of our main contributions and suggestions for future research.

5.2 Some Relevant Concepts

5.2.1 *The Relaxed Assignment Problem*

The Assignment Problem (AP) is a well-known relaxation of the TSP, which is used more often for the ATSP than for the STSP. Let $D = (V, A, C)$ be a bipartite digraph with bipartition $V = V_1 \cup V_2$, $|V_1| = |V_2| = n$, and such that $A = V_1 \times V_2$. The AP is defined as the problem to find n arcs (s_i, t_i), $1 \leq i \leq n$ of minimum total weight such that $s_i \neq s_j$ and $t_i \neq t_j$ for every $1 \leq i \neq j \leq n$, i.e., the AP is the problem to find a minimum weight perfect matching. Notice that if V_1 and V_2 are two copies of the vertex set of an TSP instance, where the arc weights of the bipartite directed graph correspond to the arc weights of the TSP and the arc weight of a vertex and its copy is set to ∞, then the AP solution can be interpreted as a collection of cycles (called subtours) for the instance.

An integer programming formulation of the AP on an ATSP instance defined on a complete digraph $G = (V, A, C)$ (where $|V| = n$, and $C = [c(i,j)]$) using variables x_{ij}, $i, j \in V$ such that $x_{ij} = 1$ when (i, j) is included in the solution and 0 otherwise, is given below.

$$\begin{aligned}
\text{Minimize} \quad & \sum_{i=1}^{n}\sum_{j=1}^{n} c(i,j)x_{ij} \\
\text{Subject to} \quad & \sum_{j=1}^{n} x_{ij} = 1 \quad i \in \{1, \ldots, n\} && (5.1) \\
& \sum_{i=1}^{n} x_{ij} = 1 \quad j \in \{1, \ldots, n\} && (5.2) \\
& x_{ij} \in \{0, 1\} \quad i, j \in \{1, \ldots, n\}
\end{aligned}$$

The Relaxed Assignment Problem (RAP) is a relaxation of the AP in which constraint set (5.2) is removed from the earlier formulation. Thus, instead of an one-to-one matching in case of the AP, in the RAP the first copy of V maps into the second copy of V. Note that a solution to the

RAP may not consist exclusively of cycles.

5.2.2 *Determining Tolerances for AP and RAP*

Extending the informal definition of tolerances in the introductory section, the upper (lower) tolerance of an arc that is included in (respectively, excluded from) an optimal solution to the AP is the maximum amount by which the weight of the arc can be increased (respectively, reduced) while keeping other arc weights unchanged, such that the current optimal solution to the AP remains optimal. Tolerances for arcs of the RAP can be defined analogously.

Computing arc tolerances for the AP involves revising the arc weight to a suitably high value if the arc is a part of the optimal solution, and a suitably low value if it is not (see [Goldengorin *et al.* (2006)]), and re-solving the AP. The AP can be solved in $\mathcal{O}(n^3)$ time, and using a shortest path based approach, all arc tolerances can also be computed in $\mathcal{O}(n^3)$ time (see [Volgenant (2006)]).

Computing arc tolerances for the RAP, on the other hand, is a more tractable problem. An optimal solution to the RAP can be characterized as a collection of arcs, one from each vertex in the graph, such that the weight of the arc is the smallest among those of all arcs from that vertex. Therefore, for each arc that belongs to an optimal solution to the RAP, its upper tolerance is the excess of the weight of the second smallest weight out-arc from the same vertex over the weight of that arc. If the arc is not in an optimal solution, then its lower tolerance would be the excess of the weight of the arc over that of the smallest weight out-arc from the same vertex. Obtaining all tolerances therefore requires finding the weights of the two least weight entries in each row of the cost matrix, and then performing a simple subtraction operation once for each arc. Both jobs can be achieved in $\mathcal{O}(n^2)$ time so that the overall complexity of determining all arc tolerances for the RAP is $\mathcal{O}(n^2)$ time.

5.2.3 *The Contraction Procedure and a Greedy Algorithm*

The (path) contraction procedure (see, e.g., [Glover *et al.* (2001)]) is a method of updating a digraph once a directed path is removed from it and replaced by a single vertex. Consider a digraph $D = (V, A, C)$ with $C = [c(i, j)]$ and a directed path $P = v_1 v_2 \cdots v_k$ in it. The contraction procedure for marking the path (and replacing it by a vertex p) replaces D

by a digraph D_p. The vertex set of D_p is $V_p = V \cup \{p\} - \{v_1, \ldots, v_k\}$. The arc set of D_p includes all arcs (i, j) from D where $i, j \notin P$. In addition for all vertices i in V_p except p, it introduces and includes arcs (i, p) with weight $c(i, v_1)$ and (p, i) with weight $c(v_k, i)$ (where c, i, v_1, and v_k are defined for digraph D). In this paper we only need a special case of the path contraction procedure, namely contracting only a single arc (say a) from a digraph D. We use a shorthand notation $CP(a, D)$ for this procedure.

Given the contraction procedure, a generic greedy algorithm can be defined as follows:

A generic greedy algorithm
Input: A weighted complete digraph $D = (V, A, C)$.
Output: A tour T.
Step 1: $G \leftarrow D, T \leftarrow \emptyset$.
Step 2: While G consists of at least three vertices, using a suitable myopic procedure, choose an arc (say $a = (u, v)$), that does not create a cycle, to include in the tour.
(For example, if arcs (1,2) and (2,3) are already contracted, the contraction of (3,1) would create a cycle.)
Set $T \leftarrow T \cup \{a\}$, $G \leftarrow CP(a, G)$.
Step 3: Set $T \leftarrow T \cup \{(v_1, v_2), (v_2, v_1)\}$ and output T.

This algorithm is generic since the myopic arc selection procedure used in Step 2 has not been defined. Typically greedy algorithms employ myopic procedures based on arc weights, choosing the least weight arc as the one to contract. Therefore, as a benchmark for tolerance-based algorithms that we present in the next section, we define the following variant.

W-GREEDY algorithm: At each iteration of the generic greedy algorithm, in Step 2, the myopic procedure chooses the least weight arc. This arc is chosen for contraction (i.e., inclusion in the tour).

5.3 Tolerance-based Greedy Algorithms

Since exploratory computations (see, e.g., [Turkensteen (2005)]) show that, given an optimal AP solution to an TSP instance, the 'probability' of the arc with the largest upper tolerance for the AP solution being in an optimal TSP solution is much higher than the 'probability' of the smallest weight arc being in an optimal TSP solution, it is interesting to create myopic procedures for the generic greedy algorithm developed in Section 5.2.3 which use tolerances instead of arc weights to choose arcs. In this section, we introduce the following three variants of such myopic procedures, leading to three greedy algorithms.

R-R-GREEDY algorithm: At Step 2 of each iteration of the generic greedy algorithm, the myopic procedure generates an optimal RAP solution on the digraph. Then the upper tolerances of each arc included in the solution are generated. The arc in the optimal RAP solution with the largest upper tolerance is chosen for contraction (i.e., inclusion in the tour).

A-R-GREEDY algorithm: At each iteration, the myopic procedure generates an optimal AP solution and an optimal RAP solution on the digraph. For each arc in the AP solution *and* in the RAP solution, the upper tolerance (w.r.t. the RAP) is computed, and for each arc in the AP solution but *not* in the RAP solution, the lower tolerance (w.r.t. the RAP) is computed, and multiplied with -1. The values thus obtained are sorted, and the arc with the largest value is chosen for contraction.

The relaxation of constraint set (5.2) in the formulation of AP to generate RAP was arbitrary. One could easily come up with another relaxation of the AP (let us call it RAP1) in which constraint set (5.1) is relaxed instead of the set (5.2). The third algorithm implements a myopic procedure that uses both the RAP and RAP1 relaxations.

A-RC-GREEDY algorithm: Optimal solutions are generated for AP as well as for RAP and RAP1. The myopic procedure described in the A-R-GREEDY algorithm is carried out twice, once with the optimal solutions to AP and RAP, and the second time with the optimal solutions to AP and RAP1. In the second case, the tolerances are computed with respect to the RAP1 relaxation. Of the two candidates that emerge from the two procedures, the one which has a larger value is chosen for contraction.

Note that for A-R-GREEDY and A-RC-GREEDY we only approximate the upper tolerances for the AP. The reason is, that in practice, solving an AP in $\mathcal{O}(n^3)$ time by the Hungarian algorithm and then computing approximate tolerances in $\mathcal{O}(n^2)$ time is much faster than using the Hungarian algorithm and then Volgenant's method (see [Volgenant (2006)]) for exactly computing the tolerances in $\mathcal{O}(n^3)$, even though both methods have an overall $\mathcal{O}(n^3)$ time complexity.

The greedy algorithms described above can be speeded up considerably using book-keeping techniques. For example, in R-R-GREEDY, if in an iteration, the end vertex of the contracted arc does not contain a smallest or a second smallest weight arc from any of the vertices, then in the next iteration, both the RAP solution and the upper tolerances remain unchanged. Even otherwise, the changes in the RAP solution and upper tolerance at the next iteration involve only those vertices from which the smallest or second smallest weight arcs were directed to the end vertex of the contracted arc in the previous iteration. Furthermore, in the A-R-GREEDY and A-RC-GREEDY algorithms, if the arc contracted does not belong to a subtour with two arcs only, the optimal AP solution before and after the contraction operation differ only by the arc contracted.

Our extensive computational experiments with the W-GREEDY and R-R-GREEDY applied to a wide set of the AP instances with $n \geq 100$ (see [Dell'Amico and Toth (2000)] for a description of the instances) show that the quality of R-R-GREEDY solutions is at least 10 times better than the quality of W-GREEDY solutions and these results are further supported by domination analysis. The domination number of a heuristic $\mathcal{H}$ for a combinatorial optimization problem $\mathcal{P}$ is the maximum number of solutions that are not better than the solution found by $\mathcal{H}$ for any instance of size n. The domination number of W-GREEDY for the AP equals 1 [Gutin and Yeo (2005)], i.e., for every n there are instances of AP for which W-GREEDY finds the *unique* worst solution. It can be shown that the domination number of R-R-GREEDY for the AP is exponential.

In the next section, we compare the three tolerance-based greedy algorithms introduced in this section with each other on benchmark TSP instances using the W-GREEDY algorithm to calibrate the algorithms. Since the performance of the W-GREEDY algorithm has been compared with other well-known algorithms for the TSP (see, e.g., [Glover *et al.* (2001)]), the next section also provides an indirect comparison of the three algorithms proposed here with those algorithms.

5.4 Computational Experience

The four greedy algorithms mentioned in the paper were implemented in order to observe their performance on benchmark instances of the TSP.

The implementations were done in C under Linux on a GenuineIntel Intel® Xeon™ 3.2GHz machine with 4 GB RAM. In our implementations we use the Jonker and Volgenant's (see [Jonker and Volgenant (1987)]) code for solving the AP.

Out of the four algorithms, only the W-GREEDY algorithm is known in the literature (see the GR algorithm in [Glover *et al.* (2001)] and [Gutin *et al.* (2002b)]). Therefore, we report our computational results using W-GREEDY as a base. Assume that for a particular TSP instance, W-GREEDY finds a tour of length L_W and in T_W time, while another algorithm $\mathcal{A}$ takes execution time T_A, and finds a tour of length L_A. Then for that instance we define the solution quality parameter q_A and time parameter τ_A for $\mathcal{A}$ as

$$q_A = \frac{L_A - L^\star}{L_W - L^\star} \times 100 \qquad \tau_A = \frac{T_A \times 100}{T_W}$$

Clearly, the smaller the values of q_A and τ_A the better the algorithm. We tested the algorithms on nine classes of instances. Classes 1 through 7 were taken from [Glover *et al.* (2001)], Class 8 is the class of GYZ instances introduced in [Gutin *et al.* (2002b)] for which the domination number of the W-GREEDY algorithm for the ATSP is 1 (see Theorem 2.1 in [Gutin *et al.* (2002b)]) and Class 9 is the amalgamation of several classes of instances from [Johnson *et al.* (2002)]. The nine classes are described below.

Class 1: All asymmetric instances from TSPLIB [Reinelt (1991)] (26 instances).

Class 2: All symmetric instances from TSPLIB [Reinelt (1991)] with less than 3000 vertices (99 instances).

Class 3: Asymmetric instances with $c(i, j)$ randomly and uniformly chosen from $\{0, 1, \cdots, 100000\}$ for $i \neq j$. 10 instances are generated for dimensions $100, 200, \cdots, 1000$ and three instances for dimensions $1100, 1200, \cdots, 3000$ (160 instances).

Class 4: Asymmetric instances with $c(i, j)$ randomly and uniformly chosen from $\{0, 1, \cdots, i \cdot j\}$ for $i \neq j$. 10 instances are generated for dimensions $100, 200, \cdots, 1000$ and three instances for dimensions $1100, 1200, \cdots, 3000$ (160 instances).

Class 5: Symmetric instances with $c(i, j)$ randomly and uniformly chosen from $\{0, 1, \cdots, 100000\}$ for $i < j$. 10 instances are generated for dimensions $100, 200, \cdots, 1000$ and three instances for dimensions $1100, 1200, \cdots, 3000$ (160 instances).

Class 6: Symmetric instances with $c(i,j)$ randomly and uniformly chosen from $\{0, 1, \cdots, i \cdot j\}$ for $i < j$. 10 instances are generated for dimensions $100, 200, \cdots, 1000$ and three instances for dimensions $1100, 1200, \cdots, 3000$ (160 instances).

Class 7: Sloped plane instances with given x_i, x_j, y_i, y_j randomly and uniformly chosen from $\{0, 1, \cdots, i \cdot j\}$ for $i \neq j$ and $c(i,j) = \sqrt{(x_i - x_j)^2 + (y_i - y_j)^2} - \max\{0, y_i - y_j\} + 2 \cdot \max\{0, y_j - y_i\}$ for $i \neq j$. 10 instances are generated for dimensions $100, 200, \cdots, 1000$ and three instances for dimensions $1100, 1200, \cdots, 3000$ (160 instances).

Class 8: GYZ instances (see Theorem 2.1 in [Gutin *et al.* (2002b)]) in which the arc weights $c(i,j)$ are defined as

$$c(i,j) = \begin{cases} n^3 & \text{for } i = n, j = 1; \\ in & \text{for } j = i + 1, i = 1, 2, \ldots, n - 1; \\ n^2 - 1 & \text{for } i = 3, 4, \ldots, n - 1; j = 1; \\ n \min\{i, j\} + 1 & \text{otherwise.} \end{cases}$$

One instance is generated for each $n = 5, 10, \cdots, 1000$ (200 instances).

Class 9: There are 12 problem generators from Johnson et al. [Johnson *et al.* (2002)], called *tmat, amat, shop, disc, super, crane, coin, stilt, rtilt, rect, smat,* and *tsmat.* Each of these generators yields 24 instances, 10 of dimensions 100, 10 of dimension 316, three of dimension 1000, and one of dimension 3162 (288 instances).

Note that for Classes 1 and 2 we use as L^* the known optima (see [Reinelt (1991)]), for the symmetric and almost-symmetric Classes 3, 4, 7, and 8 the AP lower bound and for the asymmetric Classes 5, 6, and 9 the HK (Held-Karp) lower bound ([Held and Karp (1970)]).

It is clear from Table 5.1 that the usual weight-based greedy algorithm is comprehensively outperformed by tolerance-based greedy algorithms in terms of solution quality, although it takes much less execution time than two of the tolerance-based algorithms. It is also clear that A-RC-GREEDY, and to a lesser extent, A-R-GREEDY are greedy algorithms of choice if one desires good-quality solutions. Even the extremely simplistic R-R-GREEDY algorithm generates better quality solutions for all except two classes (Classes 1 and 2) in nearly the same time. This fact is seen most starkly in Classes 4 and 7.

Table 5.1 Performance of tolerance-based algorithms

Algorithm	Problem Class	q values mean	q values std. dev.	τ values mean	τ values std. dev.
R-R-GREEDY	1	92.02	7.70	101.24	32.19
	2	103.80	14.99	63.76	48.59
	3	48.33	4.92	32.51	22.70
	4	9.30	2.28	34.07	20.85
	5	53.65	4.39	35.28	23.56
	6	12.08	2.74	36.66	21.21
	7	9.25	3.32	43.25	24.45
	8	33.23	0.01	153.68	41.46
	9	80.61	24.58	76.09	27.06
A-R-GREEDY	1	87.56	7.07	100.24	26.61
	2	91.16	12.15	275.99	152.25
	3	31.02	5.11	37.68	23.06
	4	7.75	1.85	41.13	20.98
	5	37.17	4.86	609.72	234.95
	6	10.52	2.43	1189.16	502.16
	7	6.31	3.30	3150.58	1259.05
	8	16.75	0.01	149.70	39.96
	9	69.17	26.07	373.18	543.07
A-RC-GREEDY	1	84.48	9.21	120.21	32.02
	2	89.67	12.91	292.76	159.29
	3	27.55	4.45	52.16	22.25
	4	7.92	1.88	56.90	19.88
	5	33.44	4.53	497.98	178.01
	6	11.38	2.62	746.48	295.36
	7	6.19	3.19	3396.10	1289.19
	8	16.75	0.01	174.78	69.03
	9	67.39	26.84	391.89	565.70

An interesting observation is that AP relaxation based algorithms require very long execution times on average for instances in Classes 7 and 9. For instances in these classes, experiments show that the optimal solutions to the AP relaxation for the digraphs in several iterations have many cycles of length 2, and the arc to be contracted usually comes from one of these cycles. Consequently, in the next step of the algorithm, the AP relaxation needs to be solved again, and the tolerances recalculated, thus leading to long execution times (refer to the discussion on book-keeping techniques in Section 5.3).

5.5 Summary and Future Research Directions

In this paper, we examine in detail the idea of using arc tolerances instead of arc weights as a basis for making algorithmic decisions on whether or not to include an arc in an optimal solution. Such methods have only been studied in passing in the literature (see [Helsgaun (2000)]) and deserve more attention. In order to evaluate the usefulness of the concept, three tolerance-based greedy algorithms are proposed (see Section 5.3) for the traveling salesman problem. Two of these (A-R-GREEDY and A-RC-GREEDY) are based on an AP relaxation of the original problem, while the third one (R-R-GREEDY) is based on a new relaxation of the AP relaxation itself. With the purpose of investigating the usefulness of the relaxed AP (RAP), we made extensive computational experiments with our R-R-GREEDY heuristic applied to the AP (not reported here in detail due to the space limitation). The computational results for the TSP show that the R-R-GREEDY outperforms a weight-based greedy (W-GREEDY) in quality at least 10 times on average, while for AP the corresponding domination numbers for R-R-GREEDY and W-GREEDY are 2^{n-1} and 1, respectively.

Our experiments show that the quality of solutions produced by tolerance-based greedy algorithms are overall significantly better than those found by the arc weight-based greedy algorithm. We measure quality using the ratio $\frac{L_A - L^\star}{L^\star}$ where L_A is the length of the tour returned by the heuristic, and $L^\star$ is the length of the optimal tour or of a good lower bound where the optimal tour is not known. This measure is called the "excess over the length of an optimal tour or lower bound". Unfortunately, A-R-GREEDY and A-RC-GREEDY are often slower than W-GREEDY, but R-R-GREEDY, being superior to W-GREEDY in quality, is nearly as fast as W-GREEDY. Overall, the simplest tolerance-based greedy, R-R-GREEDY, is the best algorithm for solving the STSP, while the A-RC-GREEDY algorithm could be suggested for the ATSP.

It is worth mentioning that the construction heuristics in [Glover *et al.* (2001)] (see from Table 1 in [Glover *et al.* (2001)]) have the following average excesses (taken over seven families of instances) over the length of an optimal tour or lower bound: GR= 580.35%, RI= 710.95%, KSP= 135.08%, GKS= 98.09%, RPC= 102.02%, COP= 23.01%. Computational experiments reported in [Goldengorin and Jäger (2005)] for our algorithms give R-R-GREEDY= 67.14%, A-R-GREEDY=34.75%, and A-RC-GREEDY= 29.19%. We know that the domination number of ATSP-R-R-GREEDY is $2(n-3)!$ and it would be interesting to find a non-trivial lower bound.

Another question is whether a 1-tree-based relaxation of the traveling salesman problem would generate tolerance-based greedy algorithms that are better for the STSP. Also it would be interesting to replicate the success of tolerance-based algorithms on the TSP to other combinatorial optimization problems.

Bibliography

Balas, E. and Saltzman, M. J. (1991). An algorithm for the three-index assignment problem, *Operations Research* **39**, pp. 150–161.

Dell'Amico, M. and Toth, P. (2000). Algorithms and codes for dense assignment problems: The state of the art, *Discrete Applied Mathematics* **100**, pp. 17–48.

Gamboa, D., Rego, C., and Glover, F. (2006). Implementation analysis of efficient heuristic algorithms for the traveling salesman problem, *Computers & Operations Research* **33**, pp. 1154–1172.

Glover, F., Gutin, G., Yeo, A., and Zverovich, A. (2001). Construction heuristics for the asymmetric TSP, *European Journal of Operational Research* **129**, 555–568.

Goldengorin, B. and Jäger, G. (2005). How To Make a Greedy Heuristic for the Asymmetric Traveling Salesman Problem Competitive. SOM Research Report 05A11, University of Groningen, The Netherlands(`http://som.eldoc.ub.rug.nl/reports/themeA/2005/05A11`).

Goldengorin, B., Jäger, G., and Molitor, P. (2006). Some Basics on Tolerances. To appear in the Proceedings of AAIM 2006, Lecture Notes in Computer Science (Springer).

Goldengorin, B. and Sierksma, G. (2003). Combinatorial optimization tolerances calculated in linear time. SOM Research Report 03A30, University of Groningen, The Netherlands (`http://som.eldoc.ub.rug.nl/reports/themeA/2003/03A30/`).

Goldengorin, B., Sierksma, G., and Turkensteen, M. (2004). Tolerance based algorithms for the ATSP, in Graph-Theoretic Concepts in Computer Science. 30th International Workshop, WG 2004, Bad Honnef, Germany, June 21-23, 2004.

Gutin, G., and Yeo, A. (2005). Domination analysis of combinatorial optimization algorithms and problems, in M. Golumbic, I. Hartman (eds.),*Graph Theory, Combinatorics and Algorithms: Interdisciplinary Applications* (Springer).

Gutin, G., Yeo, A., and Zverovich, A. (2002). Traveling salesman should not be greedy: domination analysis of greedy type heuristics for the TSP, *Discrete Applied Mathematics* **117**, pp. 81–86.

Held, M. and Karp, R. (1970). The traveling-salesman problem and minimum spanning trees, *Operations Research*, **18**, pp. 1138–1162.

Helsgaun, K. (2000). An effective implementation of the Lin-Kernighan traveling salesman heuristic, *European Journal of Operational Research* **126**, pp. 106–130.

Johnson, D. S., Gutin, G., McGeoch, L. A., Yeo, A., Zhang, W., and Zverovich, A. (2002). Experimental analysis of heuristics for the ATSP, Chapter 10 in: G. Gutin, A.P. Punnen (eds.), *The Traveling Salesman Problem and Its Variations*, pp. 445–489 (Kluwer).

Jonker, R. and Volgenant, A. (1987). A shortest augmenting path algorithm for dense and sparse linear assignment problems, *Computing* **38**, pp. 325–340.

Libura, M. (1991). Sensitivity analysis for minimum Hamiltonian path and traveling salesman problems, *Discrete Applied Mathematics* **30**, pp. 197–211.

Reinelt, G. (1991). TSPLIB — a traveling salesman problem library. *ORSA Journal of Computing* **3**, pp. 376–384.

Turkensteen, M., Ghosh, D., Goldengorin, B., and Sierksma, G. (2005). Tolerance-based branch and bound algorithms, A EURO conference for young OR researches and practitioners, ORP3 2005, 6 – 10 September 2005, Valencia, Spain. Proceedings Edited by C. Maroto et al., ESMAP S.L., pp. 171–182.

Volgenant, A. (2006). An addendum on sensitivity analysis of the optimal assignment, *European Journal of Operational Research* **169**, pp. 338–339.

Chapter 6

On the Membership Problem of the Pedigree Polytope

T. S. Arthanari
Department of Information Systems & Operations Management.
University of Auckland,
Private Bag 92019, Auckland, New Zealand
e-mail: t.arthanari@auckland.ac.nz

Abstract

Given a polytope P and a in the interior of P and $x \notin P$, to identify a violated facet of P, whose supporting hyperplane separates x from P constitutes the separation problem for P. In [Grötscel, Lovász and Schrijver (1988)] a construction found in [Yudin and Nemirovskii (1976)] is used to establish conditions for the existence of a polynomial separation algorithm for a bounded convex body. This proof uses Ellipsoid algorithm twice. Recently [Maurras (2002)] has given under certain conditions, a simple construction for the separation problem for P. This uses a polynomial number of calls to an oracle checking membership in P. In this paper we consider an alternative polytope $conv(A_n)$ different from the standard polytope, Q_n associated with the symmetric traveling salesman problem and verify whether Maurras's construction is possible for this polytope. $conv(A_n)$ is obtained by a projection of the pedigree polytope defined and studied in [Arthanari (2006)]. This leads us to the study of the membership problem for the pedigree polytope. A necessary and sufficient condition for membership in the pedigree polytope is given in [Arthanari (2006)]. In this paper we show that a necessary condition for membership in the pedigree polytope is the existence of a multicommodity flow with value equal to unity, in a layered network. This network is recursively constructed adding one layer at a time, and checking it is well-defined. An ill-defined network at any stage automatically precludes membership of the solution in the polytope. Future research will focus on the consequences of this result and the complexity of checking the condition.

Key Words: Hamiltonian cycles, symmetric traveling salesman problem, pedigree polytope, multistage insertion formulation, membership problem

6.1 Introduction

Polyhedral Combinatorics, bridges two very important research topics, namely, computational complexity(efficiency?) and representation of combinatorial optimisation problems using linear programs. Given a polyhedral convex set $\mathbf{C}$ and a point $\mathbf{x}$, deciding whether the point is a member of the convex set is called the membership problem and if $\mathbf{x}$ is not in the convex set, identifying a violated defining inequality is called the separation problem. A linear optimisation problem over a polyhedral convex set, gives raise to a linear programming problem. However to use [Khachiyan (1979)]'s finite precision polynomial algorithm to solve that linear programming problem, we require to solve the separation problem efficiently. Khachiyan's work generated considerable enthusiasm to study polytopes corresponding to combinatorial optimisation problems. A major work in this area is [Grötscel, Lovász and Schrijver (1988)]. This brings out the connections between the efficiencies of solving linear optimisation, separation and membership problems. In this paper we study the membership problem of the pedigree polytope defined and studied in [Arthanari (2006)].

6.1.1 *Computational Complexity, Polytopes and Efficiency*

Theoretical computer science, among other things, deals with the design and analysis of algorithms. When one wants to solve a problem efficiently using an algorithm, the amount of storage space and computational time required are considered and compared. The class of problems solvable in polynomial time by a Turing machine is designated as the class $\mathcal{P}$ (see e.g. [Garey and Johnson (1979)], [Korte and Vygen (2002)].) Another class of problems is the class $\mathcal{NP}$, which consists of problem that can be solved by a nondeterministic Turing machine in polynomial time. Polynomial time algorithms received their prominence against slow exponential time, often brute force, algorithms. [Edmonds (1965)] called them *good algorithms*, and presented one for the matching problem, which has a linear programming formulation involving exponentially many constraints.

After the seminal work [Cook (1971)] and the immediate recognition by [Karp (1972)] of its importance to combinatorial optimisation and integer programming, the so-called $\mathcal{NP}$ – *complete* problems as opposed to polynomially solvable problems received renewed attention. The problem of whether $\mathcal{P} = \mathcal{NP}$ is one of the most outstanding problems in mathematics.

Since the $\mathcal{NP}$ – *complete* subclass of $\mathcal{NP}$ is known to consist of difficult combinatorial problems, the popular belief in the fields of mathematics, theoretical computer science and operations research is the conjuncture $\mathcal{P} \neq \mathcal{NP}$. Optimisation problems are of our interest. Their complexity can be related to that of decision problems studied. If a problem is not in $\mathcal{NP}$, like in the case of an optimisation problem, but it is as hard as some $\mathcal{NP}-complete$ problem, then it is called $\mathcal{NP}-hard$. A typical problem of this kind is the Symmetric Traveling Salesman Problem ($STSP$) and is about finding a minimum cost *Tour* of n cities that starts from the home city and visits every city once and returns back to the home city, and the cost of traveling from city i to city j is the same as that of traveling from city j to city i.

In [Grötscel, Lovász and Schrijver (1988)] the construction of [Yudin and Nemirovskii (1976)] is used to establish conditions for the existence of a polynomial separation algorithm for a bounded convex body. The approach in [Grötscel, Lovász and Schrijver (1988)], uses Ellipsoid algorithm twice. Given a polytope P and a in the interior of P and $x \notin P$, recently [Maurras (2002)] has given under certain conditions, a simple construction to identify a violated facet of the polytope, P, whose supporting hyperplane separates x from P. This uses a polynomial number of calls to an oracle checking membership in P. This paper we consider an alternative polytope associated with $STSP$ and verify whether Maurras's construction is possible for this polytope. This leads us to the study of the membership problem for the pedigree polytope.

Section 6.2 introduces the preliminaries and notation used, and introduces the concepts required to study an alternative polytope $conv(A_n)$ associated with $STSP$. The pedigree is a combinatorial object defined and studied in [Arthanari (2006)] and [Arthanari (2005)]. Pedigrees are in $1-1$ correspondence with n-tours or Hamiltonian cycles. A_n is defined and used in [Arthanari (2007)] proving the dimension of the pedigree polytope. Section 6.3 checks the conditions for the existence of a polynomial separation algorithm for the alternative polytope $conv(A_n)$ and the implications for studying the membership problem of the pedigree polytope. Section 6.4 develops a layered network used to prove a necessary condition for membership in the pedigree polytope in Section 6.5. In Section 6.6 conditions for having pedigree paths in the layered network to bring a specified flow along an arc in the last layer is studied. This deals with the concept of pedigree packability. Section 6.7 defines a multicommodity flow problem which is used to check a necessary condition for membership in the pedigree polytope. Section 6.8 discusses the computational complexity of verifying the necessary condition for membership in the pedigree polytope, and shows this can be done efficiently. Section 6.9 concludes the paper indicating future research.

6.2 Preliminaries & Notations

We repeat some notations and preliminaries from [Arthanari (2006)] for convenience. Let R denote the set of reals. Similarly Q, Z, N denote the rationals, integers and natural numbers respectively, and B stands for the binary set of $\{0, 1\}$. Let R_+ denote the set of non negative reals. Similarly the subscript $_+$ is understood with rationals. Let R^d denote the set of d -tuples of reals. Similarly the superscript ^{d} is understood with rationals, etc. Let $R^{m\times n}$ denote the set of $m \times n$ real matrices.

Let n be an integer, $n \geq 3$. Let V_n be a set of *vertices*. Assuming, without loss of generality, that the vertices are numbered in some fixed order, we write $V_n = \{1, \ldots, n\}$. Let $E_n = \{(i,j)|i,j \in V_n, i < j\}$ be the set of *edges*. The cardinality of E_n is denoted by $p_n = n(n-1)/2$. Let $K_n = (V_n, E_n)$ denote the complete graph of n vertices.

We denote the elements of E_n by e where $e = (i,j)$. We also use the notation ij for (i,j). Notice that, unlike the usual practice, an edge is assumed to be written with $i < j$.

Definition 6.1. [Edge Label] Let the elements of E_n be labelled as follows: $(i,j) \in E_n$, has the label, $l_{ij} = p_{j-1} + i$.

This means, edges $(1,2), (1,3), (2,3) \in E_3$ are labelled, 1, 2, and 3 respectively. Once the elements in E_{n-1} are labelled then the elements of $E_n \setminus E_{n-1}$ are labelled in increasing order of the first coordinate, namely i.

For a subset $F \subset E_n$ we write the *characteristic* vector of F by $x_F \in R^{p_n}$ where

$$x_F(e) = \begin{cases} 1 \text{ if } e \in F, \\ 0 \text{ otherwise.} \end{cases}$$

We assume that the edges in E_n are ordered in increasing order of the edge labels.

For a subset $S \subset V_n$ we write

$$E(S) = \{ij|ij \in E, i,j \in S\}.$$

Given $u \in R^{p_n}$, $F \subset E_n$,we define,

$$u(F) = \sum_{e\in F} u(e).$$

For any subset S of vertices of V_n, let $\delta(S)$ denote the set of edges in E_n with one end in S and the other in $S^c = V_n \setminus S$. For $S = \{i\}$, we write $\delta(\{i\}) = \delta(i)$.

A subset H of E_n is called a *Hamiltonian cycle* in K_n if it is the edge set of a simple cycle in K_n, of length n. We also call such a Hamiltonian cycle a $n - tour$ in K_n. At times we represent H by the vector $(1 i_2 \ldots i_n 1)$ where $(i_2 \ldots i_n)$ is a permutation of $(2 \ldots n)$, corresponding to H.

In addition to the notations and preliminaries introduced in [Arthanari (2006)], we required a few definitions and concepts with respect to bipartite flow problems. For details on graph related terms see any standard text on graph theory such as [Bondy and Murthy (1985)].

Definition 6.2. [Diconnected Components] Given a digraph $G = (V, A)$, we say $u, v \in V$ are *diconnected* if there is a directed path from u to v and also there is a directed path from v to u in G. We write $u \leftrightarrow v$. $\leftrightarrow$ is an equivalence relation and it partitions V into equivalence classes, called diconnected components of G.

Diconnected components are also called *strongly connected* components.

Definition 6.3. [Interface] Given a digraph $G = (V, A)$, consider the diconnected components of G. An arc $e = (u, v)$ of G is called an *interface* if there exist two different diconnected components C_1, C_2 such that $u \in C_1$ and $v \in C_2$.

The set of all interfaces of G is denoted by $I(G)$.

Definition 6.4. [Bridge] Given a graph $G = (V, E)$, an edge of G is called a *bridge* if $G - e$ has more components than G, where by component of a graph we mean a maximal connected subgraph of the graph.

Definition 6.5. [Mixed Graph] A *mixed graph* $G = (V, E \cup A)$ is such that it has both directed and undirected edges. A gives the set of directed edges (arcs). E gives the set of undirected edges(edges). If A is empty G is a graph and if E empty G is a digraph.

In general, we call the elements of $E \cup A$, edges. Finding the diconnected components of a digraph can be achieved using a depth first search method in $O(|G|)$ where $|G|$ is the size of G given by $|V| + |A|$. Similarly bridges in a graph can be found in $O(|G|)$.

6.2.1 Rigid, Dummy arcs in a Capacited Transportation Problem

Consider a balanced transportation problem, in which, some arcs called the *forbidden* arcs are not available for transportation. We call the problem of finding whether a feasible flow exists in such an incomplete bipartite network, a Forbidden Arcs Transportation (FAT) problem [Murty (1992)]. This could be viewed as a capacited transportation problem, as well. In general a FAT problem is given by $O = \{O_\alpha, \alpha = 1, \ldots, n_1\}$, the set of origins, with availability a_α at O_α, $D = \{D_\beta, \beta = 1, \ldots, n_2\}$, the set of destinations with requirement b_β at D_β and $\mathcal{A} = \{(O_\alpha, D_\beta)|$ arc (O_α, D_β) is *not forbidden* $\}$, the set of arcs. We may also use (α, β) to denote an arc.

We state Lemma 6.1 from [Arthanari (2006)] on such a flow feasibility problem arising with respect to non empty partitions of a finite set.

Lemma 6.1. *Suppose $\mathcal{D} \neq \emptyset$ is a finite set and $g : \mathcal{D} \rightarrow Q_+$, is a nonnegative rational function such that, $g(\emptyset) = 0$, and $g(\mathcal{D}) = 1$. Let $\mathcal{D}^1 = \{D^1_\alpha, \alpha = 1, \ldots, n_1\}, \mathcal{D}^2 = \{D^2_\beta, \beta = 1, \ldots, n_2\}$ be two non empty partitions of $\mathcal{D}$. (That is, $\bigcup_{\alpha=1}^{n_1} \mathcal{D}^1_\alpha = \mathcal{D}$ and $\mathcal{D}^1_s \bigcap \mathcal{D}^1_r = \emptyset, r \neq s$. Similarly $\mathcal{D}^2$ is understood.) Consider the FAT problem defined as follows:*

Let the origins correspond to D^1_α, with availability $a_\alpha = g(D^1_\alpha), \alpha = 1, \ldots, n_1$ and the destinations correspond to D^2_β, with requirement $b_\beta = g(D^2_\beta), \beta = 1, \ldots, n_2$. Let the set of arcs be given by

$$\mathcal{A} = \{(\alpha, \beta)|D^1_\alpha \cap D^2_\beta \neq \emptyset\}.$$

Then $f_{\alpha\beta} = g(D^1_\alpha \cap D^2_\beta) \geq 0$ is a feasible solution for the FAT problem considered.

Several other FAT problems are defined and studied in the later sections of this paper. FAT problems can be solved using any efficient bipartite maximal flow algorithm (see [Korte and Vygen (2002)]). If the maximal flow is equal to the maximum possible flow, namely $a(O)$, we have a feasible solution to the problem.

Definition 6.6. [Rigid Arcs] Given a FAT problem with a feasible solution f we say $(\alpha, \beta) \in \mathcal{A}$ is a *rigid* arc in case $f_{\alpha,\beta}$ is same in all feasible solutions to the problem. Rigid arcs have *frozen flow.*

Definition 6.7. [Dummy Arc] A rigid arc with zero frozen flow is called a *dummy* arc.

The set of rigid arcs in a FAT problem is denoted by $\mathcal{R}$. Identifying $\mathcal{R}$ is the *frozen flow finding* problem (FFF problem). Interest in this arises in various contexts.(see [Ahuja, Magnanti and Orlin (1996)]) Application in statistical data security is discussed by [Gusfield (1988)]. The problem of protecting sensitive data in a two way table, when some not sensitive data and the marginal sums are made public is studied there. A sensitive cell is unprotected if its exact value can be identified by an adversary. This corresponds to finding rigid arcs and their frozen flows.

Even though this problem can be posed as a linear programming problem we describe the graph algorithm developed in [Gusfield (1988)].

Definition 6.8. $[G_f]$ With respect to a feasible flow f for a FAT problem, we define a mixed graph $G_f = (\mathcal{V}, A \cup E)$ where $\mathcal{V}$ is as given in the FAT problem, and

$$A = \{(O_\alpha, D_\beta)|f_{\alpha,\beta} = 0, (\alpha,\beta) \in \mathcal{A}\} \cup \{(D_\beta, O_\alpha)|f_{\alpha,\beta} = c_{\alpha,\beta}, (\alpha,\beta) \in \mathcal{A}\},$$

$$E = \{(O_\alpha, D_\beta)|0 < f_{\alpha,\beta} < c_{\alpha,\beta}, (\alpha,\beta) \in \mathcal{A}\}.$$

Definition 6.9. [Flow change Cycle] A simple cycle in G_f is called a *flow change cycle* (fc-cycle), if it is possible to trace it without violating the direction of any of the arcs in the cycle. Undirected edges of G_f can be oriented in one direction in one fc-cycle and in the other direction in another fc-cycle.

Theorem 6.1. *[Characterisation of Rigid Arcs] Given a feasible flow, f, to a FAT problem, an arc is rigid if and only if it's corresponding edge is not contained in any fc-cycle in G_f.*

The proof of this is straight forward from the definitions (see [Gusfield (1988)]). It is also proved there that the set $\mathcal{R}$ is given by the algorithm, that we call, **Frozen Flow Finding** (FFF)algorithm , stated below:

Algorithm 6.1 (Frozen Flow Finding).
Given: *A Forbidden arcs transportation problem with a feasible flow f.*
Find: *The set of rigid arcs $\mathcal{R}$, in the bipartite graph of the problem.*
  Construct *The mixed graph G_f as per Definition 6.8.*
  Find *The diconnected components of G_f (say $C_1, \ldots, C_q$).*
  Find *The set of interfaces, $I(G_f)$.*
  Find *The set of all bridges $B(G_f)$ in the underlying graphs, treating each C_r as an undirected graph.*
  Output $\mathcal{R} = I(G_f) \cup B(G_f)$. *Stop.*

In fact we have a linear time algorithm, as each of steps 1 - 3 can be done in $O(|G_f|)$.

Definition 6.10. [Layered Network] A network $N = (\mathcal{V}, \mathcal{A})$ is called a *layered network* if the node set of N can be partitioned into l sets $V_{[1]}, \ldots, V_{[l]}$ such that if $(u, v) \in \mathcal{A}$ then $u \in V_{[r]}, v \in V_{[r+1]}$ for some $r = 1, \ldots, l-1$. We say N has l layers. Nodes in $V_{[1]}$, $V_{[l]}$ are called sources and sinks respectively.

In the cases we are interested we have *capacities both on arcs and nodes.* Any flow in the layered network should satisfy apart from nonnegativity and flow conservation, the capacity restrictions on the nodes. Of course any such problem can be recast as a flow problem with capacities only on arcs.(see [Ford and Fulkerson (1962)])

6.2.2 *Definition of the Pedigree Polytope*

In this sub section we present an alternative polyhedral representation of the $STSP$, using the definition of pedigrees.

Let Q_n denote the standard $STSP$ polytope, given by

$$Q_n = conv(\{X_H : X_H \text{ is the characteristic vector of } H \in \mathcal{H}n\})$$

where $\mathcal{H}_n$ denotes the set of all *Hamiltonian cycles* (or $n - tours$) in K_n.

In polyhedral combinatoric approaches, generally, Q_n is studied while solving $STSP$ (see [Lawler et. al. (1985)]). However, in this paper we consider an alternative polytope $conv(A_n)$ for this purpose. The required notations and concepts follow.

Given $H \in \mathcal{H}_{k-1}$, the operation *insertion* is defined as follows: Let $e = (i, j) \in H$. Inserting k in e is equivalent to replacing e in H by $\{(i, k), (j, k)\}$ obtaining a $k - tour$. When we denote H as a subset of E_{k-1}, then inserting k in e gives us a $H' \in \mathcal{H}_k$ such that,

$$H' = (H \cup \{(i, k), (j, k)\}) \setminus \{e\}.$$

We write $H \overrightarrow{\ e, k\ } H'$.

Similarly the inverse operation shrinking can be defined.

Definition 6.11. [Pedigree] The vector $W = (e_4, \ldots, e_n) \in E_3 \times \ldots \times E_{n-1}$ is called a *pedigree* if and only if there exists a $H \in \mathcal{H}_n$ such that H is obtained from the $3 - tour$ by the sequence of insertions, viz.,

$$3 - tour \overrightarrow{\ e_4, 4\ } H^4 \ \ldots\ H^{n-1} \overrightarrow{\ e_n, n\ } H.$$

The pedigree W is referred to as the pedigree of H. Pedigree is a compact way of writing H. The pedigree of H can be obtained by shrinking H sequentially to the $3-tour$ and noting the edge created at each stage. We then write the edges obtained in the reverse order of their occurrence. Motivation for the definition of the pedigree and its connection to MI-formulation [Arthanari and Usha (2000)] are given in [Arthanari (2006)].

Let the set of all pedigrees, corresponding to $H \in \mathcal{H}_n$ be denoted by $\mathcal{P}_n$. For any $4 \leq k \leq n$, given an edge $e \in E_{k-1}$, with edge label l, we can associate a $0-1$ vector, $\mathbf{x}(e) \in B^{p_{k-1}}$, such that, $\mathbf{x}(e)$ has a 1 in the lth coordinate, and zeros else where. That is, $\mathbf{x}(e)$ is the indicator of e.

Similarly, we can associate a $X = (\mathbf{x}_4, \ldots, \mathbf{x}_n) \in B^{\tau_n}$, the characteristic vector of the pedigree W, where $(W)_k = e_k$, the $(k-3)^{rd}$ component of W, $4 \leq k \leq n$ and $\mathbf{x}_k$ is the indicator of e_k. The number of coordinates in X, is $\sum_{k=4}^{n} p_{k-1}$, and is denoted by τ_n

Let $P_n = \{X \in B^{\tau_n} \ : X$ is the characteristic vector of the pedigree $W \in \mathcal{P}_n\}$.

Thus there is a one to one correspondence between $H \in \mathcal{H}_n$ and $X \in P_n$. Consider the convex hull of P_n. We call this the *pedigree polytope*, denoted by $conv(P_n)$. An interesting property of $X = (\mathbf{x}_4, \ldots, \mathbf{x}_n) \in P_n$ is that, for any $4 \leq k \leq n$, X restricted to the first $k-3$ stage(s), written as

$$X/k = (\mathbf{x}_4, \ldots, \mathbf{x}_k)$$

is in P_k. Similarly, $X/k-1$ and $X/k+1$ are interpreted as restrictions of X. We use this notation for any $X \in R^{\tau_n}$ as well.

Definition 6.12. [Generators of an edge] Given $e_\beta = (i,j) \in E_k$, we say $G(e_\beta)$ is the set of generators of e_β in case

$$G(e_\beta) = \begin{cases} E_3 \setminus \{e_\beta\} & \text{if } e_\beta \in E_3 \\ \delta_i \cap E_{j-1} & \text{otherwise.} \end{cases}$$

Since an edge $e = (i,j), j > 3$ is generated by inserting j in any e' in the set $G(e)$, the name *generator* is used to denote any such edge.

Lemma 6.2 proved in [Arthanari (2006)]gives an equivalent definition of a pedigree.

Lemma 6.2. *Given n, consider $W = (e_4, \ldots, e_n)$, where $e_k = (i_k, j_k)$ for $1 \leq i_k < j_k \leq k-1, 4 \leq k \leq n$. W corresponds to a pedigree in $\mathcal{P}_n$ if and only if*

(1) $e_k, 4 \leq k \leq n$, are all distinct,

(2) $e_k \in E_{k-1}, 4 \leq k \leq n$, *and*

(3) for every $k, 5 \leq k \leq n$, *there exists a* $e' \in G(e_k)$ *such that,* $e_q = e'$, *where* $q = max\{4, j_k\}$.

This lemma allows us to define a pedigree without explicitly considering the corresponding Hamiltonian cycle.

Definition 6.13. [Extension of a Pedigree] Let $y(e)$ be the indicator of $e \in E_k$. Given a pedigree, $W = (e_4, \ldots, e_k)$ (with the characteristic vector, $X \in P_k$) and an edge $e \in E_k$, we call $(W, e) = (e_4, \ldots, e_k, e)$ an *extension* of W in case $(X, y(e)) \in P_{k+1}$.

Using Lemma 6.2, observe that given W a pedigree in $\mathcal{P}_k$ and an edge $e = (i, j) \in E_k$, (W, e) is a pedigree in $\mathcal{P}_{k+1}$ if and only if, 1] $e_l \neq e, 4 \leq l \leq k$ and 2] there exists a $q = max(4, j)$ such that e_q is a generator of $e = (i, j)$.

6.2.3 *Multistage Insertion and Related Results*

Here we present excerpts from papers on MI-formulation and related issues. Here x_{ijk} denotes $x_k(e)$ where $e = (i, j)$.

Problem 6.1 (MI- **Relaxation).**

$$\sum_{1 \leq i < j \leq k-1} x_{ijk} = 1, \ 4 \leq k \leq n \tag{6.1}$$

$$\sum_{k=4}^{n} x_{ijk} \leq 1, \ 1 \leq i < j \leq 3 \tag{6.2}$$

$$-\sum_{r=1}^{i-1} x_{rij} - \sum_{s=i+1}^{j-1} x_{isj} + \sum_{k=j+1}^{n} x_{ijk} \leq 0, \ 4 \leq j \leq n-1; \ 1 \leq i < j \tag{6.3}$$

$$x_{ijk} = 0 \text{ or } 1, \ 1 \leq i < j \leq k-1; \ 4 \leq k \leq n \tag{6.4}$$

Relaxing the integer constraints 6.4 with just non-negativity constraints (as constraints $x_{ijk} \leq 1$ *are implied by equation 6.1) and adding the following constraints*

$$-\sum_{r=1}^{i-1} x_{rin} - \sum_{s=i+1}^{n-1} x_{isn} \leq 0 \ \ i = 1, \ldots, n-1. \tag{6.5}$$

we obtain the MI *- relaxation of the* $STSP$.

Notice that constraints 6.5 are redundant and are added only because of their slack variables have special meaning.

Definition 6.14. [$P_{MI}(n)$ Polytope] The polytope corresponding to MI-relaxation is called the $P_{MI}(n)$ polytope, where n refers to the number of cities.

All X in P_n are extreme points of $P_{MI}(n)$ polytope. But $P_{MI}(n)$ has fractional extreme points as well. Thus the pedigree polytope $conv(P_n)$ is contained in $P_{MI}(n)$ polytope. We introduce the following notation (from [Arthanari and Usha (2001)]).

Definition 6.15. In general, let $E_{[n]}$ denote the matrix corresponding to equation (6.1); let $A_{[n]}$ denote the matrix corresponding to the inequalities (6.2, 6.3 & 6.5). Let $\mathbf{1}_r$ denote the row vector of r $1's$. Let I_r denote the identity matrix of size $r \times r$. Then we can write recursively,

$$E_{[n]} = \begin{pmatrix} \mathbf{1}_{p_3} & \dots & 0 & 0 \\ \vdots & \vdots & \vdots & \vdots \\ 0 & \dots & \mathbf{1}_{p_{n-2}} & 0 \\ 0 & 0 & \dots & \mathbf{1}_{p_{n-1}} \end{pmatrix} = \begin{pmatrix} E_{[n-1]} & 0 \\ \mathbf{0} & \mathbf{1}_{p_{n-1}} \end{pmatrix}.$$

To derive a similar expression for $A_{[n]}$ we first define

$$A^{(n)} = \begin{pmatrix} I_{p_{n-1}} \\ -M_{n-1} \end{pmatrix}$$

where M_i is the $i \times p_i$ *node-edge incidence* matrix.

Then

$$A_{[n]} = \left(\begin{array}{c|c|c|c} A^{(4)} & A^{(5)} & & A^{(n)} \\ & & \ddots & \\ \mathbf{0} & \mathbf{0} & & \end{array}\right) = \left(\begin{array}{c|c} A_{[n-1]} & A^{(n)} \\ \mathbf{0} & \end{array}\right).$$

Observe that $A^{(n)}$ is the submatrix of $A_{[n]}$ corresponding to $\mathbf{x}_n$. The number of rows of $0's$ is decreasing from left to right.

Lemma from [Arthanari and Usha (2001)] is useful in checking the membership of an $X \in P_{MI}(n)$.

Lemma 6.3. *Let $U^{(k-3)}$ denote the slack variable vector obtained from the MI - relaxation for $n = k$ by substituting (X/k) in the inequalities (6.2, 6.3 & 6.5). Let $U^{(0)} = \mathbf{1}_3$. Also assume $U_{ij}^{(k-3)} = 0$ for $1 \le i < j,\ j > k$. Given X and $U^{(l)}$ as defined above, we have*

$$U^{(k-4)} - A^{(k)}\mathbf{x}_k = U^{(k-3)},\ \textit{for all } k,\ 4 \le k \le n. \tag{6.6}$$

Observe that we can reformulate MI- relaxation, using both $X = (\mathbf{x}_4, \ldots, \mathbf{x}_n)$ and $U = (U^{(0)}, \ldots, U^{(n-3)})$ in matrix notation as follows:

$$\begin{aligned} E_{[n]}X &= \mathbf{1}_{n-3} \\ U^{(0)} &= \mathbf{1}_3 \\ U^{(k-3)} - A^{(k+1)}\mathbf{x}_{k+1} &= U^{(k-2)}, \ for\ all\ k,\ 3 \le k \le n-1. \\ X &\ge 0. \end{aligned} \tag{6.7}$$

In [Arthanari and Usha (2001)] the connection between MI- relaxation and cycle shrink relaxation, CS-relaxation, given by [Carr (1997)] are brought out using this reformulation. So if $X \in P_{MI}(n)$ then X satisfies Equation 6.7.

Definition 6.16. [Weight Vector] Given $X \in P_{MI}(n)$ and $X/k \in conv(P_k)$, consider $\lambda \in R_{+}^{|P_k|}$ that can be used as a weight to express X/k as a convex combination of $X^r \in P_k$. Let $I(\lambda)$ denote the index set of positive coordinates of λ. Let $\Lambda_k(X)$ denote the set of all possible *weight vectors*, for a given X and k.

Definition 6.17. [Active Pedigree] Given $X \in conv(P_k)$, we call a $X^* \in P_k$ *active* for X/k, in case there exists a $\lambda \in \Lambda_k(X)$ and an $r \in I(\lambda)$ such that $X^* = X^r$. In other words X^* receives positive weight in at least one convex combination expressing X.

Definition 6.18. [Link] Let $l \in V_{n-2} \setminus V_3$ and $(e, e') \in E_l \times E_{l+1}$. Given $X \in P_{MI}(l+1)$, we say (e, e') is a *link* in case

- $x_l(e) > 0$ and $x_{l+1}(e') > 0$
- either $e' \in E_{l+1} \setminus E_l$ and $e \in G(e')$, or $e, e' \in E_l$ and $e \neq e'$.

A link can be used to extend a pedigree from P_l to a pedigree in P_{l+1}. We make use of links in Section 6.4 to construct recursively a layered network.

6.3 Polytopes and Efficiency

$P \subset R^d$ is called a ν*-polytope*, if P is the convex hull of finitely many points $X_1, \ldots, X_r$ in R^d. $P \subset R^d$ is called a $\mathcal{H}$*-polyhedron*, if P is the intersection of finitely many half-spaces, $\mathbf{a}_i X \le a_0$, for $(\mathbf{a}_i, a_0) \in R^{d+1}$, for $i = 1, \ldots, s$. It is well known that a bounded $\mathcal{H}$-polyhedron is indeed a ν -polytope. The *affine rank* of a polytope P (denoted by $arank(P)$) is defined as the maximum number of affinely independent vectors in P. The *dimension* of a polytope P is (denoted $dim(P)$) and defined to be $arank(P)$ minus 1.

Let $P \subset R^d$ be a polytope. The *barycentre* of P is defined as

$$\bar{X} = 1/p \sum_{X^i \in vert(P)} X^i,$$

where p is the cardinality of $vert(P)$, the vertex set of P. [See [Zeigler (1995)] for introduction to polytopes.]

Given $n \in Z$ the *input size* of n is the number of digits in the binary expansion of the number n plus 1 for the sign if n is non zero. We write,

$$\langle n \rangle = 1 + \lceil \log_2(n+1) \rceil.$$

Input size of n, $\langle n \rangle$, is also known as the *digital size* of n.

Given $r = p/q$ a rational number, where p and q are mutually prime, that is $gcd(p,q) = 1$, we have input size of r given by

$$\langle r \rangle = \langle p \rangle + \langle q \rangle.$$

For every rational r we have $|r| \leq 2^{\langle r \rangle - 1} - 1$.

Definition 6.19. [Rationality Guarantee] Let $P \subset R^d$ be a polytope and ϕ and ν positive integers. We say that P has *facet complexity* at most ϕ if P can be described as the solution set of a system of linear inequalities each of which has input size $\leq \phi$. We say, P has vertex complexity at most ν if P can be written as $P = conv(V)$, where $V \subset Q^d$ is finite and each vector in V has input size $\leq \nu$.

We have Lemma 6.4 from [Grötscel, Lovász and Schrijver (1988)] connecting facet and vertex complexities.

Lemma 6.4. *Let $P \subset R^d$ be a non empty, full dimensional polytope. If P has vertex complexity at most ν, then P has facet complexity at most $3d^2\nu$.*

6.3.0.1 *Problems Related to Polytopes*

Given a polytope $P \subset Q^d$, and a $Y \in Q^d$, the problem to decide whether $Y \in P$ or not, is called the *membership problem* for P. Let $P \subset Q^d$, be a polytope with facet complexity at most ϕ. Let $MemAl(P, Y, Answer)$ be an algorithm[1] to solve the membership problem, where P is known to $MemAl$ not necessarily explicitly, and on input of $Y \in Q^d$ having input size $\langle Y \rangle$,

[1]The term *oracle* or *subroutine* is generally used to mean that this algorithm is called by another algorithm as a procedure.

$MemAl$ halts with $Answer = yes$ if $Y \in P$ and $Answer = no$ otherwise. If the membership checking time of $MemAl$ is polynomially bounded above by a function of $(d, \phi, \langle Y \rangle)$ we say $MemAl$ is an efficient oracle.

Given a polytope $P \subset Q^d$, and a $Y \in Q^d$, the problem to decide whether $Y \in P$, and if $Y \notin P$ then identifying a hyperplane that separates P and Y, is called the *separation problem* for P. Identifying a hyperplane is achieved through finding a vector $a \in Q^d$ such that $aX < aY$ for all $X \in P$

Given a non empty polytope $P \subset Q^d$, and a $C \in Q^d$, the problem of finding a $X^* \in P \ni CX^* \leq CX$ for all $X \in P$ is called the *linear optimisation problem* for P.

Recently [Maurras (2002)] shows that an intuitively appealing construction is possible for the separation problem of a polytope, by finding a hyperplane separating the polytope and a point not in the polytope, after a polynomial number of calls of to a membership oracle. The conditions under which this is possible are same as that of [Yudin and Nemirovskii (1976)], namely,

Assumption 6.1 (Maurras's Conditions).

1 *The polytope P is well defined in the d-dimensional space of Q^d of rational vectors. (There is a bound on the encoding length of any vertex of P. The polytope is rationality guaranteed.)*
2 *P has non-empty interior.*
3 *$a \in int(P)$ is given.*

[Grötscel, Lovász and Schrijver (1988)] use a construction due to [Yudin and Nemirovskii (1976)] to devise a polynomial algorithm for finding a separating plane using a membership oracle, when a convex set K instead of the polytope P is considered. But this algorithm requires in addition the radii of the inscribed and circumscribed balls, also uses Ellipsoid algorithm twice. Next we check that the Assumption 6.1 is met for a polytope closely related to the $STSP$ polytope.

6.3.1 *Properties of the Polytope, conv(A_n)*

Here we consider a compact representation of a pedigree by removing some redundancy present. Let $e^k = (k-2, k-1)$, and $E'_{k-1} = E_{k-1} \setminus \{e^k\}$, for $k \in V_n \setminus V_3$. Let $\tau'_n = \tau_n - (n-3)$.

Definition 6.20. [Projection M] Given $X \in P_n$, consider the transformation $Y = MX$, where M deletes the p^{th}_{k-1} component of $\mathbf{x}_k$ in X, giving a vector $Y \in B^{\tau'_n}$.

The projection M is given by the matrix

$$M = \begin{bmatrix} I_{p_3-1} & 0 & 0 & 0 & \cdots & 0 \\ 0 & 0 & I_{p_4-1} & \vdots & & \vdots \\ \vdots & \vdots & & \cdots & 0 & 0 \\ 0 & 0 & \cdots & & I_{p_{n-1}-1} & 0 \end{bmatrix}.$$

Notice that Y is the compact string that has the information contained in X, but there is some redundancy in X, namely, for any k, the last component of $\mathbf{x}_k$ does not say anything more than what is already said in the $p_{k-1}-1$ preceding components. This is so because, for each k we have a unique edge $e \in E_{k-1}$ that is in the pedigree X or $x_k(e) = x_k(E_{k-1}) = 1$. (We have in fact, $x_k(e^k) = 1 - x_k(E'_{k-1})$.

Definition 6.21 (A_n). *Let* $A_n = \{Y \in B^{\tau'_n} | Y = MX, X \in P_n\}$.

Lemma 6.5. *There is a* $1-1$ *correspondence between* A_n *and* $\mathcal{H}_n$.

Proof. Given any $T \in \mathcal{H}_n$, the corresponding characteristic vector X of the pedigree is unique. Thus $Y \in A_n$, given by the transformation M of X is unique as well. On the other hand, given a $Y \in A_n$, we can uniquely define,

$$x_k(e) = \begin{cases} y_k(e) \; if \; e \in E'_{k-1} \\ 1 - y_k(E'_{k-1}) \text{ for } e = e^k. \end{cases}$$

Hence the lemma. □

Theorem 6.2. *[dim(conv(A_n))] Given* $n \geq 4$, *and* A_n *as defined above, we have,*

(1) The polytope $conv(A_n)$ *is full dimensional, that is* $dim(conv(A_n)) = \tau'_n$
(2) The barycentre of $conv(A_n)$ *is given by,*

$$\bar{Y} = (\underbrace{1/p_3, 1/p_3}_{2-times}, \ldots, \underbrace{1/p_{n-1}, \ldots, 1/p_{n-1}}_{p_{n-1}-1 \; times}).$$

(3) $\bar{Y} \in int(conv(A_n))$.

Proof. Part 1 of the theorem is proved recently in [Arthanari (2007)].

Part 2 can be verified by noticing that the cardinality of $vert(conv(A_n))$, is $(n-1)!/2$, and in any X in P_n, the $(k-3)^{rd}$ component has p_{k-1} coordinates, and exactly one of the coordinates is a 1. In P_n for any component the 1 appears equally likely among the coordinates. And for any $Y \in A_n$ we have deleted the last coordinate in each component of the corresponding X.

Proof of part 3 of the theorem follows from the fact that $\bar{Y}$ does not lie on any facet defining hyperplane $CY = c_0$, for $(C, c_0) \in Q^{\tau'_n+1}$. Suppose, it lies on some facet defining hyperplane $CY = c_0$ (that is $CY \leq c_0$ for all $Y \in conv(A_n)$). Then

$$C\bar{Y} - c_0 = [2/(n-1)!] \sum_{X \in P_n} (CY^X - c_0) = 0,$$

where Y^X is the element of A_n corresponding to a $X \in P_n$. Thus for all $X \in P_n$, we have $CY^X = c_0$,

$$\Longrightarrow dim(conv(A_n)) \leq \tau'_n - 1.$$

This contradicts the fact $dim(conv(A_n))$ is τ'_n.

Therefore $\bar{Y} \in int(conv(A_n))$. Hence the theorem. □

Theorem 6.3. *[Facet - Complexity of $conv(A_n)$] $conv(A_n)$ has facet complexity at most $\phi = 3\tau'^3_n + 3\tau'^2_n(n-3)$. That is $conv(A_n)$ is rationality guaranteed.*

Proof. Each vertex Y of $conv(A_n)$ is a $0-1$ vector of length τ'_n. So Y can be encoded with input size

$$\langle Y \rangle \leq \tau'_n + (n-3) = \nu.$$

(This follows from the fact that there are at most $n-3$ $1's$ in any Y and $\langle 0 \rangle = 1$ & $\langle 1 \rangle = 1 + \lceil \log_2 2 \rceil = 2$.)

Therefore, $conv(A_n)$ has vertex complexity $\leq \nu$.

Using lemma(6.4), we have, facet complexity of $conv(A_n)$,

$$\begin{aligned} &\leq 3\tau'^2_n \nu \\ &= 3\tau'^2_n(\tau'_n + (n-3)) \\ &= 3\tau'^3_n + 3\tau'^2_n(n-3). \end{aligned}$$

Hence, $conv(A_n)$ is rationality guaranteed. □

Thus we find $conv(A_n)$ satisfies all the requirements of Maurras's conditions (Assumption 6.1). Therefore if we have a membership oracle for $conv(A_n)$ we can call that a polynomial number of times to separate a $Y \in Q^{\tau'_n}$ from $conv(A_n)$. With this in view we direct our attention to the membership problem of the pedigree polytope, since Y is in $conv(A_n)$ if and only if the corresponding pedigree X is in $conv(P_n)$.

6.4 Construction of the Layered Network N_k

In this section we define the layered network $N_k(X)$ with respect to a given $X \in P_{MI}(n)$ and for $k \in V_{n-1} \setminus V_3$. Given $X/k \in conv(P_k)$, this network is used in showing whether $X/k+1 \in conv(P_{k+1})$ or not. Since X is fixed throughout this discussion we drop the X from the notation for the network and write simply N_k.

With respect to a given X we define for each k, a layered network, N_k, with $(k-2)$ layers.

We denote the node set of N_k by $\mathcal{V}(N_k)$ and the arc set by $\mathcal{A}(N_k)$. Let $v = [k : e]$ denote a node in the $(k-3)^{rd}$ layer corresponding to an edge $e \in E_{k-1}$. Let $x(v) = x_k(e)$ for $v = [k : e]$.

Let

$$V_{[r]} = \{v | v = [r+3 : e], e \in E_{r+2}, x(v) > 0\}.$$

Notice that the node name $[r+3 : e]$ alludes to insertion decision corresponding to the stage r; that is, the edge e used for insertion of $r+3$. First we define the nodes in the network N_k, for $k = 4$.

$$\mathcal{V}(N_4) = V_{[1]} \bigcup V_{[2]}.$$

And

$$\mathcal{A}(N_4) = \{(u,v) | u \in V_{[1]}, v \in V_{[2]}, e_\alpha \in G(e_\beta)\}$$

where $u = [4 : e_\alpha]$ and $v = [5 : e_\beta]$.

Let $x(v)$ be the capacity on a node $v \in V_{[r]}, r = 1, 2$. Capacity on an arc $(u,v) \in \mathcal{A}(N_4)$ is $x(u)$. Given this network we consider a flow feasibility problem of finding a nonnegative flow defined on the arcs that saturates all the node capacities and violates no arc capacity. We refer to this problem F_4.

Notice that the problem F_4 is one and the same as the problem $FAT_4(\mathbf{x}_4)$ defined in [Arthanari (2006)]. Therefore, F_4 feasibility is equivalent to $FAT_4(\mathbf{x}_4)$ feasibility. So $X/5 \in conv(P_5)$. If F_4 is infeasible we do not proceed further.(We conclude that $X \notin conv(P_{n})$ as shown in [Arthanari (2006)].) Otherwise if $k < n-1$ we continue to define the next network for $k+1$.

If F_4 is feasible we use the FFF algorithm (or any such) and identify $\mathcal{R}$. If there are any dummy arcs in $\mathcal{R}$ we delete them from $\mathcal{A}(N_4)$ and update $\mathcal{A}(N_4)$. For rigid arc with positive frozen flow we update the capacity of the arc as the frozen flow and colour the arc *'green'*. Therefore in every feasible solution to the updated F_4 the green arcs are saturated.

Now we say N_4 is *well-defined.*

Given N_{k-1} is well-defined, we proceed to define N_k recursively. Firstly we define,

$$\mathcal{V}(N_k) = \mathcal{V}(N_{k-1}) \bigcup V_{[k-2]}. \tag{6.8}$$

Now consider the links between layers $k-3$ and $k-2$. Any of the links, $L = (e, e')$ can give raise to an arc in the network N_k depending on the solution to a max flow problem defined on a sub network derived from N_{k-1} and the link L. If the maximal flow in the sub network is zero we can not use the link (e, e').

Next we define the restricted network which is induced by deletion of a subset of nodes from $\mathcal{V}(N_{k-1})$.

Definition 6.22. [Restricted Network $N_{k-1}(L)$] Given $k \in V_{n-1} \setminus V_4$, a link $L = (e_\alpha, e_\beta) \in E_{k-1} \times E_k$, with $e_\alpha = (r, s)$ and $e_\beta = (i, j)$. $N_{k-1}(L)$ is the sub network induced by the subset of nodes $\mathcal{V}(N_{k-1}) \setminus \mathcal{D}(L)$, where $\mathcal{D}(L)$, the set of deleted nodes is constructed as follows: Let $\mathcal{D}(L) = \emptyset$.

(a) Include $[l : e_\beta]$ in $\mathcal{D}(L)$, for $max(4, j) < l < k$.
(b) Include $[l : e_\alpha]$ in $\mathcal{D}(L)$, for $max(4, s) < l < k$.
(c) Include $[j : e], e \notin G(e_\beta)$ in $\mathcal{D}(L)$, if $e_\beta \in E_k \setminus E_3$; otherwise include $[4 : e_\beta]$ in $\mathcal{D}(L)$.
(d) Include $[s : e], e \notin G(e_\alpha)$ in $\mathcal{D}(L)$, if $e_\alpha \in E_{k-1} \setminus E_3$; otherwise include $[4 : e_\alpha]$ in $\mathcal{D}(L)$.
(e) Include all nodes $V_{[k-3]} \setminus \{[k : e_\alpha]\}$ in $\mathcal{D}(L)$.

Set $\mathcal{V}(N_{k-1}(L)) = \mathcal{V}(N_{k-1}) \setminus \mathcal{D}(L)$. The sub network induced by $\mathcal{V}(N_{k-1}(L))$ is called the *Restricted Network* $N_{k-1}(L)$.

Remark 6.1.

1 Deletion rule [a] ([b]) ensures that the edge e_β (e_α) does not appear earlier in a path from source(s) in layer 1 to $(k-3)^{rd}$ layer. Deletion rule [c] ([d]) ensures that the edge(s) not in the generator of the edge e_β (e_α) are deleted from $(j-3)^{rd}((s-3)^{rd})$ layer. Finally [e] ensures that the only sink in $(k-3)^{rd}$ layer is $[k : e_\alpha]$.
2 Deletion of a node can be equivalently interpreted as imposing an upper bound of zero on the flow through a node with respect to a given link (treated as a commodity). This interpretation is useful in considering a multicommodity flow through the network N_k.
3 A multi commodity flow problem is solved to answer the question: Given $X/k \in conv(P_k)$, does $X/k+1$ belong to $conv(P_{k+1})$ in Section 6.7.

We consider the problem of finding the maximal flow in $N_{k-1}(L)$ satisfying all the restrictions on nonnegativity, flow conservation and capacity on the available nodes and arcs.

The only sink in the network is $[k : e_\alpha]$ and the sources are the undeleted nodes in $V_{[1]}$. Let $C(L)$ be the value of the maximal flow in the restricted network $N_{k-1}(L)$. We find $C(L)$ for each link L.

Now we are in a position to define the FAT problem, called F_k.

Definition 6.23 (F_k). *Consider a forbidden arc transportation problem with*

$O --$ *Origins*$] : u = [k : e_\alpha] \in V_{[k-3]}$
$D --$ *Sinks*$] : v = [k+1 : e_\beta] \in V_{[k-2]}$
$\mathcal{A} --$ *Arcs*$] : \{(u, v)$ *such that* $L = (e_\alpha, e_\beta)$ *is a link and* $C(L) > 0\}$
$C --$ *Capacity*$] : C_{u,v} = C(L)$.

If F_k is feasible and $k < n-1$ we apply the *Frozen Flow Finding* algorithm and identify $\mathcal{R}$ and the dummy subset of arcs in that. Update the capacity of the rigid arcs with positive flow equal to the frozen flow. Update $\mathcal{A}$ by deleting the dummy arcs. Rigid arcs are marked green. We finally have

$$\mathcal{A}(N_k) = \mathcal{A}(N_k) \bigcup \mathcal{A}. \tag{6.9}$$

If $k = n-1$ we stop.

This completes the construction of N_k given by Equations 6.8 and 6.9. Next task is to check that N_k is well-defined.

We need the following definitions, which are used in the sections that follow.

Definition 6.24. [Pedigree path] Consider the network, N_l. Let $path(X^r)$ denote the path corresponding to a $X^r \in P_{l+1}$, given by

$$[4 : e_4^r] \to [5 : e_5^r] \ldots \to [l+1 : e_{l+1}^r]$$

where X^r is the characteristic vector of $(e_4^r, \ldots e_{l+1}^r)$.

Definition 6.25. Consider any feasible flow, f in $N_{l-1}(L)$ for a link $L = (e_\alpha, e_\beta) \in E_l \times E_{l+1}$. Let v_f be the value of the flow f, that is, v_f reaches the sink in $N_{l-1}(L)$. We say v_f is *pedigree packable* in case there exists a subset $P(L) \subset P_l$ such that

(1) $\lambda_r (\geq 0)$ is the flow along $path(X^r)$ for $X^r \in P(L)$,
(2) $e_l^r = e_\alpha$, $X^r \in P(L)$,

(3) $\sum_{r \ni x^r(v)=1} \lambda_r \leq x(v), v \in \mathcal{V}(N_{l-1}(L))$, and
(4) $\sum_{X^r \in P(L)} \lambda_r = v_f$.

We refer to $P(L)$ as a *pedigree pack* of v_f.

Definition 6.26. [Extension Operation] Given a pedigree pack corresponding to a flow f in $N_{l-1}(L)$ for a link $L = (e_\alpha, e_\beta)$ with $v_f > 0$, we call $\overrightarrow{X^r L} = (X^r, \mathbf{x}^r_{l+1})$ the *extension* of $X^r \in P(L)$ in case

$$x^r_{l+1}(e) = \begin{cases} 1 \text{ if } e = e_\beta \\ 0 \text{ otherwise.} \end{cases} \tag{6.10}$$

That is, $X^r = (e^r_4, \dots, e^r_l = e_\alpha)$ and this pedigree can be extended to $(e^r_4, \dots, e^r_l = e_\alpha, e_\beta)$. And the corresponding characteristic vector, $(X^r, \mathbf{x}^r_{l+1}) \in P_{l+1}$ (see Figure 6.1).We denote the subset of P_{l+1} thus obtained by $\overrightarrow{P(L)}$, and call it the extension of $P(L)$. Notice that $v_f > 0$ implies $e_\beta \in T^r$, the $l-$*tour* corresponding to $X^r \in P(L)$.

Characteristic Vector of Pedigree:

$$X^r = (\mathbf{x}_4, \dots, \mathbf{x}_l)$$

$$\Downarrow$$

$$Pedigree:$$

$$(e^r_4, \dots, e^r_l = e_\alpha)$$

$$+$$

$$Link: L = (e_\alpha, e_\beta) \text{ such that } e_\beta \in T^r \in \mathcal{H}_l.$$

$$\Downarrow$$

$$Extended\ Pedigree:$$

$$(e^r_4, \dots, e^r_l = e_\alpha, e^r_{l+1} = e_\beta)$$

$$[4 : e^r_4] \rightarrow [5 : e^r_5] \ \dots \ \rightarrow [l : e^r_l = e_\alpha] \rightsquigarrow [l+1 : e^r_\beta]$$

Fig. 6.1 Extendable Pedigree path

6.5 Necessity of F_k Feasibility for Membership

We define yet another FAT problem for which an obvious feasible flow is available, and this flow can be very useful in proving the feasibility of F_k.

Definition 6.27. Given $X/k+1 \in conv(P_{k+1}), \lambda \in \Lambda_{k+1}(X)$ we define a FAT problem obtained from $I(\lambda)$ for a given $l \in \{4, \ldots, k\}$ as follows:

Partition $I(\lambda)$ in two different ways according to $\mathbf{x}_l^r, \mathbf{x}_{l+1}^r$, resulting in two partitions S_O and S_D. We have

$$S_O^q = \{r \in I(\lambda) | x_l^r(e_q) = 1\}, e_q \in E_{l-1} \text{ and } x_l(e_q) > 0$$

and

$$S_D^s = \{r \in I(\lambda) | x_{l+1}^r(e_s) = 1\}, e_s \in E_l \text{ and } x_{l+1}(e_s) > 0.$$

Let $|y|_+$ denote the number of positive coordinates of any vector y. Let $n_O = |\mathbf{x}_l|_+$ and $n_D = |\mathbf{x}_{l+1}|_+$. Let $a_q = \sum_{r \in S_O^q} \lambda_r = x_l(e_q), q = 1, \ldots, n_O$. Let $b_s = \sum_{r \in S_D^s} \lambda_r = x_{l+1}(e), s = 1, \ldots, n_D$. Let the set of forbidden arcs, F, be given by

$$F = \{(q, s) | S_O^q \cap S_D^s = \emptyset\}.$$

The problem with an origin for each q with availability a_q, a sink for each s with demand b_s and the forbidden arcs given by F is called the FAT *problem induced by* (λ, l).

Remark 6.2.

1 From Lemma 6.1 we know that such a problem is feasible and a feasible flow is given by

$$f(q, s) = \sum_{r \in S_O^q \cap S_D^s} \lambda_r.$$

We call such an f the *instant flow* for the FAT problem induced by (λ, l).

2 The availabilities and demands in the above problem are same as that of F_l.

3 Any arc in F_l and its capacity were given by solving the max flow problem in $N_{l-1}(L)$ for each link L. But there are no capacity restrictions on the arcs of the FAT problem induced by (λ, l).

We shall show that the instant flow for the FAT problem induced by any (λ, l) is indeed feasible for the problem F_l. Let L_l^r denote (e, e') such that $x_l^r(e) = x_{l+1}^r(e') = 1$. That is L_l^r is the $(l-3)^{rd}$ and $(l-2)^{nd}$ elements of the pedigree given by X^r.

Lemma 6.6. *Given $\lambda \in \Lambda_{k+1}$ and l if $path(X^r/l)$ is available in $N_{l-1}(L_l^r)$, $\forall\, r \in I(\lambda)$, then the instant flow f in the FAT problem induced by (λ, l) is feasible for F_l.*

Proof. Let $path(X^r/l)$ be available in $N_{l-1}(L_l^r)$, for every $r \in I(\lambda)$. Now consider $r \in S_O^q \cap S_D^s \neq \emptyset$ for some q, s. Let $L = (e_q, e_s)$. For all these r, $L_l^r = L$, and is available in the FAT problem induced by (λ, l).

Along the $path(X^r/l)$ we can have a flow of λ_r into $[l : e_q]$ in the restricted network $N_{l-1}(L)$. So we are ensured that the maximum flow, (say $C(L)$) in $N_{l-1}(L)$ is positive. According to the construction of problem F_l we have an arc $([l : e_q], [l+1 : e_s])$ with capacity $C(L)$. Now the definition of the instant flow f and the maximality of $C(L)$ imply that f satisfies all the capacity restrictions. (Since a flow of at least $\sum_{r \in S_O^q \cap S_D^s} \lambda_r$ along the paths, $path(X^r/l)$ can reach $[l : e_q]$ for $r \in S_O^q \cap S_D^s \neq \emptyset$. So the maximum flow $C(L)$ in $N_{l-1}(L)$ should be at least this, which is precisely $f(q, s)$.) Thus we have shown that the instant flow of the FAT problem induced by (λ, l) is feasible for F_l. Hence the lemma. □

Next we address the question: Is the condition stated by Lemma 6.6 always met? Towards this we first show that the $path(X^*/5)$ is available in $N_4(L_5^*)$, for any X^* active for $X/k+1 \in conv(P_{k+1})$.

Lemma 6.7. *Every $X^* = (e_4^*, \ldots, e_{k+1}^*)$ active for $X/k+1$, is such that $path(X^*/5)$ is available in $N_4(L_5^*)$, where $L_5^* = (e_5^*, e_6^*)$.*

Proof. $Path(X^*/5)$ is given by $[4 : e_4^*] \to [5 : e_5^*]$. We have $[5 : e_5^*]$ in $N_4(L_5^*)$ as the lone sink.

Case 1. $[4 : e_4^*]$ is not a node in $N_4(L)$

This implies there exists a deletion rule among the rules (a) through (e) (see Definition 6.22 that deleted $[4 : e_4^*]$ from $\mathcal{V}(N_4)$. Notice that rules (a), (b) and (e) are not applicable as they delete a node $[l : e]$ with $l > 4$.

Claim 6.1. *Rule (c) does not delete* $[4 : e_4^*]$.

Proof. [Proof of Claim] Suppose $e_6^* = (i, j) \in E_5 \setminus E_3$ and $j = 5$. Then rule (c) deletes a node $[5 : e]$. Suppose $e_6^* = (i, 4)$ for some $1 \leq i < 4$, then rule (c) deletes $[4 : e], e \notin G(e_6^*)$. So if $[4 : e_4^*]$ is one such node then e_4^* is not a generator of e_6^*. And so $X^*/6$ can not be in P_6. Contradiction. This leaves the possibility $e_6^* \in E_3$. Then $[4 : e_6^*]$ is deleted. Hence the claim. □

Similarly we can check that rule (d) is not deleting $[4 : e_4^*]$. Thus we have seen the impossibility of Case 1.

Case 2. $[4 : e_4^*]$ exists but $([4 : e_4^*], [5 : e_5^*])$ is not an arc in $N_4(L_5^*)$.

Suppose $([4 : e_4^*], [5 : e_5^*])$ is not an arc in $N_4(L_5^*)$. X^* being the characteristic vector of a pedigree, implies that $e_4^* \in G(e_5^*)$. So this arc exists in F_4 with capacity $x_4(e_4^*) > 0$. We shall show that F_4 is feasible. Since X^* is active for $X/6$ we have a $r \in I(\lambda)$ such that $X^* = X^r$, for some $\lambda \in \Lambda_6(X)$. Consider the FAT problem induced by $(\lambda, 4)$. So we have the instant flow f that is feasible for this FAT problem. From feasibility of f, we have

$$\sum_s f_{qs} = x_4(e_q), [4 : e_q] \in V_{[1]} \tag{6.11}$$

$$\sum_q f_{qs} = x_5(e_s), [5 : e_s] \in V_{[2]} \tag{6.12}$$

Recall that in F_4 the capacity of any arc, $([4 : e], [5 : e'])$ is defined to be $x_4(e)$. We see from equation 6.11 that f meets these capacity restrictions. Thus, from equations 6.11 and 6.12 and the observation made above, we have shown that f, the instant flow, is feasible for F_4.

So we are eligible to apply FFF algorithm to find the dummy arcs in F_4, towards constructing N_4.

Notice that the flow along $([4 : e_4^*], [5 : e_5^*])$ is positive as the corresponding set $S_O^q \cap S_D^s \neq \emptyset$, has at least r corresponding to X^* in it. This ensures that the arc $([4 : e_4^*], [5 : e_5^*])$ is not a dummy. And so it exists in $N_4(L_5^*)$. Hence Case 2 is also not possible.

This completes the proof of the lemma. □

Lemma 6.7 forms the basis to prove Lemma 6.8 that is crucial in showing that infeasibility of F_k implies that $X/k+1 \notin conv(P_{k+1})$.

Lemma 6.8. *[Existence of Pedigree Paths] Every X^* active for $X/k+1$, is such that $path(X^*/l)$ is available in $N_{l-1}(L_l^*)$, for $5 \le l \le k$, where L_l^* denotes (e, e') such that $x_l^*(e) = x_{l+1}^*(e') = 1$.*

Proof. [Proof by induction on l] From Lemma 6.7 we have the result for $l = 5$. Assume that the result is true for $l \le k-1$. We shall show that the result is true for $l = k$.

By hypothesis we have $path(X^*/k-1)$ available in $N_{k-2}(L_{k-1}^*)$ ending in $[k-1 : e_{k-1}^*]$. Now consider $N_{k-1}(L_k^*)$. Suppose $path(X^*/k)$,that is,

$$[4 : e_4^*] \to \ldots [k-1 : e_{k-1}^*] \to [k : e_k^*]$$

is not available in $N_{k-1}(L_k^*)$. But $[4:e_4^*] \to \ldots [k-1:e_{k-1}^*]$ is available in $N_{k-2}(L_{k-1}^*)$. This implies that it is also available in N_{k-1}. (Recall the construction of N_{k-1} via $N_{k-2}(L')$, for L' a link.) Also $[k:e_k^*]$ is available in $N_{k-1}(L_k^*)$ as it is the lone sink. So our assumption really means that the arc $W = ([k-1:e_{k-1}^*],[k:e_k^*])$ does not exist in $N_{k-1}(L_k^*)$.

Case 1. W does not exist in N_{k-1}.

We shall show that this case is not possible. The existence of $path(X^*/k-1)$ in $N_{k-2}(L_{k-1}^*)$ implies that the maximum flow into the sink, $[k-1:e_{k-1}^*]$ is positive. So W is an arc in F_{k-1}. Since X^* is active for $X/k+1$ we have a $r_0 \in I(\lambda^*)$ such that $X^* = X^{r_0}$, for some $\lambda^* \in \Lambda_{k+1}(X)$. Consider the FAT problem induced by $(\lambda^*, k-1)$ for such a λ^*. Notice that every $r \in I(\lambda^*)$ is active for $X/k+1$. So from the hypothesis $path(X^r/k-1)$ is available in $N_{k-2}(L_{k-1}^r)$ for every $r \in I(\lambda^*)$. We have the condition of Lemma 6.6 met here. Consider the instant flow f that is feasible for the FAT problem induced by $(\lambda^*, k-1)$. Lemma 6.6 asserts that f is indeed feasible for F_{k-1}.

But if W does not exist in N_{k-1} it implies that W has been subsequently declared dummy by the FFF algorithm. Since the flow along W as per f is at least equal to $\lambda_{r_0}^*$ corresponding to X^*, which agrees with $L_{k-1}^* = (e_{k-1}^*, e_k^*)$. But $\lambda_{r_0} > 0$ as X^* is active for $X/k+1$. Therefore, $W = ([k-1:e_{k-1}^*],[k:e_k^*])$ can not be declared as dummy by FFF algorithm. Thus, Case 1 is impossible.

Case 2. W does not exist in $N_{k-1}(L_k^*)$.

This implies that as a consequence of the deletion rules (a) through (e) W has been deleted. Notice that no arc with both of its ends available in $N_{k-1}(L_k^*)$ is deleted from the network. Since nodes $[k-1:e_{k-1}^*]$ and $[k:e_k^*]$ have been shown to be in $N_{k-1}(L_k^*)$, Case 2 is impossible.

This completes the proof of the lemma. □

Theorem 6.4. *[Theorem on non-Membership] Given $X \in P_{MI}(n)$, and for a $k \in V_{n-1} \setminus V_3$, if $X/k \in conv(P_k)$, then*

$$F_k \textit{ infeasible implies } X/k+1 \notin conv(P_{k+1}).$$

Proof. Suppose $X/k+1 \in conv(P_{k+1})$. Consider any $\lambda \in \Lambda_{k+1}(X)$. Then from Lemma 6.8 we have the path corresponding to X^r/l available in $N_{l-1}(L_l^r)$ for each $r \in I(\lambda)$. Now conditions of Lemma 6.6 are met and so the instant flow in the FAT problem induced by (λ, l) is feasible for F_l, for $5 \leq l \leq k$. We have a contradiction. Hence the theorem. □

Remark 6.3.

1 With this theorem we have a procedure to check $X/k+1 \notin conv(P_{k+1})$ by solving F_k.
2 However if F_k is feasible we can not, in general, conclude that $X/k+1 \in conv(P_{k+1})$. See Example 6.1.
3 As a corollary to this theorem we have, given $X/k \in conv(P_k)$ implies for any $\lambda \in \Lambda_k(X)$ we have all the paths corresponding to $\{X^r | r \in I(\lambda)\}$ in N_{k-1}.

Example 6.1. Consider X given by

$$\mathbf{x}_4 = (0, 3/4, 1/4);$$
$$\mathbf{x}_5 = (1/2, 0, 0, 1/2, 0, 0);$$
$$\mathbf{x}_6 = (0, 1/4, 1/2, 0, 1/4, 0, 0, 0, 0, 0).$$

It can be verified that $X \in P_{MI}(6)$. And F_4 is feasible and f given by

$$f_{([4:1,3],[5:1,2])} = 1/4, f_{([4:2,3],[5:1,2])} = 1/4, f_{([4:1,3],[5:1,4])} = 1/2$$

does it. Also
$X/5 = 1/4(0,1,0;1,0,0,0,0,0) + 1/4(0,0,1;1,0,0,0,0,0)$
$+1/2(0,1,0;0,0,0,1,0,0)$.
Next via the restricted networks $N_4(L)$ for the links in $\{(1,2),(1,4)\} \times \{(1,3),(2,3)),(2,4)\}$ we obtain the bipartite network given in Figure 6.2.

Notice that F_5 is feasible, with f given by

$$f_{([5:1,2],[6:1,3])} = 1/4, f_{([5:1,2],[6:2,4])} = 1/4, f_{([5:1,4],[6:2,3])} = 1/2.$$

Suppose $X/6$ is in P_6, consider any $\lambda \in \Lambda_6(X)$. Then there are pedigrees $X^r, r \in I(\lambda)$ such that $x_6^r(2,3) = 1$. The total weight for these pedigrees is $x_6(2,3) = 1/2$. But these pedigrees can not have $x_4^r(2,3) = 1$ as they all have $x_6^r(2,3) = 1$. So this forces the alternative $x_4^r(1,3) = 1$ for all these pedigrees. However no pedigree X^r with $x_6^r(e) = 1$ for $e = (1,3)$ or $(2,4)$ can also have $x_4^r(1,3) = 1$, as $(1,3) \notin G(e)$ for both the e's. Hence

$$\sum_{r \in I(\lambda), x_4^r(1,3)=1} \lambda_r = \sum_{r \in I(\lambda), x_6^r(2,3)=1} \lambda_r = 1/2 < 3/4 = x_4(1,3).$$

Thus λ can not belong to $\Lambda_6(X)$. Contradiction. Therefore $X \notin conv(P_6)$.

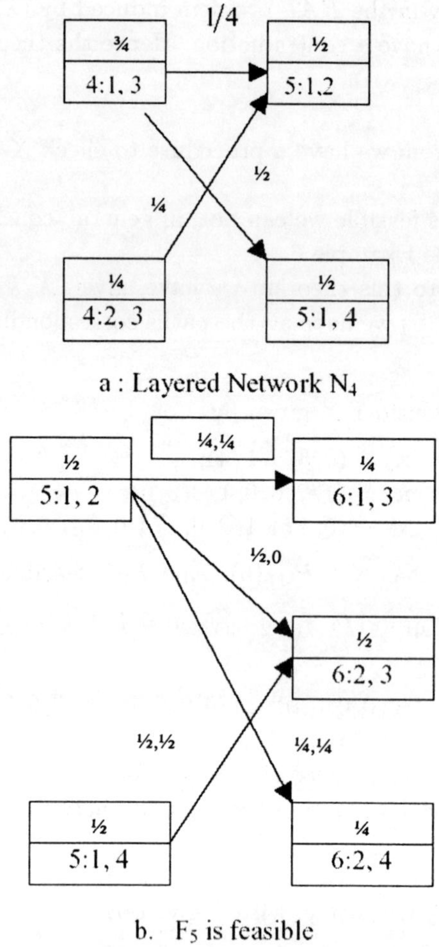

Fig. 6.2 Layered Network for Example 6.1

6.6 Pedigree Packability

Explicit use of the fact $X \in P_{MI}(n)$ is made in this section to establish the pedigree packability of any node capacity at layer $k-2$ given that $X/k \in conv(P_k)$.

Theorem 6.5. *Given $X \in P_{MI}(n)$ and $X/k \in conv(P_k)$, consider the network N_{k-1}. For any $[k+1:e] \in V_{[k-2]}$, we have a flow f_α in $N_{k-1}(L_\alpha)$*

for a link $L_\alpha = (e_\alpha, e)$, *and such that,*

(1) The value of the flow f_α, *given by* v_α, *is pedigree packable for each link* L_α,
(2) $\sum_\alpha v_\alpha = x_{k+1}(e)$, *and*
(3) $\sum_\alpha \sum_{X^r \in \mathcal{P}(L_\alpha),\ x^r(u)=1} \mu_r \le x(u), u \in \mathcal{V}(N_{k-1})$,

where μ_r *is the flow along the path*(X^r).

In other words, Theorem 6.5 assures the existence of pedigree paths in N_{k-1} bringing in a flow of v_α into the sink, $[k : e_\alpha]$, for some e_α and all these paths can be extended to pedigree paths in N_k bringing in a total flow of $x_{k+1}(e)$ into $[k+1 : e] \in V_{[k-2]}$.

Proof. Notice that $X/k+1 \in P_{MI}(k+1)$. Recall the definition of U^{k-3} we have Equation 6.7, that is,

$$U^{k-3} - A^{(k+1)}\mathbf{x}_{k+1} = U^{k-2} \ge 0.$$

In fact, U^{k-3} is the slack variable vector corresponding to X/k. So, $x_{k+1}(e) \le U^{k-3}(e)$. Now $[k+1 : e] \in V_{[k-2]}$ means $x_{k+1}(e) > 0$, and so $U^{k-3}(e) > 0$.

Since $X/k \in conv(P_k)$, consider any weight vector $\lambda \in \Lambda_k(X)$. We have

$$\sum_{X^r \in I(\lambda),\ x^r(u)=1} \lambda_r = x(u), u \in \mathcal{V}(N_{k-1}). \tag{6.13}$$

From Lemma 4.1 from [Arthanari and Usha (2000)] we have if (X, U) is an integer solution to MI- relaxation then U is the edge-tour incident vector of the n-tour corresponding to X. Applying this with $n = k$ and noticing $X/k \in conv(P_k)$, we find the same λ can be used to write U^{k-3} as a convex combination of $\dot{T}^r, r \in I(\lambda)$, where T^r is the $k-tour$ corresponding to X^r and $\dot{T}$ denotes the edge-tour incident vector of T.

Let

$$J = \{r | X^r \in P_k \text{ and } e \in T^r\}.$$

Thus,

$$U^{k-3}(e) = \sum_{r \in I(\lambda) \cap J} \lambda_r \dot{T}^r. \tag{6.14}$$

Now partition $I(\lambda)$ with respect to $\mathbf{x}_k^r$ as follows: Let I_α denote the subset of $I(\lambda)$ with $x_k^r(e_\alpha) = 1$.

$$\sum_{r \in I_\alpha} \lambda_r = x_k(e_\alpha), \text{ for } e_\alpha \in E_k.$$

We have,

$$I(\lambda) \cap J = \cup \alpha | I_{\alpha \neq \emptyset} I_\alpha \cap J. \tag{6.15}$$

As I_α's are disjoint, we a have a partition of $I(\lambda) \cap J$. Notice that the $path(X^r)$ corresponding to $r \in I_\alpha \cap J$, is available in $N_{k-1}(L_\alpha)$, since any $X^r \in P_k$ active for X/k is such that the $path(X^r)$ is available in N_{k-1} (see remark 6.3.3 following the proof of Theorem 6.4).

Now let

$$\mathcal{P}(L_\alpha) = \{X^r | r \in I_\alpha \cap J\}.$$

Any $path(X^r)$ for an $X^r \in \mathcal{P}(L_\alpha)$ can be extended to a path in N_k, ending in $[k+1 : e]$, using the arc $([k : e_\alpha], [k+1 : e])$. Thus we have a subset of pedigrees in P_{k+1}, corresponding to these extended paths. We see from equations 6.14 and 6.15 that we can do this for each e_α, and a maximum of $U^{k-3}(e)$ can flow into $[k+1 : e]$.

Since $x_{k+1}(e) \le U^{k-3}(e)$, we can choose nonnegative $\mu_r \le \lambda_r$, so that we have exactly a flow of $x_{k+1}(e)$ into $[k+1 : e]$ along the paths corresponding to $X^r \in \cup_\alpha \mathcal{P}(L_\alpha)$. Now we have part 3 of the theorem, from

(1) $\cup_\alpha \mathcal{P}(L_\alpha)$ is a subset of $\{X^r | r \in I(\lambda)\}$,
(2) $\mu_r \le \lambda_r$ and
(3) the expression for $x(u)$, given by equation 6.13.

Letting $v_\alpha = \sum_{X^r \in \mathcal{P}(L_\alpha)} \mu_r$ we have the parts 1 and 2 of the result. Hence the theorem. □

Remark 6.4. Even though we can apply this theorem for any $x_{k+1}(e) > 0$, the simultaneous application of this theorem for more than one e, in general, may not be correct. This is so because, for some paths the total flow with respect to the different e_β may violate the node capacity, $x_l(e)$, at some layer l for some e. Example 6.2 illustrates this point.

Corollary 6.1. *Given $X \in P_{MI}(n)$ and $X/k \in conv(P_k)$, if $x_{k+1}(e) = 1$ for some e, then $X/k+1 \in conv(P_{k+1})$.*

Proof. $X/k+1$ as given, means that e is available for insertion of $k+1$ with certainty. In other words, every pedigree active for X/k is such that the corresponding $k-tour$ contains e. Essentially, the proof lies in seeing the fact that given a $\lambda \in \Lambda_k(X)$, we can extend every $X^r, r \in I(\lambda)$ to $(X^r, y(e))$ with the same weight λ_r, where $y(e)$ is the indicator of e.

Now refer to the proof of Theorem 6.5. Since $x_{k+1}(e) = 1$, we have $x_{k+1}(e) = U^{k-3}(e)$. So $\mu_r = \lambda_r, r \in I(\lambda) \cap J$. It follows from an application of Theorem 6.5 and the above observation that the part 3 of the theorem yields strict equalities for each node in each layer of N_{k-1}. This with part 2 of the theorem implies the required result. □

Example 6.2. Consider X as given below:

$$\begin{aligned}\mathbf{x}_4 &= (0, 3/4, 1/4);\\ \mathbf{x}_5 &= (1/2, 0, 0, 1/2, 0, 0);\\ \mathbf{x}_6 &= (0, 1/4, 1/2, 0, 1/4, 0, 0, 0, 0, 0).\end{aligned}$$

It can be verified that $X \in P_{MI}(6)$ and $X/5 \in conv(P_5)$. In fact,
$X/5 = 1/4(0,1,0;1,0,0,0,0,0) + 1/4(0,0,1;1,0,0,0,0,0)$
$+1/2(0,1,0;0,0,0,1,0,0)$.

Now $x_6(1,3) = 1/4$ and the path $[4:2,3] \rightarrow [5:1,2] \rightarrow [6:1,3]$ brings that flow to $[6:1,3]$. The path $[4:1,3] \rightarrow [5:1,4] \rightarrow [6:2,3]$ brings the flow $1/2$ to $[6:2,3]$, as required. Also for the node $[6:2,4]$, we have the corresponding path $[4:2,3] \rightarrow [5:1,2] \rightarrow [6:2,4]$ with the flow $1/4$. Notice that these paths correspond to the extensions of the pedigrees active for $X/5$. However, we can not satisfy the requirements at nodes $[6:2,4]$ and $[6:1,3]$ simultaneously, using the respective paths, as at $[4:2,3]$ the node capacity is violated.

Instead, consider X' as given below:

$$\begin{aligned}\mathbf{x}'_4 &= \mathbf{x}_4;\\ \mathbf{x}'_5 &= \mathbf{x}_5;\\ \mathbf{x}'_6 &= (0, 0, 0, 0, 0, 1, 0, 0, 0, 0).\end{aligned}$$

We can check that the paths

$$[4:2,3] \rightarrow [5:1,2] \rightarrow [6:3,4],$$

$$[4:1,3] \rightarrow [5:1,2] \rightarrow [6:3,4],$$

and

$$[4:1,3] \rightarrow [5:1,4] \rightarrow [6:3,4]$$

bring the flows of $1/4$, $1/4$ and $1/2$, respectively to $[6:3,4]$, totalling up to $1 = x_6(3,4)$. However, these paths correspond to the extensions of all the pedigrees active for $X/5$. And we can verify that $X' \in conv(P_6)$, as assured by Theorem 6.5.

6.7 A Multicommodity Flow Problem to Check Membership

Recall the construction of the network N_k. We solve several restricted network flow problems (in $N_{k-1}(L)$, for each link L) to obtain the capacities of the arcs in F_k. After ensuring the feasibility of F_k, rigid arcs are identified. Then we declare the network N_k to be well defined, if we can have evidence that $X/k \in conv(P_k)$. This was easy for $k = 4$. As seen in the Example 6.1, even though there are pedigree paths bringing the flow along each arc in F_k, there could be conflicts arising out of the simultaneous capacity restrictions on these flows in the network N_{k-1}. We need to ensure that these restrictions are not violated. The multicommodity flow problem defined in this section does precisely this.

Definition 6.28 (Commodities). *Consider the network N_k and focus on the last two layers, consider the arcs in $\mathcal{A}$ obtained by solving F_k to feasibility and then deleting the dummy arcs. These arcs are in N_k by construction. For every arc $a \in \mathcal{A}$ designate a unique* commodity *s. Let L_s be the link corresponding to commodity s. Let $\mathcal{S}$ denote the set of commodities. We write $a \leftrightarrow s$ and read a designates s.*

Definition 6.29. Given a pedigree $X^r \in P_{k+1}$, we say that it agrees with an arc $a = (u, v) \in F_l, 4 \leq l \leq k$ in case $u = [l : e^r_l], v = [l + 1 : e^r_{l+1}]$. We denote this by, $X^r \parallel a$ and read X^r agrees with a. For any $\lambda \in \Lambda_{k+1}(X)$ and $r \in I(\lambda)$, let $I_s(\lambda)$ denote the subset of $I(\lambda)$ such that the corresponding pedigrees agree with $a \in F_k$ such that $a \leftrightarrow s$.

Next we describe an enlarged network using which we define the multicommodity problem. To the layered network N_k, we add a single source o in layer 0, with one unit availability of each commodity, and we add a layer $(k - 1)$, of sinks, one for each commodity $s \in \mathcal{S}$. We have arcs connecting the source o to the nodes in $V_{[1]}$. And we have an arc between a node $[k + 1 : e_\beta] \in V_{[k-2]}$ and a sink s, if $L_s = (e, e_\beta)$ corresponds to an arc $a \in \mathcal{A}$. We denote the set of newly added arcs by A_{new}. They all have unit capacity. The demand at sink s corresponding to a green arc is denoted by b_s and is equal to the frozen flow of arc L_s. We do not have any specified demands at other sinks. We call the network thus obtained as the *enlarged network* N. $\mathcal{V}(N) = \mathcal{V}(N_k) \cup \{o\} \cup \mathcal{S}$ and $\mathcal{A}(N) = \mathcal{A}(N_k) \cup \{(o, v)|v \in V_{[1]}\} \cup \{(v, s)|v \in V_{[k-2]}, s$ corresponds to an arc, $a = (u, v) \in \mathcal{A}$, for some $u \in V_{[k-3]}\}$. The enlarged network has k

layers in all, numbered from 0 to $k-1$. Figure 6.3 gives the schematic diagram of the enlarged network.

Let a denote an arc in the enlarged network. Let c_a denote the capacity of any arc a. Let $f_a^s \geq 0$ be the flow through arc a for commodity s. Let v^s be the total flow into sink s.

We have the following restrictions on the commodity flow through any arc $a \in \mathcal{A}(N_{k-1})$. For any s, we allow this flow to be positive only for the arcs in the restricted network $N_{k-1}(L_s)$. Let u_a^s be the upper bound on f_a^s, that is

$$0 \leq f_a^s \leq u_a^s, \; s \in \mathcal{S}, \; a \text{ an arc in } \mathcal{A}(N). \tag{6.16}$$

where,

$$u_a^s = \begin{cases} 1 & \text{if } a \in \mathcal{A}_{new} \\ c_a & \text{if } a \in N_{k-1}(L_s) \text{ or} \\ & \quad a \text{ defines } s \\ 0 & \text{otherwise.} \end{cases} \tag{6.17}$$

For each commodity, at each node $v \in \mathcal{V}(N_k)$ we conserve the flow. That is,

$$\sum_{u \ni a=(u,v)} f_a^s = \sum_{w \ni a=(v,w)} f_a^s, \; v \in \mathcal{V}(N_k), s \in \mathcal{S}. \tag{6.18}$$

We have the so called bundle capacity restriction for each arc that is,

$$\sum_{s \in \mathcal{S}} f_a^s \leq c_a, \; a \text{ an arc in } \mathcal{A}(N). \tag{6.19}$$

In addition, at each node $v \in \mathcal{V}(N_k)$ we have the node capacity restriction on the total flow through the node as well. Recall that the node capacity $x(v)$ denotes $x_l(e) > 0$ for the node $v = [l : e]$, in layer l for some $e \in E_{l-1}$.

$$\sum_{s \in \mathcal{S}} \sum_{u \ni a=(u,v)} f_a^s \leq x(v), \; v \in \mathcal{V}(N_k). \tag{6.20}$$

The flow into any sink s, denoted by, v^s is defined by,

$$v^s = \sum_{s \in \mathcal{S}} \sum_{u \ni a=(u,s) \in \mathcal{A}_{new}} f_a^s, \; s \in \mathcal{S}. \tag{6.21}$$

At the sinks corresponding to green arcs with frozen flow in F_k, we require that the demand restrictions be met.

$$v^s = b_s, \; s \ni \text{the arc has frozen flow in } F_k. \tag{6.22}$$

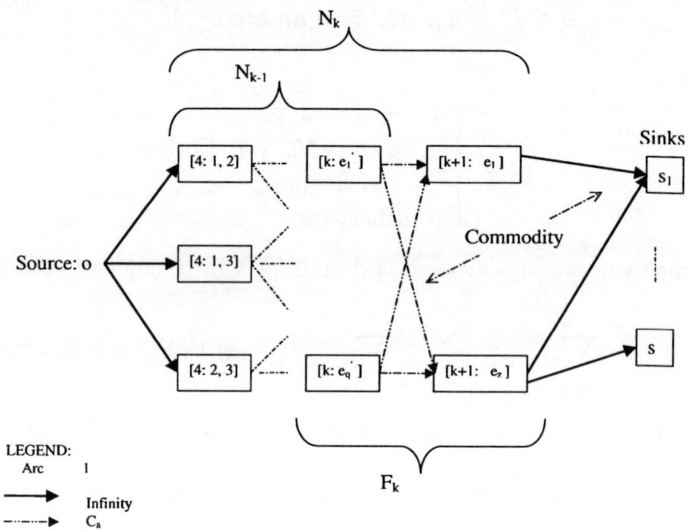

Fig. 6.3 Schematic diagram of the enlarged network, N

The problem now can be stated as:

Problem 6.2. *[Multi-flow Problem]*

$$\text{maximise} \qquad (6.23)$$

$$\sum_{s \in \mathcal{S}} v^s$$

$$\text{subject to} \qquad (6.24)$$

$$\text{constraints} \quad (6.16) \quad (6.22).$$

Next we observe some easy to prove facts about the enlarged network and Problem 6.2.

Let z^* denote the objective function value for an optimal solution to Problem 6.2. Let $f = (f^1, \ldots, f^{|\mathcal{S}|})$ denote any feasible multicommodity flow for Problem 6.2., where f^s gives the flow vector for commodity s, for a fixed ordering of the arcs of N.

Remark 6.5.

1 In the enlarged network each node $v \in V_{[k-2]}$ has an even degree. For each $s \in \mathcal{S}$, we have a pair of arcs , one emanating from and one entering v. The arc $a = (u, v)$ for some $u \in V_{[k-3]}$ enters v and $a \leftrightarrow s$. Arc $a' = (v, s)$ leaves v.

2 The node capacities at each layer in N_k add to 1, and so z^* is at most 1.

3 If $z^* = 1$, then for any optimal solution to Problem 6.2, the bundle capacity $x(v)$ at each node $v \in V_{[l-3]}, 4 \leq l \leq k+1$ is saturated.

4 Every feasible path bringing a positive flow to a sink s passes through a node in each layer of the enlarged network, satisfying the commodity flow restrictions for s. N is a layered network without any cycles, and recall that every arc a in N_k is such that $a = (u, v), u \in V_{[l]}$ and $v \in V_{[l+1]}$ for some $l, 4 \leq l \leq k$.

5 If $z^* = 1$ then any optimal solution f to Problem 6.2, is such that the solution restricted to the portion of the network corresponding to arcs in F_k, constitutes a feasible solution to F_k. This follows from Remark 6.5.3. Now f saturates the node capacities at layers $k-3$ and $k-2$ and $f_a^s, a \in F_k$ can be positive only for the arc a designating s. So letting $f_a^s = g_a$, we can check that g is feasible for F_k as all the restrictions of problem F_k are also present in Problem 6.2.

6 If $z^* = 1$ then the demand restriction of any sink s is non binding at the optimum. This is so because, the fusibility of g for F_k obtained as per the previous remark, implies that the flow along the rigid arcs are equal

to the respective frozen flows. Suppose the rigid arc is $a = (u, v)$. Now consider the corresponding arc a' mentioned in Remark 6.5.1, a' leaves v and ends up in s designated by a. Since the commodity flow is conserved at v and this is the unique arc entering s, and no other commodity can flow through this arc, (from the commodity flow upper bound restrictions 6.16) we have the flow along this arc equal to that of a. Hence the flow into s is exactly equal to the frozen flow for arc a, given by b_s. Thus we see that the demand restrictions are automatically met for any optimal solution with $z^* = 1$. So these restrictions 6.22 can be dropped.

Interestingly, the results proved earlier (Theorem 6.4 and Lemma 6.8) on the necessity of feasibility of F_k for $X/k+1$ to be in the pedigree polytope and the fact that for any $\lambda \in \Lambda_{k+1}(X)$, the pedigree paths, $path(X^r/l), r \in I(\lambda)$, for $4 \leq l \leq k$ are all available in $N_{l-1}(L_l^r)$ can be used to prove Theorem 6.6.

Theorem 6.6. *Given $X/k+1 \in conv(P_{k+1})$ then there exists a f, feasible for the multicommodity flow problem (Problem 6.2), with $z^* = \sum_{s \in \mathcal{S}} v^s = 1$.*

Proof. Since $X/k+1$ is in $conv(P_{k+1})$ we have from the proof of Theorem 6.4 for any $\lambda \in \Lambda_{k+1}(X)$, a feasible solution to F_k is given by the instant flow for the FAT problem induced by (λ, k).

Define f as follows for $a \notin \mathcal{A}_{new}$:

$$f_a^s = \begin{cases} \sum_{r \in I_s(\lambda) | X^r \parallel a} \lambda_r & \text{if} \quad a \in N_{k-1}(L_s) \\ \sum_{r \in I_s(\lambda)} \lambda_r & \text{if } a \leftrightarrow s \\ 0 & \text{otherwise.} \end{cases} \tag{6.25}$$

For $a \in \mathcal{A}_{new}$, f is defined in the obvious manner to conserve the flow at the nodes in layer 1 and layer $k-2$. The net flow into $s \in \mathcal{S}$ is same as the flow along the arc a designating s. That is,

$$v^s = \sum_{r \in I_s(\lambda)} \lambda_r.$$

We shall show that this f is feasible for the problem.

Nonnegativity and capacity on nodes are all met as λ_r are positive and add up to $x(v)$ for each node v, as

$$\sum_{r \in I(\lambda), x_l^r(e)=1} \lambda_r = x(v), v = [l : e],$$

as $X/k+1$ can be written as a convex combination of the pedigrees in $I(\lambda)$. Since the instant flow of the FAT problem induced by (λ, l) is feasible for F_l, bundle capacity restrictions on arcs are all met as well. This is so because the arcs in the network other than the new ones correspond to the arcs in $F_l, 4 \le l \le k-2$. And Lemma 6.8 ensures that we have not violated any of the upper bound restrictions on f_a^s.

For each commodity s, for $r \in I_s(\lambda)$, notice that X^r represents a path in N_k. Thus we have paths with respect to $o - X^r - s$ in N. For any of these paths, commodity flow is conserved at each node along the path. And a node not in any of these paths, does not have a positive flow of this commodity through that node.
We have,

$$\sum_{u \ \ni \ a=(u,v)} f_a^s = \sum_{u \ \ni \ a=(u,v)} \ \sum_{r \in I_s(\lambda) | X^r \ \| \ a} \lambda_r \tag{6.26}$$

$$= \sum_{r \in I_s(\lambda) | x_l^r(e)=1} \lambda_r, \tag{6.27}$$

for $v = [l : e] \in \mathcal{V}(N_k), s \in \mathcal{S}$.

Similarly

$$\sum_{w \ \ni \ a=(v,w)} f_a^s = \sum_{w \ \ni \ a=(v,w)} \ \sum_{r \in I_s(\lambda) | X^r \ \| \ a} \lambda_r \tag{6.28}$$

$$= \sum_{r \in I_s(\lambda) | x_l^r(e)=1} \lambda_r, \tag{6.29}$$

for $v = [l : e] \in \mathcal{V}(N_k), s \in \mathcal{S}$.

Hence commodity flow conservation restrictions are all met. As noticed in Remark 6.5.6, the flow into any sink s given by v^s is equal to the flow along the defining arc $a \in F_k$. Hence the total flow in the network is $\sum_{s \in \mathcal{S}} v^s = \sum_{s \in \mathcal{S}} f_a^s = \sum_{s \in \mathcal{S}} \sum_{r \in I_s(\lambda)} \lambda_r = 1$. Thus we have verified that f is feasible and the objective function value is 1. Hence the theorem. □

6.8 Computational Complexity of Checking the Necessary Condition

In this section we show that the necessary condition, given in the previous section, for $X/k+1$ to be in $conv(P_{k+1})$ can be checked efficiently. Given $X \in P_{MI}(n)$, if $X/k \in conv(P_k)$ we have the network N_{k-1} well-defined.

Constructing N_k involves solving at most $p_{k-1} \times p_k$ ($< k^4$) maximal flow problems in $N_{k-1}(L)$ for each link L to find the capacity $C(L)$. Each of this can be solved in time polynomial in k. Next solving the FAT problem F_k can also be done in time polynomial in k. If F_k is infeasible we stop. Otherwise we use Frozen Flow Finding (FFF) algorithm to identify rigid and dummy arcs. This as stated in Subsection 6.2.1 can be done in linear time in the size of the graph G_f corresponding to the problem F_k. The size of G_f is at most $p_{k-1} \times p_k + p_{k-1} + p_k$. Thus N_k can be constructed in time polynomial in k. The next task is to construct the enlarged network N corresponding to N_k and find whether a feasible multicommodity flow exists with value unity. Sine we only need to solve a linear programming problem for answering this, this can be done in polynomial time in the input size of the corresponding linear programming problem. Thus the necessary condition can be verified in time polynomial in the input size of the multicommodity flow problem 6.2. Here we have not gone for tight bounds for the computational requirements, as the purpose is to provide a remark that the necessary condition can be checked efficiently.

If solving the Problem 6.2 results in a maximal flow less than unity, we can conclude $X/k + 1 \notin conv(P_{k+1})$. Current research is directed towards the important issue, of studying the complexity of declaring N_k to be well-defined once we have shown the existence of a multicommodity flow with unit value in the recursively constructed layered network.

6.9 Concluding Remarks

In this paper an alternative polytope $conv(A_n)$ that is closely related to the pedigree polytope is studied. We verify that the conditions ([Yudin and Nemirovskii (1976)]) for the existence of a separation algorithm that calls a polynomial number of times a membership oracle for the polytope, are satisfied for $conv(A_n)$. Hence the recent polynomial construction of [Maurras (2002)] could be applied for solving the separation problem of the polytope $conv(A_n)$. Thus the membership problem of the pedigree polytope (defined and studied in [Arthanari (2006)] and [Arthanari (2005)]) becomes relevant. A necessary condition for membership in the pedigree polytope is shown as the existence of a multicommodity flow with unit value in a recursively constructed layered network. The complexity of checking this necessary condition is polynomial in input size of the linear multicommodity flow problem. Thus this condition may not be sufficient, unless $\mathcal{P} = \mathcal{NP}$.

Hence an interesting future research area is to discover evidence that the condition is not sufficient.

Acknowledgements

The author thanks his family for their patience and support during the preparation of this paper. Department of ISOM is thanked for the support to attend the Symposium. The organisers of the Symposium are thanked for the invitation.

Bibliography

Ahuja, R. K., Magnanti, T. L., and Orlin, J. B. (1996). *Network Flows Theory, Algorithms and Applications*, (Prentice Hall, Englewood Cliffs, NJ).

Arthanari, T. S. (2005). *Pedigree polytope is a Combinatorial Polytope*, In: *Operations Research with Economic and Industrial Applications: Emerging Trends*, (eds.) Mohan, S.R. and Neogy, S.K., pp. 1–17, (Anamaya Publishers, New Delhi, India).

Arthanari, T. S. (2006). On Pedigree Polytopes and Hamiltonian Cycles, *Discrete Mathematics* **306**, pp. 1474–1492.

Arthanari, T. S. (2007). *A Comparison of the Pedigree Polytope with the Symmetric Traveling Salesman Polytope*, invited paper presented at Computational, Mathematical and Statistical Methods 2007, The Fourteenth International Conference of the FIM, January 6 - 8, 2007, Chennai, India.

Arthanari, T. S. and Usha, M. (2000). An Alternate Formulation of the Symmetric Traveling Salesman problem and its Properties, *Discrete Applied Mathematics* **98**, pp. 173–190.

Arthanari, T. S. and Usha, M. (2001). On the Equivalence of the Multistage-Insertion and Cycle Shrink Formulations of the Symmetric Traveling Salesman Problem, *Operations Research Letters* **29**, pp. 129–139.

Bondy, J., and Murthy, U.S.R. (1985). *Graph Theory and Applications*, (North-Holland, New York).

Carr, B. (1997). Separating Clique Tree and Bipartition Inequalities having a fixed number of handles and teeth in Polynomial Time, *Mathematics of Operations Research* **22**, pp. 257–265.

Cook, S.A. (1971). *The Complexity of Theorem-proving Procedures*, in Proceedings of the Third Annual ACM Symposium on the Theory of Computing, pp. 151–158.

Edmonds, J. (1965). *Paths, Trees, and Flowers*, Canadian Journal of Mathematics, **17**, pp. 449–467.

Ford, L.R., and Fulkerson, D.R. (1962). *Flows in Networks*, (Princeton Univ. Press, Princeton, NJ).

Garey, M.R., and Johnson, D.S. (1979). *Computers and intractability: a Guide to the Theory of NP-completeness*, (W. H. Freeman, San Francisco).

Grötschel, M. Lovász, L. and Schrijver, A. (1988). *Geometric Algorithms and Combinatorial Optimization*, (Springer-Verlag, Berlin).

Gusfield, D. (1988). A Graph Theoretic Approach to Statistical data Security, *SIAM. J. Comput.* **17**, pp. 552–571.

Khachiyan, L.G. (1979). *A Polynomial Algorithm in Linear Programming*, (in Russian) Doklady Akademii Nauk SSSR, **244**, 1093-1096, (English translation: *Soviet Mathematics Doklady*, **20**, pp. 191–194.

Karp, R.M. (1972). *Reducibility among Combinatorial Problems*, in *Complexity of Computer Computations*, Miller, R. E. and Thatcher, J. W. (eds.), pp. 85-103, (Plenum Press, New York).

Murty, K.G. (1992). *Network Programming*, (Prentice-Hall, NJ).

Korte, B., and Vygen, J. (2002) *Combinatorial Optimization, Theory and Algorithms*, (Springer, Second edition, New York).

Lawler, E., Lenstra, J.K., Rinooy Kan, A.H.G., and Shmoys, D.B. (1985)(eds.), *The Traveling Salesman Problem*, (Wiley, New York).

Maurras, J.E. (2002). *From Membership to Separation, a Simple Construction*, Combinatorica,**22**, pp. 531-536.

Yudin, D.B. and Nemirovskiĭ, A.S.(1976) *Information complexity and Efficient Methods for Solution of Convex Extremal Problems* (in Russian), Ekonomika i Mathematicheskie Metody, **12**, pp. 357–369.(Translated in English, *Metekon*, **13**, **3**, 1977, pp. 25–45.

Ziegler, G.M. (1995) *Lectures on Polytopes*, Grad. Texts in Maths., (Springer-Verlag, Berlin).

Chapter 7

Exact Algorithms for a One-defective Vertex Colouring Problem

Nirmala Achuthan , N. R. Achuthan and R. Collinson
Department of Mathematics and Statistics
Curtin University of Technology
GPO BOX U1987, Perth, Australia - 6845

Abstract

Many real life scheduling problems involve the use of a graph colouring problem where the vertices of a graph $G(V, E)$ are coloured such that the coloured graph satisfies certain desired properties. This paper discusses one such graph colouring problem. A graph is $(m, k) - colourable$ if its vertices can be coloured with m colours such that the maximum degree of the subgraph induced on vertices receiving the same colour is at most k. The $k-defective$ *chromatic number* $\chi_k(G)$ of a graph G is the least positive integer m for which G is $(m, k) - colourable$. In this chapter, we develop exact algorithms based on partial enumeration methods to determine the one defective chromatic number $\chi_1(G)$, of a graph G. Furthermore, we assess the computational performance of the algorithms by determining the one defective chromatic number of several simulated graphs.

Key Words: Chromatic number of a graph, 1-defective chromatic number, partial enumeration methods, optimization, scheduling.

7.1 Introduction

All graphs considered in this paper are undirected, finite, loopless and have no multiple edges. For the most part we follow the notation of Chartrand and Lesniak (1986). For a graph G, we denote the vertex set and the edge set by $V(G)$ and $E(G)$ respectively. The degree of a vertex $v \in G$ is the number of adjacent vertices of v and it is denoted by $d(v)$. The maximum degree of a graph is denoted by $\Delta(G)$. For a subset U of $V(G)$, the subgraph of G induced on the set U is denoted by $G[U]$.

Several real life optimization problems with resource constraints involve scheduling activities and/or resources over certain time intervals. Some examples of such problems are: course scheduling [Dowsland (1990)] ; school timetabling [de Werra (1985); Mehta (1981)]; operational timetable [Costa (1994); Jagota (1996); Hertz (1991)]. The following vertex colouring problem (VCP) appears in disguise as part of several of the optimization problems listed above.

Given a graph $G(V, E)$, find an assignment of colours to the vertices in V of the graph G using the least number of colours so that two vertices that are adjacent are assigned different colours.

The minimum number of colours used in the VCP is called the *chromatic number* $\chi(G)$ of the graph G. The decision version of VCP is known to be NP-complete [Garey and Johnson (1978)]. For a good survey on heuristic and exact algorithms for VCP see [de Werra (1990)].

The notion of colouring of a graph has been generalised in many ways, see [Frick (1993)], for a survey. One interesting extension is the $k-defective$ colouring of a graph and we consider the $k-defective$ vertex colouring problem ($k-DVCP$), in this paper.

Let k be a non-negative integer. A subset U of $V(G)$ is said to be $k-$ *independent* if the maximum degree of $G[U]$ is at most k. A graph is $(m, k)-$ *colourable* if its vertices can be coloured with m colours such that the set of vertices receiving the same colour is $k-independent$. Sometimes we refer to an $(m, k)-$ *colouring* of G as a $k-defective$ *colouring* of G. Note that any $(m, k)-$ *colouring* of a graph G partitions the vertex set of G into m subsets $V_1, V_2, ..., V_m$, such that every V_i is $k-independent$. The $k-defective$ *chromatic number* $\chi_k(G)$ of G is the least positive integer m for which G is $(m, k)-$ *colourable*. Note that $\chi_0(G)$ is the usual chromatic number of G. Clearly $\chi_k(G) \leq \lceil \frac{n}{k+1} \rceil$, where n is the order of G. If $\chi_k(G) = m$ then G is said to be an $(m, k)-$ *chromatic* graph. The $k-defective$ vertex colouring problem ($k-DVCP$) is to assign colours to the vertices of G using the minimum number of colours such that the set of vertices receiving the same colour is $k-$ *independent*.

The concepts of $k-$ *independent* sets and $k-defective$ chromatic numbers have been studied by several authors under different names, see [Frick and Henning (1994); Achuthan, Achuthan and Simanihuruk (1996); Simanihuruk, Achuthan and Achuthan (1997)]

In the next section we present a sequential colouring heuristic for the k-DVCP and it is a generalisation of a heuristic proposed for the 0-DVCP.

In this paper we develop exact algorithms for 1-DVCP based on partial enumeration methods.

7.2 Sequential Colouring Heuristics for k-DVCP

The literature has several sequential colouring heuristics proposed for the vertex colouring of a graph, see [de Werra (1990); Brown (1972)]. Most of them can be easily generalised to a corresponding heurestic for the $k-defective$ colouring of a graph. In the following we discuss the DSATUR graph colouring heuristic to provide a $k-defective$ colouring of the graph.

DSATUR graph colouring: This graph colouring heuristic was first proposed by Brėlaz (1979) for the $0-defective$ colouring problem. This heuristic, at every stage, partitions the set of vertices into the set of coloured vertices and the set of uncoloured vertices. For each uncoloured vertex v, define the saturation degree as follows:

$$Satdeg(v) = \sum_{c=1}^{j} x_{cv}$$

where j is the total number of colours used up to the current stage, C_c is the set of vertices that are coloured by the colour c, $1 \leq c \leq j$ and if $\Delta(G[C_c \cup v]) \leq k$ then $x_{cv} = 0$. Otherwise, $x_{cv} = 1$. In other words $x_{cv} = 1$ if and only if the uncoloured vertex v cannot be assigned the colour c. Thus the saturation degree of v is the current number of colours that cannot be assigned to v among the available j colours.

(a) Order the vertices $v_1, ..., v_n$ such that $d(v_1) \geq d(v_2) \geq ... \geq d(v_n)$.

(b) Assign colour 1 to v_1, define $C_1 = \{v_1\}$, $r = 2 =$ the index of the next vertex to be coloured, $j = 1 =$ the number of colours used up to now, $U =$ the set of current uncoloured vertices $= V - \{v_1\}$.

(c) Determine $Satdeg(v)$ for $v \in U$. Define

$$PV = \{v' : Satdeg(v') = max\{satdeg(v) : v \in U\}\}$$

Choose the next vertex v' to be coloured if $d(v') = max\{d(v) : v \in PV\}$

(d) Let

$$i' = \begin{cases} \infty, & \text{if } Satdeg(v') = j \\ min\{i : x_{iv'} = 0, 1 \leq i \leq j\}, & \text{if } otherwise \end{cases}$$

and $i* = min\{i', j+1\}$. Colour the vertex v' with colour $i*$, add v' to C_{i*} and update r, j and U.

(e) Repeat steps (c) and (d) until all the vertices are coloured.

7.3 Implicit Enumeration Algorithms for $1 - DVCP$

Brown (1972) developed an exact implicit enumeration algorithm for the $0 - DVCP$. Subsequently some variations of this algorithm were developed and certain errors were corrected, see [Brown (1972); Brèlaz (1979); Kubale and Jackowski (1985); Korman (1979)]. In this section we extend these concepts to develop exact implicit enumeration algorithms for $1 - DVCP$.

The first algorithm is a simple implicit enumeration algorithm and it is referred to as Algorithm 1. The salient features of this algorithm are detailed in the following.

Let the vertices of the graph G be ordered say, $v_1, v_2, ..., v_n$. Let q denote an upper limit of the 1-defective chromatic number of G at any stage of the algorithm. Initially q is fixed as $\lceil \frac{n}{2} \rceil$ since $\chi_1(G) \leq \chi_1(K_n) = \lceil \frac{n}{2} \rceil$, where K_n is a complete graph on n vertices. The value of q is updated whenever a better solution is encountered by the implicit enumeration procedure. When the algorithm terminates, the $1 - defective$ chromatic number of the graph G is q.

Note that, $\chi_1(H) \leq \chi_1(G)$ for any subgraph H of G. In particular, if K_ω is a subgraph of G then $\lceil \frac{\omega}{2} \rceil = \chi_1(K_\omega) \leq \chi_1(G)$. Let lb denote a lower bound on the $1 - defective$ chromatic number of G at any stage of the algorithm. Initially the lb is fixed as $\lceil \frac{\omega}{2} \rceil$ where the subgraph of G induced by the vertices $v_1, v_2, ..., v_\omega$ forms a K_ω.

At any intermediate stage of the algorithm let $v_1, v_2, ..., v_\omega, v_{\omega+1}, ..., v_{r-1}$ be the vertices of G that are coloured using the colours $1, 2, ..., u_r$ such that it is a valid $1 - defective$ colouring for the subgraph of G induced by these $r - 1$ vertices. Let C_c denote the set of vertices that are assigned the colour c. Then the algorithm uses a Forward-scheme to colour the next vertex v_r by the least indexed colour from the set of feasible colours, $FC(v_r)$, for the vertex v_r where

$$FC(v_r) = \{c : 1 \leq c \leq min\{u_r + 1, q - 1\}; \Delta(G[C_c \cup \{v_r\}]) \leq 1\}.$$

From the definition of $FC(v_r)$, it is clear that any new $1 - defective$ colouring of G would use at most $(q - 1)$ colours. This Forward-scheme of colouring is repeated to colour the remaining $(n - r + 1)$ vertices namely $v_r, v_{r+1}, ..., v_n$ and it stops in one of the following two cases:

(a) The vertices $v_r, ..., v_{i-1}$ are coloured such that $G[\{v_1, v_2, ..., v_{i-1}\}]$ has a valid $1 - defective$ colouring and $FC(v_i)$ is empty.

(b) All the remaining $(n - r + 1)$ vertices are coloured such that G has a valid $1-defective$ colouring using q' colours where $q' \leq (q-1)$. In this case the best known solution and the value of q are updated. Further determine the vertex v_r such that

$$r = min\{j : v_j \text{ has colour } q\}.$$

The vertices $v_r, ..., v_n$ are converted as uncoloured vertices. Note that with the updated q, the set $FC(v_r)$ is empty.

If the Forward-scheme of colouring fails to colour the vertex v_r then by a Backtrack-scheme a vertex v_i that can be recoloured (with another feasible colour) is chosen from the set of coloured vertices $v_1, v_2, ..., v_{r-1}$. The vertex v_i is chosen such that $i = max\{j : j \in Nb(r)\}$ where $Nb(r)$ includes the index j if $1 \leq j \leq (r-1)$, $(v_j, v_r) \in E$ and j is the least rank among vertices with same colour.

Then the Forward-scheme of colouring is continued starting with a new feasible colouring of v_i.

The algorithm terminates either during the Forward-scheme when the upper bound q equals the lower bound lb or during the Backtrack-scheme when it fails to locate any vertex that can be recoloured. When the algorithm terminates the $1 - defective$ chromatic number of of G is q and the best known solution provides a valid $1 - defective$ chromatic colouring of G.

In the following we provide a justification for the fact that Algorithm 1 terminates in finite number of steps with an optimal $k-defective$ colouring of the graph G.

For a specified number of colours q, a complete enumeration of all possible colourings of the vertices $v_1, v_2, ..., v_n$ may be visualised through a corresponding complete enumeration tree that has $n+1$ levels. Level 0 of the tree represents the root node and level i represents the vertex v_i, $1 \leq i \leq n$. Thus, level i has q^i nodes of the tree corresponding to the q^i possible colourings of the vertices $v_1, v_2, ..., v_i$ using q colours, $1 \leq i \leq n$. Such a complete enumeration tree with q colours is denoted by $q - complete - Etree$. The following observations are easy to verify:

1. Note that every valid $1 - defective$ colouring using less than or equal to q colours will correspond to a unique path from the root node to an appropriate node in level n of the $q - complete - Etree$.

2. For every positive integer q, the $q - complete - Etree$ is a subtree of the $(q+1) - complete - Etree$.
3. Let q be the number of colours used in the best known $1 - defective$ colouring. The Forward scheme of the Algorithm 1 generates a subtree of $(q-1) - complete - Etree$, that is using $(q-1)$ colours. Hence $lb \leq \chi_1(G) \leq q$. The subtree thus generated is denoted by $(q-1) - S - tree$. Note that every path of length n from the root node of the $(q-1) - complete - Etree$ has a nonempty intersection with the $(q-1) - S - tree$.
4. In fact if the Forward scheme stops in case(a) it leads to pruning certain paths of the subtree that will not yield a valid $1 - defective$ colouring of the vertices. On the other hand, if the Forward scheme stops in case(b) it locates a better known solution using q' colours such that $q' \leq q$ and again prunes the subtree further to generate a subtree of $(q'-1) - complete - Etree$. At this stage note that $lb \leq \chi_1(G) \leq q' \leq q \leq \lceil \frac{n}{2} \rceil$.
5. Algorithm 1 starts with $q = \lceil \frac{n}{2} \rceil$, that is, with an $\lceil \frac{n}{2} \rceil - complete - Etree$. For each run of the Forward scheme of the algorithm, the corresponding $q - complete - Etree$ is pruned further or reduced to a $q' - complete - Etree$ where $q' \leq q$.
6. If Algorithm 1 stops with $lb = q$ then $\chi_1(G) = lb$, since $lb \leq \chi_1(G) \leq q$. In this case the best known solution provides an optimal $1 - defective$ colouring of G.
7. If Algorithm 1 terminates during the Backtrack-scheme for not finding a vertex that can be recoloured, then for every uncoloured vertex v we have $FC(v) = [IMAGE]$ and the best known $1 - defective$ colouring of G uses q colours. At this stage, the incomplete colouring using $(q-1)$ colours generates, $(q-1) - S - tree$, a subtree of the $(q-1) - complete - Etree$. We claim that $\chi_1(G) = q$. Suppose that the claim is not true. Then $\chi_1(G) \leq (q-1)$. Hence there exists a valid $1 - defective$ colouring of G that corresponds to a unique path P from root node to a node in level n of the $(q-1) - complete - Etree$. This path P has a nonempty intersection with the $(q-1) - S - tree$ and contradicts the fact that for every uncoloured vertex v the set $FC(v)$ is empty. Hence the claim is proved.

Using the observations 6 and 7 it is easy to see that Algorithm 1 terminates in finite number of steps providing an optimal $1-defective$ colouring of G.

The flow diagram in Fig. 7.1 presents the main steps of Algorithm 1 in terms of certain key words that represent suitable procedures.

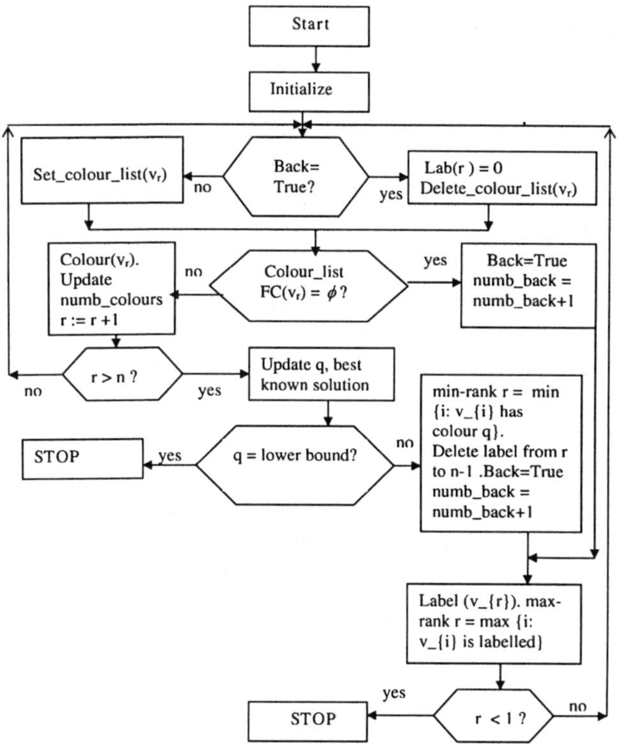

Fig. 7.1 Flow chart of Algorithm 1

In the following we briefly explain the key words used in Fig. 7.1.

- **Initialize**
This procedure initializes all the variables used in the algorithm. Fix an ordering among the vertices say, $v_1, ..., v_n$. Let ω be such that the first ω vertices, $v_1, v_2...v_\omega$ form a clique. Provide a $1-defective$ colouring for the first ω vertices, that is, for the vertices of the clique. Let $colv[0][i]$ denote the colour assigned to the vertex v_i, $1 \leq i \leq \omega$. Fix $colv[1][i] = j$ and $colv[1][j] = i$ if vertices v_i and v_j received the same colour. Fix $colv[0][i] = -1$ and $colv[1][i] = -1$

to indicate that the vertices v_i for $\omega+1 \leq i \leq n$ are not yet coloured. Fix $q = \lceil \frac{n}{2} \rceil$. Set $bkcolv[i]$ = the colour assigned to vertex v_i by the best known solution. Fix $numb-colours = \lceil \frac{\omega}{2} \rceil$ and $r = \omega+1$. Fix $BACK = False$; $Numb - back = 0$ and $lab[i] = 0$ for $1 \leq i \leq n$.

- **Set colour list** (v_r)
  This procedure defines the set of feasible colours, $FC(v_r)$ for the vertex v_r.
- **Colour**(v_r)
  This procedure assigns the least indexed colour available in the set $FC(v_r)$ to the vertex v_r. Furthermore, it appropriately fixes the colv[0][r] and colv[1][r] as required.
- **Delete colour list** (v_r)
  This procedure deletes the current colour of the vertex v_r and appropriately updates $FC(v_r)$; sets $Colv[0][r] = -1$ and $Colv[1][r] = -1$ to indicate that the vertex v_r is currently not coloured. If a vertex v_t is adjacent to v_r and they both were given the same colour then reset $Colv[1][t] = -1$.
- **Label**(v_r)
  This procedure gives a label r to every vertex that is adjacent to v_r, has smaller rank than r, and has minimal rank among all the vertices of the same colour which are adjacent to v_r.
- **max rank**
  This procedure finds the vertex that has the maximum rank among all the labelled vertices.

The second algorithm is a simple variation of Algorithm 1. The initialize step of Algorithm 2, includes determining the best known solution and the upper bound q for the $1 - defective$ chromatic number through the DSATUR graph colouring heuristic for the $1-defective$ colouring of graph G.

The third algorithm is a variation of Algorithm 2 that incorporates a look ahead feature while selecting a colour for the vertex v_r. In other words the feasible colours in $FC(v_r)$ are ordered according to certain criteria that may help in reducing the number of backtracks. This modified algorithm is referred to as Algorithm 3.

More precisely, for every colour $c \in FC(v_r)$ define the **number of preventions of** c as $|\{i : r+1 \leq i \leq n; c \in FC(v_i)\}|$.

Similarly, for every colour $c \in FC(v_r)$ define the **number of blockings of** c as $|\{i : r+1 \leq i \leq n; FC(v_i) = c\}|$.

The colours of $FC(v_r)$ are ordered first by the ascending order of the number of blockings and then by the ascending order of the number of preventions. Then the procedure colour(v_r) is modified to colour the vertex v_r with the first colour in the ordered list $FC(v_r)$. This scheme of looking at the blockings enables Algorithm 3 to seek a reduction in the number of colours used before successfully completing a $1 - defective$ colouring of all the vertices.

7.4 Computational Performance of the Algorithms

Computational performance of the three algorithms were studied by solving several simulated problems. For this purpose the number of vertices, n, ranged from 20 to 70 with an increment of 5. The computational effort required in solving a problem is likely to depend on the density of the graph. Hence the simulated problems were generated using three distinct graph density intervals, namely 0.26 - 0.34; 0.44 - 0.54 and 0.61- 0.72. For each choice of n and every graph density interval, thirty graphs were generated. All the three algorithms were coded in c++ and implemented in the environment of SGI Altix 1.6Ghz Itanium2 processor available with the Australian participation for advanced computing (APAC) national facility.

Each of the three algorithms was used to solve every simulated problem. Every algorithm's attempt on each problem was limited to a maximum CPU time of 2 hours.

Table 7.1 presents a comparison of the three algorithms applied to the simulated problems with graph density in the range 0.26 - 0.34. The table provides the number of vertices (n), average number of edges ($\overline{m}$), the number of solved problems, average number of backtracks generated, and average CPU time (seconds) over the solved problems.

From Table 7.1 note that Algorithm 2 solves problems with 60 vertices more easily as compared to Algorithm 3. Algorithm 3 is computationally slow as compared to Algorithm 2. Furthermore, both these algorithms perform better than the simple Algorithm 1.

Similarly Table 7.2 and Table 7.3 provide the comparison of the algorithms for problems with graph density in the ranges 0.44 - 0.54 and 0.61 - 0.72 respectively.

Table 7.1 Performance of the algorithms on graphs with density 0.26-0.34.

n	$\overline{m}$	Algorithm 1: Number of solved problems	Algorithm 1: Number of backtracks (average)	Algorithm 1: Avg CPU time
20	57.23	30	944.17	0.0
25	90.43	30	27420.37	0.05
30	127.97	30	51959.1	0.11
35	181.17	30	1817766.37	5.33
40	234.6	30	71121437.3	244.05
45	299.07	27	428011526.33	910.07
50	367.7	16	1162151215.07	2705.60

n	$\overline{m}$	Algorithm 2: Number of solved problems	Algorithm 2: Number of backtracks (average)	Algorithm 2: Avg CPU time
20	57.23	30	64.60	0.01
25	90.43	30	290.03	0.01
30	127.97	30	382.17	0.02
35	181.17	30	3713.23	0.17
40	234.6	30	89704.5	0.66
45	299.07	30	285236.97	1.51
50	367.7	30	4261015.43	23.49
55	465.1	29	116405123.20	399.83
60	535.87	28	187018279.43	736.04
65	645.07	24	351818115.23	1143.19
70	741.5	11	666355940.77	2302.51

n	$\overline{m}$	Algorithm 3: Number of solved problems	Algorithm 3: Number of backtracks (average)	Algorithm 3: Avg CPU time
20	57.23	30	19.63	0.00
25	90.43	30	141.23	0.03
30	127.97	30	188.93	0.05
35	181.17	30	1580.53	0.39
40	234.6	30	20473.43	1.04
45	299.07	30	80128.83	3.65
50	367.7	30	621615.83	37.59
55	465.1	29	13386717.97	782.47
60	535.87	29	16696369.63	1260.89
65	645.07	22	3402351953	2281.55
70	741.5	7	43160233.37	1764.04

Table 7.2 Performance of the algorithms on graphs with density 0.44 - 0.54.

n	$\overline{m}$	Algorithm 1		
		Number of		Avg CPU time
		solved problems	backtracks (average)	
hline 20	93.83	30	4660.33	0.01
25	149.7	30	110160.93	0.26
30	214.07	30	2459035.60	7.92
35	289.4	30	25928052.07	90.91
40	385.77	25	554591635.57	1740.29
45	482.9	8	1095988002.03	2346.68

n	$\overline{m}$	Algorithm 2		
		Number of		Avg CPU time
		solved problems	backtracks (average)	
20	93.83	30	322.70	0.01
25	149.7	30	1655.10	0.05
30	214.07	30	12484.13	0.35
35	289.4	30	123260.37	0.92
40	385.77	30	3139283.67	17.21
45	482.9	30	15852250.27	89.16
50	621.13	26	366755445.80	1931.82
55	733.87	12	700459753.43	2634.73

n	$\overline{m}$	Algorithm 3		
		Number of		Avg CPU time
		solved problems	backtracks (average)	
20	93.83	30	162.67	0.03
25	149.7	30	812.63	0.18
30	214.07	30	6090.13	0.62
35	289.4	30	42497.43	2.01
40	385.77	30	729137.93	40.58
45	482.9	30	3412853.47	230.61
50	621.13	21	39359356.40	1967.65
55	733.87	9	50511037.87	2538.46

From these tables we observe that all the three algorithms perform well on smaller dimension problems (that is problems with less than or equal to 35 vertices). For instance, problems with 35 vertices and graph density in the range 0.61-0.72, needed CPU times on an average around 20mt, 16sec and 50sec respectively by Algorithms 1, 2 and 3. The number of backtracks used by Algorithm 3, on average is just about a tenth of that used by Algorithm 2. But the look-ahead scheme of Algorithm 3 seems to increase the CPU time considerably as compared to Algorithm 2. When the graph density is small (that is the graph is sparse) Algorithms 2 and 3

Table 7.3 Performance of the algorithms on graphs with density 0.61- 0.72.

		Algorithm 1		
		Number of		Avg
n	$\overline{m}$	solved problems	backtracks (average)	CPU time
20	127.67	30	23964.87	0.05
25	201.4	30	341645.40	0.98
30	291.03	30	16608999.00	63.3
35	398.13	27	378197102.67	1165.81
40	523.4	7	1127533972.73	2849.79

		Algorithm 2		
		Number of		Avg
n	$\overline{m}$	solved problems	backtracks (average)	CPU time
20	127.67	30	1883.73	0.05
25	201.4	30	18024.10	0.42
30	291.03	30	257514.57	1.36
35	398.13	30	3346560.87	15.20
40	523.4	30	103806685.57	589.57
45	656.9	21	485207911.00	1592.46

		Algorithm 3		
		Number of		Avg
n	$\overline{m}$	solved problems	backtracks (average)	CPU time
20	127.67	30	1057.23	0.21
25	201.4	30	11246.73	0.77
30	291.03	30	77317.90	2.82
35	398.13	30	980539.70	41.92
40	523.4	30	22679796.27	1013.34
45	656.9	30	60643828.03	1504.85

are able to solve, within a reasonable time, problems of larger size, that is up to 65 vertices. In fact Algorithm 2 performs better than the other two algorithms for problems of size less than or equal to 70 vertices.

Acknowledgement

The authors thank the Australian Participation for Advanced Computing (APAC) for computing facilities provided for this work under the project code g18.

Bibliography

Achuthan, N., Achuthan, N. R. and Simanihuruk, M.(1996). *On Defective Colourings of Complementary Graphs, The Australasian Journal of Combinatorics*, **13**, pp.175 - 196.

Brèlaz, D.(1979). *New methods to color the vertices of a graph, Communications of ACM*, **22**, pp. 251 - 256.

Brown, R.J. (1972). *Chromatic scheduling and the chromatic number problem, Management science*, **19**, pp. 451 - 463.

Chartrand, G. and Lesniak, L. (1986). *Graphs and Digraphs*, 2nd edition, (Wadsworth and Brooks/Cole, Monterey, California).

Costa, D. (1994). *A Tabu Search algorithm for computing an operational timetable, European J. of Operational Research*, **76**, pp. 98–110.

Dowsland,K.A. (1990). *A Timetabling Problem in which Clashes are Inevitable, J. Operational Research Society*, **41**, pp. 907–918.

Frick, M. , (1993). *A Survey of (m,k)-Colourings, Annals of Discrete Mathematics*, **55**, pp.45 - 58.

Frick, M and Henning, M. A., (1994). *Extremal Results on Defective Colourings of Graphs, Discrete Mathematics*, **126**, pp. 151 - 158.

Garey,M.R. and Johnson, D.S. (1978). *Computers and Intractability: a Giude to the Theory of NP-Completeness*(W.H. Freemann, San Francisco)

Hertz, A. (1991) *Tabu search for large scale timetabling problems, European J. of Operational Research*, **54**, pp. 39–47.

Jagota, A. (1996). *An adaptive, multiple restarts neural network algorithm for graph coloring , European J. of Operational Research*, **93**, pp. 257–270.

Korman, S.M. (1979). *The graph-colouring problem, in Combinatorial Optimization*, Christofides, N., Mingozzi, A. , Toth, P. and Sandi, C. Eds., pp. 211–235 (Wiley, New York).

Kubale, M. and Jackowski, B. (1985). *Ageneralised Implicit Enumeration Algorihm for Graph Coloring, Communications of ACM*, **28**, pp. 412 – 418.

Mehta, N. (1981). *The application of graph coloring method to an examination scheduling problem, Interfaces*, **11**, pp. 57–64.

Simanihuruk, M., Achuthan, N. and Achuthan, N.R. (1997). *On Defective Colourings of Triangle-Free Graphs, The Australasian Journal of Combinatorics*, **16**, pp. 259 – 283.

Simanihuruk, M., Achuthan, N. and Achuthan, N.R. (1997). *On Minimal Triangle-Free Graphs With Prescribed 1- Defective Chromatic Number", The Australasian Journal of Combinatorics*, **16**, pp. 203 – 227.

de Werra, D.(1985). *An introduction to timetabling , European J. of Operational Research*, **19**, pp. 151–162.

de Werra, D.(1990). *Heuristics for Graph Coloring , Computing Suppl. Computational graph theory*, **7**, pp. 191–208.

Chapter 8

Complementarity Problem involving a Vertical Block Matrix and its Solution using Neural Network Model

S. K. Neogy
Indian Statistical Institute, New Delhi-110016
e-mail: skn@isid.ac.in.
A. K. Das
Indian Statistical Institute, Kolkata-700108
e-mail: akdas@isical.ac.in.
P. Das
Indian Statistical Institute, Kolkata-700108
e-mail: dasprasun@rediffmail.com.

Dedicated to the memory of Professor S. R. Mohan

Abstract

In this paper, we consider several classes of vertical block matrices and characterize these matrix classes in the context of vertical linear complementarity problem (VLCP). We develop a neural network dynamics for solving VLCP. We have shown that the performance of our proposed dynamics is quite encouraging and it performs well for various classes of vertical block matrices. This seems to be a good alternative of Cottle-Dantzig algorithm for solving VLCP.

Key Words: Vertical block matrix, equivalent LCP, VLCP, Cottle-Dantzig algorithm, neural network dynamics

8.1 Introduction

The *linear complementarity problem(LCP)* is an important problem in mathematical programming and it has several applications in other fields.

The problem is stated as follows:

Given a square matrix M of order n with real entries and an n dimensional vector q, find n dimensional vectors w and z satisfying

$$w - Mz = q,\ w \geq 0,\ z \geq 0, \tag{8.1}$$

$$w^t z = 0 \tag{8.2}$$

or show that no solution exists.

This problem is denoted as LCP(q, M). If a pair of vectors (w, z) satisfies (8.1), then the problem LCP$(q.M)$ is said to have a feasible solution. A pair (w, z) of vectors satisfying (8.1) and (8.2) is called a solution to the LCP(q, M). In LCP(q, M), $F(q, M)$ denotes the feasible region and $S(q, M)$ denotes the solution set of LCP(q, M). For a detailed discussion on this problem and applications see [Cottle, Pang, and Stone (1992)] and [Murty (1988)].

The concept of a vertical block matrix was introduced by [Cottle and Dantzig (1970)] in connection with the generalization of the linear complementarity problem and it is defined as follows.

Consider a rectangular matrix A of order $m \times k$ with $m \geq k$. Suppose A is partitioned row-wise into k blocks in the form

$$A = \begin{bmatrix} A^1 \\ A^2 \\ \vdots \\ A^k \end{bmatrix}$$

where each $A^j = ((a^j_{rs})) \in R^{m_j \times k}$ with $\sum_{j=1}^{k} m_j = m$. The block matrix considered above is called a *vertical block matrix of type* $(m_1, \ldots, m_k)$. The generalization of the linear complementarity problem by [Cottle and Dantzig (1970)] involving a vertical block matrix is known as vertical linear complementarity problem and it is stated as follows:

Given a vertical block matrix A of type $(m_1, \ldots, m_k)$ and a vector $q \in R^m$, the vertical linear complementarity problem (VLCP(q, A)) is to find $w \in R^m$ and $z \in R^k$ such that

$$w - Az = q,\ w \geq 0,\ z \geq 0 \tag{8.3}$$

$$z_j \prod_{i=1}^{m_j} w^j_i = 0, \text{ for } j = 1, 2, \ldots, k. \tag{8.4}$$

For details on vertical linear complementarity problem see [Mohan, Neogy and Sridhar (1996a)], [Mohan and Neogy (1996a,b)] and the references therein.

The vertical block matrix arises naturally in the literature of stochastic games where the states are represented by the columns and actions in each state are represented by rows in a particular block. See [Mohan, Neogy and Parthasarathy (1997a,b, 2001)] and [Mohan, Neogy, Parthasarathy and Sinha (1999)]. Neural network approach for computing VLCP solution will be extremely useful for computation in Stochastic game problem. The rest of the paper is organized as follows.

In Section 8.2, we present the notations, definitions and the results which are used for obtaining subsequent results. In Section 8.3, we present the main results which extend several results in LCP setting to VLCP setting. In Section 8.4, we propose a neural network model for solving VLCP described by the nonlinear dynamic system. Finally, in Section 8.5, we present a number of numerical experiments for finding solutions of VLCP(q, A) to demonstrate the effectiveness and efficiency of the proposed neural network dynamics.

8.2 Preliminaries

For a matrix $A \in R^{m\times k}$, $A_{\cdot j}$ denotes the jth column of A and $A_{i\cdot}$, the ith row of A. For any positive integer n, if $\alpha \subseteq \{1, 2, \ldots, n\}$, $\bar{\alpha}$ denotes the complement of α in $\{1, 2, \ldots, n\}$. If M is a square matrix of order n and α, β are two nonempty subsets of $\{1, 2, \ldots, n\}$ then $M_{\alpha\beta}$ denotes the submatrix of M consisting of only the rows and columns of M whose indices are in α and β respectively. $M_{\cdot\beta}$ denotes the submatrix of those columns of M whose indices are in β. Similarly $M_{\alpha\cdot}$ denotes the submatrix of the rows of M whose indices are in α. Let $J_1 = \{1, 2, \ldots, m_1\}$ be the set of row indices in A corresponding to A^1 and let $J_r = \{\sum_{j=1}^{r-1} m_j + 1, \sum_{j=1}^{r-1} m_j + 2, \ldots, \sum_{j=1}^{r} m_j\}$ be the set of row indices in A corresponding to A^r, $r = 2, 3, \ldots, k$.

We say that $M \in R^{n\times n}$ is

- $P\,(P_0)$-matrix if all its principal minors are positive (nonnegative).
- $N\,(N_0)$-matrix if all the principal minors of M are negative (nonpositive).
- $\bar{N}$-*matrix* if there exists a sequence $\{N^{(k)}\}$ where $N^{(k)} = [m_{ij}^{(k)}]$

are N-matrices such that $m_{ij}^{(k)} \to m_{ij}$ for all $i, j \in \{1, 2, \ldots, n\}$.

- *copositive* (C_0) (*strictly copositive* (C)) if $x^t M x \geq 0 \ \forall \ x \geq 0$ ($x^t M x > 0 \, \forall \, 0 \neq x \geq 0$).
- *copositive-plus* (C_0^+) if $M \in C_0$ and the implication $[x^t M x = 0, x \geq 0] \Rightarrow (M + M^t)x = 0$ holds.
- a star matrix if $x \in S(0, M) \Rightarrow M^t x \leq 0$.
- *copositive-star* (C_0^*) if M is copositive and star matrix.
- a Q matrix if LCP(q, M) has a solution $\forall \ q \in R^n$
- a Q_0-matrix if for all $q \in R^n$, $F(q, M) \neq \emptyset \Rightarrow S(q, M) \neq \emptyset$.
- a S-matrix if there exists a vector $z \in R^n$ such that $Mz > 0, \ z > 0$.
- $\mathcal{L}_1$*-matrix* if for every $0 \neq y \geq 0$, $y \in R^n$ $\exists$ an i such that $y_i > 0$ and $(My)_i \geq 0$.
- $\mathcal{L}_2$*-matrix* if for each $0 \neq \xi \geq 0$, $\xi \in R^n$ satisfying $\eta = M\xi \geq 0$ and $\eta^t \xi = 0 \ \exists$ a $0 \neq \hat{\xi} \geq 0$ satisfying $\hat{\eta} = -M^t \hat{\xi}, \ \ \eta \geq \hat{\eta} \geq 0$, $\xi \geq \hat{\xi} \geq 0$.
- $\mathcal{L}$*-matrix* if it is in both $\mathcal{L}_1$ and $\mathcal{L}_2$.
- a matrix with T-property if for every nonempty set $\alpha \subseteq \{1, 2, \ldots, n\}$, the existence of a solution z_α to the system

  $$z_\alpha > 0, \ M_{\alpha\alpha} z_\alpha \leq 0, \ M_{\bar{\alpha}\alpha} z_\alpha \geq 0,$$

  implies that there exists a nonzero vector $y_{\alpha_0} \geq 0$ such that

  $$y_{\alpha_0}^t M_{\alpha_0 \alpha} = 0 \text{ and } y_{\alpha_0}^t M_{\alpha_0 \bar{\alpha}} \leq 0$$

[Flores-Bazan and Lopez (2005)] considers a matrix class F_1 which extends the class $\mathcal{L}_2$. We say that $M \in R^{n \times n}$ is a F_1 matrix if for any nonempty set $\alpha \subseteq \{1, 2, \ldots, n\}$, the following implication holds ($\bar{\alpha} = \{1, 2, \ldots, n\} \setminus \alpha$) :

$$z_\alpha > 0, \ M_{\alpha\alpha} z_\alpha = 0, \ M_{\bar{\alpha}\alpha} z_\alpha \geq 0,$$

implies that there exists a nonzero vector $x_\alpha \geq 0$ such that

$$x_\alpha^t M_{\alpha\alpha} = 0 \text{ and } x_\alpha^t M_{\alpha\bar{\alpha}} \leq 0.$$

A vertical block matrix A of type $(m_1, \ldots, m_k)$ is called a *vertical block P (P_0, C_0, C, C_0^+, $\mathcal{L}_1$, N, N_0)-matrix* if all its representative submatrices are P (P_0, C_0, C, C_0^+, $\mathcal{L}_1$, N, N_0)-matrices.

A vertical block matrix A of type $(m_1, \ldots, m_k)$ is called a

- Q_0-matrix if for any $q \in R^m$, (8.3) has a solution implies that the VLCP(q, A) has a solution.
- Q-matrix if for any $q \in R^m$, VLCP(q, A) has a solution.

- vertical block N-matrix of the first category if A is a vertical block N-matrix and A has at least one positive entry.
- vertical block N-matrix of the second category if A is a vertical block N-matrix with all entries negative.
- vertical block matrix with T property if every representative submatrix has T property.
- vertical block R_0-matrix if VLCP$(0, A)$ has the unique solution $w = 0,\ z = 0$.
- vertical block matrix with F_1 (copositive-star, $\mathcal{L}_1$, $\mathcal{L}_2$, R_0) property if every representative submatrix is F_1 (copositive-star, $\mathcal{L}_1$, $\mathcal{L}_2$, R_0).

The concept of equivalent matrix introduced in [Mohan, Neogy and Sridhar (1996a)] is defined as follows.

Consider a vertical block matrix A of type $(m_1, \ldots, m_k)$ where m_j is the size of the jth block. We construct a matrix M by copying $A_{.j}$, m_j times for $j = 1, 2, \ldots, k$. Thus $M_{.p} = A_{.s}\ \forall\ p \in J_s$. This construction leads to a square matrix M of order m. We call the matrix M obtained in this manner the *equivalent square matrix* of A and we call the problem LCP(q, M) as **equivalent LCP** of the VLCP(q, A). The following lemma due to [Mohan, Neogy and Sridhar (1996a)] which establishes a connection between the solution of the equivalent LCP(q, M) and VLCP(q, A).

Lemma 8.1. *Given the VLCP(q, A), let M be the equivalent square matrix of A. VLCP(q, A) has a solution if and only if LCP(q, M) has a solution.*

We make use of the following results to prove our main results.

Theorem 8.1. *([Ebiefung and Kostreva (1993)][p. 167]) VLCP(q, A) has a complementary feasible solution if and only if there exists a representative submatrix A_G and a corresponding subvector q_G of q so that LCP(q_G, A_G) is solvable with a solution z and $Az + q \geq 0$.*

Theorem 8.2. *([Valiaho (1986)]) If $M = M^t \in R^{n \times n}$ is copositive of exact order $(n-1)$, then*
(i) M is positive definite of order $(n-2)$;
(ii) all the principal minors of order ≥ 2 of M^{-1} are negative;
(iii) $M^{-1} \leq 0$ with negative off-diagonal elements.

Lemma 8.2. *([Mohan, Neogy and Sridhar (1996a)]) A is a vertical block $Q(Q_0)$ matrix iff the equivalent square matrix $M \in Q(Q_0)$.*

8.3 Main Results

Given a vertical block matrix A of type $(m_1, \ldots, m_k)$, let $u^1, \ldots, u^k$ be a collection of row vectors such that $0 \neq u^j \geq 0$ has m_j coordinates and $\sum_{j=1}^{k} m_j = m$. Then

$$U = \begin{bmatrix} u^1 \ldots 0 \\ \vdots \;\; \ldots \;\; \vdots \\ 0 \;\ldots\; u^k \end{bmatrix} \tag{8.5}$$

is of order $k \times m$. The jth row of the matrix UA is the u^j-weighted sum of the rows in A^j.

Theorem 8.3.
If A is a vertical block N-matrix of type $(m_1, \ldots, m_k)$ and U is given by (8.5), then UA is an N-matrix.

Proof. It is easy to check that

$$UA = \begin{bmatrix} \sum_{i \in J_1} u_i^1 A_{i.}^1 \\ \vdots \\ \sum_{i \in J_k} u_i^k A_{i.}^k \end{bmatrix}.$$

The determinant of a matrix is a multilinear function of its rows. Therefore

$$\det UA = \sum_{i \in J_1} \cdots \sum_{i \in J_k} \prod_{j=1}^{k} u_i^j \det \begin{bmatrix} A_{i.}^1 \\ \vdots \\ A_{i.}^k \end{bmatrix}. \tag{8.6}$$

All the terms in (8.6) are nonpositive and atleast one is negative since for every $j = 1, \ldots, k$ there exists an index i_0 such that $u_{i_0}^j > 0$. □

[Parthasarathy and Ravindran (1990)] proved that an N-matrix has exactly one real negative eigenvalue. We generalize this result.

Theorem 8.4. *Let $A \in R^{m \times k}$ be a vertical block N-matrix of type $(m_1, \ldots, m_k)$. Then the equivalent matrix M of A has exactly one real negative eigenvalue and have atleast $(m-k)$ zero eigenvalues. Also all nontrivial principal submatrices have exactly one real negative eigenvalue.*

Proof. Let $A \in R^{m\times k}$ be a vertical block matrix of type $(m_1, \ldots, m_k)$ and $U \in R^{k\times m}$ with $k \leq m$. Let M be the equivalent matrix of A. Then, by Theorem 1.3.20 in [Horn and Johnson (1985)] [p. 53], the equivalent matrix $M = AU$ has the same eigenvalues as UA, counting multiplicity with an additional $(m-k)$ eigenvalues equal to zero. Since by Theorem 8.3, UA is an N-matrix, so UA has exactly one negative eigenvalue. Hence, for the equivalent matrix $M = AU$ among k eigenvalues, exactly one real eigenvalue is negative and the additional $(m-k)$ eigenvalues are equal to 0. Also, all nontrivial submatrices of M have exactly one real negative eigenvalue. □

The following theorem was observed by [Mohan, Neogy and Sridhar (1996b)]. We provide the proof by [Mohan, Neogy and Sridhar (1996b)] for the sake of completeness.

Theorem 8.5. *Let A be an $m\times k$ vertical block matrix of type $(m_1, \ldots, m_k)$ and let M be the equivalent matrix of A of order m. The following statements are equivalent:*
(i) Every representative submatrix of A is copositive;
(ii) M is copositive.

Proof. **(ii)⇒ (i).** This follows from the inheritance property of a copositive matrix. See [Cottle, Habetler and Lemke (1970)][p. 296].
(i)⇒ (ii). We shall prove this by showing that every principal submatrix of M including M itself is copositive using induction on the order of the principal submatrices of M.

First we show that any 2×2 principal submatrix of M is copositive. Suppose i_1, i_2 are the row and column indices of the 2×2 principal submatrix G of M. Suppose $i_1 \in J_r$ and $i_2 \in J_s$ with $r \neq s$. Then G is a 2×2 principal submatrix of a representative submatrix of A and hence is copositive. Suppose $i_1, i_2 \in J_r$. Then G is of the form $\begin{bmatrix} a & a \\ b & b \end{bmatrix}$. Now the copositivity of the representative submatrices of A implies that all the diagonal entries of M are nonnegative and hence both a and b are nonnegative. It follows that G is copositive.

Now we show that any 3×3 principal submatrix of M is copositive.

Let i_1, i_2, i_3 be the row and column indices of a 3×3 principal submatrix G of M. If i_1, i_2, i_3 are from 3 distinct sets J_r, J_s, J_t, then G is a principal submatrix of a representative submatrix of A and therefore is copositive. If on the other hand, i_1, i_2, i_3 are from the same set J_r, then G is of type

$\begin{bmatrix} a\,a\,a \\ b\,b\,b \\ c\,c\,c \end{bmatrix}$ with a, b and c as nonnegative. Thus G is a nonnegative matrix and hence it is copositive.

The only other type of cases to be considered is $i_1, i_2 \in J_r$ and $i_3 \in J_s$ with $r \neq s$. In this case G is of the form $G = \begin{bmatrix} a\,a\,e \\ b\,b\,f \\ c\,c\,g \end{bmatrix}$. Let $B = G + G^t = \begin{bmatrix} 2a & a+b & c+e \\ a+b & 2b & c+f \\ c+e & c+f & 2g \end{bmatrix}$. Note that all the 2×2 submatrices of G are copositive. Hence G and therefore $G + G^t$ are copositive of order 2.

Consider the submatrix
$B_{\alpha\alpha} = \begin{bmatrix} 2a & a+b \\ a+b & 2b \end{bmatrix}$ where $\alpha = \{1, 2\}$.
Note that $\det B_{\alpha\alpha} = 4ab - (a+b)^2 = -(a-b)^2 \leq 0$.

Suppose B is not copositive. Note that all 2×2 submatrices are copositive. Hence B is copositive matrix of exact order 2. From Theorem 8.2 part (ii), it follows that $\det(B) < 0$ and $B^{-1} \leq 0$ with off-diagonal entries strictly negative by Theorem 8.2 part (iii).

However $(3,3)^{th}$ element of $B^{-1} = \frac{\det B_{\alpha\alpha}}{\det(B)} \geq 0$ which contradicts Theorem 8.2 part(iii). Therefore $B = G + G^t$ is copositive. So, any principal submatrix of order 1, 2 and 3 of M is copositive.

Let us make the induction hypothesis that every principal submatrix of M of order p or less is copositive. Now suppose G is a principal submatrix of order $(p+1)$. If G is a submatrix of a representative submatrix of A then by hypothesis G is copositive. Otherwise, there are at least two columns of G which are identical. Suppose the rows $i_1, \ldots, i_{p+1}$ of M are the rows and columns of G. Then there are at least two indices i_r, i_t such that $i_r, i_t \in J_s$ for some s.

Suppose G is not a copositive matrix. Then $(G + G^t)$ is also not copositive. By our induction hypothesis G is a copositive matrix of exact order p. By Theorem 8.2 part(i), $(G + G^t)$ is positive definite of order $(p-1)$. Note that the 2×2 submatrix containing the row and column indices i_r, i_t of $(G + G^t)$ is of the form $\begin{bmatrix} 2a & a+b \\ a+b & 2b \end{bmatrix}$ and its determinant is $-(a-b)^2 \leq 0$. This is a contradiction to Theorem 8.2 part(i). This completes the proof of the theorem. □

Lemma 8.3. *Suppose A is an $m \times k$ vertical block matrix of type $(m_1, \ldots, m_k)$ and M is its equivalent matrix. If M is copositive-plus, then every representative submatrix of A is copositive-plus.*

Proof. This follows from the inheritance property of copositive-plus matrices. See [Cottle, Habetler and Lemke (1970)]. □

The converse of the above lemma is not true. This is illustrated in the following example given in [Mohan, Neogy and Sridhar (1996b)].

Example 8.1. *Let* $A = \begin{bmatrix} 1 & -1 \\ 0 & -1 \\ 1 & 0 \end{bmatrix}$ *be a vertical block matrix of type* $(2, 1)$. *Both the representative submatrices of A are clearly copositive-plus. However the equivalent matrix M is not copositive-plus.*

However, if all the representative submatrices of a vertical block matrix A are copositive-plus, then A is a vertical block Q_0 matrix, even though its equivalent matrix M need not be a copositive-plus matrix. A constructive proof of this is given by [Cottle and Dantzig (1970)].

Lemma 8.4. *[Mohan, Neogy and Sridhar (1996a)][Lemma 6.1, p.213]*

Let A be a given vertical block matrix of type $(m_1, m_2, \ldots, m_k)$. Let M be the equivalent square matrix of order m. VLCP$(0, A)$ has a unique solution if and only if the equivalent LCP$(0, M)$ has a unique solution.

Thus essentially A is a vertical block R_0-matrix if and only if M is an R_0-matrix. Note that a vertical block R_0-matrix need not guarantee its representative submatrices as R_0.

Recall that a vertical block matrix A of order $m \times k$ and type $(m_1, \ldots, m_k)$ has $F_1(\mathcal{L}_2,\ R_0$, copositive-star)-property if every representative submatrix of A is $F_1(\mathcal{L}_2,\ R_0$, copositive-star).

[Mohan and Neogy (1996a)] observes that if a vertical block matrix A of type $(m_1, \ldots, m_k)$ has $\mathcal{L}_1$ property then the equivalent matrix $M \in \mathcal{L}_1$. Further, [Mohan, Neogy and Sridhar (1996b)] proved that if a vertical block matrix A of type $(m_1, \ldots, m_k)$ has $\mathcal{L}_2$ property then the equivalent matrix $M \in \mathcal{L}_2$.

The following theorem generalizes this result.

Theorem 8.6. *Suppose a vertical block matrix A of order $m \times k$ and type $(m_1, \ldots, m_k)$ has F_1-property. Then the equivalent matrix $M \in F_1$.*

Proof. Suppose (w^*, z^*) solves LCP$(0, M)$. We first construct a slightly different solution $(\bar{w}, \bar{z})$ to LCP$(0, M)$ as follows:

We take $\bar{w} = w^*$. Suppose for some $r_1, r_2 \in J_r$, we have $z^*_{r_1} > 0$, $z^*_{r_2} > 0$. It follows that $w^*_{r_1} = w^*_{r_2} = 0$. Hence also $\bar{w}_{r_1} = \bar{w}_{r_2} = 0$. Now choose $r_t \in J_r$ as any index, $1 \leq t \leq m_r$ such that $\bar{w}_{r_t} = \min_{1 \leq s \leq m_r} \bar{w}_{r_s}$. Note that $\bar{w}_{r_t} = 0$. We define $\bar{z}_{r_t} = \sum_{i=1}^{m_r} z^*_{r_i}$ and $\bar{z}_{r_i} = 0$ for $r_i \neq r_t$. Consider the set $L = \{1_t, 2_t, \ldots, k_t\}$. Note that $L \cap J_i$ is a singleton set for each $1 \leq i \leq k$ and M_{LL} is a representative submatrix of A. Note also that $(\bar{w}_L, \bar{z}_L)$ solves LCP$(0, M_{LL})$ where M_{LL} is a square matrix of order k.

Let $\alpha = \{i \;:\; \bar{z}^i_L > 0\}$. Since by hypothesis $M_{LL} \in F_1$, therefore for any nonempty set $\alpha \subseteq \{1, 2, \ldots, k\}$ with $\bar{z}^\alpha_L \in R^{|\alpha|}$, $\bar{z}^\alpha_L > 0$, $M^{\alpha\alpha}_{LL} \bar{z}^\alpha_L = 0$, $M^{\bar{\alpha}\alpha}_{LL} \bar{z}^\alpha_L \geq 0 \Rightarrow \exists\, 0 \neq x^\alpha_L \in R^{|\alpha|}, x^\alpha_L \geq 0$ such that $x^\alpha_L M^{\alpha\alpha}_{LL} = 0, x^\alpha_L M^{\bar{\alpha}\alpha}_{LL} \leq 0,$

Now define $\bar{x} \in R^m$ by taking $\bar{x}_L = x_L$ and $\bar{x}_{\bar{L}} = 0$. It follows that $M \in F_1$. □

Corollary 8.1. *If a vertical block matrix A of order $m \times k$ and type $(m_1, \ldots, m_k)$ has copositive-star (T)-property. Then the equivalent matrix M is a copositive-star matrix(a matrix with T-property).*

Corollary 8.2. *If a vertical block matrix A of order $m \times k$ and type $(m_1, \ldots, m_k)$ has $\mathcal{L}_2$ (R_0)-property, then the equivalent matrix $M \in \mathcal{L}_2$ (R_0).*

The following example shows that the converse of the above theorem is not true.

Example 8.2. *Let* $A = \begin{bmatrix} 0 & 0 & 1 \\ 0 & 0 & 1 \\ 1 & -1 & 1 \\ -2 & 1 & 1 \end{bmatrix}$ *be a vertical block matrix of type* $(1, 1, 2)$. *The equivalent matrix is given by* $M = \begin{bmatrix} 0 & 0 & 1 & 1 \\ 0 & 0 & 1 & 1 \\ 1 & -1 & 1 & 1 \\ -2 & 1 & 1 & 1 \end{bmatrix}$. *It is easy to verify that $M \in R_0$ and $M \in F_1$ whereas none of the representative matrix is an F_1-matrix.*

Lemma 8.5. *A is a vertical block N-matrix of the first category if and only if every representative submatrix is an N-matrix of the first category.*

Definition 8.1. A matrix $A \in R^{n\times n}$ is said to be an $\bar{N}$*-matrix* if there exists a sequence $\{A^{(k)}\}$ where $A^{(k)} = [a_{ij}^{(k)}]$ are N-matrices such that $a_{ij}^{(k)} \to a_{ij}$ for all $i, j \in \{1, 2, \ldots, n\}$.

Definition 8.2. We say that A is a *vertical block* $\bar{N}$*-matrix* of type $(m_1, \ldots, m_k)$ if every representative submatrix is an $\bar{N}$-matrix.

Example 8.3. *Let* $A = \begin{bmatrix} 0 & -1 & 0 \\ 0 & 0 & 1 \\ 0 & 1 & 0 \\ 1 & 0 & 0 \end{bmatrix}$ *be a vertical block matrix of type* $(1, 1, 2)$. *It is easy to see that* $A \in$ *vertical block* $\bar{N}$ *of type* $(1, 1, 2)$ *since we can get* A *as a limit point of the sequence* $A^{(k)} = \begin{bmatrix} -\frac{1}{k} & -1 & \frac{2}{k} \\ -\frac{1}{k} & -\frac{1}{k} & 1 \\ \frac{4}{k} & 1 & -\frac{1}{k} \\ 1 & \frac{2}{k} & -\frac{1}{k} \end{bmatrix}$ *of vertical block* N*-matrices of type* $(1, 1, 2)$ *which converges to* A *as* $k \to \infty$.

Lemma 8.6. *Suppose* A *is a vertical block* $\bar{N}$*-matrix of type* $(m_1, \ldots, m_k)$. *Let* M *be the equivalent matrix. Then there exists a nonempty subset* ν *of* $\{1, 2, \ldots, n\}$ *such that* M *can be written in the partitioned form as (if necessary, after a principal rearrangement of its rows and columns)*

$$M = \begin{bmatrix} M_{\nu\nu} & M_{\nu\bar{\nu}} \\ M_{\bar{\nu}\nu} & M_{\bar{\nu}\bar{\nu}} \end{bmatrix}$$

where $M_{\nu\nu} \leq 0, \ M_{\bar{\nu}\bar{\nu}} \leq 0, \ M_{\nu\bar{\nu}} \geq 0, \ M_{\bar{\nu}\nu} \geq 0.$

Proof. This follows from Remark 3.1 in [Mohan, Sridhar and Parthasarathy (1994)][p. 623] and from the definition of vertical block $\bar{N}$-matrices. □

We make use of the following result due to [Murty (1972)] and [Saigal (1972)].

Theorem 8.7. *If* $M \in R_0$ *and* $|S(q, M)|$*=odd, then* $A \in Q$.

Theorem 8.8. *Let* A *be a vertical block* $\bar{N}$*-matrix of type* $(m_1, \ldots, m_k)$ *with* R_0*-property and* M *be the equivalent matrix. Assume* $v(M) > 0$. *Then* A *is a vertical block* Q*-matrix.*

Proof. Since A is a vertical block matrix with R_0-property then by Corollary 8.2 M is a R_0-matrix.

Note that M can be written in the partitioned form as (if necessary, after a principal rearrangement of its rows and columns)

$$M = \begin{bmatrix} M_{\nu\nu} & M_{\nu\bar{\nu}} \\ M_{\bar{\nu}\nu} & M_{\bar{\nu}\bar{\nu}} \end{bmatrix}$$

where $M_{\nu\nu} \leq 0,\ M_{\bar{\nu}\bar{\nu}} \leq 0,\ M_{\nu\bar{\nu}} \geq 0,\ M_{\bar{\nu}\nu} \geq 0$.

Note that M is an N_0-matrix. Let $q > 0$, $q \in R^m$ be nondegenerate with respect to M (i.e., (w, z) is a solution to LCP(q, M) implies that $(w+z) > 0$.

Then clearly LCP(q, M) has exactly 3 solutions; $w^1 = q,\ z^1 = 0$,
$w^2 = \begin{bmatrix} 0 \\ w^2_{\bar{\nu}} \end{bmatrix};\ z^2 = \begin{bmatrix} z^2_{\nu} \\ 0 \end{bmatrix}$ where $z^2_{\nu} > 0$ and $w^2_{\bar{\nu}} = q_{\bar{\nu}} + M_{\bar{\nu}\nu} z^2_{\nu} > 0$ and
$w^3 = \begin{bmatrix} w^3_{\nu} \\ 0 \end{bmatrix};\ z^3 = \begin{bmatrix} 0 \\ z^3_{\bar{\nu}} \end{bmatrix}$ where $z^3_{\bar{\nu}} > 0$ and $w^3_{\nu} = q_{\nu} + M_{\nu\bar{\nu}} z^3_{\bar{\nu}} > 0$.

Since $M \in R_0$ and LCP(q, M) has an odd number of solutions, by Theorem 8.7, it follows that $A \in Q$. □

The following approach is proposed by [Mohan, Neogy and Sridhar (1996a)] which makes it easier to calculate the VLCP degree of a vertical block matrix A. Given the vertical block matrix A of type $(m_1, \ldots, m_k)$ consider the mapping $F_A : R^m \to R^m$ defined as follows:

Given $x \in R^m$, let x^+ and x^- be the positive and negative parts of x. Let

$$F_A(x) = x^+ - \sum_{i=1}^{k} A_{\cdot i} \big(\sum_{j \in J_i} x_j^- \big).$$

It is easy to see that given a $q \in R^m$, if there is a x such that $F_A(x) = q$, then defining $w = x^+$ and $z \in R^k$ by taking $z_i = \sum_{r \in J_i} x_r^-$, we see that (w, z) solves VLCP(q, A). Actually, it is easy to see that the VLCP map $F_A(x)$ defined above is the same as LCP map $F_M(x)$ where M is the equivalent matrix of A.

If we define the VLCP degree of A to be the degree of the piecewise linear map $F_A(x)$, then it turns out that this is also the LCP degree of the equivalent square matrix of order m.

Theorem 8.9. *Suppose, A is a vertical block $\bar{N}$-matrix of type $(m_1, \ldots, m_k)$ with $v(M) > 0$. Then $|deg(A)| =$ odd.*

Proof. This follows from the fact that LCP(q, M) has odd number of solutions for any non-degenerate $q > 0$ with respect to M. Therefore, LCP degree of the equivalent matrix M is odd. Since, VLCP degree of A is also same as LCP degree of the equivalent square matrix M of order m, therefore $|\deg(A)| = \text{odd}$. □

8.4 Computing VLCP Solution Using the Neural Network Dynamics

Neural networks approaches in optimization were introduced in early 80's (see [Tank and Hopfield (1986)],[Kennedy and Chua (1988)]). It is basically to establish a nonnegative energy function and a dynamic system that represents an artificial neural network. Normally, the dynamic system is in the form of first order differential equation. The concept behind the neural network based optimization techniques is that the objective function and constraints are mapped into a closed-loop network so that when a constraint violation occurs, the magnitude and direction of the violation are fed back to adjust the states of the neurons in the network. The energy function of the network decreases until it attains a minimum and the states of the neurons of the network are taken to be the minimizer of the original problem. Mainly pivoting algorithms are used for solving complementarity problem with a vertical block matrix apart from complete enumeration procedure. But pivoting algorithms are heavily dependent on matrix classes. The neural network approach seems to be promising for solving complementarity problems.

8.4.1 *Proposed Neural Network Dynamics*

We propose the following recurrent neural network model which is described by the following nonlinear dynamic system.

$$\frac{dx_j}{dt} = \max_{i=1,\ldots,m_j} (-q - A(x + k\frac{dx}{dt})),\ x_j > 0 \tag{8.7}$$

Theorem 8.10. *If the neural network whose dynamics is described by the differential equations (8.7) converges to a stable state then the convergence state is a solution for VLCP.*

Proof. Consider a *vertical block matrix A of type* $(m_1, \ldots, m_k)$. Equation (8.7) can be written as

$$\frac{dx_j}{dt} = \max_{i=1,\ldots,m_j} (-q - A * (x + dx))_i^j, \quad \text{if } x_j > 0 \tag{8.8}$$

$$\frac{dx_j}{dt} = \max\{(-q - A*(x+dx))_1^j, \ldots, (-q - A*(x+dx))_{m_j}^j, 0\}, \quad \text{if } x_j = 0 \tag{8.9}$$

Note that (8.9) ensures that x will be bounded from below by 0. Let $\lim_{t\to\infty} x(t) = x^*$. By stability of convergence $\frac{dx^*}{dt} = 0$. So, (8.8) and (8.9) become

$$(-q - Ax^*)_i^j \leq 0 \tag{8.10}$$

$$x_j^* \prod_{i=1}^{m_j} (-q - Ax^*)_i^j = 0 \tag{8.11}$$

Therefore,

$$[-q - Ax^*] \leq 0, \quad x^* \geq 0 \tag{8.12}$$

$$x_j^* \prod_{i=1}^{m_j} (-q - Ax^*)_i^j = 0 \tag{8.13}$$

Therefore, we get the inequalities (8.3)-(8.4). By definition, x^* is a solution of VLCP(q, A). □

Remark 8.1. In order to solve the differential equations (8.7), the Euler's method may be used. The following Matlab code describes the discrete implementation of our neural network. Coefficient k is set to equal to the time step dt to simplify the calculations.

while $\|dx\| > \epsilon$
for $j = 1 : k$;
for $i = (m_j + 1) : (m_j + m_{j+1})$;
$dx_j = dt * max_i(-q - A*(x+dx))_i^j$;
end;
end;
$dx = max(x + dx, 0) - x$; %(to make $x \geq 0$)
$x = x + dx$;
end.

8.5 Simulation Results

We conduct a number of numerical experiments for finding solutions of VLCP(q, A) to demonstrate the effectiveness and efficiency of the proposed neural network dynamics. The simulation is carried out on Matlab to solve the differential equations using Euler's method. To start with, we initialize x and dx at $t = 0$. We take small positive values for step length dt and ϵ. The dynamics stops if $\|dx\| < \epsilon$. The simulation runs on a Compaq PC with intel pentium 4 processor 1.99 GHz 248 MB RAM. We mention two examples below.

Example 8.4. Consider the VLCP(q, A) where A is a vertical block P-matrix of type (1,1,1,1,2):

$$A = \begin{bmatrix} 1 & -8 & -10 & -1 & -1 \\ 0 & 1 & 1 & 1 & 1 \\ 0 & -1 & 1 & 1 & 1 \\ 0 & -1 & 1 & 1 & 1 \\ 0 & 1 & 1 & 1 & 1 \\ 0 & 1 & 1 & 1 & 1 \end{bmatrix} \quad q = \begin{bmatrix} -1 \\ -1 \\ -1 \\ -1 \\ -1 \\ -1 \end{bmatrix}$$

The dynamics converges to the point $(6.4907, 0.0001, 0.5000, 0.5000, 0.0001)$ after 62 iterations with a step of 0.1 and ϵ equal to 0.001. The solution for this problem is $(6.5, 0, 0.5, 0.5, 0)$.

Example 8.5. [Mohan, Neogy, Parthasarathy and Sinha (1999)] consider an example to formulate a discounted zero-sum stochastic game with ARAT structure as VLCP(q, A), where A is vertical block matrix of type (2,2,2,2). Though, the matrix A is not an R_0-matrix, Cottle-Dantzig algorithm processes this VLCP(q, A) and provides a solution. In this paper, we consider the same VLCP(q, A) and show that the proposed dynamics is able to converge to the same solution.

$$A = \begin{bmatrix} -\frac{1}{4} & 0 & \frac{3}{4} & 0 \\ -\frac{1}{4} & 0 & \frac{3}{4} & 0 \\ 0 & -\frac{1}{4} & 0 & \frac{3}{4} \\ 0 & -\frac{1}{4} & 0 & \frac{3}{4} \\ -\frac{3}{4} & 0 & \frac{1}{4} & 0 \\ -1 & \frac{1}{4} & 0 & \frac{1}{4} \\ 0 & -\frac{3}{4} & 0 & \frac{1}{4} \\ \frac{1}{4} & -1 & \frac{1}{4} & 0 \end{bmatrix} \quad q = \begin{bmatrix} -4 \\ -5 \\ -3 \\ -4 \\ 3 \\ 6 \\ 6 \\ 2 \end{bmatrix}$$

The complementary solution as obtained by using Cottle-Dantzig algorithm is $(7, 6, 9, 7.33)$. The dynamics converges to $(7, 6.0076, 9, 7.3538)$ just after 669 iterations with a step of 0.5 and ϵ as 0.01.

Computational experience on the performance of the proposed model is reported in the following table and figures.

Example No.	dt	Norm	Iteration	Solution	Optimal Solution
8.4	0.1	0.001	62	(6.4907, 0.0001, 0.5000, 0.5000, 0.0001)	(6.5, 0, 0.5, 0.5, 0)
8.5	0.5	0.01	669	(7, 6.0076, 9, 7.3538)	(7, 6, 9, 7.33)

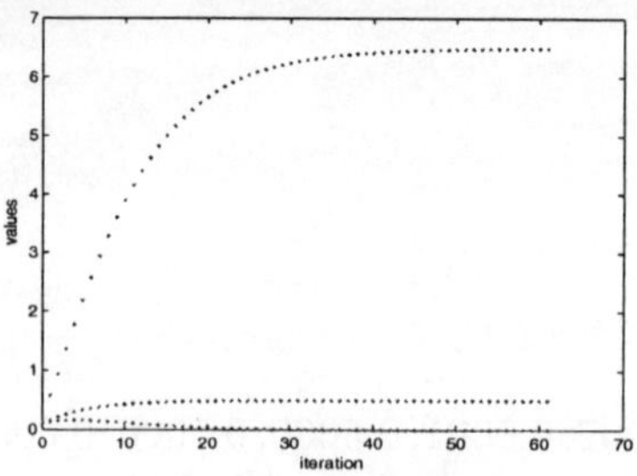

Figure 1 Example No. 8.4

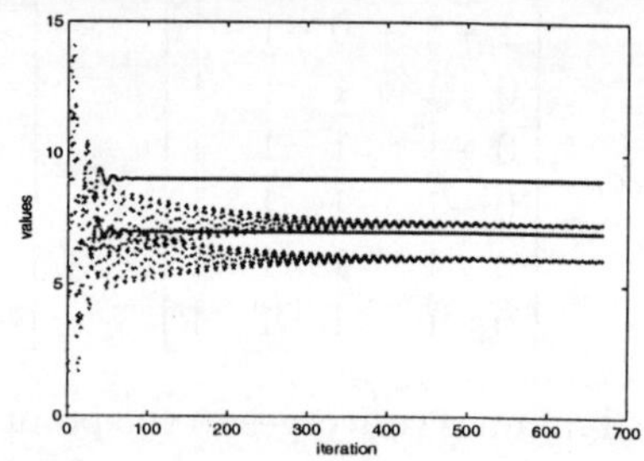

Figure 2 Example No. 8.5

From the above tables and figures (see Figure 1-2), we observe that the proposed model computes solution for VLCP involving different classes of vertical block matrices. In fact, the proposed neural network model was tried for a large number of test problems and it shows that the proposed model has the faster convergence to compute a solution of a vertical LCP which is very encouraging.

Bibliography

Cottle, R. W. and Dantzig, G. B. (1970). A generalization of the linear complementarity problem, *Journal of Combinatorial Theory*, **8**, pp. 79–90.

Cottle, R. W., Pang, J. S. and Stone, R. E. (1992). *The Linear Complementarity Problem*, (Academic Press, New York).

Cottle, R. W., Habetler, G. J. and Lemke, C. E. (1970). On classes of copositive matrices, *Linear Algebra and Its applications*, **3**, pp. 295–310.

Ebiefung, A. A. and Kostreva, M. M. (1993). The generalized Leontief input-output model and its application to the choice of new technology, *Annals of Operations Research*, **44**, pp. 161–172.

Flores-Bázan, F. and López, R. (2005). Characterizing Q-matrices beyond L-matrices, *Journal of Optimization theory and applictions*, **127**, pp. 447–457.

Garcia, C. B. (1973). Some classes of matrices in linear complementarity theory, *Mathematical Programming*, **5**, pp. 299–310.

Horn, R. A. and Johnson, C. R. (1985). *Matrix Analysis*, (Cambridge University Press, Cambridge).

Kennedy, M. P. and Chua, L. O. (1988). Neural networks for non-linear programming, *IEEE transactions on Circuits System*, **35**, pp. 554–562.

Lloyd, N. G.(1978). *Degree Theory*, (Cambridge University Press, Cambridge).

Mohan, S. R. and Sridhar, R. (1992). On characterizing N-matrices using linear complementarity, *Linear Algebra and Its Applications*, **160**, pp. 231–245.

Mohan, S. R., Neogy, S. K. and Sridhar, R. (1996). The generalized linear complementarity problem revisited, *Mathematical Programming* **74**, pp. 197–218.

Mohan, S. R., Neogy, S. K. and Sridhar, R., (1996b). Copositive, Sufficient and Semimonotone matrices in Vertical Linear Complementarity, *Technical Report # 9608*, (Indian Statistical Institute, Delhi Centre, India).

Mohan, S. R. and Neogy, S. K. (1996). The role of representative submatrices in vertical linear complementarity theory, *Linear and Multilinear Algebra* **41** pp. 175–187.

Mohan, S. R. and Neogy, S. K.(1996). Generalized linear complementarity in a problem of n person games, *OR Spektrum* **18**, pp. 231–239.

Mohan, S. R., Neogy, S. K. and Parthasarathy, T. (1997a). Linear complementarity and discounted polystochastic game when one player controls transitions, in *Complementarity and Variational Problems,* eds: M.C. Ferris and Jong-Shi Pang, pp. 284–294, (SIAM, Philadelphia).

Mohan, S. R., Neogy, S. K. and Parthasarathy, T. (1997b). Linear complementarity and the irreducible polystochastic game with the average cost criterion when one player controls transitions, in *Game theoretical applications to Economics and Operations Research*, eds. T. Parthasarathy, B. Dutta, J. A. M. Potters, T. E. S. Raghavan, D. Ray and A. Sen, pp. 153–170, (Kluwer Academic Publishers, Dordrecht, The Netherlands).

Mohan, S. R., Neogy, S. K. and Parthasarathy, T. (2001). Pivoting algorithms for some classes of stochastic games: A survey, *International Game Theory Review*, **3**, pp. 253–281.

Mohan, S. R., Neogy, S. K., Parthasarathy, T. and Sinha, S. (1999). Vertical linear complementarity and discounted zero-sum stochastic games with ARAT structure, *Mathematical Programming*, Series A, pp. 637–648.

Murty, K. G.(1972). On the number of solutions of the complementarity problem and the spanning properties of complementary cones, *Linear Algebra and Its Applications*, **5**, pp. 65–108.

Murty, K. G. 1988. *Linear complementarity, Linear and Nonlinear Programming*, (Heldermann Verlag, West Berlin).

Mohan, S. R., Sridhar R. and Parthasarathy, T. (1992). $\bar{N}$ matrices and the class Q, in B. Dutta et al. (Eds), *Lecture Notes in Economics and Mathematical Systems*, **389**, pp. 24–36, (Springer Verlag, Berlin).

Mohan, S. R., Sridhar R. and Parthasarathy, T. (1994). The linear complementarity problem with exact order matrices, *Mathematics of Operations Research*, **19**, pp. 618–644.

Murthy, G. S. R. and Parthasarathy, T. (1995). Some properties of fully semi-monotone matrices, *SIAM Journal on Matrix Analysis and Applications*, **16**, pp. 1268–1286.

Murthy, G. S. R., Parthasarathy, T. and Ravindran, G. (1995). On copositive semi-monotone Q matrices, *Mathematical Programming*, **68**, pp. 187–203.

Murthy, G. S. R., Parthasarathy, T. and Ravindran, G. (1995). A copositive Q-matrix which is not R_0, *Mathematical Programming*, **61**, pp. 131–135.

Parthasarathy, T. and Ravindran, G., (1990). N-Matrices, *Linear Algebra and its Applications*, **139**, pp. 89–102.

Saigal, R. (1972). A characterization of the constant parity property of the number of solutions to the linear complementarity problem, *SIAM Journal on Applied Mathematics*, **23**, pp. 40–45.

Tank, D. W. and Hopfield, J. J. (1986). Simple neural optimization networks: An A/D converter, signal decision network, and a linear programming circuit, *IEEE transactions on Circuits System*, **33**, pp. 533–541.

Väliaho, H. (1986). Criteria for Copositive matrices, *Linear Algebra and Its Applications*, **81**, pp. 19–34.

Chapter 9

Fuzzy Twin Support Vector Machines for Pattern Classification

Reshma Khemchandani
Department of Mathematics
Indian Institute of Technology, Hauz Khas
New Delhi-110016, India.
e-mail: reshmaiitd@gmail.com
Jayadeva
Department of Electrical Engineering
Indian Institute of Technology, Hauz Khas
New Delhi-110016, India
e-mail: jayadeva@ee.iitd.ac.in
Suresh Chandra
Department of Mathematics
Indian Institute of Technology, Hauz Khas
New Delhi-110016, India.
e-mail: chandras@maths.iitd.ac.in

Abstract

We propose a fuzzy extension to twin support vector machines for binary data classification. Here, a fuzzy membership value is assigned to each pattern, and points are classified by assigning them to the nearest of two non parallel planes that are close to their respective classes. Fuzzy Twin Support Vector Machines determine two non-parallel planes by solving two related support vector machines-type problems, each of which is smaller than conventional fuzzy support vector machines. The approach can be used to obtain an improved classification when one has an estimate of the fuzziness of samples in either class.

Key Words: Support vector machines, pattern classification, machine learning, fuzzy, generalized eigenvalues, eigenvalues and eigenvectors.

9.1 Introduction

The last decade has witnessed the evolution of Support Vector Machines (SVMs) as a powerful paradigm for pattern classification and regression [Burges (1998)]-[Cherkassky and Mulier (1998)]. SVMs emerged from research in statistical learning theory on how to regulate the tradeoff between structural complexity and empirical risk. One of the most popular SVM classifiers is the "maximum margin" one, that attempts to reduce generalization error by maximizing the margin between two disjoint half planes [Burges (1998)]-[Cherkassky and Mulier (1998)]. The resultant optimization task involves the minimization of a convex quadratic function subject to linear inequality constraints.

Taking motivation from [Mangasarian and Wild (2006)], recently, the present authors [Jayadeva, Khemchandani and Chandra (2007)] have proposed a non-parallel plane classifier for binary data, termed as the Twin Support Vector Machine(TWSVM). In this approach, data points of each class are proximal to one of two non-parallel planes. The non-parallel planes are obtained by solving a pair of small sized quadratic programming problems (QPPs) compared to SVM where we solve a single large size QPP. In SVM, the QPP has all data points in the constraints, but in TWSVM they are distributed in the sense that the patterns of one class give the constraints of the other QPP, and vice-versa. This strategy of solving two smaller sized QPPs rather than one large QPP, makes TWSVMs work faster than standard SVMs.

In practice, there are often situations where patterns belonging to one class play a more significant role in classification. Traditionally such problems have been solved by fuzzy SVMs, e.g. [Lin and Wang (2002)], and fuzzy proximal SVMs [Jayadeva, Khemchandani and Chandra (2004)], where patterns of the more important class are assigned higher membership values.

In this paper, we propose a fuzzy extension to twin support vector machines, termed as the Fuzzy Twin Support Vector Machine (FTWSVM) for binary data classification. Similar to TWSVMs, FTWSVMs also aim at generating two non-parallel planes such that each plane is closer to one of the two classes and is as far as possible from the other. The introduction of fuzzy memberships allow us to improve the overall error rate since each of the two problems being solved can be associated with a different set of fuzzy memberships, thereby improving the accuracy for each problem independent of the other.

The paper is organized as follows: Section 9.2 briefly dwells on SVMs, and also introduces the notations used in the rest of the paper. Section 9.3 discusses linear Twin Support Vector Machines for binary data classification. Section 9.4 introduces linear Fuzzy Twin Support Vector Machines. Section 9.5 deals with experimental results and Section 9.6 contains concluding remarks.

9.2 Support Vector Machines

Let the patterns to be classified be denoted by a set of m row vectors $A_i, (i = 1, 2, \ldots, m)$ in the n-dimensional real space $\mathbf{R}^n$, where $A_i = (A_{i1}, A_{i2}, \ldots, A_{in})^T$. Also, let $y_i \in \{1, -1\}$ denote the class to which the ith pattern belongs. We first consider the case when the patterns belonging to the two classes are linearly separable. Then, we need to determine $w \in \mathbf{R}^n$ and $b \in \mathbf{R}$ such that

$$\begin{aligned} A_i w &\geq 1 - b \quad \mathit{for} \quad y_i = 1, \text{ and} \\ A_i w &\leq -1 - b \quad \mathit{for} \quad y_i = -1. \end{aligned} \tag{9.1}$$

The plane described by

$$w^T x + b = 0 \tag{9.2}$$

lies midway between the bounding planes given by

$$w^T x + b = 1 \quad \mathit{and} \quad w^T x + b = -1, \tag{9.3}$$

and separates the two classes from each other with a margin of $\frac{1}{||w||_2}$ on each side. In other words, the margin of separation between the two classes is given by $\frac{2}{||w||_2}$. Here, $||w||_2$ denotes the L_2 norm of a vector w. Data samples which lie on the planes given by (9.3) are termed as support vectors. The maximum margin classifier, which is the standard SVM, is obtained by maximizing this margin, and is equivalent to the following problem

$$(SVM1) \qquad \begin{aligned} &\underset{w,b}{Min} \quad \frac{1}{2} w^T w \\ &\mathit{subject\ to} \\ &A_i w \geq 1 - b \quad \mathit{for} \quad y_i = 1, \\ &A_i w \leq -1 - b \quad \mathit{for} \quad y_i = -1. \end{aligned} \tag{9.4}$$

When the two classes are not linearly separable, there will be an error in satisfying the inequalities (9.1) for some patterns, and we can modify (9.1) to

$$\begin{aligned} A_i w + q_i &\geq 1 - b \quad for \quad y_i = 1, \\ A_i w - q_i &\leq -1 - b \quad for \quad y_i = -1, \\ q_i &\geq 0, i = 1, 2 \ldots, m, \end{aligned} \tag{9.5}$$

where q_i denotes the error variable associated with the ith data sample. In this case, the classifier is termed as a "soft margin" one, and it approximately classifies points into two classes with some error. The classification of a given test sample x is obtained by determining the sign of $w^T x + b$. The soft margin depends on the value of the non-negative error variables q_i. In this case, one needs to choose a trade-off between the margin and the error, and the standard SVM formulation for classification of the data points with a linear kernel is given by

$$\begin{aligned} (SVM2) \qquad & \underset{w,\, b,\, q}{Min} \quad c\, e^T q + \frac{1}{2} w^T w \\ & subject\ to \\ & \quad A_i w + q_i \geq 1 - b \qquad for \quad y_i = 1, \\ & \quad A_i w - q_i \leq -1 - b \quad for \quad y_i = -1, \\ & \quad q_i \geq 0, \qquad i = 1, 2 \ldots, m. \end{aligned} \tag{9.6}$$

Here, c denotes a scalar whose value determines the trade-off; a larger value of c emphasizes the classification error, while a smaller one places more importance on the classification margin.

In practice, rather than solving (SVM1) and (SVM2), we solve their dual problems to get the appropriate hard or soft margin classifier. The case of nonlinear kernels is handled on lines similar to linear kernels [Gunn (1998)].

9.3 Twin Support Vector Machines

In this section, we give a brief outline of Twin Support Vector Machines (TWSVMs)[Jayadeva, Khemchandani and Chandra (2007)]. As mentioned earlier, TWSVM classifier is obtained by solving the two QPPs, which have the formulation of a typical SVM, except that not all patterns appear in the constraints of either problem at the same time.

Let the number of patterns in classes 1 and -1 be given by m_1 and m_2 and are represented by matrices A and B, respectively. Therefore, the sizes of matrices A and B are $(m_1 \times n)$ and $(m_2 \times n)$, respectively. The TWSVM classifier is obtained by solving the following pair of quadratic programming problems

$$(TWSVM1) \quad \underset{w^{(1)},\, b^{(1)},\, q}{Min} \quad \frac{1}{2}(Aw^{(1)} + e_1 b^{(1)})^T (Aw^{(1)} + e_1 b^{(1)}) + c_1 e_2^T q$$
$$subject\ to$$
$$-(Bw^{(1)} + e_2 b^{(1)}) + q \geq e_2,$$
$$q \geq 0, \tag{9.7}$$

and,

$$(TWSVM2) \quad \underset{w^{(2)},\, b^{(2)},\, q}{Min} \quad \frac{1}{2}(Bw^{(2)} + e_2 b^{(2)})^T (Bw^{(2)} + e_2 b^{(2)}) + c_2 e_1^T q$$
$$subject\ to$$
$$(Aw^{(2)} + e_1 b^{(2)}) + q \geq e_1,$$
$$q \geq 0, \tag{9.8}$$

where $c_1,\ c_2 > 0$ are parameters, and e_1 and e_2 are vectors of ones of appropriate dimensions.

In a nutshell, TWSVMs comprise of a pair of quadratic programming problems, such that in each QPP the objective function corresponds to a particular class, and the constraints are determined by patterns of the other class. Thus, TWSVMs give rise to two smaller sized QPPs. In TWSVM1, patterns of class 1 are clustered around the plane $x^T w^{(1)} + b^{(1)} = 0$. Similarly in TWSVM2, patterns of class -1 cluster around the plane $x^T w^{(2)} + b^{(2)} = 0$.

The Wolfe dual [Mangasarian (1994)] of (TWSVMs) is obtained by considering Karush-Kuhn-Tucker conditions and is given by

$$(DTWSVM1) \ \underset{\alpha}{Max} \quad e_2^T \alpha - \frac{1}{2}\alpha^T G (H^T H)^{-1} G^T \alpha$$
$$subject\ to$$
$$0 \leq \alpha \leq c_1. \tag{9.9}$$

Similarly, the Wolfe Dual of (TWSVM2) is given by

$$(DTWSVM2) \ \underset{\gamma}{Max} \quad e_1^T \gamma - \frac{1}{2}\gamma^T P (Q^T Q)^{-1} P^T \gamma$$
$$subject\ to$$
$$0 \leq \gamma \leq c_2, \tag{9.10}$$

where, $H = [A \;\; e_1]$, $G = [B \;\; e_2], P = [A \;\; e_1]$, $Q = [B \;\; e_2]$, and the augmented vectors $u = \begin{bmatrix} w^{(1)} \\ b^{(1)} \end{bmatrix}$, and $v = \begin{bmatrix} w^{(2)} \\ b^{(2)} \end{bmatrix}$, are given by

$$u = (H^T H)^{-1} G^T \alpha, \tag{9.11}$$

and

$$v = (Q^T Q)^{-1} P^T \gamma, \tag{9.12}$$

respectively. In the above discussion, matrices $H^T H$ and $Q^T Q$ are of size $(n+1) \times (n+1)$, where in general, n is much smaller than the number of patterns of classes 1 and -1.

Once vectors u and v are known from (9.11) and (9.12), the separating planes

$$x^T w^{(1)} + b^{(1)} = 0 \text{ and } \quad x^T w^{(2)} + b^{(2)} = 0 \tag{9.13}$$

are obtained. A new data sample $x \in \mathbf{R}^n$ is assigned to class r $(r = 1, 2)$, depending on which of the two planes given by (9.13) it lies closest to, i.e

$$x^T w^{(r)} + b^{(r)} = \min_{l=1,2} |x^T w^{(l)} + b^{(l)}|, \tag{9.14}$$

where $|\cdot|$ is the perpendicular distance of point x from the plane $x^T w^{(l)} + b^{(l)} = 0, \; l = 1, 2$.

9.4 Fuzzy Twin Support Vector Machines

In this section, we introduce fuzzy extension of twin support vector machines (FTWSVMs), which incorporates the information of fuzziness in the data. FTWSVMs obtain non-parallel planes around which the data points of the corresponding class get clustered.

The FTWSVM classifier is obtained by solving the following pair of quadratic programming problems

$$
\begin{aligned}
(FTWSVM1) \quad & \underset{w^{(1)},\, b^{(1)},\, q}{Min} \quad \tfrac{1}{2}(S_1 A w^{(1)} + e_1 b^{(1)})^T (S_1 A w^{(1)} + e_1 b^{(1)}) + c_1 e_2^T q \\
& \textit{subject to} \\
& \qquad -(S_2 B w^{(1)} + e_2 b^{(1)}) + q \geq e_2, \\
& \qquad q \geq 0,
\end{aligned} \tag{9.15}
$$

and,

$$
\begin{aligned}
(FTWSVM2) \quad & \underset{w^{(2)},\, b^{(2)},\, q}{Min} \quad \tfrac{1}{2}(S_2 B w^{(2)} + e_2 b^{(2)})^T (S_2 B w^{(2)} + e_2 b^{(2)}) + c_2 e_1^T q \\
& \textit{subject to} \\
& \qquad (S_1 A w^{(2)} + e_1 b^{(2)}) + q \geq e_1, \\
& \qquad q \geq 0,
\end{aligned} \tag{9.16}
$$

where $c_1,\ c_2 > 0$ are parameters, e_1 and e_2 are vectors of ones of appropriate dimensions, and S_1, S_2 are the matrices of membership values of two classes, respectively.

The algorithm finds two hyperplanes, one for each class, and classifies points according to which hyperplane a given point is closest to. The first term in the objective function of (9.15) or (9.16) is the weighted sum of squared distances from the hyperplane to points of one class. Therefore, minimizing it tends to keep the hyperplane close to points of one class (say class 1). The constraints require the hyperplane to be at a weighted distance of at least 1 from points of the other class (say class -1); a set of error variables is used to measure the error wherever the hyperplane is closer than this minimum distance of 1. The second term of the objective function minimizes the sum of error variables, thus attempting to minimize mis-classification due to points belonging to class -1.

Further, FTWSVM is approximately four times faster than the usual fuzzy SVM. This is because the complexity of the usual FSVM is no more than m^3, and FTWSVM solves two problems viz. (9.15) and (9.16), each of roughly half of the size.

The Lagrangian corresponding to the problem FTWSVM1 (9.15), is given by

$$\begin{aligned} L(w^{(1)}, b^{(1)}, q, \alpha, \beta) = &\frac{1}{2}(S_1Aw^{(1)} + e_1b^{(1)})^T(S_1Aw^{(1)} + e_1b^{(1)}) \\ &+c_1e_2^Tq - \alpha^T(-(S_2Bw^{(1)} + e_2b^{(1)}) + q - e_2)) - \beta^Tq \quad (9.17) \end{aligned}$$

where $\alpha = (\alpha_1, \alpha_2 \dots \alpha_{m_2})^T$, and $\beta = (\beta_1, \beta_2 \dots \beta_{m_2})^T$ are the vectors of Lagrange multipliers. The Karush-Kuhn-Tucker (K. K. T.) necessary and sufficient optimality conditions [Mangasarian (1994)] for (FTWSVM1) are given by

$$S_1A^T(S_1Aw^{(1)} + e_1b^{(1)}) + S_2B^T\alpha = 0 \quad (9.18)$$

$$e_1^T(S_1Aw^{(1)} + e_1b^{(1)}) + e_2^T\alpha = 0 \quad (9.19)$$

$$c_1e_2 - \alpha - \beta = 0 \quad (9.20)$$

$$-(S_2Bw^{(1)} + e_2b^{(1)}) + q \geq e_2, \quad q \geq 0 \quad (9.21)$$

$$\alpha^T(-(S_2Bw^{(1)} + e_2b^{(1)}) + q - e_2) = 0, \quad \beta^Tq = 0 \quad (9.22)$$

$$\alpha \geq 0, \quad \beta \geq 0. \quad (9.23)$$

Since $\beta \geq 0$, from (9.20) we have

$$0 \leq \alpha \leq c_1. \quad (9.24)$$

Next, combining (9.18) and (9.19) leads to

$$[(S_1A)^T \quad e_1^T][S_1A \quad e_1]\begin{bmatrix} w^{(1)} \\ b^{(1)} \end{bmatrix} + [(S_2B)^T \quad e_2^T]\alpha = 0. \tag{9.25}$$

We define

$$H = [S_1A \quad e_1], \quad G = [S_2B \quad e_2], \tag{9.26}$$

and the augmented vector $u = \begin{bmatrix} w^{(1)} \\ b^{(1)} \end{bmatrix}$. With these notations, (9.25) may be rewritten as

$$H^T Hu + G^T\alpha = 0 \quad \text{i.e.} \quad u = -(H^T H)^{-1}G^T\alpha. \tag{9.27}$$

Although $H^T H$ is always positive semidefinite, it is possible that it may not be well conditioned in some situations. On the lines of the regularization term introduced in Ridge Regression approaches such as [Saunders, Gammerman and Vovk (1998)], we introduce a regularization term ϵI, $\epsilon > 0$, to take care of problems due to possible ill-conditioning of $H^T H$. Here, I is an identity matrix of appropriate dimensions. Therefore, (9.27) gets modified to

$$u = -(H^T H + \epsilon I)^{-1}G^T\alpha. \tag{9.28}$$

However in the following, we shall continue to use (9.27) with the understanding that, if the need be, (9.28) is to be used for the determination of u.

Using (9.17) and the above K.K.T. conditions, we obtain the Wolfe dual [Mangasarian (1994)] of FTWSVM1 as follows

$$(FDTWSVM1)\ \underset{\alpha}{Max} \quad e_2^T\alpha - \tfrac{1}{2}\alpha^T G(H^T H)^{-1}G^T\alpha$$
$$\textit{subject to}$$
$$0 \le \alpha \le c_1. \tag{9.29}$$

Similarly, we consider FTWSVM2 and obtain its dual as

$$(FDTWSVM2)\ \underset{\gamma}{Max} \quad e_1^T\gamma - \tfrac{1}{2}\gamma^T P(Q^TQ)^{-1}P^T\gamma$$
$$\textit{subject to}$$
$$0 \le \gamma \le c_2. \tag{9.30}$$

Here, $P = [S_1A \quad e_1]$, $Q = [S_2B \quad e_2]$, and the augmented vector $v = \begin{bmatrix} w^{(2)} \\ b^{(2)} \end{bmatrix}$, which is given by

$$v = (Q^TQ)^{-1}P^T\gamma. \tag{9.31}$$

Once (FDTWSVM1) and (FDTWSVM2) are solved to obtain the planes (9.13), a new pattern $x \in \mathbf{R}^n$ is assigned to class 1 or class -1 in a manner similar to the linear TWSVM case (9.14).

9.5 Experimental Results

Fuzzy Twin Support Vector Machine (FTWSVM), TWSVM, Fuzzy SVM (FSVM), and SVM data classification methods were implemented by using MATLAB 7 [Matlab] running on a PC with an Intel P4 processor (3 GHz) with 1 GB RAM. The methods were evaluated on datasets from the UCI Machine Learning Repository [Blake and Merz]. Generalization error was determined by following the standard ten fold cross-validation methodology [Duda, Hart and Strok (2001)].

Let us consider a situation where the patterns belonging to a particular class are much more important, or their membership is less ambiguous, in comparison to patterns of the other class. In such situations, we would like to classify patterns of this particular class preferentially. For example, when screening patients who are clearly healthy from those who require further examination, it is desirable to err on the side of caution. In such situations, we may assign a membership value of 1 to patterns of the class for which a higher generalizability is desired, and assign relatively smaller membership values to patterns belonging to the remaining class.

As an illustration, we now consider the case where the membership of patterns of class 1 is less ambiguous and patterns of class -1 are associated with a higher degree of ambiguity. Therefore, data samples belonging to class 1 are assigned a membership value s_1, while samples belonging to class -1 are assigned a membership value s_2, i.e. in this case $s_1 < s_2$. Hence, while implementing (FTWSVM1), we assign a membership value of 1 to patterns of class 1 and s_2 to patterns of class -1, and while implementing (FTWSVM2), we assign a membership value of 1 to patterns of class -1, and s_1 to patterns of class 1.

Tables 1, 2, 3 and 4 summarize FTWSVM performance on some benchmark datasets available at the UCI machine learning repository [Blake and Merz]. The table compares the performance of the FTWSVM classifier with that of SVM [Gunn (1998)], FSVM and TWSVM and illustrate that an appropriate assignment of the fuzzy membership values can be used to improve the classification accuracy. Optimal values of c_1 and c_2 were obtained by using a tuning set comprising of 10% of the dataset.

Table 5 compares the training time of FSVM with that of FTWSVM for ten folds. The table indicates that FTWSVM is not just effective, but is almost four times faster than a conventional FSVM, because it solves two quadratic programming problems of a smaller size instead of a single QPP of a very large size.

Table I: Percentage Test Set Accuracy of Heart-statlog with a Linear Kernel

Data Set	FTWSVM	TWSVM	FSVM	SVM
Class 1	83.15±8.91	74.07±13.61	82.27±9.42	80.01±9.25
Class -1	84.64±8.07	90.82±9.77	84.48±8.64	85.64±9.21
Overall	83.70±7.07	83.33±7.81	83.70±6.87	83.33±7.08

here($s_1 = 0.9,\ s_2 = 0.7$))

Table II: Percentage Test Set Accuracy of Sonar with a Linear Kernel

Data Set	FTWSVM	TWSVM	FSVM	SVM
Class 1	78.70±14.83	72.62±14.24	77.54±11.81	75.91±12.95
Class -1	76.42±10.36	80.90±11.15	79.07±5.95	80.51±6.91
Overall	76.86±8.41	75.40±7.82	76.86±5.51	76.38±5.75

here($s_1 = 0.8,\ s_2 = 1$)

Table III: Percentage Test Set Accuracy of Ionosphere with a Linear Kernel

Data Set	FTWSVM	TWSVM	FSVM	SVM
Class 1	72.41±11.75	61.40±11.33	67.40±9.31	62.92±12.24
Class -1	94.72±3.30	99.47±1.58	94.14±3.93	96.33±3.95
Overall	87.17±3.45	86.33±4.39	85.19 ±3.30	85.18 ±2.50

here($s_1 = 1,\ s_2 = 0.7$)

Table IV: Percentage Test Set Accuracy of Cleveland Heart with a Linear Kernel

Data Set	FTWSVM	TWSVM	FSVM	SVM
Class 1	84.74±9.29	87.85±4.53	84.67±5.96	87.07±7.93
Class -1	81.84±7.52	78.14±10.6	80.27±9.85	77.38±10.65
Overall	83.48±5.20	83.49±4.91	82.83±5.87	82.83±5.11

here($s_1 = 0.1,\ s_2 = 1$)

Table V: Training Times (in seconds)

Data Set	FTWSVM	FSVM
Sonar (208×60)	6.64	24.9
Heart-statlog (270×14)	11.3	50.9
Heart-c (303×14)	14.92	68.2
Ionosphere (351×34)	25.9	102.2

9.6 Concluding Remarks

In this paper, we have proposed a fuzzy extension to TWSVM approach for data classification, termed as FTWSVM. In FTWSVM, we solve two quadratic programming problems of a smaller size instead of a large sized one as we do in traditional Fuzzy SVMs. This makes FTWSVM almost four times faster than a standard Fuzzy SVM classifier. Furthermore, in contrast to a single hyperplane as given by traditional Fuzzy SVMs, FTWSVMs yield two non-parallel planes such that each plane is close to one of the two datasets, and is distant from the other dataset. The incorporation of fuzzy memberships into each of the twin problems allows us to improve the error on both subsets. In terms of generalization, FTWSVM compares favourably with Fuzzy SVM and TWSVM.

Another line of work in FTWSVMs which immediately suggests itself is to use a nonlinear kernel to perform the classification task in a feature space, with obvious applications in the case of non-linearly separable data sets and its extension to multicategory classification.

Acknowledgements

The first author acknowledges the financial support of the Council of Scientific and Industrial Research (India) in the form of a scholarship for pursuing her Ph.D. The authors are extremely thankful to the learned referees whose valuable comments have helped to improve the content and presentation of the paper.

Bibliography

Blake C. L. and Merz, C. J. UCI Repository for Machine Learning databases Irvine, CA:University of California, Department of Information and Computer Sciences. On-line at http:// www.ics.uci.edu/ mlearn/ MLRepository.html

Bradley, P. S., and Mangasarian, O. L. (2000). Massive Data Discrimination via Linear Support Vector Machines, *Optimization Methods and Software*, **13**, pp. 1–10.

Burges, C. (1998). A Tutorial on Support Vector Machines for Pattern Recognition, *Data Mining and Knowledge Discovery*, 2.

Cortes, C. and Vapnik, V. N. (1995). Support Vector Networks, *Machine Learning*, **20**, pp. 273–297.

Cherkassky V. and Mulier, F (1998). *Learning from Data - Concepts, Theory, and Methods*, (John Wiley and Sons, New York).

Duda, R. O., Hart, P. R. and Stork, D. G. (2001). *Pattern Classification*, 2nd edition, (John Wiley and Sons, Inc, New York).

Fung, G., Mangasarian, O. L. (2001). Proximal Support Vector Machines, Proc. KDD-2001, San Francisco, pp. 77–86.

G.H. Golub and C.F. Van Loan (1996). *Matrix Computations*, 3rd Ed, (The John Hopkins Univ. Press, Maryland).

Gunn, S. R. (1998). *Support Vector Machines for Classification and Regression*, Technical Report, School of Electronics and Computer Science, University of Southampton, Southampton, U.K. On-line at http://www.isis.ecs.soton.ac.uk/resources/svminfo/

Jayadeva, Khemchandani, R. and Chandra, S. (2004). Fast and Robust Learning Through Fuzzy Linear Proximal Support Vector Machines, *Neurocomputing*, **61**, pp. 401–411.

Jayadeva, Khemchandani, R. and Chandra, S. (2007). Twin Support Vector Machines for Pattern Classification, *IEEE Transactions on Pattern Analysis and Machine Intelligence*, **29**, pp. 905–911.

Lin C-F. and Wang, S-D. (2002). Fuzzy Support Vector Machines, *IEEE Trans. Neural Networks*, **13** (2), pp. 464–471.

Mangasarian, O. L. (1994). *Nonlinear Programming*, SIAM, Philadelphia, PA.

Mangasarian O. L. and Wild, E. W. (2006). Multisurface Proximal Support Vector Classification via Generalized Eigenvalues", *IEEE Transactions on Pattern Analysis and Machine Intelligence*, **28** (1), pp. 69–74.

Saunders, C., Gammerman, A. and Vovk, V. (1998). Ridge regression learning algorithm in dual variables In ICML '98, Proceedings of the 15th International Conference on Machine Learning, Madison, WI, pp. 515–521. http://citeseer.csail.mit.edu/saunders98ridge.html

Matlab. http://www.mathworks.com

Chapter 10

An Overview of the Minimum Sum of Absolute Errors Regression

Subhash C. Narula
School of Business
Virginia Commonwealth University
Richmond, Virginia 23284-4000, USA
e-mail: snarula@vcu.edu
John F. Wellington
School of Business and Management Sciences
Indiana U. - Purdue U. Fort Wayne
Fort Wayne, Indiana 46805-1499
e-mail: wellingj@ipfw.edu

Abstract

Our objective is to provide an overview of the minimum sum of absolute errors (MSAE) regression. Although proposed fifty years before the concept of least squares regression, the MSAE regression did not receive much attention until the second half of the last century. During this period, several very effective and efficient algorithms to compute the MSAE estimates of the unknown parameters of the multiple linear regression model were proposed and studied. Today a number of very good computer programs are available in the open literature and in publicly available computer packages. Efficient algorithms and computer programs for the selection of models with fewer variables are also available.

The asymptotic and small sample properties of the MSAE estimators have been studied. Based on these results, formulae for confidence intervals and procedures for testing hypotheses have been developed.

Some of these results have appeared in survey articles in the literature. However, since their publication, a number of new results have appeared in he literature that makes the MSAE regression procedure more attractive. For example, an R^2 like measure for the MSAE regression is now available. We now understand how special characteristics unique to the MSAE regression explain its robustness to certain type of outliers. And unlike least squares regression, variations in the values of the response and predictor variables within definable limits do not change the fitted MSAE regression. We highlighted those features of MSAE regression in this paper.

Key Words: Algorithms, coefficient of determination, computer programs, goodness-of-fit, robustness, r square, statistical inference, statistical properties, variable selection

10.1 Introduction

It is interesting to note that minimization of absolute error regression (MSAE) was introduced about mid-eighteenth century when Boscovich (1757) proposed that a straight line should be fitted to three or more non-collinear points in a plane so as to satisfy two conditions, namely that: (i) the sum of the positive errors and the sum of absolute negative errors of the given points from the fitted line be equal in magnitude and (ii) the sum of the absolute errors be minimum. Boscovich (1760) developed accordingly a geometric algorithm for the simple linear regression model. The algebraic formulation of the algorithm was given by Laplace (1792).

Although proposed a half century before the publication of Legendre's "Principle of Least Squares," the development of the MSAE regression has been slow. Eisenhart (1961) suggested that this might have been due to the following factors: the uniqueness of the least squares solution; the relative computational simplicity of least squares regression; and the thorough reformulation and development of the method of least squares by Gauss (1809, 1823, and 1828) and Laplace (1818).

The MSAE regression reappeared in the 1880's largely due to the work of Edgeworth. His contributions included that condition (i) of Boscovich should be dropped so that the minimum in condition (ii) can obtain its smallest value (Edgeworth, 1887); a discussion of the non-uniqueness of the MSAE regression; and a demonstration that the MSAE estimator is maximum likelihood estimator when the random error follows a Laplace distribution (Edgeworth, 1888). He also developed an algorithm for the simple linear MSAE regression model. The interested reader may refer to Farebrother (1987) for further historical details.

Until fifty years ago, the computational problems associated with MSAE regression in any but the simple regression model effectively prevented its use. Charnes, Cooper and Ferguson (1955) formulated the MSAE regression problem as a linear goal programming problem and solved it using the simplex method. This development has been of the utmost importance for its role in making the MSAE regression computationally available to a broad audience of researchers and practitioners. Since then a number of very efficient algorithms have been developed for the MSAE regression. Now it is possible to solve the problem in slightly more computational time than that required for the least squares regression.

Besides computational difficulties, a serious limitation to wider application of the MSAE regression in the past has been the limited development and knowledge of procedures for statistical inference. However, results from Monte Carlo studies of small sample properties of the MSAE estimator have been reported and inference procedures are now available, Dielman (2005).

The MSAE regression is considered a robust alternative to least squares regression by a number of authors. For example, Huber (1974, p. 927) stated that with regard to L_p estimators in regression, "$p = 1$ (minimum sum of absolute errors regression) gives robustness in a technical sense (Hampel, 1971), i.e., resistance against arbitrary outliers." Unlike other robust regression procedures, the MSAE regression does not require a 'tuning constant.' Because the MSAE regression is resistant to outliers, it provides a good starting solution for one step and iteratively weighted multi-step least squares procedures.

The MSAE regression is less sensitive to outliers than the least squares regression. To some extent, this happens because the inconsistent observations are treated quite differently by the two procedures. The least squares regression is drawn to unusual observation(s). The MSAE regression is not generally so affected. Unusual data points often stand out in MSAE regression. It is useful to observe that the MSAE regression is to the least squares regression what the sample median is to the sample mean. Both the sample mean and the least squares regression estimator are determined and influenced by all the observations whereas the sample median and the MSAE regression estimator are also influenced by all the observations but are determined by only a subset of observations. It is well known that the value of a sample median is unaffected if the magnitude of an observation is changed such that it remains on the same side of the sample median. A similar result is true for the MSAE regression.

Since the successful application of the MSAE regression by Charnes, Cooper, and Ferguson (1955) to determine executive compensation, the MSAE regression has been used in many situations. Successful applications of the MSAE regression include estimating investment functions; setting time standards for work elements; detecting data errors in search for subsurface formations where mineral resources may exist; modeling data related to orbiting space objects; in astronomy; modeling a variety of geophysical data; estimating cost functions; analyzing seismic data; estimating pharmokinetic parameters; obtaining estimates of the state of power systems; and assessing the market value of unsold residential real estate properties. The use of the MSAE regression has been recommended in economic studies where errors with non-finite variance are more representative of random disturbance than errors with finite variance. Giloni and Padberg (2002) cited use of the MSAE regression model in predicting stock market prices. In comparative studies, the MSAE regression has performed as well as, if not better than, the least squares regression.

It is worth pointing out that the MSAE regression has been studied in several contexts under a variety of names including the least sum of absolute errors (LSAE); minimum (or least) absolute deviations, errors or values (MAD, MAE, LAD, LAV); L_1-norm, and others.

Giloni and Padberg (2002) have noted that textbooks in statistics nearly ignore the discussion of regression by MSAE. We attribute the phenomenon to the positioning of the subject. It is at the interface of statistics (regression analysis) and optimization methods (mathematical programming) and as such draws upon an extensive body of knowledge from both areas. Except for constrained least squares, seldom is linear regression by least squares presented as an optimization problem whereas regression by MSAE is always framed as a linear optimization problem. However, most students of statistics are unfamiliar with methods of mathematical programming. Consequently, textbook writers may be unwilling to allocate competing page space to the requisite background material; feel the subject is not statistically developed to warrant discussion beyond a superficial treatment as an alternative to the least squares methodology; or is an advanced topic more suitable to a monograph. Given the dearth of treatment to regression by MSAE in textbooks, literature reviews and updates to MSAE regression such as this paper become important learning resources to the student, researcher, and practitioner.

The rest of the material is organized as follows. In the next section, we discuss some of the computational algorithms and computer programs that are currently available to solve the simple and multiple MSAE regression models. In Section 10.3, we provide some results relating to MSAE statistical properties and inference procedures. In Section 10.4, we discuss variable selection procedures for MSAE regression. In Section 10.5, we describe the goodness-of-fit measure that has been developed for the MSAE regression model followed by other results in Section 10.6. In Section 10.7, we discuss the likelihood displacement function for the MSAE regression and in Section 10.8 we discuss the robustness of the MSAE fit. A recent proposal for regression under multiple criteria including MSAE is discussed in Section 10.9. We conclude the paper with a few remarks in Section 10.10.

10.2 Computational Algorithms

Let y be an $n \times 1$ vector of value of the response variable corresponding X, the $n \times k$ matrix of regressor (predictor) variable values that may include a columns of ones for the intercept term. Consider the multiple linear regression model

$$y = X\beta + \varepsilon \tag{10.1}$$

where β is the $k \times 1$ vector of the unknown parameters and ε is the $n \times 1$ vector representing the unobservable random errors (disturbances). The components of ε are independent and identically distributed random variables with density function $f(\cdot)$.

Simple Linear Regression: Let y_i denote the value of the response variable corresponding to x_i, the value of a regressor (or predictor) variable for the i-th observation, $i = 1, 2, \ldots, n$, where n is the number of observations. The simple linear regression model is

$$y_i = \beta_0 + \beta_i x_i + \varepsilon_i,\ i = 1, \ldots, n \tag{10.2}$$

where β_o and β_1 are the unknown intercept and slope parameters of the model and ε_i represents the unobservable random error.

The objective is to determine the estimators $\hat{\beta}_0$ and $\hat{\beta}_1$ of β_0 and β_1 such that $\sum_{i=1}^{n} |y_i - \hat{y}_i|$ is minimum, where $\hat{y}_i = \hat{\beta}_0 + \hat{\beta}_1 x_i$. Edgeworth (1888) proposed an algorithm to compute the MSAE estimates for the simple linear regression model; however, his method was not widely used. Seventy years later, Karst (1958) developed an intuitively appealing iterative algorithm for the problem. Since then, a number of algorithms have been proposed. They include Abdelmalek (1980), Armstrong and Kung (1978), Barrodale and Roberts (1973), Josavanger and Sposito (1983), Klingman and Mote (1982), and Wesolowsky (1981). Narula, Sposito, and Gentle (1991) reported results of a computational study of computer programs for simple linear MSAE regression.

Multiple Linear Regression: In an effort to determine compensation for executives (i.e., salary plus fringe benefits), Charnes, Cooper, and Ferguson (1955) pointed out that the MSAE regression problem is essentially a linear goal programming problem. Wagner (1959) formulated it as the following linear programming problem:

$$\begin{aligned} &\text{Minimize } 1'(e^+ + e^-) \quad \text{(LP)} \\ &\text{Subject to } X\hat{\beta} + e^+ - e^- = y, \\ &\qquad e^+, e^- \geq 0, \\ &\qquad \hat{\beta} \text{ unrestricted in sign} \end{aligned}$$

where 1 is the $n \times 1$ vector of ones, and e^+ and e^- are $n \times 1$ vectors of residuals corresponding to under- and over-prediction of y, respectively. Wagner (1959) stated that a simplex algorithm for bounded variables might solve more efficiently the dual formulation of (LP). Barrodale and Roberts (1973) proposed a special purpose algorithm for solving (LP). At present, several very efficient and effective algorithms, namely, Armstrong, Frome, and Kung (1979), Bartels, Conn, and Sinclair (1976,1978), Bloomfield and Steiger (1983), Coleman and Li (1992), Madsen and Nielsen (1993), Ruzinsky and Olsen (1989), Wesolowsky (1981), and Zhang (1993) may be used to solve the simple and multiple linear MSAE regression models.

Computer Programs: A number of computer programs, Bartels, Conn, and Sinclair (1976), Armstrong and Kung (1978), Armstrong, Frome, and Kung (1979), and Josavanger and Sposito (1983) may be used to compute the MSAE estimates of the parameters of the simple as well as the multiple linear regression models. For the simple linear regression problem, the computer program by Josavanger and Sposito (1983) based on the modification of the algorithm of Wesolowsky (1981) performed well in a comparative study conducted by Gentle, Sposito, and Narula (1988). In a study of the relative performance of computer programs for the multiple linear regression model, Gentle, Sposito, and Narula (1987, 1988) reported that within the limitations of the study, the program of Armstrong, Frome, and Kung (1979) performed best.

One may also fit the MSAE model using the robust regression package ROBSYS. Furthermore, one can write a short FORTRAN program to calculate the MSAE estimates of β in (10.1) using the IMSL (1980) subroutine *RLLAV*. At present, computer programs are also available in popular statistical packages such as S-Plus (L_1-*fit* function) and SAS (1983) (proc

IML). Therefore, it is reasonable to claim that the computational difficulties associated with the use of the MSAE regression no longer exist.

10.3 Statistical Properties and Inference

In the past, besides computational difficulties, a serious limitation to a greater application of the MSAE regression has been the lack of a well known and widely distributed statistical inference apparatus comparable to what exists for regression by least squares. However, a number of Monte Carlo studies of the small sample properties of the MSAE estimator have been conducted and reported in the literature. Asymptotic distributional results and inference procedures have also been developed.

Statistical Properties: The MSAE estimator of β in (10.1) is maximum likelihood and hence asymptotically unbiased and efficient when errors follow the Laplace distribution. Small sample properties of the MSAE estimator have been investigated extensively via Monte Carlo methods. These studies supported the thesis that the estimators are unbiased (or nearly so). However, for symmetric error distributions, Sielken and Hartley (1973) give two linear programming formulations for unbiased estimators of β under MSAE.

Based on the Monte Carlo studies, use of the MSAE regression is recommended whenever errors follow the Laplace or the Cauchy distributions, a mixture of normal and uniform distributions, or contaminated normal distribution. The efficiency of the MSAE estimator is about 80 percent even when the assumptions for the least squares procedure are satisfied.

In an extensive Monte Carlo study, Rosenberg and Carlson (1977) studied the small sample properties for the MSAE estimators in a multiple linear regression model for symmetric error distributions. They concluded that: (i) when errors followed a symmetric distribution with high kurtosis, the MSAE estimator followed an almost normal distribution and had a significantly smaller standard error than the least squares estimator and (ii) for symmetric error distributions, the MSAE estimator followed approximately multi-normal distribution with mean β and variance-covariance matrix $\tau^2(X'X)^{-1}$, where X is the design matrix and τ^2/n is the variance of the median of a sample of size n from the error distribution.

Bassett and Koenker (1978) have proven analytically that for the general linear model with independently and identically distributed errors, the

MSAE estimator is unbiased, consistent, and asymptotically follows a multi-normal distribution with variance-covariance matrix $\tau^2(X'X)^{-1}$. An important implication of this result is that the MSAE estimator has a strictly smaller confidence ellipsoid than the least squares estimator of β for any error distribution for which the sample median is a more efficient estimator of location than the sample mean. The median is a more efficient estimator of location than mean for the distributions in Table 1.

Table 1: Distributions for which the sample median is more efficient estimator of location than the sample mean

Distribution	*pdf*[1]	Var(median)	Range
Cauchy	$f(x) = 1/[\pi c\{1 + (\frac{x}{c})^2\}], c > 0$	$\frac{\pi^2 c^2}{(4n)}$	$-\infty < x < \infty$
Laplace	$f(x) = (\frac{c}{2})\ \exp\ [(-\lvert x - b\rvert)/c], c > 0$	$\frac{1}{(4nc^2)}$	$-\infty < x < \infty$
Logistic	$f(x) = \text{sech}^2(\frac{x}{c})/(2c), c > 0$	$\frac{\pi c^2}{n}$	$-\infty < x < \infty$
Symmetric stable	$f(x)$ does not exist except for the Cauchy and Normal distributions	finite[2]	$-\infty < x < \infty$

1. Although the results apply to more general cases, the pdf's reported here are such that the distributions are centered at zero.
2. This result applies to symmetric stable distributions with the characteristic exponent less than 2. Thus, the normal distribution is excluded, but the Cauchy distribution is included. The var $(\cdot)$ is finite whenever $n > \frac{4}{\alpha} + n$ where α is the characteristic exponent.

Statistical Inference: Based on the asymptotic distributional results, formulae for constructing confidence intervals and procedures for testing hypotheses related to β of (10.1) have been developed by Dielman and Pfaffenberger (1982) and Narula (1987). We give a few formulae for confidence intervals and tests of hypotheses for element β_i of β and for the linear combination $r'\beta$ of the regression parameters of (10.1) where r is a $k \times 1$ vector of known constants.

• For a single component of β, say β_i, the $(1 - \alpha)100\%$ confidence interval is

$$\hat{\beta}_i \pm z_{\alpha/2}\hat{\tau}\sqrt{(X'X)_{ii}^{-1}},$$

where $(X'X)_{ii}^{-1}$ is the i-th diagonal element of $(X'X)^{-1}$, $\hat{\beta}_i$ is the i-th component of $\hat{\beta}$, z_p denotes the $(1-p)$-th percentile of the standard normal distribution, and $\hat{\tau}$ is a consistent estimator of τ, $i = 1, 2, \ldots, k$. A number of estimators of τ have been proposed. Birkes and Dodge (1993) and McKean and Schrade (1987), estimator is

$$\hat{\tau} = \sqrt{n^*}(e_{(n^*-m+1)} - e_{(m)})/4,$$

where $m = (n^* + 1)/2 - \sqrt{n^*}, n^*$ is the number of nonzero MSAE residuals and $e_{(1)}, e_{(2)}, \ldots, e_{(n^*)}$ are the nonzero MSAE residuals arranged in ascending order.

• For $r'\beta$, the $(1 - \alpha)100\%$ confidence interval is

$$r'\hat{\beta} \pm z_{\alpha/2}\hat{\tau}\{r'(X'X)^{-1}r\}^{\frac{1}{2}}.$$

• To test the null hypothesis $H_0 : \beta_i = 0$ versus $\beta_i \neq 0$ for the single component β_i of β at the α level of significance, the decision rule is to reject H_0 whenever

$$z^* = \left| \frac{\hat{\beta}_i}{\hat{\tau}\sqrt{(X'X)^{-1}_{ii}}} \right| > z_{\alpha/2}.$$

• To test the null hypothesis, $H_0 : r'\beta = \rho$ versus the alternative hypothesis $H_1 : r'\beta \neq \rho$ at the α level of significance, the decision rule is to reject H_0 whenever

$$z^* = \left| \frac{r'\hat{\beta} - \rho}{\hat{\tau}\sqrt{r'(X'X)^{-1}r}} \right| > z_{\alpha/2}.$$

In a Monte-Carlo study, Stangenhaus and Narula (1991) found that the sampling distribution of the estimator followed a normal distribution for a sample of size as small as 10 (the smallest sample size used in the study) if the errors followed a normal distribution; for sample size 20 when the errors followed a contaminated normal distribution; and for sample sizes 100 and 200 when the errors follow the Cauchy and the Laplace distributions, respectively. They also reported results for studies of interval estimates.

The results of the study are clearly encouraging. Although the sampling distribution of the estimator may converge to normality very slowly (for certain error distributions), we can use the asymptotic properties of the estimator to construct confidence intervals and tests of hypotheses for sample sizes as small as 10.

Statistical inference procedures for small sample size have also been investigated by Dielman and Pfaffenberger (1990a, 1990b) and Dielman and Rose (1995) using Monte Carlo studies. Their results also show that one can use the statistical inference procedures based on the normal distribution for small sample sizes. Stangenhaus, Narula, and Ferreira (1993) have proposed bootstrap procedures for statistical inference purposes. Dielman (2005) provided a good discussion of bootstrap methodology as an alternative to the likelihood ratio and other tests.

10.4 Variable Selection

It is generally tacitly assumed that the k regressor of the linear model (10.1) include all relevant variables and their functions and, at times, may include a few extraneous variables and their functions. Often it is possible to select a model with $m(< k)$ variables without essentially losing any information about the response variable contained in the k predictor variables. A simplified model may also lead to a better understanding of the phenomenon under investigation. If prediction is the analyst's objective, it is well known that a model with fewer variables is more desirable. Moreover, models with fewer variables are easier to understand, to explain, and are less expensive to maintain. In fact, for economic, computational, and statistical reasons, it may be desirable to include fewer than k variables in the model.

Narula and Wellington (1979) proposed an efficient implicit enumeration algorithm to find the best model with $m(= 1, 2, \ldots, k-1)$ regressor variables. The best model with m variables is the model with the smallest value of the sum of absolute errors among all models with m variables. Their procedure guarantees the best model of $m(= 1, 2, \ldots, k-1)$ regressor variables without examining all the models. A computer program based on their algorithm appears in Wellington and Narula (1981). Suggestions for accelerating the search so that all $2^k - 2$ possible regressions of size $m = 1, \ldots, k-1$ are implicitly but not explicitly examined in the search for best model of size $m(< k)$ predictor variables appeared in Narula and Wellington (1983). Sklar (1988) also provided a procedure to find the best regressions of size $m(< k)$.

Andre, Narula, Elian, and Tavares (2003) proposed stepwise procedures for selection of variables. Their proposed automatic selection procedures included forward selection, backward elimination, and stepwise. The forward selection procedure begins with one predictor variable in the model and proceeds by adding one variable at a time until no further additions are indicated. A backward elimination procedure starts with all the variables in the model and eliminates one variable at a time until no further eliminations are indicated. The stepwise procedure adds and eliminates a variable at each step until no further additions or deletions to the model are indicated.

In most practical situations, as a rule, there does not exist a single 'best' model but rather many 'equally good' models. One possible method to select a model among a few good models is to compute the sum of predictive absolute errors (SPAE) for each model in the following way.

Omit observation i from the data set to be used for fitting the model. Fit the model using the remaining $n-1$ observations. Use this model to predict the value of the response variable for the omitted observation and calculate its residual. Return the omitted observation to the data set to be used for fitting, remove the next observation, and repeat the operation until each observation $i(=1,\ldots,n)$ has been so treated. Sum successively the residual generated at each iteration, that is, compute

$$SPAE = \sum_{i=1}^{n} |y_i - \hat{y}_{(i)}|$$

where y_i is the observed value of the response variable for omitted observation i and $\hat{y}_{(i)}$ is its predicted counterpart derived from the MSAE model fitted without observation i. Choose the model associated with the minimum SPAE. We hasten to add that this process of computing SPAE is computationally very intensive. The reader may note the similarity in producing the SPAE to jackknifing and bootstrap methods.

10.5 Coefficient of Determination

To measure the goodness of the MSAE fit, McKean and Sievers (1987) proposed the coefficient of determination R_1. Let RSAE denote the reduction in the sum of absolute errors due to the fitting of a p-variable model and observe that

$$RSAE = \sum_{i=1}^{n} |y_i - \text{ median } (y)| - SAE \tag{10.3}$$

where $\sum_{i=1}^{n} |y_i$ - median $(y)|$ denotes the sum of absolute errors for the fitted model $\hat{y}_i =$ median (y) and SAE is the sum of absolute errors for the p-variable model. Then the coefficient of determination R_1 is:

$$R_1 = RSAE|(RSAE + (n-p-1)\hat{\tau}/2) \tag{10.4}$$

where $\hat{\tau}$ is a consistent estimator of τ. It is well known that when the errors follow Laplace distribution, then

$$\hat{\tau} = SAE/n \tag{10.5}$$

the mean of the sum of absolute MSAE residuals, is the maximum likelihood and consistent estimator of τ, Engelhardt and Bain (1973). Andre, Elian, Narula, and Tavares (2000) proposed the use of $\hat{\tau} = SAE/n$ in (10.4). Because the sum of absolute residuals is a non-increasing function of p, the use of $\hat{\tau}$ in (10.4) assures that the value of R_1 increases in moving from a reduced to a full model. Thus, R_1 has all the desirable properties of a coefficient of determination, Kvalseth (1985). Large values of R_1 are desirable.

Eline, Narula, and Tavares (2000) proposed the following adjusted coefficient of determination R_{1a} for the MSAE regression model with p-variables:

$$R_{1a} = 1 - (1 - R_1)\left(\frac{n}{n-p-1}\right) \tag{10.6}$$

where R_1 is defined in (10.4). See Elian, Narula, and Tavares (2000) for other results related to use of the adjusted coefficient of determination R_{1a}.

10.6 Other Results

It is well known that the MSAE regression hyperplane passes through at least k observations, Appa and Smith (1973). These observations with zero residual value are known as the defining (or basic) observations. Observations with nonzero residual values are the nondefining (or nonbasic) observations.

The MSAE estimate of β is completely determined by the defining observations. In particular, the system of equations corresponding to MSAE regression may be written as

$$\begin{bmatrix} y_{(1)} \\ y_{(2)} \end{bmatrix} = \begin{bmatrix} X_{(1)} & 0 \\ X_{(2)} & I^* \end{bmatrix} \begin{bmatrix} \hat{\beta} \\ e_{(2)} \end{bmatrix},$$

where subscripts (1) and (2) denote respectively sets of indices for the defining and nondefining observations. With this ordering, the k observations with subscripts in (1) lie on the MSAE regression hyperplane. Further, let $X_{(1)}$ denote the $k \times k$ matrix containing the values of the predictor variables for the defining observations; $X_{(2)}$ indicate the $(n-k) \times k$ matrix containing the values of the predictor variables for the nondefining observations; $y_{(1)}$ be the $k \times 1$ vector of values of the response variable for the defining observations; $y_{(2)}$ be the $(n-k) \times 1$ vector of values of the response variable for the nondefining observations; vector $e_{(2)}$ be the $(n-k) \times 1$ vector with component i $(= |y_i - \hat{y}_i|$ if $y_i - \hat{y}_i \neq 0$, $i = 1, \ldots, n)$ and I^* is the $(n-k) \times (n-k)$ diagonal matrix with diagonal elements that are either $+1$ (if $y_i - \hat{y}_i > 0$) or -1 (if $y_i - \hat{y}_i < 0$); and y_i is the i-th value of the response variable and $\hat{y}_i$ is its MSAE predicted counterpart, $i = 1, \ldots, n$.

It then follows that

$$\hat{\beta} = X_{(1)}^{-1} y_{(1)}$$

and

$$e_{(2)} = I^* y_{(2)} - I^* X_{(2)} X_{(1)}^{-1} y_{(1)}.$$

The hat matrix for the MSAE regression is:

$$\begin{bmatrix} \hat{y}_{(1)} \\ \hat{y}_{(2)} \end{bmatrix} = \begin{bmatrix} I \\ X_{(2)} X_{(1)}^{(-1)} \end{bmatrix} y_{(1)} \equiv H y_{(1)}$$

where I is a $k \times k$ identity matrix, Narula and Wellington (1985). For the defining observations, the $h_{ii} = 1$, $i = 1, \ldots, k$ of the hat matrix H and may be considered influential. However, they are not influential in the same way as the observations associated with the large diagonal elements of the hat matrix in the least squares analysis, Hoaglin and Welsch (1978).

Although diagnostic statistics are available for the least squares model, few techniques have been developed for the MSAE model.

Using the Box-Cox transformation, Parker (1988) proposed a method to assess the need for transformation in MSAE regression. It has been shown that the MSAE regression is more sensitive to leverage points, i.e., outliers in the direction of the predictor (or regressor) variable, than the least squares regression, Rousseeuw and Leroy (1987).

It is also known that the fitted MSAE regression model is invariant to changes in the value of a response variable for a nondefining observation as long as it remains on the same side of the fitted MSAE regression hyperplane, Narula and Wellington (1985). The fitted model is also invariant to certain changes in the value of a predictor variable for a nondefining observation. Because the fitted MSAE regression model is determined by a subset of $k < n$ defining observations, certain variations in the values of the nondefining observations leave the parameter estimates unchanged. This feature is discussed and illustrated in Section 10.8.

Non-linear regression by MSAE is reviewed in Dielman (2005) and treated in Gonin and Money (1989). The latter treated the subject within the context of L_p-norm estimation with MSAE as the special case ($p = 1$) and much of the discussion is given to algorithm development.

10.7 Likelihood Displacement

Ellis and Morgenthaler (1992) point out that, at present, leverage does not have a precise meaning. Vaguely stated, a design point far from the bulk of the others is called a leverage point. It is important to distinguish leverage points from *influential points.* An observation taken at a leverage point has the potential to influence the fit, but it does not necessarily do so.

Cook, Pea, and Weisberg (1988) proposed the likelihood displacement function as a unifying principle for influence measure. They pointed out that if desirable, this displacement can be transformed to a more familiar scale and compared to percentiles of a chi-squared distribution with k degrees of freedom.

To determine the influence of the i-th observation, $i = 1, \dots, n$, Cook, Pea, and Weisberg (1988) suggested the following likelihood displacement function

$$LD_i(\theta) = 2[L(\hat{\theta}, y) - L(\hat{\theta}_{(i)}, y)],$$

where $\hat{\theta}$ is the maximum likelihood estimator of based on all the observations and $\hat{\theta}_{(i)}$ is the maximum likelihood estimator of θ based on all the observations except observation i. According to Cook, Pea, and Weisberg (1988), if this function is large, observation i is influential and deleting it may cause a substantial change in important results.

It is well known that MSAE estimators are maximum likelihood estimators of the parameter β, when the errors ε_i/s in (10.1) follow Laplace distribution with mean equal to zero and variance equal to $2\tau^2$, i.e., the probability density function of y_i is given by

$$f(y_i) = 1/(2\tau) \exp\left(-|y_i - x_i\beta|/\tau\right), \quad -\infty < y_i < \infty, \quad i = i, \dots, n.$$

The log likelihood function L is

$$L(\beta, \tau) = -n \ln(2\tau) - \sum_{i=1}^{n} |y_i - x_i\beta|/\tau,$$

and the maximum likelihood estimator $\hat{\tau}$ of τ is

$$\hat{\tau} = \sum_{i=1}^{n} |y_i - x_i\hat{\beta}|/n = MSAE/n$$

For our problem $\theta = (\beta, \tau)$ and the likelihood displacement function is

$$LD_i(\beta, \tau) = 2[L(\hat{\beta}, \hat{\tau}) - L(\hat{\beta}_{(i)}, \hat{\tau}_{(i)})],$$

where $\hat{\beta}_{(i)}$ is the MSAE estimator of β and $\hat{\tau}_i$ of τ based on all the observations except observation i.

The likelihood displacement function for the i-th observation can be written as:

$$LD_i(\beta, \tau) = 2(n \ln(\hat{\tau}_{(i)}/\hat{\tau}) + |y_i - x_i\hat{\beta}_{(i)}|/\hat{\tau}_{(i)} - 1). \qquad (10.7)$$

The measure in (10.7) will be large if $|y_i - x_i\hat{\beta}_{(i)}|/\hat{\tau}_{(i)}$ is large, or $(\hat{\tau}_{(i)}/\hat{\tau})$ is large or both.

When the i-th observation is not influential, then the estimate $\hat{\tau}_{(i)}$ without the i-th observation should be very similar to $\hat{\tau}$, i.e., $\hat{\tau}_{(i)}/\hat{\tau} \cong 1$, and $|y_i - x_i\hat{\beta}_{(i)}|$ should be close to the mean of the absolute errors, $\hat{\tau}_{(i)}$, i.e., $|y_i - x_i\hat{\beta}_{(i)}| \cong \hat{\tau}_{(i)}$, and so $LD_i \cong 0$. That is, when the i-th observation is not influential, the likelihood displacement function will be close to zero. However, large values of the function suggest that the observation might be influential. Clearly, $LD_i(\beta, \tau)$ takes both β and τ into consideration.

To determine if the i-th observation is influential only for the estimation of β, a measure based on the method proposed by Cook et al (1988) is

$$LD_i(\beta||\tau = 2n \ln \left(\frac{\sum_{j=1}^{n} |y_j - x_j\hat{\beta}_{(i)}|}{\sum_{j=1}^{n} |y_j - x_j\hat{\beta}|} \right). \tag{10.8}$$

That is, we compare the sum of absolute value of the residuals when they are calculated using the MSAE estimator of β computed with and without the i-th observation. If the deletion of the i-th observation changes the estimates such that the sum of the absolute residuals does not change much then $LD_i(\beta|\tau)$will be close to zero.

It is interesting to note that

$$LD_i(\beta|\tau) = 2n \ln(1 + \lambda_{(i)}),$$

where $\lambda_{(i)} = \frac{\sum_{j=1}^{n} |y_j - x_j\hat{\beta}_{(i(}| - \sum_{j=1}^{n} |y_j - x_j\hat{\beta}}{\sum_{j=1}^{n} |y_j - x_j\hat{\beta}|}$, which may be interpreted as relative increase in the MSAE when $\hat{\beta}$ is substituted for $\hat{\beta}_{(i)}$.

Cook, Pena, and Weisberg (1988) pointed that the values of $LD_i(\beta, \tau)$, and $LD_i(\beta|\tau)$ may be compared with the percentiles of the chi-squared distribution with degrees of freedom k and $k - 1$, respectively, to decide whether an observation is influential or not. Although the necessary regularity conditions are not satisfied for Laplace distribution, Cox and Hinkley (1974) and Basset and Koenker (1978) have shown that the likelihood ratio behaves in the usual way.

Elian, Andre, and Narula (2000) used the Laplace distribution to develop the preceding influence measures since it has been assumed for the error distribution in many applications, Engelhardt and Bain (1973). Further the use of MSAE regression has been recommended whenever one suspects outliers in the response variable, that is, if the error distribution has thick tails. Therefore, even if the error distribution is not exactly Laplace, it may be close enough so that the proposed influence measures would be useful.

10.8 Robustness

It is useful to observe that the MSAE regression is to the least squares regression what the sample median is to the sample mean. Both the sample mean and the least squares estimates of β are determined and influenced by all the observations whereas the sample median and the MSAE estimates are also influenced by all observations but determined by only a subset of observations. Just as the value of the sample median is unaffected if the magnitude of an observation changes such that it remains on the same side (either above or below) of the sample median, a similar result holds true for the MSAE regression, Narula and Wellington (1985). The fitted MSAE regression model remains unchanged if the value of the response variable for a nondefining observation changes such that the observation remains on the same side of the fitted MSAE regression hyperplane and no other change occurs in the original data. This is unlike least squares regression where any change in the value of the response variable for any observation changes the values of the parameter estimates.

The MSAE regression estimate of β also remains unchanged if the value of a predictor variable for a nondefining observation varies within certain interval(s) and no other change occurs in any of the original data, the ceteris paribus assumption. Procedures to compute such intervals for the simple linear MSAE regression model are given in Narula, Sposito, and Wellington (1993). These intervals give the analyst useful information about the variation in the value of each predictor variable within each nondefining observation that could be accommodated without changing the MSAE regression estimate of β, ceteris paribus. Recently, Narula and Wellington (2002) have extended the investigation, called interval analysis, to the multiple linear regression model under MSAE. We illustrate interval analysis with data from Draper and Stoneman (1966) that is further analyzed in Hoaglin and Welsch (1978) and Narula and Wellington (1985). The fitted MSAE regression model for the data is

$$\hat{y}_i = 9.084 + 9.189x_{i1} - 0.171x_{i2},\ i = 1, \ldots, 10 \tag{10.9}$$

The MSAE parameter estimates $9.084, 9.189$, and -0.171 are determined by the x-data and y-data of the defining observations $\{3, 8, 9\}$. Table 2 is a display of the variation in each x-datum of each nondefining observation $\{1, 2, 4, 5, 6, 7, 10\}$ that maintains the values of the MSAE parameter estimates, assuming no other changes in the original data. For x_{12}, the value of predictor variable 2 in observation 1, if an alternate value of interest under any condition were in the interval [10.956, 11.832],

Table 2: Alternate Values of the Predictor Variables for the Nondefining Observations That Maintain the MSAE Fit for the Wood Beam Data

Obs. i	Lower Bound	X_1 Datum	Upper Bound	Lower Bound	X_2 Datum	Upper Bound	MSAE Residual	Y Datum
1	0.488	0.499	0.584	10.956	11.1	11.832	-0.632	11.14
2	0.473	0.558	0.563	8.609	8.8	9.044	0.050	12.74
3[1]	-	0.604	-	-	8.8	-	0.	13.13
4	0.431	0.441	0.526	8.756	8.9	9.515	-0.105	11.51
4				11.285	8.9	12.032	-0.105	11.51
5	0.539	0.550	0.635	8.656	8.8	9.532	-0.254	12.38
6	0.443	0.528	0.539	9.168	9.9	10.044	0.356	12.60
7	0.333	0.418	0.422	10.502	10.7	10.844	0.034	11.13
8[1]	-	0.480	-	-	10.5	-	0.	11.70
9[1]	-	0.406	-	-	10.5	-	0.	11.02
10	0.326	0.467	0.361	10.556	10.7	11.432	-0.136	11.41
10	0.456	0.467	0.552				-0.136	11.41

1 Defining observation. Hence, variation in any datum would change the MSAE parameter estimates.

the MSAE parameter estimates given in (10.9) would be unchanged, ceteris paribus. For alternate values of predictor variable 2 in observation 4, there are two disjointed intervals. Ceteris paribus, an alternate value for x_{42} could be in the interval [8.756, 9.515] or [11.285, 12.032] and produce the MSAE model given in (10.9). Alternate values for predictor variable 1 in observation 10 also occur in two disjointed intervals.

The robust features of the fitted MSAE regression model discovered in interval analysis are useful when the analyst is interested in assessing the stability of the MSAE regression model, that is, identifying where small variations in the data used for fitting could or could not produce large changes in the model. Narrow intervals should move the analyst to check that the associated data were accurately collected, recorded, and transmitted. Small variations in any such datum could alter the MSAE regression. Wide intervals give some assurance that small variations in the associated x-datum would not affect the model. Interval analysis assists in this examination by providing the analyst with the specificity of allowable variation. For example, the value $x_{12}(= 11.1)$ was used to produce the model given in (10.9). If the alternate value $x_{12} = 10.6$ were of interest to the analyst, the model of (10.9) would change to

$$\hat{y}_i = 8.243 + 9.864x_{i1} - 0.122x_{i2}, \; i = 1, \ldots, 10 \qquad (10.10)$$

and the analyst would have to determine the significance of the changes in the MSAE parameter estimates. However, note that the change of -0.5 in the original value of $x_{12}(= 11.1)$ changed the model whereas the change of $+0.5(x_{12} = 11.6)$ would leave the result given in (10.9) unchanged. If the analyst discovered that the collection, recording, or transmitting of the datum x_{12} allowed variation from whatever source as great as 0.5, he would know through the results of interval analysis in which case the error would be problemsome. In general, the intervals are not symmetric with respect to the original value of the datum. Observe also that the value $(= 9.9)$ of predictor variable 2 in observation 6 is close to the upper bound $(= 10.044)$ of variation in x_{62} that maintains the result given in (10.9). The analyst may want to confirm its accuracy; otherwise the MSAE parameter estimates could be different.

Table 3: Alternate Values of the Response Variable for the Nondefining Observations That Maintain the MSAE Fit of the Wood Beam Data

Index i	Lower Bound	y_i	Upper Bound	MSAE Residual
1	$-\infty$	11.14	11.772	-0.632
2	12.690	12.74	∞	0.050
4	$-\infty$	11.51	11.615	-0.105
5	$-\infty$	12.38	12.634	-0.254
6	12.244	12.60	∞	0.356
7	11.096	11.13	∞	0.034
10	$-\infty$	11.41	11.546	-0.136

Table 3 is taken from Narula and Wellington (2002) and displays the results of interval analysis for the values of the response variable in the Wood beam data. Note the relationship between the finite value of the lower or upper bounds and the value as well as sign of the MSAE residual for each $y_i, i = 1, 2, 4, 5, 6, 7, 10$, of the nondefining observations. In each case, the finite bound is the predicted value of the response variable obtained from the fitted MSAE model in (10.9), that is $y_i - e_i = \hat{y}_i$, $i = 1, 2, 4, 5, 6, 7, 10$. The other bound has the same sign as the MSAE residual. This feature holds in general for MSAE regression and accounts for the property that the value of the response variable for each nondefining observation can change indefinitely as long as its residual value does not change sign, ceteris paribus. This means geometrically that the observation under the variation of interest does not cross to the opposite side of the fitted MSAE regression plane. Although the variation in the value of the response variable for each nondefining observation can change indefinitely in one direction, the change in the value of a predictor variable for any nondefining observation always has real bounds. The illustration helps in understanding the variation in the original data that can be accommodated without changing the MSAE parameter estimates of β in (10.1). Interval analysis provides some insight to the robust feature of MSAE regression.

Ha and Narula (1989) developed tolerance limits that indicate the maximum amount by which the values of the response variable for all non-defining observations may be simultaneously changed without affecting the MSAE parameter estimates.

The MSAE regression result also provides a good starting solution for a number of robust regression procedures. We refer the interested reader to Arthanari and Dodge (1981), Bloomfield and Steiger (1983), Dielman (2005), Dodge (1987a, 1987b, 1992, 1997), Gentle (1977), Narula and Wellington (1982), and Sposito, Smith, and McCormick (1978).

Narula, Saldiva, Andre, Elian, Ferreira, and Capelozzi (1999) discussed MSAE regression as a robust alternative to least squares regression.

10.9 Multiple Criteria Linear Regression Using MSAE

Recently, Narula and Wellington (2007) proposed a very different approach to linear regression, that is, the use of multiple criteria including MSAE. When loss due to nonzero error of prediction is related to absolute as well as relative errors, joint criteria such as MSAE and minimization of the sum of absolute relative errors (MSRE) may produce useful alternative models for the analyst's and decision maker's consideration. Narula and Wellington (2007) illustrated the approach with a model for predicting the market value of unsold residential property based on recently sold property. The model is estimated with simultaneous consideration of MSAE and MSRE. The rationalization of the approach is based on the loss/gain of tax revenues and the expected increase/decrease in homeowner complaints that arise from the under-assessments (positive residual errors) and over-assessments (negative residual errors) of property values produced by the model. Narula and Wellington (2007) related loss/gain in tax revenues to absolute errors and expected increase/decrease in homeowner complaints to relative errors. The taxing authority, the decision maker in this scenario, is interested in minimizing the loss in tax revenue due to under-assessments and in the number of complaints to be adjudicated due to over-assessments.

10.10 Concluding Remarks

Development of computational procedures, computer programs, inference, diagnostics, and understanding of the special properties of MSAE regression continues. And so does the dissemination of those results. More students, teachers, researchers, and practitioners than ever know how to compute, analyze, and infer under MSAE. Some understand the experimental circumstances in which MSAE regression model is to be preferred to least squares. However, more need to know and much remains to be done. There is an urgent and important need to develop residual and diagnostic analyses comparable to what is available for least squares regression and to assess the consequences of multicollinearity in MSAE regression. Although work is currently underway to expand interval analysis to multiple and simultaneous variations in the data used for fitting, more needs to be understood about the sensitivity/insensitivity of the MSAE fit to a wide variety of data perturbations. In time, the invariance feature of the MSAE fit may become one of the strongest attractions of researchers and practitioners to linear regression analysis under MSAE. To happen, the inference procedures and diagnostics must be simplified, more widely known and understood, and applied with good outcome.

Bibliography

Abdelmalek, N. N. (1980). L_1 solution of overdetermined system of linear equations, *ACM Transactions on Mathematical Software* **6**, pp. 220–227.

Andre, C. D. S., Elian, S. N., Narula, S. C. and Tavares,R. A. (2000). Coefficient of determination for variable selection in the MSAE regression, *Communications in Statistics-Theory and Methods* **29**, pp. 623–642.

Andre, C. D. S., Narula, S. C., Elian, S. N. and Tavares, R. A. (2003). An overview of the variable selection methods for the minimum sum of absolute errors regression, *Statistics in Medicine,* **22**, pp. 2101–2111.

Appa, G. and Smith, C. (1973). On L_1 and Chebyshev estimation, *Mathematical Programming* **5**, pp. 73–87.

Armstrong, R. D., Frome, E. L. and Kung, D. S. (1979). A revised simplex algorithm for the absolute deviation curve fitting problem, *Communications in Statistics: Simulation and Computation* **B8(2)**, pp. 175–190.

Armstrong, R. D. and Kung, M. T. (1978). Algorithm AS 132: Least absolute value estimates for a simple linear regression problem. *Applied Statistics* **27**, pp. 363–366.

Arthanari, T. S. and Dodge, Y. (1981). *Mathematical Programming in Statistics,* (John Wiley).

Barrodale, I. and Roberts, F. D. K. (1973). An improved algorithm for discrete $L1$ linear approximation, *SIAM J. Numerical Analysis* **10**, pp. 839–848.

Barrodale, I. and Roberts, F. D. K. (1974). Algorithm 498: Solution of an overdetermined system of equations in the L_1 norm, *Communications of the ACM* **17**, pp. 319–320.

Bartels, R. H., Conn, A. R. and Sinclair, J. W. (1976). *A FORTRAN program for solving overdetermined systems of linear equations in the L_1 sense*, Tech. Rep. No. 236, Math. Sc. Dept., John Hopkins University.

Bartels, R. H., Conn, A. R. and Sinclair, J. W. (1978). Minimization techniques for piecewise differentiable functions: The L_1 solution to an overdetermined linear system. *SIAM Journal of Numerical Analysis* **15**, pp. 224–241.

Basset, G. and Koenker, R. (1978). Asymptotic theory of least absolute error regression. *Journal of the American Statistical Association* **73**, pp. 618–622.

Birkes, D. and Dodge, Y. (1993). *Alternative Methods of Regression*, (John Wiley).

Bloomfield, P. and Steiger, W. (1983). *Least Absolute Deviations: Theory, Applications and Algorithms*, (Birkhauser).

Boscovich, R. J. (1757). De litteraria expeditione per pontificiam detionem, et synopsis amplioris operis, ac habentur plura ejus ex exemplaria etiam sensorum impressa, *Bon. Sci. et Art. Inst. Atque Acad. Comm.,* **4**, pp. 353–396.

Boscovich, R. J. (1760). De recentissimis graduum dimensionibus, et figura, ac magnitudine terrae inde derivanda, *Philosophiae Recentioris, a Benedicto Stai in Romano Archigynasis Publico Eloquentare Professore, vesibus traditae, Libri X, cum adnotianibus et Supplementis P. Rogerii Joseph Boscovich,* S. J., **2**, pp. 406–426.

Charnes, A., Cooper, W. W. and Ferguson, R. D. (1955). Optimal estimation of executive compensation by linear programming, *Management Science* **1**, pp. 138–151.

Coleman, T. F. and Li, Y. (1992). A globally and quadratically convergent affine scaling method for linear L_1 problems. *Mathematical Programming* **56**, pp. 189–222.

Cook, R. D., Pena, D. and Weisberg, S. (1988). The likelihood displacement: A unifying principle for influence measures, *Communications in Statistics-Theory and Method* **17**, pp. 623-640.

Cox, D. R. and Hinkley, D. V. (1974). *Theoretical Statistics,* (Chapman-Hall).

Dielman, T. and Pfaffenberger, R. (1982). LAV (Least absolute value) estimation in the regression model: A review, *TIMS Studies in the Management Sciences* **19**, pp. 31–52.

Dielman, T. and Pfaffenberger, R. (1990a). Tests of linear hypothesis in LAV regression. *Communications in Statistics: Simulation and Computation* **19**, pp. 1179–1199.

Dielman, T. and Pfaffenberger, R. (1990b). A further comparison of tests of hypotheses in LAV regression, *Computational Statistics and Data Analysis* **14**, pp. 375–384.

Dielman, T. and Rose, E. L. (1995). A bootstrap approach to hypothesis testing in least absolute value regression, *Communications in Statistics: Simulation and Computation* **20**, pp. 119–130.

Dielman, T. (2005). Least absolute value regression: Recent contributions. *Journal of Statistical Computation and Simulation* **75**, pp. 263–286.

Dodge,Y. (1987a). *Statistical Data Analysis: Based on the L_1-norm and Related Methods* (North-Holland, Amsterdam).

Dodge, Y. (l987b). An introduction to L_1-norm based statistical data analysis, *Computational Statistics and Data Analysis* **5**, pp. 239–253.

Dodge, Y. (1992). *L_1-Statistical Analysis and Related Methods*, (North-Holland).

Dodge, Y. (1997). *L_1-Statistical Analysis and Related Methods*, Institute of Mathematical Statistics, Lecture Notes and Monograph Series, 31, (Hayward).

Draper, N. R. and Stoneman, D. M. (1966). Testing for the inclusion of variables in linear regression by a randomization technique, *Technometrics* **8**, pp. 695–699.

Edgeworth, F. Y. (1887). On observations relating to several quantities. *Hermathena* **6**, pp. 279–285.

Edgeworth, F. Y. (1888). On a new method of reducing observations relating to several quantities. *Philosophical Magazine* **25**, pp. 184–191.

Elian, S. N., Andre, C. D. S. and Narula, S. C. (2000). Influence measure for the L_1 regression. *Communications in Statistics-Theory and Methods* **29**, pp. 837–849.

Eisenhart, C. (1961). *Boscovich and the combination of observations*, In Roger Joseph Boscovich, (L. L. Whyte, ed.), pp. 200-212. (Allen & Unwin, London, Reprinted in Kendall and Plackett in 1977).

Ellis, S. P. and Morgenthaler, S. (1961). Leverage and breakdown in L_1 regression, *Journal of the American Statistical Association* **87**, (1992), pp. 143–148.

Engelhardt, M. and Bain, L. (1973). Some complete and censored sampling results for the Weibull or extreme-value distribution, *Technometrics* **15, 3**, pp. 541–549.

Farebrother, R. W. (1987). The historical development of the L_1 and L_∞ estimation procedures, *Statistical Data Analysis Based on the L_1-norm and Related Methods* (Y. Dodge, editor), North-Holland, Amsterdam, pp. 37–64.

Gauss, C. F. (1809). *Theoria Motus Corporum Coelestium in Sectionibus Conicis Solum Ambientium, Hamburg,* (Frid. Perthes et. I. H. Besser).

Gauss, C. F. (1821),(1823).*Theoria Combinationis Observationum Erroribus Minimum Obnoxiae. Commentiones Societatis Regiae Scientiarum Gottingensis Recentiores,* 5, German summary, Gottingische Gelehrte Anzeigen , pp. 321–327 and pp. 313–318.

Gauss, C. F. (1826), (1828).*Supplementum Theoriae Combinationis Observati onum Erronibus Minimis Obnoxiae, Commentiones Societatis Regiae Scientiarum Gottingensis Recentiores* 6, German summary, Gottingiesche Gelehrte Anzeigen , pp. 1521–1527.

Gentle, J. E. (1977). Least absolute values estimation: An introduction, *Communications in Statistics - Simulation and Computation* **B6**, 313-328.

Gentle, J. E., Sposito, V. A. and Narula, S. C. (1987). Algorithms for unconstrained L_1 linear regression. *Statistical Data Analysis Based on the L_1-Norm and Related Methods* (Y. Dodge, editor), North-Holland, Amsterdam, Holland, pp. 83–94.

Gentle, J. E., Sposito, V. A. and Narula, S. C. (1988). Algorithms for unconstrained L_1 simple linear regression, *Computational Statistics and Data Analysis* **6**, pp. 335–339.

Giloni, A. and Padberg, M. (2002). Alternative methods of linear regression, *Mathematical and Computer Modelling* **35**, pp. 361–374.

Gonin, R. and Money, A. (1989). Nonlinear L_p-norm estimation, Marcel Dekker.

Ha, C. D. and Narula, S. C. (1989). Perturbation analysis for the minimum sum of absolute errors regression, *Communications in Statistics - Simulation and Computation* **B18, 3**, pp. 957–970.

Hampel, F. R. (1971). A general qualitative definition of robustness, *Annals of Mathematical Statistics* **42**, pp. 1887–1896.

Hoaglin, D. C. and Welsch, R. E. (1978). The hat matrix in regression and ANOVA, *The American Statistician* **32**, pp. 17–22.

Huber, P. J. (1974). Comment on "Adaptive robust procedures: A partial review and some suggestions for future applications and theory", by R.V. Hogg, *Journal of the American Statistical Association* **69**, pp. 926–927.

Josvanger, L. A. and Sposito, V. A. (1983). L_1-norm estimates for the simple linear regression. *Communications in Statistics- Simulation and Computation* **12**, pp. 215–221.

Karst, O. J. (1958). Linear curve fitting using least deviations, *Journal of the American Statistical Association* **53**, pp. 118–132.

Kvalseth, T. O. (1985). Cautionary note about R_2, *The American Statistician* **39**, 4, November, pp. 279–285.

Klingman, D. and Mote, J. (1982). Generalized network approaches for solving least absolute value and Tchebycheff regression problems, *TIMS Studies in Management Sciences* **19**, pp. 53–66.

Laplace, P. S. (1792). Sur les degres mesures des meridiens, et sur les longueurs observees sur pendule. *Histoire de L'Academic Royale des Inscriptions et Belles Lettres, avec les Memoires de Litterature Tirez des Registres de Cette Academie, Annee 1789* Paris.

Laplace, P. S. (1818). *Deuxieme Supplement de la Theorie Analytique des Probabilites Courcies*, Paris.

Madsen, K. and Nielsen, H. B. (1993). A finite smoothing algorithm for linear L_1 estimation, *SIAM Journal on Optimization* **3**, pp. 223–235.

McKean,J. W. and Schrader, R. M. (1987). Least absolute errors analysis of variance, *Statistical Data Analysis Based on the L_1-Norm and Related Methods* (Y. Dodge, editor), Elsevier Science Publishers, pp. 297–305.

McKean, J. W. and Sievers, G. L. (1987). Coefficient of determination for least absolute deviation analysis, *Statistics and Probability Letters* **5**, pp. 49–54.

Narula, S. C. (1987). The minimum sum of absolute errors regression, *Journal of Quality Technology* **19** (1), pp. 37–45.

Narula, S. C., Saldiva, P. H. N., Andre, C. D. S., Elian, S. N., Ferreira, A. F. and Capelozzi, V. (1999). The minimum sum of absolute errors regression:

A robust alternative to the least squares regression, *Statistics in Medicine* **18**, pp. 1401–1417.

Narula, S. C., Sposito, V. A. and Gentle, J. E. (1991). Comparison of computer programs for simple linear L_1 regression, *Journal of Statistical Computation and Simulation* **39** (1 & 2), pp. 63–68.

Narula, S. C., Sposito, V. A. and Wellington, J. F. (1993). Intervals which leave the minimum sum of absolute errors regression unchanged, *Applied Statistics* **42** (2), pp. 369–378.

Narula, S. C. and Wellington, J. F. (1979). Selection of variables in linear regression using the minimum sum of weighted absolute errors criterion, *Technometrics* **21** (3), pp. 299–306.

Narula, S. C. and Wellington, J. F. (1982). The minimum sum of absolute errors regression: A state of the art survey, *International Statistical Review* **50** (3), pp. 317–326.

Narula, S. C. and Wellington, J. F. (1983). Selection of variables in linear regression: A pragmatic approach, *Journal of Statistical Computation and Simulation* **17**, pp. 159–172.

Narula, S. C. and Wellington, J. F. (1985). Interior analysis for the minimum sum of absolute errors regression, *Technometrics,* **27** (2) pp. 181–188.

Narula, S. C. and Wellington, J. F. (2002). Sensitivity analysis for the predictor variable in the MSAE regression, *Computational Statistics and Data Analysis* **40**, pp. 355–373.

Narula, S. C. and Wellington, J. F. (2007). Multiple criteria regression analysis, *European Journal of Operational Research* **181** (2), pp. 767–772.

Parker, I. (1988). Transformations and influential observations in minimum sum of absolute errors regression, *Technometrics* **30**, pp. 215–220.

Rosenberg, B. and Carlson, D. (1977). A simple approximation of the sampling distribution of least absolute residuals regression estimates, *Communications in Statistics* **B6**, pp. 421–437.

Rousseeuw, P. and Leroy, A. (1987). *Robust Regression and Outlier Detection*, (John Wiley).

Ruzinsky, S. and Olsen, E. (1989). L_1 and L_∞ minimization via a variant of Karmarkar's algorithm, *IEEE Transactions on Acoustics, Speech, and Signal Processing* **37**, pp. 245–253.

SUGI Supplemental Library User's Guide, 1983 ed. (SAS Institute, Inc).

Sielken, R. H. and Hartley, H. O. (1973). Two linear programming algorithms for unbiased estimation of linear model, *Journal of the American Statistical Association* **68**, pp. 639–641.

Sklar, M. G. (1988). Extensions to a best subset algorithm for least absolute value estimation, *American Journal of Mathematical and Management Sciences* **8**, 1-58.

Sposito, V. A., Smith, W. C. and McCormick, C. (1978). *Minimizing the Sum of Absolute Deviations*, (Vandenhoeck and Ruprecht, Gottingen, Germany).

Stangenhaus, G. and Narula, S. C. (1991). Inference procedures for the L_1 regression, *Computational Statistics and Data Analysis* **12** (1), pp. 79–85.

Stangenhaus, G., Narula, S. C. and Ferreira, P. Fo. (1993). Bootstrap confidence intervals for the minimum sum of absolute errors regression, *Journal of Statistical Computation and Simulation* **48**, 127-133.

Wagner, H. M. (1959). Linear programming techniques for regression analysis, *Journal of the American Statistical Association* **54**, pp. 206–212.

Wellington, J. F. and Narula, S. C. (1981). Variable selection in multiple linear regression using the minimum sum of weighted absolute errors criterion, *Communications in Statistics - Simulation and Computation* **B10** (6), pp. 641–648.

Wesolowsky, G. O. (1981). A new descent algorithm for the least absolute value regression problem, *Communications in Statistics - Simulation and Computation* **Bl0**, pp. 479–481.

Zhang, Y. (1993). Primal-dual interior point approach for computing L_1 solutions and L_∞ solutions of overdetermined systems, *Journal of Optimization Theory and Applications* **77**, pp. 323–341.

Chapter 11

Hedging against the Market with No Short Selling

Stephen A. Clark
Department of Statistics
University of Kentucky
Lexington, Kentucky 40506-0027
e-mail: saclar@ms.uky.edu
Cidambi Srinivasan
Department of Statistics
University of Kentucky
Lexington, Kentucky 40506-0027
e-mail: srini@ms.uky.edu

Abstract

Consider a stochastic securities market model with a finite state space and a finite number of trading dates. We study how arbitrage price theory is modified by a no short-selling constraint. The principle of No Arbitrage is characterized by the existence of an equivalent supermartingale measure. If we measure present value as conditional expectations after an equivalent change of measure, then the fundamental value of a security might fall below its market value, leading to the possibility of a price bubble. We show that the Law of One Price holds for marketed claims if and only if there exists an equivalent martingale measure. The latter condition indicates that price bubbles are fragile. Given that the Law of One Price prevails, then a contingent claim has a unique fundamental value if and only if it is the difference of two marketed claims. The main tool for arbitrage analysis in this essay is finite-dimensional LP duality theory.

Key Words: Arbitrage, bubble, fundamental value, hedging prices, martingale, short-selling, supermartingale

11.1 Introduction

This essay examines the valuation problem for a contingent claim delivered in the future from the viewpoint of arbitrage price theory. In brief, an arbitrage opportunity is a simple free lunch, i.e. a trading strategy which provides a sure profit without risk of loss. This theory is built on the premise that prices are adjusted by the actions of arbitrageurs until all arbitrage opportunities are eliminated from the market. The consequences of No Arbitrage (NA) have been thoroughly analyzed in perfect markets, leading to a linear valuation theory for contingent claims [e.g. Kreps, 1981; Clark, 1993]. In the context of dynamic securities markets, a linear valuation operator typically possesses an elegant representation as an expectations operator with respect to an equivalent martingale measure [e.g. Harrison and Kreps, 1979; Harrison and Pliska, 1981]. Nevertheless, this theory is less understood in the presence of market frictions. The purpose here is to analyze the effect of just one trading constraint, no short-selling, upon a securities market.

The securities market model described in the next section has a relatively simple mathematical structure. There are only a finite number of securities available for trading at a finite number of dates. All prices and dividends are measured relative to the price of a 'riskless' treasury bill that serves as money. Furthermore, the information filtration is created by finite partitions of the state space and is the same for all market investors. The price and dividend processes are adapted. A *feasible portfolio* consists of a non-negative number of shares of each security and an arbitrary number of treasury bills. Thus, an investor cannot short-sell securities, but he is allowed to borrow or lend money. Future portfolios must be designed just before future prices and dividends are announced. Thus, they depend only on the information available in the preceding time period. In this context, the principle of NA asserts that it is impossible to design a feasible portfolio so that its earnings are non-negative in all states and positive in some states. The main result of this section is that NA holds if and only if there exists an *equivalent supermartingale measure*, i.e. an equivalent probability measure under which the price plus cumulative dividend process is a supermartingale. This result includes the Harrison-Pliska [1981] Theorem as a special case. In fact, it is easy to show that an equivalent supermartingale measure reduces to an equivalent martingale measure whenever the short-selling constraint is not binding.

The third section measures the *fundamental value* of a contingent claim as its expectation with respect to an equivalent supermartingale measure. Suppose the return from one share of a security is given by its terminal share price plus cumulative dividend. Then the supermartingale property implies that the fundamental value of the security's return does not exceed initial share price. The residual between market value and fundamental value is a *price bubble* on the security. Notice that the price bubble is zero if and only if the supermartingale property reduces to a martingale property. This conclusion is somewhat surprising in view of the literature on price bubbles. Sequential equilibrium theory typically associates the emergence of price bubbles with an infinite number of trading dates [e.g. Santos and Woodford, 1997; Loewenstein and Willard, 2000; or Montrucchio and Privileggi, 2001]. On the other hand, price bubbles are known to emerge in equilibrium across a finite number of trading dates whenever investors have asymmetric information [e.g. Allen, Morris, and Postlewaite, 1993; Morris, Postlewaite, and Shin, 1995; or Conlon, 2004]. Our model has both a finite number of trading dates and symmetric information. Thus, the source of a price bubble is found solely within the trading constraint of no short-selling.

The fourth section sets up the standard framework for the arbitrage analysis of trading strategies. By definition, a *trading strategy* is a predictable, feasible portfolio process. A trading strategy is *self-financing* provided that all earnings are rolled over into investments before the terminal date. A contingent claim is *marketed* whenever it is replicated by the terminal value of a self-financing trading strategy. The presence of a no short-selling constraint disrupts conventional thinking about self-financing trading strategies. For example, it is possible that two self-financing trading strategies have different initial costs, but the same terminal values. Thus, the Law of One Price does not necessarily apply to marketed claims. Nevertheless, the share holdings and initial cost of a self-financing trading strategy completely determine its cash holdings and, hence, its terminal value.

The fifth section fully exploits the LP structure of our model. The key is the fact that conditional expectations is a linear operator. Thus, an equivalent supermartingale measure is characterized as a strictly positive solution to an appropriate linear system of equalities and inequalities. The slack variables are readily identified as incremental price bubbles, and they vanish if and only if the feasible solution is an equivalent martingale measure. Consider an arbitrary, but fixed, contingent claim. The primal LP problem consists of selecting an equivalent supermartingale measure to maximize (or minimize) the fundamental value of the contingent claim. The dual LP problem consisting of selecting a marketed claim equal to or less than (resp., equal to or greater than) the contingent claim to minimize (resp., maximize) the initial cost. The optimal values of these dual programs are called hedging prices. Thus, the LP Duality Theorem shows that these hedging prices are the endpoints of the interval of possible fundamental values for the contingent claim. Finally, LP methods lead to the following interesting result. The Law of One Price for marketed claims prevails if and only if there exists an equivalent martingale measure. In turn, the latter condition indicates that price bubbles are fragile [e.g. Santos and Woodford, 1997; or Montrucchio and Privileggi, 2001]. Given that the Law of One Price holds, we conclude that a contingent claim has a unique fundamental value if and only if it is the difference between two marketed claims.

11.2 Securities Market Model

We study a stochastic model of the securities market with a finite number of trading dates. Time t is measured discretely by the non-negative integers up to a finite horizon $T > 0$. Information is described by a state space Ω and a sequence of finite partitions $\{\Omega_t\}_{t=0}^T$ of Ω, where Ω_t represents the information at date t. We assume that $\Omega_0 = \{\Omega\}$ and that Ω_{t+1} is a *refinement* of Ω_t for every date $0 \leq t < T$, i.e. Ω_{t+1} is a collection of partitions of the events in Ω_t. Uncertainty is measured by a probability measure P on the σ-algebra $\sigma\{\Omega_T\}$. We presume that Ω_t consists of non-null events for every date $0 \leq t \leq T$ without loss of generality. We say that a random variable x is Ω_t*-measurable* provided that x is constant on the events in Ω_t. Let X_t denote the vector space of all Ω_t-measurable random variables for every date $t \geq 0$. A sequence of random variables $\{x_t\}_{t=0}^T$ is *adapted* provided that $x_t \in X_t$ for every date $0 \leq t \leq T$, and $\{x_t\}_{t=0}^T$ is *predictable* provided that x_0 is constant and $x_t \in X_{t-1}$ for every date $0 < t \leq T$. All equalities and inequalities are presumed to hold almost surely in the following development.

The market itself consists of a finite number n of securities paying dividends and a short-term treasury bill which serves as numeraire. The ith security is characterized by an adapted price process $\{z_t^i\}_{t=0}^T$ and an adapted dividend process $\{d_t^i\}_{t=0}^T$ such that $d_0^i = 0$ for every $1 \leq i \leq n$. Let $z_t := [z_t^1, z_t^2, ..., z_t^n]$ and $d_t := [d_t^1, d_t^2, ..., d_t^n]$ for brevity. It is customary to assume these vector processes are positive, but this extra assumption is not necessary in the following analysis. On the other hand, the government bill has a current price of 1 at every date, simply because it serves as the unit of account, and it pays no coupons. Thus, all future prices and dividends are implicitly discounted to the date 0 by measuring values relative to treasury bill prices.

A portfolio at date $t > 0$ consists of a random number θ_t^i of shares of the ith security for every $1 \leq i \leq n$ and a number ξ_t of treasury bills. We presume that each θ_t^i and ξ_t are Ω_{t-1}-measurable for every date $t > 0$. The time lag indicates that an agent must select a portfolio just before date t without knowledge of the prevailing prices and dividends at date t. Let $\theta_t := [\theta_t^1, \theta_t^2, ...\theta_t^n]$. A *feasible* portfolio (θ_t, ξ_t) at date $t > 0$ must satisfy the additional condition that $\theta_t \geq 0$, indicating there is no short-selling of the long-lived securities. Since there are no sign restrictions on ξ_t in a feasible portfolio, this model allows unlimited borrowing of treasury bills. Notice that the market value of the portfolio (θ_t, ξ_t) at date $t-1$ is given by

$$\theta_t \cdot z_{t-1} + \xi_t := \sum_{i=1}^{n} \theta_t^i z_{t-1}^i + \xi_t \tag{1}$$

for any $t > 0$.

The only axiom of market consistency that we will evoke in this essay is a sequential version of the principle of NA. For any date $t > 0$, let $\Delta z_t^i := z_t^i - z_{t-1}^i$ denote the capital gains from holding one share of the ith security from date $t-1$ to date t. We may also write $\Delta z_t := z_t - z_{t-1}$ in vector notation. By definition, the condition NA holds at date $t > 0$ provided that

$$\theta_t \cdot (\Delta z_t + d_t) \geq 0 \Longrightarrow \theta_t \cdot (\Delta z_t + d_t) = 0 \tag{2}$$

for any feasible portfolio (θ_t, ξ_t) at date t. Finally, we say that (sequential) NA holds provided that NA holds at every date $0 < t \leq T$. The random variable $\theta_t \cdot (\Delta z_t + d_t)$ measures the earnings, i.e. capital gains plus dividends, from holding the portfolio (θ_t, ξ_t) from date $t-1$ to date t. So NA eliminates the possibility of riskless profit. The condition of NA is presumed to hold throughout this essay.

The development of martingale methods, beginning with Harrison and Kreps [1979], signifies one of the great achievements of financial economics. By definition, a *state-price deflator* consists of a strictly positive, real-valued martingale $\{\lambda_t\}_{t=0}^T$ such that $\lambda_0 = 1$ and that

$$E_P[\lambda_t(z_t + d_t)\,|\Omega_{t-1}] \leq \lambda_{t-1} z_{t-1} \tag{3}$$

for every date $0 < t \leq T$. It is easy to verify that equation (3) is equivalent to the condition that the deflated price plus cumulative deflated dividend process $\{\lambda_t z_t + \sum_{s=0}^{t} \lambda_s d_s\}_{t=0}^{T}$ is a supermartingale. Clearly, a finite-horizon state-price deflator $\{\lambda_t\}_{t=0}^{T}$ is uniquely determined by its terminal variable λ_T. In turn, λ_T is the Radon-Nikodym derivative of an *equivalent supermartingale measure* Q, which is a probability measure with the same null events as P and which satisfies the supermartingale property

$$E_Q(z_t + d_t \,|\Omega_{t-1}) \leq z_{t-1} \tag{4}$$

for every date $0 < t \leq T$. Since equivalent supermartingale measures are in one-to-one correspondence with state-price deflators, the computational device selected is just a matter of convenience. On the other hand, state-price deflators are more general than equivalent supermartingale measures when working with an infinite number of trading dates [e.g. Santos and Woodford, 1997; or Montrucchio and Privileggi, 2001].

The fact that there is no sign restriction upon prices and dividends in the above model is important to the formal development of the theory. Suppose the ith and jth securities satisfy the conditions that (i) $z_t^j = -z_t^i$ and (ii) $d_t^j = -d_t^i$ at every date $0 \leq t < \infty$. Then the above model with no short-selling logically reduces to a model that allows short-selling of the ith security. In brief, the short-selling constraint is not binding on the ith security. Furthermore, the supermartingale property (3) immediately simplifies to the martingale property

$$E_P[\lambda_t(z_t^i + d_t^i) \,|\Omega_{t-1}] = \lambda_{t-1} z_{t-1}^i \tag{5}$$

at every date $0 < t \leq T$. Therefore, our model includes the possibility of unlimited short-selling on some of the securities as a special case. A rationale for studying state-price deflators as we have defined them is found in the following result.

Proposition 11.1. *There exists a state-price deflator if and only if NA holds.*

Proof. Suppose there exists a state-price deflator, say $\{\lambda_t\}_{t=0}^{T}$. Then $\theta_t \cdot (\Delta z_t + d_t) \geq 0$ implies $E_P[\lambda_t \theta_t \cdot (\Delta z_t + d_t)] \geq 0$.On the other hand, the supermartingale property (3) implies $E_P[\lambda_t \theta_t \cdot (\Delta z_t + d_t)] \leq 0$. Thus, $E_P[\lambda_t \theta_t \cdot (\Delta z_t + d_t)] = 0$. Since $\theta_t \cdot (\Delta z_t + d_t) \geq 0$, we obtain $\theta_t \cdot (\Delta z_t + d_t) = 0$. Thus, NA holds at every date $0 < t \leq T$. Conversely, suppose that NA holds. Let

$$K_t = \{x \in X_t : \exists \theta_t \geq 0 \; s.t. \; x = \theta_t \cdot (\Delta z_t + d_t)\}$$

Notice that K_t is a polyhedral cone in X_t. Let $X_t^+ := \{x \in X_t : x \geq 0\}$ and $X_t^{++} := \{x \in X_t^+ : x \neq 0\}$. Then NA holds at date t if and only if $X_t^{++} \cap K_t = \phi$. It follows from previous work [Clark, 1993] that there exists a non-zero linear functional $p_t : X_t \to \mathbb{R}$ strictly separating X_t^{++} from K_t, i.e. $p_t(x) > 0$ for every $x \in X_t^{++}$ and $p_t(x) \leq 0$ for every $x \in K_t$. We may assume $p_t(1) = 1$ without loss of generality. The Riesz representation of p_t is denoted by δ_t, so that $p_t(x) = E_P(\delta_t x)$. We obtain $E_P(\delta_t) = 1$ from $p_t(1) = 1$, and we obtain $\delta_t > 0$ from the strict positivity of p_t. Furthermore, $p_t(x) \leq 0$ for every $x \in K_t$ implies $E_P[\delta_t(\Delta z_t + d_t) \,|\Omega_{t-1}] \leq 0$. An adapted process $\{\lambda_t\}_{t=0}^T$ is recursively defined by $\lambda_0 = 1$ and $\lambda_t = \delta_t \lambda_{t-1}$ for every $0 < t \leq T$. It is straightforward to verify that $\{\lambda_t\}_{t=0}^T$ is a state-price deflator. □

Consider the special case when there is unlimited short-selling on all securities. Then there are no sign restrictions on a feasible portfolio and the supermartingale property (3) of a state-price deflator reduces to the martingale property (5) for every $1 \leq i \leq n$. Alternatively, the associated equivalent supermartingale measure reduces to an equivalent martingale measure. Harrison and Pliska [1981] originally demonstrated that there exists an equivalent martingale measure if and only if NA holds in the finite-dimensional setting. King [2002] has also devised a proof of the Harrison-Pliska Theorem using LP methods. Equilibrium and optimal portfolio models with a no short-selling constraint in an infinite-dimensional context have also been studied by Lucas [1978], He and Pearson [1991], and Montrucchio and Privileggi [2001].

11.3 Fundamental Value

We now regard $X := X_T$ as a *contingent claims space*, i.e. the vector space of all possible payoffs contingent upon the state of nature. A state-price deflator $\{\lambda_t\}_{t=0}^T$ provides a natural measurement of the present value of a contingent claim. Indeed, we propose to measure the *fundamental value* (at date 0) of the contingent claim $x \in X$ as $E_P(\lambda_T x)$. Suppose the contingent claim x has the representation $x = \sum_{t=0}^T x_t$, where $x_t \in X_t$ for every date $0 \leq t \leq T$. Since $\{\lambda_t\}_{t=0}^T$ is a martingale, we obtain the familiar formula

$$E_P(\lambda_T x) = \sum_{t=0}^{T} E_P(\lambda_t x_t) \tag{6}$$

from the Law of Iterated Expectations. If x_t is delivered at date t, then this formula expresses the fundamental value of a contingent claim directly in terms of the present value of the cash flows it actually produces. More generally, the *fundamental value at date t* of the contingent claim $x \in X$ is measured as $E_P(\lambda_T x \,|\Omega_t)$.

Consider the fundamental value from the return of a marketed security. Let $\mu_{t-1} := -E_P[\lambda_t((\Delta z_t + d_t)\,|\Omega_{t-1}]$. Then the supermartingale property (3) can be rewritten as

$$E_P[\lambda_t(z_t + d_t)\,|\Omega_{t-1}] + \mu_{t-1} = \lambda_{t-1} z_{t-1} \tag{7}$$

The ith component of this vector equation has the following interpretation. The contingent claim $x = z_t^i + d_t^i$ is the one-period return of buying one share of the ith security at date $t-1$ and holding it until date t. Its fundamental value at date $t-1$ is given by $E_P[\lambda_t(z_t^i + d_t^i)\,|\Omega_{t-1}]$. Since $\lambda_{t-1} z_{t-1}$ measures the (deflated) market value at date $t-1$ of one share of the security, the residual term μ_{t-1}^i is readily identified as an incremental price bubble on the ith security. Notice that this type of price bubble is always non-negative, and it is zero if and only if the martingale property (5) prevails. We next extend this concept to the time horizon T. We now measure the return from buying one share of the ith security at date $t < T$ and holding it until date T as $x = z_T^i + \sum_{s=t+1}^{T} d_s^i$. The fundamental value at date t of this contingent claim is given by

$$E_P[\lambda_T(z_T^i + \sum_{s=t+1}^{T} d_s^i)\,|\Omega_t] = E_P(\lambda_T z_T^i\,|\Omega_t) + \sum_{s=t+1}^{T} E_P(\lambda_s d_s^i\,|\Omega_t) \tag{8}$$

Let

$$b_t^i := E_P(\sum_{s=t}^{T-1} \mu_s\,|\Omega_t) \tag{9}$$

for every date $0 \le t < T$. Then the algebraic identity

$$E_P(b_t\,|\Omega_{t-1}) = b_{t-1} - \mu_{t-1} \tag{10}$$

for every date $0 < t < T$ implies that the process $\{b_t\}_{t=0}^{T-1}$ is a supermartingale. The next result identifies b_t^i as a *price bubble* on the ith security at date t.

Proposition 11.2. *The formula*

$$E_P[\lambda_T(z_T + \sum_{s=t+1}^{T} d_s)\,|\Omega_t] + b_t = \lambda_t z_t \tag{11}$$

holds for every date $0 \le t < T$.

Proof. We proceed by backwards induction on the date t. First notice that

$$E_P[\lambda_T(z_T + d_T)\,|\Omega_{T-1}] + \mu_{T-1} = \lambda_{T-1} z_{T-1}$$

by virtue of equation (7) with $t = T - 1$ and $b_{T-1} = \mu_{T-1}$ from definition (9). We immediately obtain

$$E_P[\lambda_T(z_T + d_T)\,|\Omega_{T-1}] + b_{T-1} = \lambda_{T-1} z_{T-1}$$

For the inductive step, assume that formula (11) holds with $0 < t \leq T - 1$. Computing conditional expectations with respect to Ω_{t-1}, we obtain

$$E_P[\lambda_T(z_T + \sum_{s=t}^{T} d_s)\,|\Omega_{t-1}] + E_P(b_t\,|\Omega_{t-1}) = E_P[\lambda_t(z_t + d_t)\,|\Omega_{t-1}]$$

Substituting from equation (10) and applying equation (7), the above equation reduces to

$$E_P[\lambda_T(z_T + \sum_{s=t}^{T} d_s)\,|\Omega_{t-1}] + b_{t-1} = \lambda_{t-1} z_{t-1}$$

Thus, formula (11) is verified at date $t - 1$, which completes the backwards induction. □

Equation (11) decomposes market value, represented component-wise by the right-hand side, into fundamental value, represented component-wise by the first term on the left-hand side, and a bubble, represented component-wise by the second term on the left-hand side. Hence, it is a generalization of the Fundamental Equation of Asset Pricing. Indeed, the above equation can be rewritten as

$$E_P(\lambda_T z_T^i) + \sum_{s=t+1}^{T} E_P(\lambda_s d_s^i) + b_0^i = z_0^i \tag{12}$$

for the ith security at date 0. Notice that the fundamental value of one share of a security is equal to its market value if and only if it has no price bubble. Furthermore, the price bubble b_t^i is zero if and only if the martingale property (5) holds for all dates past time t.

11.4 Trading Strategies

By definition, a *trading strategy* is the selection of a feasible portfolio (θ_t, ξ_t) at every date $t > 0$. The trading strategy $\{(\theta_t, \xi_t)\}_{t=1}^{T}$ is *self-financing* provided that

$$\theta_{t+1} \cdot z_t + \xi_{t+1} = \theta_t \cdot (z_t + d_t) + \xi_t \tag{13}$$

for every date $0 < t < T$. This condition asserts that there is no outflow or inflow of funds at intermediate dates. So the contingent claim $\theta_T \cdot (z_T + d_T) + \xi_T$ is replicated at date T by the self-financing trading strategy $\{(\theta_t, \xi_t)\}_{t=1}^T$ at an initial cost of $\theta_1 \cdot z_0 + \xi_0$. More briefly, we say that a contingent claim $x \in X$ is *marketed* provided that there exists a self-financing trading strategy with terminal value x. The next result is a useful accounting identity.

Proposition 11.3. *If $\{(\theta_t, \xi_t)\}_{t=1}^T$ is a self-financing trading strategy, then the condition*

$$\theta_T \cdot (z_T + d_T) + \xi_T = (\theta_1 \cdot z_0 + \xi_0) + \sum_{t=1}^{T} \theta_t \cdot (\Delta z_t + d_t) \tag{14}$$

holds.

Proof. Consider the algebraic identity

$$\sum_{t=1}^{T} \theta_t \cdot (\Delta z_t + d_t) = \sum_{t=1}^{T} [\theta_t \cdot (z_t + d_t) + \xi_t] + \sum_{t=1}^{T} (\theta_t \cdot z_{t-1} + \xi_t)$$

The self-financing condition (13) implies that this sum telescopes into

$$\sum_{t=1}^{T} \theta_t \cdot (\Delta z_t + d_t) = [\theta_T \cdot (z_T + d_T) + \xi_T] - (\theta_1 \cdot z_0 + \xi_0)$$

Hence, equation (14) immediately follows. □

This result plainly asserts that the terminal value of a self-financing trading strategy is equal to its initial value plus its cumulative earnings. It also indicates that the cash holdings $\{\xi_t\}_{t=1}^T$ of a self-financing trading strategy are implicitly determined by the security holdings $\{\theta_t\}_{t=1}^T$ and the initial cost $\gamma = \theta_1 \cdot z_0 + \xi_0$. Since there is no short-selling, the supermartingale property (3) yields the following additional conclusion.

Corollary 11.1. *If $\{(\theta_t, \xi_t)\}_{t=1}^T$ is a self-financing trading strategy, then the condition*

$$E_P[\lambda_T \theta_T \cdot (z_T + d_T) + \lambda_T \xi_T] \leq \theta_1 \cdot z_0 + \xi_0 \tag{15}$$

holds for any state-price deflator $\{\lambda_t\}_{t=0}^T$.

Proof. The supermartingale property (3) implies

$$E_P[\lambda_T \theta_t \cdot (\Delta z_t + d_t)] \leq 0$$

for every date $0 < t \leq T$. Thus, inequality (15) immediately follows from formula (14). □

This corollary implies that NA operates across time in the following way. If a marketed claim is positive, then it must incur a non-negative cost; if it is both positive and incurs zero cost, then it must be zero. Nevertheless, we cannot always assert that the Law of One Price holds in this context. Such a Law would require the initial cost of a marketed claim to be unique.

Example: Let $\Omega = \{\alpha, \omega\}$ be the state space with a uniform probability distribution P, and let $T = 1$ be the time horizon. The information partitions are given by $\Omega_0 = \{\Omega\}$ and $\Omega_1 = \{\{\alpha\}, \{\omega\}\}$. The number of securities is $n = 2$. The first security pays no dividends and has a (discounted) price process $\{z_t^1\}_{t=0}^1$ given by $z_0^1 = 1$, $z_1^1(\alpha) = 1.5$, and $z_1^1(\omega) = 0.5$. The second security also pays no dividends and has a (discounted) price process $\{z_t^2\}_{t=0}^1$ given by $z_0^1 = 2.25$, $z_1^1(\alpha) = 2.5$, and $z_1^1(\omega) = 1.5$. It is easy to verify that the process $\{\lambda_t\}_{t=0}^1$ given by $\lambda_0 = 1$ and $\lambda_1 = 1$ is a state-price deflator such that

$$E_P(\lambda_1 z_1^1 \,|\Omega) = E_P(z_1^1) = 1 = z_0^1$$
$$E_P(\lambda_1 z_1^2 \,|\Omega) = E_P(z_1^2) = 2 < z_0^2$$

Proposition 11.1 implies that NA holds. Notice that a trading strategy reduces to a single portfolio $(\theta_1^1, \theta_1^2, \xi_1)$ in this case. Consider the contingent claim x defined by $x = z_1^2$. This claim is obviously marketed by the portfolios $(0, 1, 0)$ and $(1, 0, 1)$. Since the initial cost of these portfolios are 2.25 and 1.5, respectively, the Law of One Price is violated. This state of affairs would never happen if short-selling were allowed, because the portfolio $(1, -1, 1.25)$ would become a simple free lunch.

11.5 Hedging Prices

We next study the LP formulation of this model. Define a (continuous) linear operator $A_t : X_T \to X_{t-1}^n$ by the condition that

$$A_t\lambda = E_P[\lambda(\Delta z_t + d_t)\,|\Omega_{t-1}] \tag{16}$$

for every date $t > 0$. The adjoint operator $A_t^* : X_{t-1}^n \to X_t$ is given by the formula

$$A_t^*\theta_t = \theta_t \cdot (\Delta z_t + d_t) \tag{17}$$

as confirmed by verifying the adjoint identity

$$\langle A_t^*\theta_t, \lambda\rangle = E_P[\lambda\theta_t \cdot (\Delta z_t + d_t)] = \langle \theta_t, A_t\lambda\rangle \tag{18}$$

Trading strategies also emerge from LP duality. Indeed, the linear operator $A : X_T \to \prod_{t=0}^{T-1} X_t^n$ is well-defined in matrix form by the formula

$$A\lambda := \begin{bmatrix} A_1\lambda \\ A_2\lambda \\ \dots \\ A_T\lambda \end{bmatrix} \tag{19}$$

The adjoint operator $A^* : \prod_{t=0}^{T-1} X_t^n \to X_T$ is given by the formula

$$A^*[\theta_1, \theta_2, ..., \theta_T] = \sum_{t=1}^{T} \theta_t \cdot (\Delta z_t + d_t) \tag{20}$$

Clearly, this expression is the cumulative earnings from the trading strategy $\{(\theta_t, \xi_t)\}_{t=1}^T$.

Suppose $\lambda_T \in X_T$ is a strictly positive random variable normalized so that $E_P(\lambda_T) = 1$. Let $\lambda_t = E_P(\lambda_T \,|\Omega_t)$ for every date $0 \leq t < T$. Then $\{\lambda_t\}_{t=0}^T$ is a strictly positive martingale such that $\lambda_0 = 1$. Notice that λ_T solves the linear inequality $A\lambda \leq 0$ if and only if λ_t solves the linear inequality $A_t\lambda \leq 0$ for every $0 < t \leq T$. Therefore, λ_T is the terminal variable of a state-price deflator if and only if it solves the linear system

$$\begin{aligned} &A\lambda \leq 0 \\ &E_P(\lambda) = 1 \\ &\lambda > 0 \end{aligned} \tag{21}$$

The corresponding linear operator $B : X_T \to \mathbb{R} \times \prod_{t=0}^{T-1} X_t^n$ is defined by the formula

$$B\lambda := \begin{bmatrix} E_P(\lambda) \\ A\lambda \end{bmatrix} \tag{22}$$

with adjoint operator $B^* : \mathbb{R} \times \prod_{t=0}^{T-1} X_t^n \to X_T$ given by

$$B^*[\gamma; \theta_1, \theta_2, ..., \theta_T] = \gamma + A^*[\theta_1, \theta_2, ..., \theta_T] \tag{23}$$

We immediately identify γ as the initial cost of a trading strategy with security holdings $\{\theta_t\}_{t=1}^T$. Adjusting cash balances $\{\xi_t\}_{t=1}^T$ so that the self-financing condition (13) holds, it follows from Lemma (3) that $B^*[\gamma; \theta_1, \theta_2, ..., \theta_T]$ is the terminal value of the corresponding self-financing trading strategy.

It is well-known from the study of a securities market with no trading constraints that an equivalent martingale measure and, hence, a state-price deflator is not unique unless the market is complete, i.e. all contingent claims are marketed [e.g. Taqqu and Willinger, 1981]. In fact, a contingent claim has a unique fundamental value if and only if it is marketed. Similar considerations apply to a market that does not allow short-selling. Yet market completeness is an especially stringent condition when there are trading constraints upon the linear structure of the marketed claims space as we have here. On the contrary, we must admit to the possibility that the state-price deflator is not unique.

There are two interrelated linear programs relevant to this problem. Consider the evaluation of a fixed contingent claim $x \in X$. The maximal fundamental value of x is given by $\sup E_P(\lambda x)$, where the variable λ is constrained by the linear system (21); and the minimal fundamental of x is given by $\inf E_P(\lambda x)$, where the variable λ is again constrained by the linear system (21). It is easy to verify that the optimal values of these programs do not change if we weaken the constraint $\lambda > 0$ to $\lambda \geq 0$. Thus, the maximal fundamental value is the optimal value of the LP

$$\begin{aligned} &\sup E_P(\lambda x) \\ &A\lambda \leq 0 \\ &E_P(\lambda) = 1 \\ &\lambda \geq 0 \end{aligned} \tag{24}$$

and the minimal fundamental value is the optimal value of the LP

$$\begin{aligned} &\inf E_P(\lambda x) \\ &A\lambda \leq 0 \\ &E_P(\lambda) = 1 \\ &\lambda \geq 0 \end{aligned} \tag{25}$$

These programs can be written in canonical form by introducing a slack variable $\mu := [\mu_0, \mu_1, ..., \mu_{T-1}]$, where $\mu_t \in X_t$ for every date $0 \leq t < T$. The inequality constraint $A\lambda \leq 0$ is now replaced with the condition that $A\lambda + \mu = 0$ and $\mu \geq 0$. Since the conditions $A_t\lambda + \mu_t = 0$ and $\mu_t \geq 0$ hold for every date $0 \leq t < T$, it immediately follows that the slack variable μ_t is identical to the incremental price bubble defined by equation (7). In summary, the canonical maximization problem is given by

$$\begin{aligned} &\sup E_P(\lambda x) \\ &A\lambda + \mu = 0 \\ &E_P(\lambda) = 1 \\ &[\lambda, \mu] \geq 0 \end{aligned} \tag{26}$$

and the canonical minimization problem is given by

$$\begin{aligned}&\inf E_P(\lambda x)\\&A\lambda+\mu=0\\&E_P(\lambda)=1\\&[\lambda,\mu]\geq 0\end{aligned}\tag{27}$$

The fundamental value of the contingent claim $x \in X$ is any real number between these two optimal values.

Let $\gamma \in \mathbb{R}$. The dual LP for the canonical maximization problem is given by

$$\begin{aligned}&\inf \gamma\\&B^*[\gamma;\theta_1,\theta_2,...,\theta_T]\geq x\\&[\theta_1,\theta_2,...,\theta_T]\geq 0\end{aligned}\tag{28}$$

and the dual LP for the canonical minimization problem is given by

$$\begin{aligned}&\sup \gamma\\&B^*[\gamma;\theta_1,\theta_2,...,\theta_T]\leq x\\&[\theta_1,\theta_2,...,\theta_T]\geq 0\end{aligned}\tag{29}$$

These formulas yield hedging prices for the contingent claim $x \in X$. Indeed, we immediately identify $B^*[\gamma;\theta_1,\theta_2,...,\theta_T]$ as the terminal value of a self-financing trading strategy with initial cost γ and share holdings $\{\theta_t\}_{t=1}^T$. Thus, the optimal value to program (28) is the *upper hedging price* of x and the optimal value to program (29) is the *lower hedging price* of x. In view of the LP Duality Theorem, any real number between these hedging prices corresponds to a measurement of the fundamental value of x. These results also apply to a model with unlimited short-selling as a special case. LP duality theory indicates that an equality constraint in the primal LP corresponds to no sign restriction on the dual variable in the dual LP [e.g. Gale, 1960]. Therefore, replacing the inequality constraint $A\lambda \leq 0$ with $A\lambda = 0$ in the primal LP leads to the deletion of the sign restriction $[\theta_1,\theta_2,...,\theta_T] \geq 0$ in the dual LP. This complementation between the primal and dual LP irrefutably identifies the trading constraint as the source of a bubble.

Consider a version of the well-known Law of One Price, which asserts that two self-financing trading strategies with the same terminal value must also have the same initial cost. In LP terms, this Law asserts that the conditions

$$\begin{aligned}&B^*[\gamma;\theta_1,\theta_2,...,\theta_T]=B^*[\widehat{\gamma};\widehat{\theta}_1,\widehat{\theta}_2,...,\widehat{\theta}_T]\\&[\theta_1,\theta_2,...,\theta_T]\geq 0\\&[\widehat{\theta}_1,\widehat{\theta}_2,...,\widehat{\theta}_T]\geq 0\end{aligned}\tag{30}$$

imply that $\gamma = \widehat{\gamma}$.

Proposition 11.4. *The Law of One Price prevails if and only if there exists an equivalent martingale measure.*

Proof. Suppose that The Law of One Price prevails. For any random variable x, let $x^+ := x1_{\{x\geq 0\}}$ and $x^- := -x1_{\{x\leq 0\}}$. Notice that $x = x^+ - x^-$. For any random vector θ, define θ^+ by the condition that $(\theta^+)^i := (\theta^i)^+$ for every $1 \leq i \leq n$, and define θ^- by the condition that $(\theta^-)^i := (\theta^i)^-$ for every $1 \leq i \leq n$. It is clear that $\theta = \theta^+ - \theta^-$. Indeed, we have just set up the canonical lattice operations.

Assume that $B^*[\gamma; \theta_1, \theta_2, ..., \theta_T] = B^*[\widehat{\gamma}; \widehat{\theta}_1, \widehat{\theta}_2, ..., \widehat{\theta}_T]$, where there are no sign restrictions upon $[\theta_1, \theta_2, ..., \theta_T]$ and $[\widehat{\theta}_1, \widehat{\theta}_2, ..., \widehat{\theta}_T]$. Since B^* is a linear operator, we obtain

$$\begin{aligned} &B^*[\gamma^+ + \widehat{\gamma}^-; \theta_1^+ + \widehat{\theta}_1^-, \theta_2^+ + \widehat{\theta}_2^-, ..., \theta_T^+ + \widehat{\theta}_T^-] \\ &= B^*[\gamma^- + \widehat{\gamma}^+; \theta_1^- + \widehat{\theta}_1^+, \theta_2^- + \widehat{\theta}_2^+, ..., \theta_T^- + \widehat{\theta}_T^+] \end{aligned}$$

Since

$$\begin{aligned} [\theta_1^+ + \widehat{\theta}_1^-, \theta_2^+ + \widehat{\theta}_2^-, ..., \theta_T^+ + \widehat{\theta}_T^-] \geq 0 \\ [\theta_1^- + \widehat{\theta}_1^+, \theta_2^- + \widehat{\theta}_2^+, ..., \theta_T^- + \widehat{\theta}_T^+] \geq 0 \end{aligned}$$

the Law of One Price yields $\gamma^+ + \widehat{\gamma}^- = \gamma^- + \widehat{\gamma}^+$. Thus, we obtain $\gamma = \widehat{\gamma}$ upon rearranging terms. Let

$$M := \{x \in X_T : \exists\, [\gamma; \theta_1, \theta_2, ..., \theta_T] \text{ s.t. } x = B^*[\gamma; \theta_1, \theta_2, ..., \theta_T]\}$$

denote the range of the adjoint operator B^*. A linear functional $\pi : M \to \mathbb{R}$ is well-defined by the condition that $x = B^*[\gamma; \theta_1, \theta_2, ..., \theta_T]$ implies $\pi(x) = \gamma$. Since NA holds, proposition 11.1 yields the existence of an equivalent supermartingale measure with Radon-Nikodym derivative λ_T. So the corollary to proposition 11.3 immediately implies that π is strictly positive. Since X_T is a finite-dimensional vector space, there exists a strictly positive linear functional $p : X_T \to \mathbb{R}$ extending π [e.g. Kreps, 1981; or Clark, 1993]. Let λ denote the Riesz representation of p. Then $\lambda > 0$, because p is strictly positive. In brief, we say that λ is a strictly positive extension of π. We know from previous work that $\lambda \in X_T$ solves the linear system

$$B\lambda = \begin{bmatrix} 1 \\ 0 \end{bmatrix}, \lambda \geq 0 \tag{31}$$

if and only if λ is a positive linear extension of π [Clark, 2003]. Therefore, λ is the Radon-Nikodym derivative of an equivalent martingale measure.

Conversely, suppose there exists an equivalent martingale measure with Radon-Nikodym derivative $\lambda_T > 0$. Let $\{(\theta_t, \xi_t)\}_{t=1}^{T}$ denote a self-financing trading strategy. It follows from the corollary to proposition 11.3 that

$$E_P[\lambda_T \theta_T \cdot (z_T + d_T) + \lambda_T \xi_T] = \theta_1 \cdot z_0 + \xi_0$$

after strengthening the supermartingale property to the martingale property. Since $\gamma = \theta_1 \cdot z_0 + \xi_0$ and

$$B^*[\gamma; \theta_1, \theta_2, ..., \theta_T] = \theta_T \cdot (z_T + d_T) + \xi_T$$

we deduce that the value of $B^*[\gamma; \theta_1, \theta_2, ..., \theta_T]$ uniquely determines γ. Thus, the Law of One Price prevails. □

This result has several interesting ramifications. First, the Law of One Price is a cornerstone of computational finance. Consider the hedging price problems (28) and (29) when the contingent claim x is marketed. Notice that $x = B^*[\gamma; \theta_1, \theta_2, ..., \theta_T]$ implies $[\gamma; \theta_1, \theta_2, ..., \theta_T]$ is a feasible solution to either problem, so that γ is a measurement of the fundamental value of x. So if x does not have a unique initial cost, then it does not have a unique fundamental value. Second, the existence of an equivalent martingale measure is characterized by a version of NA for which there are no sign restrictions upon a feasible portfolio [e.g. Harrison and Pliska, 1981]. Yet this type of NA is implausible as a behavioral axiom when short-selling is not feasible. Third, an equivalent supermartingale measure reduces to an equivalent martingale measure if and only if it does not produce a bubble. We say that a bubble is *persistent* whenever every equivalent supermartingale measure produces a nonzero bubble. If the bubble is not persistent, then we say it is *fragile*. Therefore, a bubble is fragile if and only if the Law of One Price prevails. The problem of fragile bubbles in infinite-horizon markets has also been studied in sequential equilibrium theory [e.g. Santos and Woodford, 1997; or Montrucchio and Privileggi, 2001]. Fourth, the above result immediately leads to conditions characterizing unique valuation.

Corollary 11.2. *Suppose the Law of One Price prevails. Then a contingent claim $x \in X_T$ has a unique fundamental value if and only if it is the difference between two marketed claims.*

Proof. Since the Law of One Price holds, the proof to proposition 11.4 shows that every contingent claim $x \in M$ has a unique cost $\pi(x)$. Furthermore, $\pi : M \to \mathbb{R}$ is a strictly positive linear functional. Since X_T is a finite-dimensional vector space, it follows from previous work [Clark, 2000] that the contingent claim $x \in X_T$ has a unique fundamental value if and only if $x \in M$. Finally, the proof to proposition 11.4 reveals that $x \in M$ if and only if x is the difference between two marketed claims. □

Taqqu and Willinger [1981] originally obtained this type of result in the case when short-selling is fully permitted. In essence, they demonstrated that a contingent claim $x \in X$ has a unique fundamental value if and only if it is marketed. It is easy to verify that their result is a special case of the above corollary, corresponding to a situation where the no short-selling constraint is not binding.

Bibliography

Allen, F., Morris, S. and Postlewaite, A. (1993). Finite bubbles with short sale constraints and asymmetric information, *Journal of Economic Theory*, **61**, pp. 206–229.

Clark, S. A. (1993). The valuation problem in arbitrage price theory, *Journal of Mathematical Economics*, **22**, pp. 463–478.

Clark, S. A. (2000). Arbitrage approximation theory, *Journal of Mathematical Economics*, **33**, pp. 167–181.

Clark, S. A. (2003). An infinite dimensional LP duality theorem, *Mathematics of Operations Research*, **28**, 2, pp. 233–245.

Conlon, J. R. (2004). Simple finite horizon bubbles robust to higher order knowledge, *Econometrica*, **72** (3), pp. 927–936.

Gale, D. (1960). *The Theory of Linear Economic Models*, (McGraw-Hill: New York).

Harrison, J. M. and Kreps, D. M. (1979). Martingales and arbitrage in multiperiod securities markets, *Journal of Economic Theory*, **20**, pp. 381–408.

Harrison, J. M. and Pliska, S. R. (1981). Martingales and stochastic integrals in the theory of continuous time trading, *Stochastic Processes and their Applications*, **11**, pp. 215–260.

He, H. and Pearson, N. D. (1991). Consumption and portfolio policies with incomplete markets and short-sale constraints: the infinite-dimensional case, *Journal of Economic Theory*, **54**, pp. 259–304.

King, A. J. (2002). Duality and martingales: a stochastic programming perspective on contingent claims, *Mathematical Programming Series B*, **91**, pp. 543–562.

Kreps, D. M. (1981). Arbitrage and equilibrium in economies with infinitely many commodities, *Journal of Mathematical Economics*, **8**, pp. 15–35.

Loewenstein M. and Willard, G. A. (2000). Rational equilibrium asset-pricing bubbles in continuous trading models, *Journal of Economic Theory*, **91**, pp. 17–58.

Lucas, R. E. (1978). Asset prices in an exchange economy, *Econometrica*, **46**, pp. 1429–1445.

Montrucchio, L. and Privileggi, F. (2001). On fragility of bubbles in equilibrium asset pricing models of Lucas-type, *J. Econ. Theory*, **101**, pp. 158–188.

Morris S., Postlewaite, A. and Shin, H. S. (1995). Depth of knowledge and the effect of higher order uncertainty, *Economic Theory* **6**, pp. 453–467.

Santos M. and Woodford M. (1997). Rational asset pricing bubbles, *Econometrica* **65**, pp. 19–57.

Taqqu M. S. and Willinger, W. (1987). The analysis of finite security markets using martingales, *Adv. Appl. Probl.*, **19**, pp. 1–25.

Chapter 12

Mathematical Programming and Electrical Network Analysis II: Computational Linear Algebra through Network analysis

H. Narayanan
Department of Electrical Engineering,
Indian Institute of Technology, Bombay,
Mumbai, 400 076, India
e-mail:hn@ee.iitb.ac.in

Dedicated to the memory of Professor S. R. Mohan

Abstract

In this paper we report our experience in solving min cost flow problems approximately by transforming them to network analysis problems. In the process we solve large (of the order of a million nodes) resistive networks. The preconditioned conjugate gradient (PCG) method appears the most suitable for this problem but runs into convergence difficulties if the conductance values have the high range of $1 - 10^8$. We solve this problem by developing a variation of the PCG (which is described in the paper) and using it to solve *hybrid analysis* equations of the network. This suggests a relook at commonly used algorithms in computational linear algebra by associating an electrical network with the linear equations in question. In order to make the paper self contained we give a formal description of commonly used network analysis procedures such as nodal, loop and hybrid analysis.

Key Words: Mathematical Programming, hybrid analysis, electrical network, loop analysis

12.1 Introduction

It is well known that optimization problems of the 'primal-dual' variety can be cast as electrical network solution problems [Dennis (1959)], [Iri (1969)].

Until recently there had been no serious attempt to explore whether this electrical approach had any computational advantage. A beginning was reported in [Narayanan (2004)]. Subsequently the min-cost flow problem has been tackled through network analysis methods with some success [Trivedi, Desai and Narayanan (2006)], [Trivedi, Punglia and Narayanan (2007)]. A natural consequence of this attempt is a re-examination of some basic algorithms of computational linear algebra. The present paper reports on difficulties encountered, their resolution and a proposed program of development of this electrical approach to computational linear algebra.

Informally, the min cost flow problem may be described as follows : We are given a directed 'flow' graph. There is a source node 's', at which the net flow leaving the node is nonnegative and a sink node 't' at which the net flow entering is nonnegative. At all other nodes the flow is conserved. This implies that the flow leaving the source node and that entering the sink node are equal. Each edge 'e' has a 'capacity' cap(e) and a 'cost per unit flow' cost(e). The flow through e must be nonnegative and must not exceed cap(e). We are given a specified flow J, which has to be sent from s to t. The problem is to determine (a) if this is feasible and, if feasible, (b) to distribute the flow among the edges so that the total cost is minimum. This is a special kind of linear programming problem which has been well studied [Ahuja, Magnanti and Orlin (1993)]. Electrically, the problem is equivalent to the solution of a network in which there is a current source from t to s of value J and with every branch a composite device as shown in Figure 12.1.

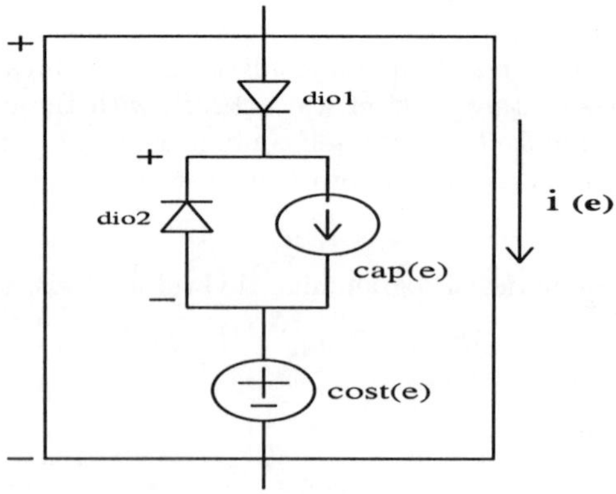

Fig. 12.1 Composite Edge

The devices $dio1, dio2$ are ideal diodes i.e.

$$v_{dio} \leq 0, \qquad i_{dio} \geq 0 \tag{12.1}$$

and

$$v_{dio}.i_{dio} = 0 \tag{12.2}$$

(For a proof of the equivalence one could refer to [Dennis (1959)], [Iri (1969)] or [Narayanan (2004)].) We can adopt one of two approaches to solve the network

(1) we could adapt the interior point method [Renegar (2001)] and solve the network exactly or
(2) we could approximate the ideal diode by a smooth approximation (say $i = I_0(e^{\frac{v}{v_T}} - 1)$) and solve this network approximately through the Newton-Raphson procedure.

The former is worth exploring but is slow in comparison with standard computer science algorithms reported in [Ahuja, Magnanti and Orlin (1993)]. The latter is approximate, can sometimes fail to converge but in many cases does provide approximate solutions at competitive speeds. When the flow network was sufficiently large (> 100,000 nodes, 300,000 edges) and the cost and capacity ranges were larger than $1 - 10^6$, we found that available implementations of standard computer science algorithms [Leda (2005)] become unreliable or too slow while with electrical network analysis, we could reach within 0.1% of the optimal cost in reasonable time (see Table 12.1 in Section 12.3).

The N-R procedure, at each iteration, converts every non-linear device with a smooth non-linear v-i characteristic into one with a straight line v-i characteristic ($i_j = G_j v_j + J_j$) which is a tangent at some point to the original non-linear characteristic. Each iteration of the N-R procedure amounts to the solution of an appropriate resistive network, which has only conductances, voltage sources and current sources, where the conductance values correspond to the slope of the straight line v-i characteristic with which the non-linear characteristic is replaced, as mentioned above. In the cases, where we have "practical diodes" ($i = I_0(e^{\frac{v}{v_T}} - 1)$) at each N-R iteration (except perhaps the first or second), we will have to solve a resistive network in which more than 30% of the resistors have high conductance and more than 30% have low conductance. By using certain scaling techniques, we can keep this ratio of conductance values within about $1 - 10^8$. This does not present any difficulty if we use sparse LU techniques for solving the linear equations. But when the networks are non-planar, sparse LU methods are too slow beyond say 10,000 nodes and 30,000 edges. When the conductance range is $1 - 10^4$, preconditioned conjugate gradient(PCG) method performs acceptably for all topologies, whether planar or non-planar. However PCG breaks down when the conductance range is $1 - 10^8$. (This is because large conductance range corresponds to large ratio of eigen values and PCG performs poorly when the 'condition number' is large). We attempted an unconventional solution for this problem. We first scaled the conductances so that they lay in the range of $10^{-4} - 10^4$. We treated the conductances in the range $10^{-4} - 1$ as resistance(1/conductance) of value $1 - 10^4$. We then divided the network into two parts corresponding to these two types of devices and wrote hybrid equations for the network. The resulting coefficient matrix has the form $\begin{bmatrix} \tilde{G} & H \\ -H^T & \tilde{R} \end{bmatrix}$, where $\tilde{G}$ and $\tilde{R}$ are positive definite. A variation of PCG works well for this kind of matrices and there is convergence within a few hundred iterations even when the matrix size is $10^6 \times 10^6$. The above experience suggests that it may be worthwhile to treat systems of linear equations with positive definite coefficient matrices as though they arise from a resistive network and solve the network through the most appropriate hybrid equations. The present paper details this suggestion and gives some experimental backing to the utility of the suggestion.

12.2 Electrical Network Analysis Procedures

An electrical network is a pair($\mathcal{G}$,$\mathcal{D}$), where $\mathcal{G}$ is a directed graph with edge set $E(G)$ and vertex set $V(G)$ and $\mathcal{D}$ is a 'device characteristic' on the edge set $E(G)$, defined to be a collection of pairs of real valued functions (v,i) on $E(G)$. For the purposes of this paper, however, $(v,i) \in \mathcal{D}$ can be treated as vectors or constant functions. Associated with $\mathcal{G}$ are two vector spaces

(1) the current space $\mathcal{V}_i(\mathcal{G})$, made up of vectors $i : E(G) \longrightarrow \Re$ which satisfy, for every node x, the condition that the net current leaving x equals zero. To illustrate, in the Figure 12.2, at node x we have

$$i(e_2) + i(e_3) + i(e_4) - i(e_1) = 0 \tag{12.3}$$

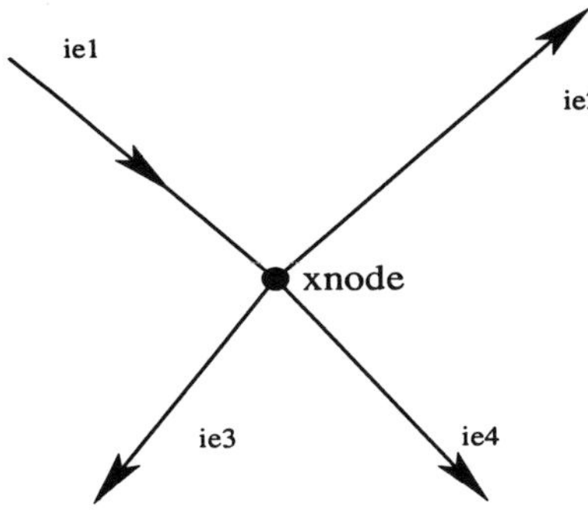

Fig. 12.2

(2) the voltage space $\mathcal{V}_v(\mathcal{G})$, made up of vectors $v : E(G) \longrightarrow \Re$, which can be derived from potential vectors $p : V(G) \longrightarrow \Re$ in the following sense : $v(e) = p(a) - p(b)$ whenever edge e is directed from a to b.

The fundamental theorem of network theory, viz. Tellegen's theorem, states that $\mathcal{V}_i(\mathcal{G})$ and $\mathcal{V}_v(\mathcal{G})$ are complementary orthogonal. To solve a network $(\mathcal{G},\mathcal{D})$ means to find all (v,i) such that $v \in \mathcal{V}_v(\mathcal{G})$, $i \in \mathcal{V}_i(\mathcal{G})$ and $(v,i) \in \mathcal{D}$. One such pair (v,i) is called a solution of the network. The device characteristics of particular interest to us are those where the network has voltage sources, current sources and resistors. Let $E(\mathcal{G}) \equiv \mathbf{V} \uplus \mathbf{I} \uplus \mathbf{T}$

Let

$\mathcal{E} : \mathbf{V} \longrightarrow \Re$,
$\mathcal{J} : \mathbf{I} \longrightarrow \Re$,
$r : \mathbf{T} \longrightarrow \Re+$.

Then $(v,i) \in \mathcal{D}$ if and only if

$v(e) = \mathcal{E}(e)$, where $e \in \mathbf{V}$,
$i(e) = \mathcal{J}(e)$, where $e \in \mathbf{I}$,
$v(e) - r(e).i(e) = 0$ or equivalently
$i(e) - g(e).v(e) = 0$ $(g(e) = \frac{1}{r(e)})$, where $e \in \mathbf{T}$.

We will call $\mathbf{V}, \mathbf{I}, \mathbf{T}$ respectively, the sets of voltage sources, current sources and resistors. The symbols for these devices are given in Figure 12.3.

Every network with voltage sources, current sources and resistors can be transformed through a linear (in the size of the network) time algorithm into another network in which each device is 'composite' as in Figure 12.3.

Here we have

$$v(e) - \mathcal{E}(e) = r(e)(i(e) - \mathcal{J}(e)), \tag{12.4}$$

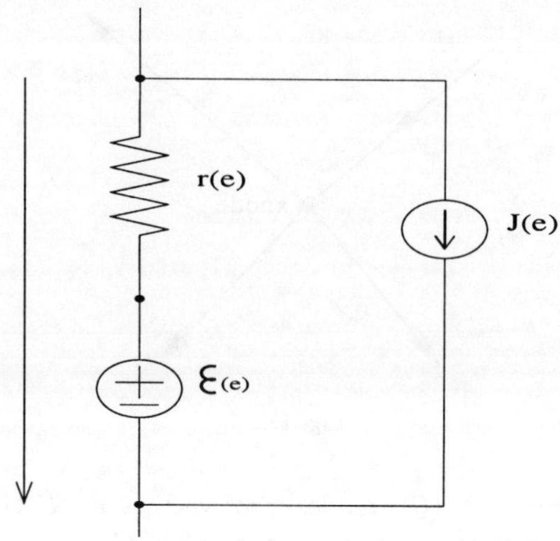

Fig. 12.3 Composite device

where $r(e)$ is non-zero. From the solution to the transformed network one can obtain that of the original, once again by a linear time process.

For network analysis, one needs convenient bases for $\mathcal{V}_v(\mathcal{G})$ and $\mathcal{V}_i(\mathcal{G})$. We next describe these. For convenience of explanation we will henceforth take $\mathcal{G}$ to be connected. This is of course not necessary for carrying out network analysis.

The incidence matrix A of a directed graph has a row corresponding to each node of the graph and a column corresponding to each edge of the graph. We have

$A(a, e) \equiv +1(-1)$ if e is incident on a and is directed away from(into) a
$\equiv 0$ otherwise

The rows of A are linearly dependent since the entries of every column add up to zero. However dropping any row, in the case of a connected graph, results in a linearly independent set of vectors.

Let p be a potential vector with a real entry corresponding to every node in $\mathcal{G}$. Clearly $p^T A$ gives the voltage vector v derived from p. Therefore by its definition, $\mathcal{V}_v(\mathcal{G})$ is the row space of A. A convenient basis of this row space is a matrix A_r, (a 'reduced incidence matrix of $\mathcal{G}$'), obtained by dropping one row of A. For $\mathcal{V}_i(\mathcal{G})$, a convenient basis is obtained through the 'fundamental circuit matrix'. This is constructed as follows. Pick a spanning tree t of the graph. Let $e' \in \mathcal{E}(\mathcal{G}) - t$. Then it is easy to see that $e' \cup t$ contains a unique loop $\mathcal{L}_{e'}$. We give this loop an orientation that agrees with that of e'.

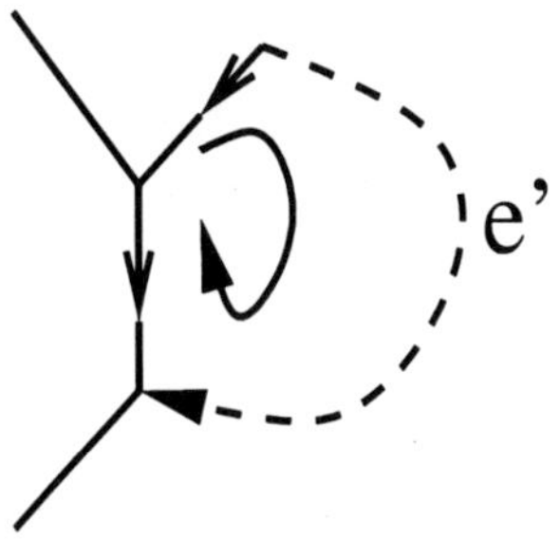

Fig. 12.4 The loop $\mathcal{L}_{e'}$

We now construct a vector $l_{e'}$, where

$l_{e'}(e) = 0$ if e is not in the loop $l_{e'}$

$l_{e'}(e) = +1(-1)$ if e is in the loop $l_{e'}$ and its orientation agrees with that of (opposes that of) the loop.

The fundamental circuit matrix B with respect to t has one row $l_{e'}$ corresponding to each $e' \in \mathcal{E}(\mathcal{G}) - t$. The rows of B are linearly independent since, if $e', e'' \in \mathcal{E}(\mathcal{G}) - t$ and $e' \neq e''$ then we have $l_{e'}(e'') = l_{e''}(e') = 0$, whereas we have $l_{e'}(e') = l_{e''}(e'') = 1$.

It is easy to see that if a current vector i (i.e., a vector in $\mathcal{V}_i(\mathcal{G})$) has zero value on all edges outside t, then it must have zero value on edges of t. From this it follows that the rows of B form a basis of $\mathcal{V}_i(\mathcal{G})$.

12.2.1 *Nodal Analysis*

This method is intended for networks which have only current sources and resistors. Let A_r be the reduced incidence matrix of the graph $\mathcal{G}$ of the network. Partition the columns of A_r into those corresponding to conductances and current sources as $\left[A_{rG} \; A_{rJ} \right]$. Let (v,i) be a solution of the network, we then have

$$A_{rG}.i_G = -A_{rJ} i_J \tag{12.5}$$

where i is partitioned in $\begin{bmatrix} i_G \\ i_J \end{bmatrix}$ corresponding to conductances and current sources. Now

$$i_G = G v_G \tag{12.6}$$

where G is a diagonal matrix.

$$\begin{bmatrix} v_G \\ v_J \end{bmatrix} = \begin{bmatrix} A_{rG}^T \\ A_{rJ}^T \end{bmatrix} x \tag{12.7}$$

since the rows of A_r form a basis of $\mathcal{V}_v(\mathcal{G})$.
It thus follows that

$$(A_{rG} G A_{rG}^T) x = -A_{rJ} i_J \tag{12.8}$$

Observe that the matrix A_r is obtained from an incidence matrix of $\mathcal{G}$ by omitting the row corresponding to some node d.

Since G is a positive diagonal and therefore positive definite matrix, $A_{rG} G A_{rG}^T$ is positive definite if rows of A_{rG} are linearly independent. The matrix AGA^T has rows and columns corresponding to nodes of $\mathcal{G}$ and $A_{rG} G A_{rG}^T$ is obtained from AGA^T by omitting the row and the column corresponding to d. The entry (i, j) of AGA^T is the negative of the conductance of the branch between i and j and the diagonal entry (i, i) is the sum of the conductances of the edges incident at i.

Any linear equation of the form $Cx = b$ where C is symmetric, can be interpreted as arising from nodal analysis of a suitable resistive network. This resistive network may be constructed as follows:
First put down one node per row/column of C and then also for an additional datum node which we will call d.
Whenever C_{ij} is non-zero join i^{th} node to j^{th} node with a conductance of value $-C_{ij}$. Connect i^{th} node to 'd' with a conductance of value $(C_{ii} + \Sigma_{i \neq j} C_{ij})$.
Next from d to each node j connect a current source of value b_j directed into j. Let $\begin{bmatrix} A_G & A_J \end{bmatrix}$ be the incidence matrix of the graph of the network, columns partitioned corresponding to conductances and current sources. Let $\begin{bmatrix} A_{rG} & A_{rJ} \end{bmatrix}$ be the reduced incidence matrix obtained from this matrix by omitting the row corresponding to d. Let the columns of A_G be according to increasing edge numbers. Let G be the diagonal matrix such that G_{kk} = conductance of kth edge. It is then clear that

$$A_{rG} G {A_{rG}}^T = C \tag{12.9}$$

and if we write nodal analysis equations for the network we will obtain the equation $Cx = b$. Let us call the resistive network corresponding to $Cx = b$ obtained by the above procedure, $\mathcal{N}_{Cb}$.

Example: Let $Cx = b$ be as given in Equation 12.10.

$$\begin{bmatrix} 5 & -3 \\ -3 & 4 \end{bmatrix} \begin{matrix} x_1 \\ x_2 \end{matrix} = \begin{bmatrix} b_1 \\ b_2 \end{bmatrix} \tag{12.10}$$

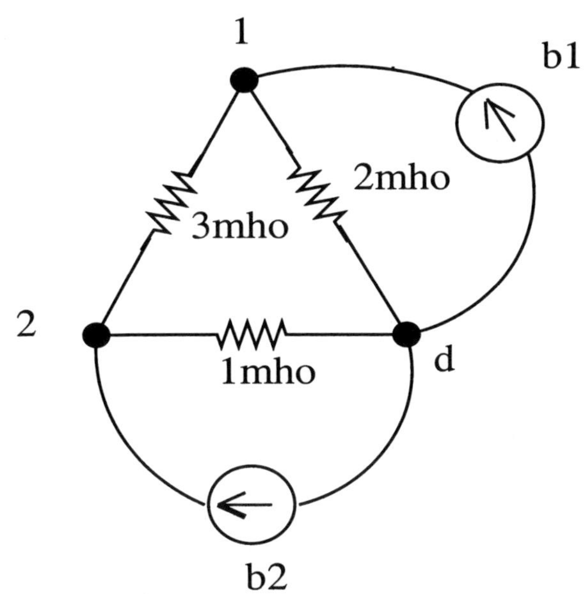

Fig. 12.5 $\mathcal{N}_{Cb}$

Then $\mathcal{N}_{Cb}$ is as in Figure 12.5 (*mho* is to indicate that we are dealing with conductance value; $2mho \equiv \frac{1}{2}\Omega$, where Ω indicates resistance value). We remark that, if all conductances in a network are positive then the matrix $A_{rG}GA_{rG}{}^T$ will be diagonally dominant with positive diagonal entries and non positive off-diagonal entries, i.e., $\sum_j C_{ij} \geq 0$ with the summation being positive for at least one node. Such a matrix is necessarily positive definite. However, even if C is positive definite, we may still have negative conductances in $\mathcal{N}_{Cb}$. For instance, if in the coefficient matrix of Equation 12.10 we replace 4 by 2, then in Figure 12.5, the $1mho$ conductance would be replaced by $-1mho$.

12.2.2 Loop Analysis

This is intended for networks which have only resistances and voltage sources. Let B be a fundamental circuit matrix of the graph $\mathcal{G}$ of the network, partitioned into $\left[B_R|B_E\right]$ corresponding to resistances and voltage sources. Note that rows of B form a basis of $\mathcal{V}_i(\mathcal{G})$.

We then have, if (v, i) is a solution of the network

$$\left[B_R B_E\right] \begin{matrix} v_R \\ v_E \end{matrix} = 0 \tag{12.11}$$

where v is partitioned as $\begin{pmatrix} v_R \\ v_E \end{pmatrix}$ corresponding to resistors and voltage sources. We have

$$v_R = R i_R \tag{12.12}$$

$$\begin{bmatrix} i_R \\ i_E \end{bmatrix} = \begin{bmatrix} B_R^T \\ B_E^T \end{bmatrix} y \tag{12.13}$$

where i is partitioned into $\begin{pmatrix} i_R \\ i_E \end{pmatrix}$, since $i \in \mathcal{V}_i(\mathcal{G})$.

Hence, $B_R R B_R^T y = -B_E v_E$.

These are the loop analysis equations of the network. It can be shown that rows of B_R would be linearly independent provided voltage sources do not form loops. Thus, if R is positive diagonal, the matrix $B_R R B_R^T$ is positive definite when the rows of B_R are linearly independent, i.e., when voltage sources do not form loops. Unlike the matrix $A_{rG} G A_{rG}{}^T$, the matrix $B_R R B_R^T$ can become very dense even when the graph has few edges. For instance, if there is a common tree branch to all the fundamental circuits, the matrix would be fully dense. However, given a vector y, computation of $\left[B_R R B_R^T\right] y$ requires only the knowledge of the spanning tree (no explicit storage of the matrix) and can be done graph theoretically in linear time (on size of the network).

It is easily verified that if every device is composite of the form $v(e) - \mathcal{E}(e) = r(e)(i(e) - \mathcal{J}(e))$, where $r(e)$ is finite and nonzero, then the nodal analysis and loop analysis equations can both be written. The former would appear as

$A_{rG} G A_{rG}{}^T x = -A_{rJ}\mathcal{J} + A_{rG} G \mathcal{E}$ and the latter as $BRB^T y = -B\mathcal{E} + BR\mathcal{J}$.

12.2.3 *Hybrid Analysis*

Hybrid Analysis methods are originally due to G.Kron ([Kron (1939)],[Kron (1963)]) as simplified by Branin [Branin Jr. (1962)]). The development presented here is however based on a topological version reported in [Narayanan (1979a)].

We will assume that every device is composite of the form $v(e) - \mathcal{E}(e) = r(e)(i(e) - \mathcal{J}(e))$.

Let the edges of the graph $\mathcal{G}$ of the network be partitioned into sets P and Q, where devices in the two sets are independent of each other. (This partition is given beforehand according to some user defined criterion.) Let t be a spanning tree that contains as many edges as possible from the set P (and therefore as few edges as possible from Q). Denote $t \cap P$ by M. Let $(E(\mathcal{G}) - t) \cap Q$ be denoted by L.

We now build two networks $\mathcal{N}_{PL}$ and $\mathcal{N}_{QM}$ as follows:- $\mathcal{N}_{PL}$ has graph $\mathcal{G}_{PL}$ with edge set $P \cup L$ built from $\mathcal{G}$ by short circuiting (fusing the end points of) edges in $t \cap Q$ and removing them. The devices in P have the same characteristics as in $\mathcal{N}$ and L has no device characteristic constraints. $\mathcal{N}_{QM}$ has graph $\mathcal{G}_{QM}$ with edge set $Q \cup M$ built from $\mathcal{G}$ by open circuiting edges (removing the edges but leaving the end points in place) in $\big(E(G) - t\big) \cap P$. The devices in Q have the same characteristics as in $\mathcal{N}$ and M has no device characteristic constraints. (Note that the L, M edges are present in both networks.) The main theorem of [Narayanan (1979a)] says that solving $\mathcal{N}$ is equivalent to solving $\mathcal{N}_{PL}$ and $\mathcal{N}_{QM}$ simultaneously keeping i_L, v_M the same in both networks.

Hybrid analysis equations can be written as follows:-

(1) Write nodal analysis equations for $\mathcal{N}_{PL}$ treating branches in L as current sources of value i_L.
(2) Write loop analysis equations for $\mathcal{N}_{QM}$ treating branches in M as voltage sources of value v_M.
(3) Force the constraints that i_L is the same in both networks and v_M is the same in both networks.

We go through the formal development below:- Let $[A_{rP} A_{rL}] = [A_{rM} A_{r(P-M)} A_{rL}]$ be a reduced incidence matrix of $\mathcal{G}_{PL}$. Let the device characteristic of the edges in P be expressible as

$$(i_P - \mathcal{J}_P) = G_P(v_P - \mathcal{E}_P). \tag{12.14}$$

We then have (since $i \in \mathcal{V}_i(\mathcal{G})$),

$$A_{rP} i_P + A_{rL} i_L = 0 \tag{12.15}$$

$$i.e., A_{rP}(i_P - \mathcal{J}_P) + A_{rL}i_L = -A_{rP}\mathcal{J}_P \tag{12.16}$$

$$i.e., A_{rP}G_P(v_P - \mathcal{E}_P) + A_{rL}i_L = -A_{rP}\mathcal{J}_P \tag{12.17}$$

$$i.e., A_{rP}G_Pv_P + A_{rL}i_L = -A_{rP}\mathcal{J}_P + A_{rP}G_P\mathcal{E}_P \tag{12.18}$$

Now,

$$\begin{bmatrix} v_P \\ v_L \end{bmatrix} = \begin{bmatrix} A_{rP}^T \\ A_{rL}^T \end{bmatrix} v_{nP}, \tag{12.19}$$

for some v_{nP} (since $\begin{bmatrix} v_P \\ v_L \end{bmatrix} \in \mathcal{V}_v(\mathcal{G}_{PL})$). We thus have,

$$(A_{rP}G_PA_{rP}^T)v_{nP} + A_{rL}i_L = -A_{rP}\mathcal{J}_P + A_{rP}G_P\mathcal{E}_P. \tag{12.20}$$

These are the nodal analysis equations of $\mathcal{N}_{PL}$. Next for $\mathcal{N}_{QM}$, we choose the spanning tree t for building the fundamental circuit matrix of $\mathcal{G}_{QM}$. Let
$\begin{bmatrix} B_M B_Q \end{bmatrix} \equiv \begin{bmatrix} B_M B_{t\cap Q} B_L \end{bmatrix} \equiv \begin{bmatrix} B_M B_{t\cap Q} I_L \end{bmatrix}$
be the fundamental circuit matrix of $\mathcal{G}_{QM}$ with respect to tree t.
Let the device characteristic in Q be expressible as $v_Q - \mathcal{E}_Q = R_Q(i_Q - \mathcal{J}_Q)$.
We then have,

$$B_Mv_M + B_Qv_Q = 0 \tag{12.21}$$

$$i.e., B_Mv_M + B_Q(v_Q - \mathcal{E}_Q) = -B_Q\mathcal{E}_Q \tag{12.22}$$

$$i.e., B_Mv_M + B_QR_Q(i_Q - \mathcal{J}_Q) = -B_Q\mathcal{E}_Q \tag{12.23}$$

$$i.e., B_Mv_M + B_QR_Qi_Q = -B_Q\mathcal{E}_Q + B_QR_Q\mathcal{J}_Q \tag{12.24}$$

We have,

$$\begin{bmatrix} i_Q \\ i_M \end{bmatrix} = \begin{bmatrix} B_Q^T \\ B_M^T \end{bmatrix} y \tag{12.25}$$

for some y, since $\begin{bmatrix} i_Q \\ i_M \end{bmatrix} \in \mathcal{V}_i(\mathcal{G}_{QM})$. Hence,

$$B_Mv_M + B_QR_QB_Q^Ty = -B_Q\mathcal{E}_Q + B_QR_Q\mathcal{J}_Q. \tag{12.26}$$

Now we impose the condition that v_M is the same in both networks and so is i_L.
But this means,

$$A_{rM}^T v_{nP} = v_M \tag{12.27}$$

and,

$$I_L^T y = i_L \tag{12.28}$$

So we get the hybrid equations,

$$A_{rP} G_P A_{rP}^T v_{nP} + A_{rL} i_L = -A_{rP} \mathcal{J}_P + A_{rP} G_P \mathcal{E}_P \tag{12.29}$$

$$B_M A_{rM}^T v_{nP} + B_Q R_Q B_Q^T i_L = -B_Q \mathcal{E}_Q + B_Q R_Q \mathcal{J}_Q \tag{12.30}$$

The matrix $\begin{bmatrix} A_{rP} G_P A_{rP}^T & A_{rL} \\ B_M A_{rM}^T & B_Q R_Q B_Q^T \end{bmatrix}$ is positive definite if G_P, R_Q are positive definite.

It can be shown that $B_M A_{rM}^T = -A_{rL}^T$ [Narayanan (1979b)].
For linear equations of the form

$$\begin{bmatrix} \tilde{G} & H \\ -H^T & \tilde{R} \end{bmatrix} \begin{bmatrix} x_1 \\ y_2 \end{bmatrix} = \begin{bmatrix} d_1 \\ d_2 \end{bmatrix} \tag{12.31}$$

where $\tilde{G}$, $\tilde{R}$ are symmetric positive definite, a variation of the preconditioned conjugate gradient ('modified PCG') method works. This is described in the Appendix.

Suppose nodal analysis equations of $\mathcal{N}$ are

$$Cx = b.$$

It is then possible to keep the currents and voltages in the resistors unchanged but make every current source appear in parallel with some resistor by using the procedure of i-shift (where we replace a current source which is across a 'path' of resistors by a sequence of current sources of the same value across each of the resistors in the path). In the resulting network, every current source only appears in parallel with some resistor. We can therefore use hybrid analysis on this network. On the face of it, hybrid analysis is more cumbersome than nodal analysis since the coefficient matrix is usually dense in the case of the former even where it is sparse in case of latter. However its behavior in the case of iterative methods such as conjugate gradient can be much better. For instance, suppose the original network has conductances of value ranging from 10^{-4} to 10^4. We can divide these into two parts - those which are in the range 1 to 10^4 can go into P and those in the range 10^{-4} to 1 can go into Q. In the hybrid analysis equations 12.29,12.30, the matrices $\left[A_{rP}G_PA_{rP}^T\right]$ and $\left[B_QR_QB_Q^T\right]$ would have condition number of the order square root of the condition number of C. However, the over all coefficient matrix has some asymmetry since the submatrices A_{rL}, $B_MA_{rM}^T$ are negative transposes of each other. Numerical experiments reported in Section 12.3 seem to support the use of hybrid analysis and the solution of the resuting equations through the modified PCG method. In the modified PCG method, the key computational step is to find Ky when given y, where K is the coefficient matrix. Suppose K is the matrix $\begin{bmatrix} A_{rP}G_PA_{rP}^T & A_{rL} \\ B_MA_{rM}^T & B_QR_QB_Q^T \end{bmatrix}$. Then it is not necessary to store K explicitly for effecting this computation. It is only necessary to store the graph $\mathcal{G}$, the spanning tree t and the conductance values. Further, all the matrix vector products can be computed either graph theoretically or, in the case of multiplication by G_P or R_Q, by scaling.

12.3 Experimental Results

In this section we first present results on the comparative performance of the circuit simulation based min-cost flow solver with respect to standard computer science based algorithms reported in [Ahuja, Magnanti and Orlin (1993)]. Next we present results on the performance of the modified PCG algorithm on randomly chosen circuits.

The graphs *fGraph* of Table 12.1 are planar flow graphs of the large 'dense grid' variety, i.e., graphs with a rectangular grid structure with an internal node in each window connected to all the peripheral nodes of the window. The cost and capacity were randomly chosen to be in the range $1-10^6$. Our present version of flow solver is based on LU factorization and cannot yet handle non planar flow graphs of size larger than about 10,000 nodes because of loss of sparsity during the factorization process. In Table 12.1, t_{cost}, t_{cap} and t_{pd} are the times taken in seconds by *Cost Scaling*, *Capacity Scaling* and *Primal-Dual* LEDA [Leda (2005)] routines for Min Cost Flow for finding the minimum cost cLR and the corresponding flow distribution of flow graph *fGraph*. (Note that cLR is far from the exact minimum cost cMs. This situation appears to arise because of the large range of values for capacity and cost). t_{fsDC} is the time taken in seconds by the DC analyzer based min cost flow simulator (DCFS) in finding the

Table 12.1 Performance of DCFS and standard computer science algorithms for large dense grid planar flow graphs with cost and capacity in the range 1-10^6

$fGraph$	t_{cost}	t_{cap}	t_{pd}	cLR $\times 10^{12}$	t_{MS}	cMS $\times 10^{12}$	t_{dcAna}	cFS $\times 10^{12}$	itr
fg10k	-	3.84	777.65	4.99	1.475	93.36	10.2	93.35	28
fg20k	-	9.86	-	8.38	16.123	96.93	22.28	96.91	28
fg30k	-	23.81	-	10.43	43.299	113.40	35.18	113.37	26
fg40k	-	39.09	-	11.73	26.347	106.15	53.30	106.11	29
fg50k	-	1468.19	-	13.50	32.968	114.44	78.08	114.41	32
fg100k	-	-	-	-	178.831	112.38	194.19	112.33	31
fg200k	-	-	-	-	985.602	117.61	644.73	117.53	29

approximate minimum cost cFS and the corresponding flow distribution of $fGraph$ in itr N-R iterations when we use cost and capacity scaling. t_{MS} is the time taken in seconds by the public domain min cost flow simulator *MCF-1.3* [Opt. software (2004)] in finding min cost flow solution cMs of the same flow graph. Here "-" in the column t_{cost} indicates that the experiment could not be performed due to the *overflow* error and in the column t_{pd} "-" indicates that the experiment was not performed because it took too long. Experiments were performed on a 3 GHz PIV processor with 1 GB RAM. It can be seen that circuit simulation based flow solver solution cFS comes within 0.1% of the exact solution cMs.

Tables 12.2, 12.3 [Trivedi (2006)] show experimental results for hybrid analysis of networks using the modified PCG method (see the Appendix). *In these cases, direct nodal analysis using preconditioned conjugate gradient method failed to converge.* Approximately 30% of the conductances were chosen to be 10^4, and equal number to be 10^{-4} and the remaining to be in between at random. The resistors were divided as discussed earlier. The set P contained conductances in the range 1 to 10^4, Q those in range 10^{-4} to 1 and hybrid analysis was performed with nodal analysis for $\mathcal{N}_{PL}$ and loop analysis for $\mathcal{N}_{QM}$, the devices in Q being treated as resistors, or more generally of the form $v_Q = R_Q(i_Q - \mathcal{J}_Q)$.

iterations is the iteration count taken by modified CG routine and $t_{Solution}$ is the time in seconds in solving matrix of form given in Equation 12.31. Experiment has been performed on a 3.2 GHz machine having 1 GB RAM. Both the planar and nonplanar graphs were generated randomly with average degree between 3 and 4. (Usually PCG performs better for random nonplanar networks than for random planar networks.)

Table 12.2 Planar circuit analysis using modified CG for conductance range 1 mho to 10000 mho and resistance range 1 ohm to 10000 ohm

Size of matrix	*iterations*	$t_{Solution}$ (in secs)
100k	1896	57.02
200k	2589	165.66
300k	3244	309.89
400k	3661	466.24
500k	4303	688.37
600k	4486	853.00
700k	5130	1141.80
800k	5260	1332.18
900k	5594	1612.74
1Million	6236	2001.06

Table 12.3 Nonplanar circuit analysis using modified CG for conductance range 1 mho to 10000 mho and resistance range 1 ohm to 10000 ohm

Size of matrix	*iterations*	$t_{Solution}$ (in secs)
100K	134	6.56
200K	141	14.98
300K	162	29.66
400K	146	39.56
500K	179	63.13
600K	164	73.56
700K	189	101.56
800K	179	112.14
900K	178	126.13
1 Million	180	144.71

12.4 A Proposal

The experimental results of the previous section appear to us to be favorable towards adopting hybrid analysis as the main method for linear network analysis if we are to use iterative methods for solution of linear equations. In our case all the resistors are positive so that variants of conjugate gradient method could be used for solution. Suppose the equation $Cx = b$ has the coefficient matrix, symmetric but not positive definite. It would be interesting to explore whether the ideas of this paper are useful even in this more general case. The steps are clear:-

(1) Use the ideas in Subsection 12.2.1 to decompose C in the product form $A_r G A_r^T$. The diagonal matrix G may have both positive and negative entries.
(2) Store the graph for which A_r is the reduced incidence matrix, the diagonal entries of G and the column vector b.
(3) Divide the resistors into two groups corresponding to any criterion relevant to the problem. Call these sets P and Q.
(4) Store the spanning tree t (containing as many edges of P as possible) described in Subsection 12.2.3. Let the resulting hybrid equations be

$$\begin{bmatrix} A_{rG_P} G_P A_{rG_P}^T & H \\ -H^T & B_Q R_Q B_Q^T \end{bmatrix} \begin{bmatrix} x_1 \\ y_2 \end{bmatrix} = \begin{bmatrix} d_1 \\ d_2 \end{bmatrix} \qquad (12.32)$$

Compute and store the vector $\begin{bmatrix} d_1 \\ d_2 \end{bmatrix}$. By storing the graph, the spanning tree and the conductances, in effect we have then stored $\begin{bmatrix} A_{rP} G_P A_{rP}^T & A_{rL} \\ B_M A_{rM}^T & B_Q R_Q B_Q^T \end{bmatrix}$. (Since C may not be positive definite, this matrix also may not be).
(5) Modify a standard iterative scheme valid for equations with general symmetric coefficient matrices along the lines of the development in the Appendix.

12.5 Conclusion

In this paper we have described our computational experience with solving networks with resistors, voltage sources and current sources. We have given a formal self contained treatment of methods of analysis such as nodal, loop and hybrid analysis. We have presented numerical evidence that hybrid analysis has some advantages while using iterative methods, since it permits us to work with matrices with better condition numbers than nodal or loop analysis. We have presented in the appendix a variation of the preconditioned conjugate gradient method which is suited for the equations that result when we use hybrid analysis. We indicate that this approach may be used with equations more general than the ones that arise in network analysis.

Appendix: A variation of PCG

In this appendix we describe a variation of the preconditioned conjugate gradient method which is particularly suited for the solution of hybrid analysis equations of electrical networks.

Let $A \equiv \begin{bmatrix} \tilde{G} & H \\ -H^\top & \tilde{R} \end{bmatrix}$ be a positive definite matrix, where $\tilde{G} \in \mathbf{R}^{m\times m}$, $H \in \mathbf{R}^{m\times(n-m)}$ *and* $\tilde{R} \in \mathbf{R}^{(n-m)\times(n-m)}$, with the sub matrices $\tilde{G}$, $\tilde{R}$ being symmetric positive definite. Such matrices arise as coefficient matrices of hybrid analysis equations, which have as unknowns some voltage and some current variables, for resistive networks. It would be convenient if we could use an algorithm similar to the Conjugate Gradient (CG) algorithm (see for instance [Greenbaum (1997)]) for solving such equations. But the CG algorithm is intended for symmetric positive definite matrices and the more general bi-CG algorithm is twice as costly in terms of computation. The elementary strategy described in this paper allows us to modify the CG algorithm to make it valid for the solution of such equations without any additional cost.

We define a pseudo-inner product $<, \tilde{>}$ on the space of all real n-tuples as follows.

Let an n-tuple $(z_1, \ldots, z_m, z_{m+1}, \ldots, z_n)$ be partitioned as $(\underset{\sim}{z}_1, \underset{\sim}{z}_2)$, where

$$\underset{\sim}{z}_1 = (z_1, \ldots, z_m) \text{ and}$$
$$\underset{\sim}{z}_2 = (z_{m+1}, \ldots, z_n).$$

Here the partition is intended to be consistent with that of the matrix $\begin{bmatrix} \tilde{G} & H \\ -H^\top & \tilde{R} \end{bmatrix}$.

Then

$$\begin{aligned} < \underset{\sim}{x}, \underset{\sim}{y} \tilde{>} &\equiv < (\underset{\sim}{x}_1, \underset{\sim}{x}_2), (\underset{\sim}{y}_1, \underset{\sim}{y}_2) \tilde{>} \\ &\equiv \sum x_{1j} * y_{1j} - \sum x_{2j} * y_{2j}. \end{aligned}$$

We note that the pseudo inner product has been defined as above specifically to deal with the matrix A. Since A is partitioned according to $m, n-m$, the number m enters the definition of the pseudo-inner product $<, \tilde{>}$.

It is clear that $< x, y \tilde{>} = < y, x \tilde{>}, < \alpha x, y \tilde{>} = \alpha < x, y \tilde{>}, < x, y + y' \tilde{>} = < x, y \tilde{>} + < x, y' \tilde{>}$.

For convenience we define another pseudo inner product

$$< \underset{\sim}{x}, \underset{\sim}{y} \tilde{>}_A \equiv < \underset{\sim}{x}, A\underset{\sim}{y} \tilde{>}$$

We say $\underset{\sim}{x}, \underset{\sim}{y}$ are A- orthogonal if

$$< \underset{\sim}{x}, \underset{\sim}{y} \tilde{>}_A = 0.$$

The following lemma states the main motivation for the definition of the pseudo inner product. This is needed for certain canonical properties, important for algorithms to go through, to hold.

Lemma 12.1. $< A\underset{\sim}{x}, \underset{\sim}{y} \tilde{>} =< \underset{\sim}{x}, A\underset{\sim}{y} \tilde{>}$.

Proof. Both LHS and RHS are equal to the expansion
$x_1{}^\top \tilde{G} y_1 - x_2{}^\top \tilde{R} y_2 + x_1{}^\top \tilde{H} y_2 + x_2{}^\top \tilde{H}^\top y_1$ □

Lemma 12.2. *Let x be linearly dependent on $y_0, y_1, \ldots y_k$ and let $y_0, y_1, \ldots y_k$ be A-orthogonal to each other.*

Then $x = \frac{<x, y_0 \tilde{>}_A}{<y_0, y_0 \tilde{>}_A} y_0 + \ldots + \frac{<x, y_k \tilde{>}_A}{<y_k, y_k \tilde{>}_A} y_k$ (assuming the denominators are non zero).

Proof. Let $x = \alpha_0 y_0 + \ldots + \alpha_k y_k$. Then

$$\begin{aligned}\frac{< x, y_j \tilde{>}_A}{< y_j, y_j \tilde{>}_A} &= \frac{< \alpha_0 y_0 + \ldots + \alpha_k y_k, y_j \tilde{>}_A}{< y_j, y_j \tilde{>}_A} \\ &= \alpha_j \frac{< y_j, y_j \tilde{>}_A}{< y_j, y_j \tilde{>}_A} \\ &= \alpha_j\end{aligned}$$

□

Equivalently, when x is linearly dependent on $y_0, y_1, \ldots y_k$, and these latter are A-orthogonal to each other, if we successively remove from x its 'A-projections' on $y_0, \ldots, y_k$ we will be left with the zero vector.
Further, let $x_r \equiv x_{r-1} - \frac{<x_{r-1}, y_{r-1} \tilde{>}_A}{<y_{r-1}, y_{r-1} \tilde{>}_A} y_{r-1}, r = 1, \cdots, k$, where $x_0 = x$.
Then it can be seen that $\frac{<x, y_j \tilde{>}_A}{<y_j, y_j \tilde{>}_A} = \frac{<x_j, y_j \tilde{>}_A}{<y_j, y_j \tilde{>}_A}$, $j = 1, 2, \ldots, k$. The modified conjugate gradient(CG) algorithm is obtained from the usual CG algorithm by replacing the usual inner product $<, >$ by the pseudo inner product $<, \tilde{>}$ wherever the former occurs. For the sake of completeness we describe the modified CG algorithm below.

A.1 *Modified CG*

This algorithm is intended for positive definite matrices of the form $A = \begin{pmatrix} \tilde{G} & H \\ -H^\top & \tilde{R} \end{pmatrix}$ where $\tilde{G}$, $\tilde{R}$ are symmetric positive definite.

Given an initial guess x_0, compute $r_0 = b - Ax_0$ and set $p_0 = r_0$.
For $k = 1, 2, \ldots,$ compute Ap_{k-1}.
Set $x_k = x_{k-1} + a_{k-1}p_{k-1}$, where $a_{k-1} = \frac{<r_{k-1}, p_{k-1}\tilde{>}}{<p_{k-1}, Ap_{k-1}\tilde{>}}$.
Compute $r_k = r_{k-1} - a_{k-1}Ap_{k-1}$.
Set $p_k = r_k + b_{k-1}p_{k-1}$, where $b_{k-1} = -\frac{<r_k, Ap_{k-1}\tilde{>}}{<p_{k-1}, Ap_{k-1}\tilde{>}}$.

Theorem 12.1. *Let* $A = \begin{bmatrix} \tilde{G} & H \\ -H^\top & \tilde{R} \end{bmatrix}$ *be a positive definite matrix with* $\tilde{G}$, $\tilde{R}$ *symmetric positive definite. Let* $e_i \equiv A^{-1}r_i, \forall i$*.If the modified CG algorithm doesn't encounter a 'A-degenerate vector' (i.e., a vector that is A- orthogonal to itself) while generating* $p_j, r_j, \quad j = 0, 1, \ldots, n-1$ *then we have*
(*a*) $< e_{k+1}, Ap_j\tilde{>} =< p_{k+1}, Ap_j\tilde{>} =< r_{k+1}, r_j\tilde{>} = 0 \quad \forall j \leq k$
(*b*) *The modified CG algorithm generates the exact solution to the linear system* $Ax = b$ *in at most* n *steps.*

Proof. (a) The standard CG proof goes through. For completeness we repeat it replacing $<,\tilde{>}$ by $<,>$.

Since we assume that no A-degenerate vector is encountered, it is clear that the coefficients in the CG algorithm are well defined unless a residual vector is zero in which case the exact solution has been found. Let $r_0, r_1, \cdots, r_k$ be nonzero. By the choice of a_0, it is clear that

$$< r_1, r_0\tilde{>} = < Ae_1, p_0\tilde{>} = < e_1, Ap_0\tilde{>} = 0. \tag{A.1}$$

and from the choice of b_0, it follows that $< p_1, Ap_0\tilde{>} = 0$. Next assume that

$$< r_k, p_j\tilde{>} =< e_k, Ap_j\tilde{>} = < p_k, Ap_j\tilde{>} = < r_k, r_j\tilde{>} = 0, \quad \forall j \leq k-1. \tag{A.2}$$

We then have

$$< p_k, Ap_k\tilde{>} = < r_k, Ap_k\tilde{>}, \tag{A.3}$$

$$< r_k, p_k\tilde{>} = < r_k, r_k\tilde{>}. \tag{A.4}$$

Since $a_k = \frac{<r_k, p_k>}{<p_k, Ap_k>}$, it follows that

$$< r_{k+1}, r_k > = < r_k, r_k > - \; a_k < r_k, Ap_k >.$$

$$= < r_k, r_k > - < r_k, p_k > = 0 \tag{A.5}$$

and

$$< e_{k+1}, Ap_k > = < r_{k+1}, p_k > = < r_k, p_k > - \; a_k < Ap_k, p_k >$$

$$= < r_k, p_k > - < r_k, p_k > = \; 0. \tag{A.6}$$

Since $b_k = - \frac{<r_{k+1}, Ap_k>}{<p_k, Ap_k>}$, we have

$$< p_{k+1}, Ap_k > = < r_{k+1}, Ap_k > + b_k < p_k, Ap_k >$$

$$=< r_{k+1}, Ap_k > - < r_{k+1}, Ap_k > = \; 0. \tag{A.7}$$

Next we have for $j \leq (k-1)$,

$$< r_{k+1}, r_j > = < r_k - a_k Ap_k, r_j > = - \; a_k < p_k, A(p_j - b_{j-1}p_{j-1}) > = \; 0, \tag{A.8}$$

$$< p_{k+1}, Ap_j > = < r_{k+1} + b_k p_k, Ap_j > =< r_{k+1}, Ap_j >$$

$$= < r_{k+1}, a_j^{-1}(r_j - r_{j+1}) > = \; 0. \tag{A.9}$$

Thus by induction the desired equalities follows.
(b) The vectors $p_0, \ldots, p_{k-1}$, $\;k < n$ are A-orthogonal to each other. Suppose we have $0 = e_k = e_{k-1} - a_{k-1}p_{k-1}$. The solution in this case is obtained in $k < n$ steps.
Next, let $k = n$. Let $p_0, \ldots, p_{k-1}$ be non A-degenerate vectors (ie., $< p_j, \, Ap_j > \neq 0$).

Claim 12.1. *$p_0, \ldots p_{k-1}$ are linearly independent.*

By induction suppose $p_0, \ldots p_j$ are independent.
If $p_{j+1} = \alpha_0 p_0 + \ldots + \alpha_j p_j$, then $< p_{j+1}, Ap_{j+1} > = {\alpha_0}^2 < p_0, \, Ap_0 > + \ldots + {\alpha_j}^2 < p_j, \, Ap_j >$, using the fact that $p_0, \ldots, p_{k-1}$ are A-orthogonal to each other. Now p_{j+1} is non A-degenerate. So LHS$\neq 0$.
Next

$$0 = < p_{j+1} - \alpha_0 p_0 - \ldots - \alpha_j p_j, \; A(p_{j+1} - \alpha_0 p_0 - \ldots - \alpha_j p_j) >$$
$$= < p_{j+1}, \; Ap_{j+1} > + {\alpha_0}^2 < p_0, \; Ap_0 > + \ldots + {\alpha_j}^2 < p_j, Ap_j >$$

But this gives $\langle p_{j+1}, Ap_{j+1}\rangle = -\langle p_{j+1}, Ap_{j+1}\rangle$ i.e., p_{j+1} is A-degenerate. This is a contradiction. We conclude that p_{j+1} is linearly independent of $p_0, \ldots, p_j$. So by induction it follows that $p_0, \ldots, p_{k-1}$ are linearly independent. This proves the claim.

Since $k = n$, e_0 is linearly dependent on $p_0, \ldots, p_{k-1}$. Now $p_0, \ldots, p_{k-1}$ are A-orthogonal to each other. Hence, $e_0 = \frac{\langle e_0, Ap_0\rangle}{\langle p_0, Ap_0\rangle} p_0 + \ldots + \frac{\langle e_0, Ap_{k-1}\rangle}{\langle p_{k-1}, Ap_{k-1}\rangle} p_{k-1}$.

We have from the algorithm, $e_k = e_0 - \frac{\langle e_0, Ap_0\rangle}{\langle p_0, Ap_0\rangle} p_0 - \ldots - \frac{\langle e_{k-1}, Ap_{k-1}\rangle}{\langle p_{k-1}, Ap_{k-1}\rangle} p_{k-1} = e_0 - \frac{\langle e_0, Ap_0\rangle}{\langle p_0, Ap_0\rangle} p_0 - \ldots - \frac{\langle e_0, Ap_{k-1}\rangle}{\langle p_{k-1}, Ap_{k-1}\rangle} p_{k-1} = 0$. Thus the algorithm must terminate in at most n steps. □

A.2 *Modified Preconditioned CG*

Lemma 12.3. *(a) There exist L_1, K_1 such that $L_1 L_1^{\top} = \tilde{G}$ and $K_1 K_1^{\top} = \tilde{R} + H^{\top}(\tilde{G})^{-1}H$.*

Hence the matrix

(b) $\begin{bmatrix} \tilde{G} & H \\ -H^{\top} & \tilde{R} \end{bmatrix}$ *can be factored in the form*

$$\begin{bmatrix} \tilde{G} & H \\ -H^{\top} & \tilde{R} \end{bmatrix} = \begin{bmatrix} L_1 & 0 \\ -H^{\top} L_1^{-\top} & K_1 \end{bmatrix} \begin{bmatrix} L_1^{\top} & L_1^{-1} H \\ 0 & K_1^{\top} \end{bmatrix}$$

Proof. (a) We note that $\tilde{G}, \tilde{R}, (\tilde{G})^{-1}$ are positive definite and $H^{\top}(\tilde{G})^{-1}H$ is positive semidefinite. Hence $\tilde{G}$ can be factored as $L_1 L_1^{\top}$ and $\tilde{R} + H^{\top}(\tilde{G})^{-1}H$, being positive definite, can be factored as $K_1 K_1^{\top}$.
(b)Routine. □

Let $A = \begin{pmatrix} \tilde{G} & H \\ -H^{\top} & \tilde{R} \end{pmatrix}$. where $\tilde{G} \in \mathbb{R}^{r \times r}$, $H \in \mathbb{R}^{r \times (n-r)}$ and $\tilde{R} \in \mathbb{R}^{(n-r)\times(n-r)}$. We define the modified transpose $M^{\tilde{\top}}$ for an $n \times n$ matrix M, partitioned as $\begin{bmatrix} M_{11} & M_{12} \\ M_{21} & M_{22} \end{bmatrix}$, where M_{11}, M_{22} are $r \times r$, $(n-r) \times (n-r)$ respectively, as $M^{\tilde{\top}} \equiv \begin{bmatrix} (M_{11})^{\top} & -M_{21}^{\top} \\ -M_{12}^{\top} & M_{22}^{\top} \end{bmatrix}$. Clearly $(M^{\tilde{\top}})^{\tilde{\top}} = M$. It is also easy to see that $\langle x, My\rangle = \langle M^{\tilde{\top}} x, y\rangle$.

We have the following desirable property for the modified transpose operation.

Lemma 12.4. *Let B be an $n \times n$ matrix. Then, $(B^{\tilde{\top}})^{-1} = (B^{-1})^{\tilde{\top}}$*

Proof. We have,

$$\left(\begin{pmatrix} B_{11} & B_{12} \\ B_{21} & B_{22} \end{pmatrix}^{\tilde{\top}}\right)^{-1} = \begin{pmatrix} B^\top{}_{11} & -B^\top{}_{21} \\ -B^\top{}_{12} & B^\top{}_{22} \end{pmatrix}^{-1}.$$

$$\text{Suppose} \begin{pmatrix} B_{11} & B_{12} \\ B_{21} & B_{22} \end{pmatrix}^{-1} = \begin{pmatrix} C_{11} & C_{12} \\ C_{21} & C_{22} \end{pmatrix}.$$

Then, it can be directly verified by multiplication of the relevant matrices that

$$\left(\begin{pmatrix} B_{11} & B_{12} \\ B_{21} & B_{22} \end{pmatrix}^{\tilde{\top}}\right)^{-1} = \begin{pmatrix} C^\top{}_{11} & -C^\top{}_{21} \\ -C^\top{}_{12} & C^\top{}_{22} \end{pmatrix},$$

which proves the lemma. □

Let M be an $n \times n$ preconditioning matrix of the form $\begin{bmatrix} M_{11} & M_{12} \\ -M_{12}{}^\top & M_{22} \end{bmatrix}$, where M_{11}, M_{22} are positive definite and are $r \times r$, $(n-r) \times (n-r)$ respectively. The modified preconditioned CG, for solving $Ax = b$ (where A is as in Theorem 12.1), is described below.

Given an initial guess x_0.
Compute $r_0 = b - Ax_0$ and solve $Mz_0 = r_0$.
Set $p_0 = z_0$ For $k = 0, 1, \ldots$.
Compute Ap_{k-1}.
Set $x_k = x_{k-1} + a_{k-1}p_{k-1}$, where $a_{k-1} = \frac{<r_{k-1}, z_{k-1}\tilde{>}}{<p_{k-1}, Ap_{k-1}\tilde{>}}$.
Compute $r_k = r_{k-1} - a_{k-1}Ap_{k-1}$.
Solve $Mz_k = r_k$.
Set $p_k = z_k + b_{k-1}p_{k-1}$, where $b_{k-1} = \frac{<r_k, z_k\tilde{>}}{<r_{k-1}, z_{k-1}\tilde{>}}$.

A.3 *Justification for Modified Preconditioned CG*

We need to solve $Ax = b$ where $A^{\tilde{\top}} = A$.
The preconditioning matrix M also satisfies $M^{\tilde{\top}} = M$. Therefore by Lemma 12.3, M can be factorized as $LL^{\tilde{\top}}$. The equation $(M^{-1}A)x = M^{-1}b$ can also be written as

$$(L^{-1}A(L^{\tilde{\top}})^{-1})y = L^{-1}b \tag{A.10}$$

where $y = L^{T}x$. By Lemma 12.4,

$$(L^{T})^{-1} = (L^{-1})^{T}.$$

$$\text{We then have}\,(L^{-1}A(L^{T})^{-1})^{T} = (L^{-1}A(L^{-1})^{T})^{T} = L^{-1}A^{T}(L^{-1})^{T} = L^{-1}A(L^{T})^{-1}.$$

So modified CG can be applied to Equation A.10.
We have the following modified CG algorithm.
Compute $\hat{r_0} = L^{-1}b - By_0$, denoting $L^{-1}A(L^{T})^{-1}$ by B.
Set $\hat{p}_0 = \hat{r}_0$.
Compute $B\hat{p}_{k-1}$.
Set $y_k = y_{k-1} + a_{k-1}\hat{p}_{k-1}$, where $a_{k-1} = \frac{<\hat{r}_{k-1},\hat{p}_{k-1}>}{<\hat{p}_{k-1},B\hat{p}_{k-1}>} = \frac{<\hat{r}_{k-1},\hat{r}_{k-1}>}{<\hat{p}_{k-1},B\hat{p}_{k-1}>}$
Compute $\hat{r}_k = \hat{r}_{k-1} - a_{k-1}B\hat{p}_{k-1}$. Set $\hat{p}_k = \hat{r}_k + b_{k-1}\hat{p}_{k-1}$, where $b_{k-1} = -\frac{<\hat{r}_k,B\hat{p}_{k-1}>}{<\hat{p}_{k-1},B\hat{p}_{k-1}>} = \frac{<\hat{r}_k,\hat{r}_k>}{<\hat{r}_{k-1},\hat{r}_{k-1}>}$

Now $r_k \equiv b - Ax_k = L(L^{-1}b - L^{-1}A(L^{T})^{-1}y_k) = L\hat{r}_k$.
Set $p_k = (L^{T})^{-1}\hat{p}_k$.
We then have

$$\begin{aligned} p_k = (L^{T})^{-1}(\hat{r}_k + b_{k-1}\hat{p}_{k-1}) &= (L^{T})^{-1}\hat{r}_k + b_{k-1}p_{k-1} \\ &= (L^{T})^{-1}L^{-1}r_k + b_{k-1}p_{k-1} \\ &= M^{-1}r_k + b_{k-1}p_{k-1} \\ &= z_k + b_{k-1}p_{k-1}. \end{aligned}$$

Next

$$\begin{aligned} a_{k-1} &= \frac{< L^{-1}r_{k-1},\, L^{-1}r_{k-1}>}{< L^{T}p_{k-1},\, BL^{T}p_{k-1}>} \\ &= \frac{< (L^{T})^{-1}L^{-1}r_{k-1},\, r_{k-1}>}{< p_{k-1},\, L^{-1}BL^{T}p_{k-1}>} \\ &= \frac{< M^{-1}r_{k-1},\, r_{k-1}>}{< p_{k-1},\, Ap_{k-1}>} \\ &= \frac{< z_{k-1},\, r_{k-1}>}{< p_{k-1},\, Ap_{k-1}>}. \end{aligned}$$

Finally, using the simplification adopted for the numerator of the above expression we have,
$b_{k-1} = \frac{<r_k,z_k>}{<r_{k-1},z_{k-1}>}$.
This completes the justification of the modified CG algorithm.

Bibliography

Ahuja, R. K., Magnanti, T. L. and Orlin, J. B. (1993). *Network flows; Theory, Algorithms and Applications*, (Prentice-Hall, Englewood Cliffs, New Jersey).

Branin Jr., F. H. (1962). The relationship between Kron's method and the classical methods of network analysis, *Matrix and Tensor Quarterly*, **12**, pp. 69–105.

Cormen, T. H., Leiserson, C. E. and Rivest, R. L. (1990). *Introduction to Algorithm*, (MIT Press, Cambridge, MA, USA).

Dennis, J. B. (1959). *Mathematical Programming and Electrical Networks*, (MIT Press, Cambridge, Massachusetts).

Greenbaum, A. (1997). *Iterative Methods for Solving Linear Systems*, (SIAM, Philadelphia).

Iri, M. (1969). Network Flows, Transportation and Scheduling, *Theory and Algorithms, Mathematics in Science and Engineering*, **57**, (Academic Press, New York).

Kron, G. (1939). *Tensor Analysis of Networks*, (J.Wiley, New York).

Kron, G. (1963).*Diakoptics - Piecewise Solution of Large Scale Systems*, (McDonald, London).

LEDA libraries, Algorithmic Solutions Software GMBH, Germany, 2005.

A. Löbel. Mcf 1.3 – a network simplex implementation, February 2004.

Narayanan, H. (1979a). A theorem on Graphs and its application to network analysis, *Proceedings of IEEE international symposium of circuit and systems*, pp. 1008–1011.

Narayanan, H. (1979b). *Submodular Functions and Electrical Networks, Annals of Discrete Mathematics*, **54**, North Holland, Amsterdam, The Netherlands.

Narayanan, H. (2004). Mathematical Programming and Electrical Networks, *Operations Research with Economic and Industrial Applications: Emerging Trends* edited by S.R. Mohan and S.K. Neogy, (Anamaya Publishers, New Delhi).

Renegar, J. (2001). *A Mathematical View of Interior-Point Methods in Convex Optimization*, SIAM.

Trivedi, G., Desai, M. P. and Narayanan, H. (2006). *Fast DC Analysis and Its Application to Combinatorial Optimization Problems*, 19th International Conference on VLSI Design, pp. 695–700.

Trivedi,G., Punglia, S. and Narayanan, H. (2007). *Application of DC Analyzer to Combinatorial Optimization Problems*, 20th International Conference on VLSI Design, Accepted.

Trivedi, G. (2006). *A Fast DC Analyzer and its Application to Combinatorial Optimization Problems*, Ph.D Thesis, submitted to EE Dept, IIT Bombay.

Bibliography

[illegible]

Chapter 13

Dynamic Optimal Control Policy in Price and Quality for High Technology Product

A. K. Bardhan
Faculty of Management Studies
University of Delhi
Delhi 110 007, India
e-mail: amit@or.du.ac.in
Udayan Chanda
Department of Operational Research,
University of Delhi, Delhi 110 007, India,
e-mail: uchanda@or.du.ac.in

Abstract

This paper studies optimal control policies of quality level and price for the introduction of a new product with two competing technology generations in a dynamic environment and also proposes a new model in this regard. Lots of work has been done to study the optimal policies pertaining to explanatory variables like price, promotional effort, quality, time etc. In comparison high technology products have received less attention. The proposed model is a combination of diffusion models and the cost function, which is capable of estimating the future profit trends. The new model uses the relationship between the repeat purchasers and the new purchasers in the overall diffusion of a new technology over multiple generations, by separately identifying the two types of adopters.

Key Words: New product diffusion, optimal control, technology substitution, marketing mix variables.

13.1 Introduction

As the global competition becomes more prevalent across product lines, a firm can succeed only through continuous innovation in its products on compressed development schedules. More new products have been launched in last two decades as compared to any time in the past. Majority of these developments have taken place in high technology sectors like information communication products. In this highly competitive environment, quality of a product plays a major role in its success. Among the various attributes of quality, information regarding reliability and utility reach the potential customers very fast. The audiences of high technology products tend to take informed decision and search for relevant data. Past buyers, trade journals and discussion forums on internet provide information comparison and cost benefit analysis on new products, enhancements etc. Therefore any quality improvement initiative taken by the developers can influence further success of the product. But at the same time effect of other marketing mix variables like price cannot be ignored in the overall diffusion of a product. As the time gap between successive generations of product reduces, the replacement of earlier technologies with the latest one occurs quite frequently, this calls for modeling of profit of the firm that includes the marketing-mix variables such as price, quality, production cost and advertisement expenditure which ultimately influences the sales figure heavily. Robinson and Lakhani (1975), modeled the consumer demand as a function of diffusion effect and price. Their dynamic price model suggests that optimal prices will decline over time. Horsky and Simon (1983) analyzed a model in which the market share response to advertisement is formulated by incorporating the diminishing return to advertising and carryover effects of advertising. Thompson and Teng (1984), proposed a general dynamic price-advertising model. In their model they have incorporated learning curve production cost. At the time of model development they have assumed that the marginal cost of production is a non increasing function of cumulative production volume, which contains as special case, the learning curve phenomenon. Badiru (1992), in his review paper discussed various univariate and multivariate learning curve models that have evolved over the past several years. Dockner and Jorgensen (1988), discussed the optimal advertising policies for diffusion model under monopolistic market situation. Thompson and Teng (1996), again derived the optimal price and quality policies and tried to establish a relationship between these two marketing strategies and the diffusion process. According to them under certain conditions higher prices imply higher quality and under the some other conditions the optimal price declines over time while the product quality improves. Lin C. *et al* (2001), proposed a general class of dynamic model by combining Dockner and Jorgensen and Thompson and Teng models. In their empirical analysis they have incorporated price, quality and advertisement in the model and have discussed the optimal control policies by using the genetic algorithm technique. Many models have been developed to study the optimal policies of different marketing variables. But most of them based on single generation framework only. In this paper we have proposed a general dynamic price-quality model for two competing product generations, in which price and quality are two control variables whose optimal values are to be determined under a finite planning-horizon.

13.2 Dynamic Diffusion of Demand

The basic mixed innovation diffusion model was proposed by Bass in 1969, since then it has become the standard for further development and modification. The model can be mathematically represented as:

$$g = \frac{dN(t)}{dt} = p(\bar{N} - N(t)) + q(\bar{N} - N(t))N(t) \tag{13.1}$$

where, $\bar{N}$ is the potential market size, 'p' is innovation coefficient and 'q' is the imitation coefficient.

Norton and Bass (1987), model is a classic example of multiple generation model, which is built upon the Bass model. During the model development they assumed that the coefficients of innovation and imitation remain unchanged from generation to generation. Islam and Meade (1997) have tested the hypothesis of coefficient constancy across generation of Norton-Bass model. Their empirical work relaxed the assumption of constant coefficient of Norton-Bass model. They proposed that the coefficients of later generation technology are constant increment/decrement over the coefficients of the first generation. Mahajan and Muller (1996), proposed a model which is an extension of Bass model to capture simultaneously both the substitution and diffusion patterns for each successive generation of technological products. Speece and MacLachlan (1995); Hardie, Danaher and Putsis Jr. (2001), developed models in a different way by incorporating price as an explanatory variable.

In this article, a more general of model under dynamic environment is proposed, which captures both diffusion and substitution processes. In our model we have assumed for second generation product that there are two groups of buyers: (a) new purchasers, who are first-time adopters of the product generation (b) repeat buyers, who had also adopted the first generation product. In almost all the marketing situations price $p(t)$ and quality level $q(t)$ play a major role in determining the total market share achieved by the firm. In addition sales figure of the product is also influenced by the word-of-mouth effect. Thus we can assume that demand of a product $(x(t))$ as a dynamic function of price, quality and cumulative sales volume $(N(t))$, and can be expressed as $x = x(p, q, N)$. For, single generation product, we can assume that the demand growth function is twice continuously differentiable and increases with quality and decreases with price and can be written as

$$\frac{\delta x}{\delta p} < 0 \; ; \quad \frac{\delta x}{\delta q} > 0 \quad \text{and} \quad \frac{\delta^2 x}{\delta p \delta q} = \frac{\delta^2 x}{\delta q \delta p} \, .$$

13.3 Model Development

The model is based on the following assumptions:

- Once an adopter adopts a new technology, he/she doesn't revert to earlier generation later.
- New adopters (first time buyers) will purchase only that particular generation product for which he/she will get the maximum utility. Utility can be expressed as a function of price and the goodwill of the product, which in turn depends on the word-of-mouth influence of the adopters.
- Sales of a second-generation durable come from two sources:
  (1) *New Purchasers* (*First time Buyers*): Those who have for the first time adopted the product.
  (2) *Repeat Purchasers*: Those adopters who have bought the earlier generation and now upgrade to latest technology.
- Each adopter can purchase exactly one product unit and she/he makes no further purchases of the product generations that they have adopted. And also each adopter after having made the first purchase may make a repeat purchase of exactly one unit in each successive generation or they can skip a generation product and can wait for more advanced one.

The process of incorporating a new technology is a process, which involves the diffusion of knowledge about the characteristic of the technology. Second generation is introduced into the market before its predecessor is withdrawn. Two components of the model are adoptions due to the new purchase and repeat sales. The basic framework behind the proposed model is:

$$\text{CumulativeAdopters}_j(t) = \text{NewPurchasers}_j(t) + \text{Re peatPurchasers}_j(t)$$

where 'j' is the index representing the generation of a particular technology and N_j is the cumulative number of adopters in the jth generation. Thus,

$$N_j(t) = N_j'(t) + R_j(t) \tag{13.2}$$

where, $N_j'(t)$ is the cumulative number of first time purchasers and $R_j(t)$ is the cumulative number of repeat purchasers of a jth generation technology product . A monopoly market situation is assumed and each adopter can adopt exactly one unit of product from each generation. The model for different market situations can be build as follows:

Case 1. When a single generation product is in the market place:

When only first generation product is in the market, the cumulative sales can be described by the following model, which is an extension of basic mixed innovation diffusion model proposed by Bass (1969):

$$x_1(t) = \frac{dN_1(t)}{dt} = (\bar{N}_1 - N_1(t))Z_1(N_1(t))g_1(p_1(t), q_1(t)) \tag{13.3}$$

where,

$$\begin{aligned} Z_i(N_i(t)) &= \text{Diffusion effect on } i\text{th generation demand at} \\ &\quad\ \text{time } t\ (i = 1, 2). \\ g_i(p_i(t), q_i(t)) &= \text{price and quality function of } i\text{th generation product} \\ &\quad\ \text{at time } t\ (i = 1, 2). \end{aligned}$$

p_i and q_i be the quoted price and quality level of the ith generation product respectively, where $p_i \geq 0$ and $q_i \geq 0$.

Case 2. When two generation products are in the market

When there are two generations of the technology in the market, the potential purchasers of first generation technology (who are yet to purchase) come under the influence of both innovation and imitation forces and price and quality factors effective for the second generation. As a result a fraction of the adopters (say, $\gamma(t)$) who would have otherwise adopted the first generation product instead adopt the latest technology and the remaining

$[1-\gamma(t)]$ will adopt the first generation product. Let us define the parameter $\gamma(t)$as the leapfrogging parameter. Demand function for two generations are:

$$\begin{aligned} x_1(t) &= \frac{dN_1(t)}{dt} \\ &= (\bar{N}_1 - N_1(t))Z_1(N_1(t))g_1(p_1(t), q_1(t))(1-\gamma(t)) \end{aligned} \tag{13.4}$$

$$\begin{aligned} x_2(t) &= \frac{dN_2(t)}{dt} \\ &= (\bar{N}_2 - N_2(t))Z_2(N_2(t))g_2(p_2(t), q_2(t)) \\ &\quad +(\bar{R}_2(t) - R_2(t))Z_2'(N_2(t))g_2'(p_2(t), q_2(t)) \\ &\quad +\gamma(t)(\bar{N}_1 - N_1(t))Z_1(N_1(t))g_1(p_1(t), q_1(t)) \end{aligned} \tag{13.5}$$

where, the leapfrogging parameter $(\gamma(t))$ can be defined as

$$\gamma(t) = \omega(t)\frac{Z_2(t_2)g_2(t_2)}{Z_1(t)g_1(t) + Z_2(t_2)g_2(t_2)} \tag{13.6}$$

where,

(1) $\omega(t)$ is a dummy variable $= \begin{cases} 0 & \text{when } 0 < t \leq t_2 \\ 1 & \text{when } t > t_2 \end{cases}$

(2) $x_i(t) =$ Rate of adoption of ith generation product $(i = 1, 2)$.

(3) $Z_2'(N_2(t)) =$ Diffusion effect on repeat purchasers of 2nd generation product at time t.

(4) $R_2(t)$ is the cumulative number of repeat purchasers of second generation technology, who have earlier purchased the earlier one.

(5) $t_2 = t - \tau$: τ is the introduction time of second generation product.

where, $\frac{\partial x_i}{\partial p_i} = x_{i_{pi}} < 0$, $x_{ip_j} > 0$ and $x_{iq_i} > 0$, $x_{iq_j} < 0$; $(i, j = 1, 2; i \neq j)$.

As we have discussed earlier the buyers of first generation product can become the potential purchasers of the second generation technology and if none of the adopters of new technology drops out of the market in the later generation, the total number of potential repeat purchasers in the second generation is equal to the summation of all prior purchasers of the first generations i.e., we can conclude that the potential repeat purchasers of the second generation product is the function of the adopters of the first generation product. Thus the possible repeat purchasers for 2nd generation technology can be expressed as a function of adoption of first generation product and can be written as follows:

$$\text{Potential repeat purchasers} = \bar{R}_2(t) = \left[\sum_{i=1}^{t} n_1(i)\right] \text{ i.e. } \bar{R}_2(t) = f(N_1(t)) \tag{13.7}$$

$n_1(i)$: Sales of first generation product due to first time purchasers at time t. The notation used in equation (4) and (5) can be defined as:

$$Z_1(N_1(t)) = \left[a_1 + b_1\frac{N_1(t)}{\bar{N}_1}\right],$$

$$Z_2(N_2(t)) = \left[a_2 + b_2\frac{N_2(t)}{\bar{N}_2 + \bar{R}_2(t)}\right]$$

and

$$Z_2'(N_2(t)) = \left[a_2' + b_2'\frac{N_2(t)}{\bar{N}_2 + \bar{R}_2(t)}\right], \quad g_i(p_i(t), q_i(t)) = e^{-d_i p_i(t) + h_i q_i(t)}$$

and

$$g_2'(p_2(t), q_2(t)) = e^{-d_2' p_2(t) + h_2' q_2(t)}$$

where,

(1) a_i and a_i' are innovation coefficients due to first time and repeat purchasers respectively.
(2) b_i and b_i' are imitation coefficient due to first time and repeat purchasers respectively.
(3) d_i and d_i' are pricing parameter due to first time and repeat purchasers respectively.
(4) h_i and h_i' are quality parameter due to first time and repeat purchasers respectively.

13.4 Dynamic Optimization

In this section the general price and quality decision model is formulated by incorporating the learning curve phenomenon as proposed by Thompson and Teng (1996). According to Thomson and Teng the marginal cost of production is a non increasing function of cumulative production volume. Following notations are used below:
r: Discount rate; τ: timing of introduction of second generation product and T be the end of the planning period.
$C(N(t), q(t))$: Total cost per unit at time t for cumulative sales volume $N(t)$ and quality level $q(t)$.
$x(t) = \dfrac{dN(t)}{dt} = x(p, q, N) =$ sales rate at time t.

Now, suppose the firm wants to maximize its total present value of profit over the finite planning horizon. Then the objective function for the firm can be given by:

$$\max_{p(t),q(t)} J = \int_0^T e^{-rt}[\{p_1(t) - C_1(N_1(t), q_1(t))\}x_1(t)]dt$$
$$+ \int_{t=\tau}^T e^{-rt}[\{p_2(t) - C_2(N_2(t), q_2(t))\}x_2(t)]dt$$

subject to (13.8)

$$x_1(t) = x_1(p_1(t), q_1(t), N_1(t), p_2(t), q_2(t), N_2(t))$$
$$x_2(t) = x_2(p_1(t), q_1(t), N_1(t), p_2(t), q_2(t), N_2(t))$$

where, $p_i(t) \geq 0$ and $q_i(t) \geq 0$.

Optimal Solution

To solve the problem, Pontryagin Maximum principle can be applied. The current value Hamiltonian is as follows:

$$\begin{aligned} H &= (p_1 - C_1(N_1, q_1))x_1 + (p_2 - C_2(N_2, q_2))x_2 + \lambda x_1 + \mu x_2 \\ &= (p_1 - C_1(N_1, q_1) + \lambda)x_1 + (p_2 - C_2(N_2, q_2) + \mu)x_2 \end{aligned} \tag{13.9}$$

where, $\lambda(t)$ and $\mu(t)$ are the current value adjoint variables (shadow prices of $x_1(t)$ and $x_2(t)$, respectively) which satisfy the following differential equations

$$\begin{aligned} \frac{d\lambda}{dt} &= \dot{\lambda} \\ &= r\lambda - H_{N_1} \\ &= r\lambda + C_{N_1}x_1 - (p_1 - C_1 + \lambda)x_{1N_1} - (p_2 - C_2 + \mu)x_{2N_1} \end{aligned} \tag{13.10}$$

with the transversality condition at $t = T$, $\lambda(t) = 0$.

$$\begin{aligned} \frac{d\mu}{dt} &= \dot{\mu} \\ &= r\mu - H_{N_2} \\ &= r\mu + C_{N_2}x_2 - (p_1 - C_1 + \lambda)x_{1N_2} - (p_2 - C_2 + \mu)x_{2N_2} \end{aligned} \tag{13.11}$$

with the transversality condition at $t = T$, $\mu(t) = 0$.

The physical interpretation of the current value Hamiltonian H can be given as follows: $\lambda(t)$ and $\mu(t)$ stand for the future benefits from first and second generation (at time 't') of having one more unit produced. Thus the current value Hamiltonian is the sum of current profit $[(p_1 + C_1)x_1 + (p_2 - C_2)x_2]$ and the future benefit $[\lambda x_1 + \mu x_2]$. In short, H represents the instantaneous total profit of the firm at time t.

The following necessary conditions hold for an optimal solution:

$$\frac{dH}{dp_1} = 0 = H_{p_1} \Rightarrow (p_1 - C_1 + \lambda)x_{1p_1} + x_1 + (p_2 - C_2 + \mu)x_{2p_1} = 0 \quad (13.12)$$

$$\frac{dH}{dp_2} = 0 = H_{p_2} \Rightarrow (p_1 - C_1 + \lambda)x_{1p_2} + (p_2 - C_2 + \mu)x_{2p_2} + x_2 = 0 \quad (13.13)$$

Necessary conditions also include

$$\frac{dH}{dq_1} = 0 = H_{q_1} \Rightarrow (p_1 - C_1 + \lambda)x_{1q_1} - x_1 C_{1q_1} + (p_2 - C_2 + \mu)x_{2q_1} = 0 \quad (13.14)$$

$$\frac{dH}{dq_2} = 0 = H_{q_2} \Rightarrow (p_1 - C_1 + \lambda)x_{1q_2} + (p_2 - C_2 + \mu)x_{2q_2} - x_2 C_{2q_2} = 0 \quad (13.15)$$

Other optimality conditions are:

$$\begin{vmatrix} H_{p_1p_1} & H_{p_1p_2} \\ H_{p_2p_1} & H_{p_2p_2} \end{vmatrix} > 0; \quad \begin{vmatrix} H_{p_1p_1} & H_{p_1p_2} & H_{p_1q_1} \\ H_{p_2p_1} & H_{p_2p_2} & H_{p_2q_1} \\ H_{q_1p_1} & H_{q_1p_2} & H_{q_1q_1} \end{vmatrix} < 0 \quad \text{and}$$

$$\begin{vmatrix} H_{p_1p_1} & H_{p_1p_2} & H_{p_1q_1} & H_{p_1q_2} \\ H_{p_2p_1} & H_{p_2p_2} & H_{p_2q_1} & H_{p_2q_2} \\ H_{q_1p_1} & H_{q_1p_2} & H_{q_1q_1} & H_{q_1q_2} \\ H_{q_2p_1} & H_{q_2p_2} & H_{q_2q_1} & H_{q_2q_2} \end{vmatrix} > 0 \quad (13.16)$$

From the above optimality conditions, we can get the following results:

From (13.12) and (13.13) we have

$$p_1^* = C_1 - \lambda - \frac{x_1 x_{2p_2} - x_2 x_{2p_1}}{x_{1p_1} x_{2p_2} - x_{2p_1} x_{1p_2}} \quad \text{and}$$

$$p_2^* = C_2 - \mu - \frac{x_2}{x_{2p_2}} + \frac{x_{1p_2}}{x_{2p_2}} \left[\frac{x_1 x_{2p_2} - x_2 x_{2p_1}}{x_{1p_1} x_{2p_2} - x_{2p_1} x_{1p_2}} \right] \quad (13.17)$$

Again, from (13.12) and (13.14)

$$C_{1q_1}^* = -\frac{x_{1q_1}}{x_{1p_1}} + (p_2 - C_2 + \mu) \left[\frac{x_{2q_1} x_{1p_1} - x_{2p_1} x_{1q_1}}{x_1 x_{1p_1}} \right] \quad (13.18)$$

from (13.13) and (13.15)

$$C_{2q_2}^* = -\frac{x_{2q_2}}{x_{2p_2}} + (p_1 - C_1 + \lambda) \left[\frac{x_{1q_2} x_{2p_2} - x_{1p_2} x_{2q_2}}{x_2 x_{2p_2}} \right] \quad (13.19)$$

Integrating equation (13.10) and (13.11) with the transversality conditions, we have the future benefit of having one more unit produced of the respective generation as

$$\lambda(t) = \int_t^T [(p_1 - C_1 + \lambda)x_{1N_1} + (p_2 - C_2 + \mu)x_{2N_1} - C_{N_1} x_1] e^{-rs} ds \quad (13.20)$$

$$\mu(t) = \int_t^T [(p_1 - C_1 + \lambda)x_{1N_2} + (p_2 - C_2 + \mu)x_{2N_2} - C_{N_2} x_2] e^{-rs} ds \quad (13.21)$$

13.5 Theoretical Results

The general formulation and characteristics of the proposed model will help in gaining some insight into the important factors influencing the optimal policies. Let,

η_i: price elasticity of demand of ith generation product with respect to price p_i, i.e., $\eta_i = -p_i \frac{x_{ip}}{x_i}$; and the cross-elasticities $\eta_{ij} = p_j \frac{x_{ip_j}}{x_i}$; $(i, j = 1, 2;\ i \neq j)$.

Equations (13.12) and (13.13) we have the following pricing policies.

$$p_1^* = \begin{cases} C_1 + \frac{C_1\eta_2 + \left(p_2\eta_{21}\left[\frac{x_2}{x_1}\right] - \lambda(\eta_1\eta_2 - \eta_{12}\eta_{21})\right)}{\eta_2(\eta_1 - 1) - \eta_{12}\eta_{21}} & \text{when } (\eta_{12}; \eta_{21}) \neq 0 \\ C_1 + \frac{C_1 - \lambda\eta_1}{\eta_1 - 1} \text{ when } \eta_{12} = \eta_{21} = 0 \end{cases} \tag{13.22}$$

$$p_2^* = \begin{cases} C_2 + \frac{C_2\eta_1 + \left(p_1\eta_{12}\left[\frac{x_1}{x_2}\right] - \mu(\eta_1\eta_2 - \eta_{12}\eta_{21})\right)}{\eta_1(\eta_2 - 1) - \eta_{12}\eta_{21}} & \text{when } (\eta_{12}; \eta_{21}) \neq 0 \\ C_2 + \frac{C_2 - \mu\eta_2}{\eta_2 - 1} \text{ when } \eta_{12} = \eta_{21} = 0 \end{cases} \tag{13.23}$$

For detail calculation see Appendix A.

For the case of zero leapfrogging or when $\eta_{12} = \eta_{21} = 0$, there will be no significant effect of price of second-generation product on the potential purchasers of the first generation. The optimal pricing policy for this situation has been discussed in detail by Bayus (1992). Now suppose $\eta_1\eta_2 - \eta_{12}\eta_{21} > 0$ and also $\eta_2(n_1 - 1) = \eta_1(n_2 - 1) = \eta_1\eta_2$:

Case 1. When $\eta_{12} > \eta_{21}$

The price path of first generation product is monotonically increasing and the price path of second-generation product is monotonically decreasing.

Case 2. When $\eta_{12} < \eta_{21}$

The price path of first generation product is monotonically decreasing and that of the second-generation product is monotonically increasing.

Case 3. When $\eta_{12} = \eta_{21}$

In this situation price-path of both the generation will follow the same direction.

13.6 Pricing and Quality Impact on Leapfrogging

The general characteristics discussed in the above section are useful to gain insight into the factors affecting the optimal price and quality. We shall now investigate the case how the price, and quality motivate a potential first time purchaser to leapfrog from an old technology to the latest one under the condition of equal diffusion effect of both the generations. The leapfrogging parameter ($\gamma(t)$) plays a major role in increasing or decreasing the sales figure of a particular generational. In majority of studies relating to the multiple generation product situation the leapfrogging behavior of consumers are either ignored or taken to be constant (e.g. Mahajan and Muller (1996)). In their empirical work Mahajan and Muller have considered the parameter 'γ' as constant over time and also not influenced by the diffusion rates of the technologies which are there in the market. This seems to be an unrealistic assumption. We suggest that $\gamma(t)$ is a time-varying component and can easily be influenced by the diffusion rate of the existing generations. In this section the cases as to how the price and quality changes influence the leapfrogging will be discussed. As defined earlier, leapfrogging parameter is

$$\gamma(t) = \omega(t)\frac{Z_2(N_2(t))g_2(p_2(t), q_2(t))}{Z_1(N_1(t))g_1(p_1(t), q_1(t)) + Z_2(N_2(t))g_2(p_2(t), q_2(t))} \tag{13.24}$$

Differentiating (13.24) with respect to price p_1, we have

$$\begin{aligned}\frac{\delta\gamma}{\delta p_1} &= \left[\frac{1}{Z_2g_2 + Z_1g_1}\right]^2 \left[(Z_2g_2 + Z_1g_1)\frac{\delta(Z_2g_2)}{\delta p_1} - Z_2g_2\frac{\delta(Z_2g_2 + Z_1g_1)}{\delta p_1}\right] \\ &= -\frac{Z_2g_2Z_1\frac{\delta g_1}{\delta p_1}}{(Z_1g_1 + Z_2g_2)} \\ &= d_1\left(\frac{Z_2g_2}{Z_1g_1 + Z_2g_2}\right)\left(\frac{Z_1g_1}{Z_1g_1 + Z_2g_2}\right) \\ &= d_1\gamma\left(\frac{Z_1g_1}{Z_1g_1 + Z_2g_2}\right)\end{aligned} \tag{13.25}$$

Similarly,

$$\frac{\delta\gamma}{\delta p_2} = -d_2\gamma\left(\frac{Z_1g_1}{Z_1g_1 + Z_2g_2}\right), \tag{13.26}$$

$$\frac{\delta\gamma}{\delta q_1} = -h_1\gamma\left(\frac{Z_1g_1}{Z_1g_1 + Z_2g_2}\right), \tag{13.27}$$

$$\frac{\delta\gamma}{\delta q_2} = -h_2\gamma\left(\frac{Z_1g_1}{Z_1g_1 + Z_2g_2}\right). \tag{13.28}$$

To examine the influence of price and quality on leapfrogging, the following can be proposed:

Proposition 1. If $Z_1 = Z_2$, $A = X$ and $B = Y$, where $A = \left[\frac{g_1}{(g_1+g_1)^2}\right]\frac{\delta g_2}{\delta p_2}$, $B = \left[\frac{g_1}{(g_1+g_1)^2}\right]\frac{\delta g_2}{\delta q_2}$, $X = \left[\frac{g_2}{(g_1+g_1)^2}\right]\frac{\delta g_1}{\delta p_1}$ and $Y = \left[\frac{g_2}{(g_1+g_1)^2}\right]\frac{\delta g_1}{\delta q_1}$ then:

	Condition	Result
Case 1.	$p_2^\bullet\,(\uparrow)$, $q_2^\bullet\,(\downarrow)$ and $p_1^\bullet\,(\downarrow)$, $q_1^\bullet\,(\uparrow)$	$\gamma^\bullet\,(\downarrow)$
Case 2.	$p_2^\bullet\,(\downarrow)$, $q_2^\bullet\,(\uparrow)$ and $p_1^\bullet\,(\uparrow)$, $q_1^\bullet\,(\downarrow)$	$\gamma^\bullet\,(\uparrow)$

Proof. See Appendix B. □

In case 1, as the firm increases the price of second generation product without increasing the standard of the product, it end up with reducing the number leapfroggers because in this situation the potential purchasers will find the first generation product more lucrative than the other one. This situation may arise when a company wants to sell out whole of the stock of first generation product. In case 2, as the price of the second generation is monotonically decreasing and the price of the first generation is increasing, hence in this situation leapfrogging parameter is also monotonically increasing. The results in Proposition 1 can be interpreted to mean that price has a higher impact than the quality in determining the rate of leapfrogging.

Now, from equation (13.25)-(13.28), following theorem results:

Theorem A. 1. $\left[\dfrac{\dfrac{\delta\gamma}{\delta p_1}}{\dfrac{\delta\gamma}{\delta p_2}}\right] = \text{Constant};$ 2. $\left[\dfrac{\dfrac{\delta\gamma}{\delta q_1}}{\dfrac{\delta\gamma}{\delta q_2}}\right] = \text{Constant};$

3. $\left[\dfrac{\dfrac{\delta\gamma}{\delta q_1}}{\dfrac{\delta\gamma}{\delta q_2}}\right] \propto \left[\dfrac{\dfrac{\delta\gamma}{\delta p_1}}{\dfrac{\delta\gamma}{\delta p_2}}\right].$

Proof. See Appendix B. □

13.7 Optimal Introduction Timing Strategy for Second Generation Product

The success of a high technology product is largely dependent on its time of introduction in the market. In multiple generation product situations the timing of introduction of a new generation technology plays a major role in the overall diffusion of new and the existing technologies. As this paper deals with two generations only, in this section the problem of optimal introduction timing of second generation product will be discussed. Suppose that a firm has introduced its first generational product at time t_1 and is still continue with it and has no intention to withdraw it from the market in near future. And by the time t_2 it has already developed the second generation product ($t_1 \leq t_s$). Now the firm has to decide the optimal time (τ) of introduction of the latest generation, in such a way that the optimal time τ, has very limited effect on the diffusion of the earlier generation product. The ideal introduction time (τ) of the second generation product should be that when the scope of leapfrogging from the first generation to the second generation will be very limited or in other words, the ideal time should be that point when a firm find a way-out to reimburse the cannibalization of sales from the first generation product due to the introduction of second generation. Differentiating equation (13.7) with respect to τ and equating to zero, we have the following proposition.

Proposition 2. The optimal time of introduction of second generations is when the profit gained from the second generation product can compensate the loss incurred on the first generation product due to leapfrogging. Otherwise the firm should wait for an ideal time to introduce the next version.

Integrating 'J' with respect to 'τ' and equating to zero, results the following equation for the optimal introduction time of second generation product:

$$\begin{aligned}
&(p_1(\tau^*) - C_1(N_1(\tau^*), q_1(\tau^*)))(\bar{N}_1 - N_1(\tau^*)) \\
&\quad \times Z_1(N_1(\tau^*))g_1(p_1(\tau^*), q_1(\tau^*))\gamma(\tau^*) \\
&\quad = [p_2(\tau^*) - C_2(N_2(\tau^*), q_2(\tau^*))][(\bar{N}_2 - N_2(\tau^*))Z_2(N_2(\tau^*)) \\
&\quad \times g_2(p_2(\tau^*), q_2(\tau^*)) + (\bar{R}_2 - R_2(\tau^*))Z_2'(N_2(\tau^*)) \\
&\quad \times g_2'(p_2(\tau^*), q_2(\tau^*)) + (\bar{N}_1 - N_1(\tau^*))Z_1(N_1(\tau^*)) \\
&\quad \times g_1(p_1(\tau^*), q_1(\tau^*))\gamma(\tau^*)]
\end{aligned} \tag{13.29}$$

(For details proof see Appendix C).

The second component of left hand side of the product in (13.29) represents the number of leapfroggers from first generation product, which has been multiplied with the marginal profit. Therefore the optimal time for introduction of the new product is when the return from the second generation (RHS of (13.29)) at least compensates the loss for the first generation. Basically the left-hand side of the equation (13.29) represents the loss figure of first generation product, which is incurred due to the leapfrogging. The right-hand side gives the profit figure on introduction of second generation product, which also consists of profit gained due to the up-graders. As long as left-hand side is greater than right-hand side, firm should wait for the ideal time to introduce the next version. The firm can launch the next version when the left-hand side becomes equal or less than right-hand side.

13.8 Conclusions

In this paper we have extended the Thompson and Teng model by considering price and quality as decision variables. In most of the earlier works properties of optimal solutions were discussed in the light of single generation only, in contrast here we tried to discuss the optimality conditions of different marketing mix variables for two product generations resulting from technological innovations. The theoretical results obtained here confirm that the optimality conditions as described in literatures for a single generation state can also hold for multiple generation situations. Finally the model can be extended in several ways, e.g. by extending the monopolistic model to a duopolistic or oligopolistic market. The model can also be extended for n-generational product situation in the dynamic market situation.

Acknowledgement

Authors are thankful to the anonymous referee for the useful and the beneficial comment. This research work is partially supported by a grant to the first author from University Grant Commission, India.

Appendix

A. Optimal Pricing

From (13.17), we have

$$p_1 = C_1 - \lambda + \frac{1}{x_1}\left[\frac{p_1\eta_2 x_1 + p_2\eta_{21}x_2}{\eta_1\eta_2 - \eta_{12}\eta_{21}}\right]$$

$$\Rightarrow p_1\left(1 - \frac{\eta_2}{\eta_1\eta_2 - \eta_{12}\eta_{21}}\right) = C_1 - \lambda + \frac{\eta_{21}x_2p_2}{x_1\left(\eta_1\eta_2 - \eta_{12}\eta_{21}\right)}$$

$$\Rightarrow p_1 = C_1 + \frac{C_1\eta_2 + \left(p_2\eta_{21}\left[\frac{x_2}{x_1}\right] \text{-}\lambda\left(\eta_1\eta_2 - \eta_{12}\eta_{21}\right)\right)}{\eta_2\left(\eta_1 - 1\right) - \eta_{12}\eta_{21}} \tag{A.1}$$

Similarly the optimal path for p_2 can be given as

$$p_2 = C_2 + \frac{C_2\eta_1 + \left(p_1\eta_{12}\left[\frac{x_1}{x_2}\right] \text{-}\mu\left(\eta_1\eta_2 - \eta_{12}\eta_{21}\right)\right)}{\eta_1\left(\eta_2 - 1\right) - \eta_{12}\eta_{21}} \tag{A.2}$$

B. Significance of Pricing and Quality in Leapfrogging

Since, $Z_1 = Z_2$; thus (13.6) $\Rightarrow \gamma = \dfrac{g_2}{g_1 + g_2}$.

Now taking the time derivative of the above equation, we have

$$\gamma^\bullet = \left[\frac{g_1}{(g_1+g_2)^2}\right]\frac{\delta g_2}{\delta p_2}p_2^\bullet + \left[\frac{g_1}{(g_1+g_2)^2}\right]\frac{\delta g_2}{\delta q_2}q_2^\bullet - \left[\frac{g_2}{(g_1+g_2)^2}\right]\frac{\delta g_1}{\delta p_1}p_1^\bullet - \left[\frac{g_2}{(g_1+g_2)^2}\right]\frac{\delta g_1}{\delta q_1}q_1^\bullet \qquad \text{(B.1)}$$

where,

$$\frac{\delta g_1}{\delta p_1},\ \frac{\delta g_2}{\delta p_2} < 0;\ \frac{\delta g_1}{\delta q_1},\ \frac{\delta g_2}{\delta q_2} > 0 \ \text{ and } \ g_1, g_2 \geq 0\,. \qquad \text{(B.2)}$$

Hence, the sign of A is positive and X is negative.

(B.I.) Corollary. There will be no leapfrogging if:

$$Ap_2^\bullet + Bq_2^\bullet = Xp_1^\bullet + Yq_1^\bullet\,.$$

Proof of Theorem A. Now,

$$(13.25)\div(13.26)\Rightarrow \frac{\delta\gamma}{\delta p_1} \div \frac{\delta\gamma}{\delta p_2} = \text{Constant}$$

Similarly, from (13.27) and (13.28) we can prove that

$$\frac{\delta\gamma}{\delta q_1} \div \frac{\delta\gamma}{\delta q_2} = -\frac{h_1}{h_2} = \text{Constant.}$$

Again,

$$\frac{\left[\frac{\frac{\delta\gamma}{\delta p_1}}{\delta\gamma\delta p_2}\right]}{\left[\frac{\frac{\delta\gamma}{\delta q_1}}{\delta\gamma\delta q_2}\right]} = \frac{d_1h_2}{d_2h_1} = \psi = \text{Constant} \Rightarrow \frac{\frac{\delta\gamma}{\delta p_1}}{\frac{\delta\gamma}{\delta p_2}} = \psi\frac{\frac{\delta\gamma}{\delta q_1}}{\frac{\delta\gamma}{\delta q_2}} \Rightarrow \left[\frac{\frac{\delta\gamma}{\delta p_1}}{\frac{\delta\gamma}{\delta p_2}}\right] \propto \left[\frac{\frac{\delta\gamma}{\delta q_1}}{\frac{\delta\gamma}{\delta q_2}}\right]$$

(B.II) *Corollary*

Case 1. If $\psi = 1 \Rightarrow \frac{\delta\gamma}{\delta p_1}\frac{\delta\gamma}{\delta q_2} = \frac{\delta\gamma}{\delta q_1}\frac{\delta\gamma}{\delta q_2}$

Case 2. If $\psi < 1 \Rightarrow \frac{\delta\gamma}{\delta p_1}\frac{\delta\gamma}{\delta q_2} < \frac{\delta\gamma}{\delta q_1}\frac{\delta\gamma}{\delta q_2}$.

Case 3. If $\psi > 1 \Rightarrow \frac{\delta\gamma}{\delta p_1}\frac{\delta\gamma}{\delta q_2} > \frac{\delta\gamma}{\delta q_1}\frac{\delta\gamma}{\delta q_2}$

C. Optimal Introduction Timing Strategy

Proof of Proposition 2.

$$
\begin{aligned}
(13.8) \Rightarrow \max_{p(t),q(t)} J = &\int_0^T e^{-rt}[(p_1(t) - C_1(N_1(t), q_1(t)))x_1 t)]dt \\
&+ \int_{t=\tau}^T e^{-rt}[(p_2(t) - C_2(N_2(t), q_2(t))x_2(t)]dt \\
= &\int_{t=0}^T e^{-rt}[p_1(t) - C_1(N_1(t), q_1(t)][(\bar{N}_1 - N_1(t)) \\
&\times Z_1(N_1(t))g_1(p_1(t), q_1(t))(1 - \gamma(t))]dt \\
&+ \int_{t=\tau}^T e^{-rt}[p_2(t) - C_2(N_2(t), q_2(t)] \\
&\times [(\bar{N}_2 - N_2'(t))Z_2(N_2(t))g_2(p_2(t), q_2(t)) \\
&+ (\bar{R}_2 - R_2(t))Z_2(N_2(t))g_2'(p_2(t), q_2(t)) \\
&+ \gamma(t)(\bar{N}_1 - N_1(t))Z_1(N_1(t))g_1(p_1(t), q_1(t))]dt \\
= &\int_0^T e^{-rt}(p_1(t) - C_1(N_1(t), q_1(t))[\bar{N}_1 - N_1(t)] \\
&\times Z_1(N_1(t))g_1(p_1(t), q_1(t))(1 - \gamma(t))dt \\
&+ \int_{t=\tau}^T e^{-rt}\{[(p_2(t) - C_2(N_2(t), q_2(t)] \\
&\times [(\bar{N}_2 - N_2'(t))Z_2(N_2(t))g_2(p_2(t), q_2(t)) \\
&+ (\bar{R}_2 - R_2'(t))Z_2'(N_2(t))g_2'(p_2(t), q_2(t))] \\
&+ \gamma(t)(\bar{N}_1 - N_1(t))Z_1(N_1(t))g_1(p_1(t), q_1(t))\}dt \\
= &\int_0^T e^{-rt}(p_1(t) - C_1(N_1(t), q_1(t))[\bar{N}_1 - N_1(t)] \\
&\times Z_1(N_1(t))g_1(p_1(t), q_1(t))dt \\
&+ \int_{t=\tau}^T e^{-rt}[(p_2(t) - C_2(N_2(t), q_2(t)] \\
&\times [(\bar{N}_2 - N_2'(t))Z_2(N_2(t))g_2(p_2(t), q_2(t)) \\
&+ (\bar{R}_2 - R_2'(t))Z_2'(N_2(t))g_2'(p_2(t), q_2(t))]dt \\
&+ \int_{t=\tau}^T e^{-rt}((p_2(t) - p_1(t)) - (C_2(N_2(t), q_2(t)) - C_1(N_1(t), q_1(t))) \\
&\times \gamma(t)(\bar{N}_1 - N_1(t))Z_1(N_1(t))g_1(p_1(t), q_1(t))dt
\end{aligned}
$$

Integrating 'J' with respect to 'τ' and equating to zero, results for the optimal introduction time of second generation can be obtained as follows:

$$\begin{aligned}&(p_1(\tau^*) - C_1(N_1(\tau^*), q_1(\tau^*)))(\bar{N}_1 - N_1(\tau^*))Z_1(N_1(\tau^*))g_1(p_1(\tau^*), q_1(\tau^*))\gamma(\tau^*)\\ &\quad = [p_2(\tau^*) - C_2(N_2(\tau^*), q_2(\tau^*))][(\bar{N}_2 - N_2(\tau^*))Z_2(N_2(\tau^*))g_2(p_2(\tau^*), q_2(\tau^*))\\ &\quad + (\bar{R}_2 - R_2(\tau^*))Z_2'(N_2(\tau^*))g_2'(p_2(\tau^*), q_2(\tau^*)) + (\bar{N}_1 - N_1(\tau^*))\\ &\quad \times Z_1(N_1(\tau^*))g_1(p_1(\tau^*), q_1(\tau^*))\gamma(\tau^*)]\end{aligned}$$

Bibliography

Badiru, A.B. (1992). *Computational Survey of Univariate and Multivariate Learning Curve Models*, IEEE Transactions of Engineering Management, **39** (2), pp. 176–188.

Bass, F.M. (1969). *A New-Product Growth Model for Consumer Durables*, Management Science, **15**, pp. 215–227.

Barry, B.L. (1992). *The Dynamic Pricing of Next Generation Consumer Durables*, Marketing Science, **11** (3), pp. 251–265.

Danaher, P.J., Hardie, B.G.S., and William Jr., P.P. (2001). *Marketing-Mix Variables and the Diffusion of Successive Generations of a Technology Innovation*, Journal of Marketing Research, **XXXVIII**, pp. 501–514.

Dockner, E. and Jorgensen, S. (1988), *Optimal advertising policies for diffusion models of new product innovation in monopolistic situations, Management Science*, **34**, pp. 119–131.

Dorfman, R. and Steiner, P.O. (1954). *Optimal advertising and optimal quality*, American Economic Review, **44**, pp. 826–836.

Horsky, D. and Simon, L.S. (1983). *Advertising and the diffusion of a new products*, Marketing Science, **2**, pp. 1–18.

Islam, T. and Meade, N. (1997). *The diffusion of successive generation of a technology - a more general model*, Technological Forecasting and Social Change, **56**, pp. 49–60.

Lin, C., Shen, S., Yeh, Y. and Ding J. (2001), *Dynamic optimal control policy in advertising price and quality*, International Journal of System Science, **32** (2), pp. 175–184.

Mahajan, V., and Muller, E. (1996). *Timing, Diffusion, And Substitution Of Successive Generations Of Technological Innovations: The IBM Mainframe Case*, Technological Forecasting and Social Change, **51**, pp. 109–132.

Norton, J.A. and Bass, F.M. (1987). *A diffusion theory model of adoption and substitution for successive generation of high-technology products*, Management Science, **33** (9), pp. 1069–1086.

Robinson, B. and Lakhani, C. (1975). *Dynamic price models for new product planning*, Management Science, **21**, pp. 1113–1122.

Speece, M.W. and MacLachlan, D.L. (1995). *Application of A Multi-Generation Diffusion Model to Milk Container Technology*, Technological Forecasting and Social Change, **49**, pp. 281–295.

Thompson, G.L. and Teng, J.T. (1984). *Optimal pricing and advertising policies for new product oligopoly market*, Marketing Science, **3**, pp. 148–168.

Thompson, G.L. and Teng, J.T. (1996). Optimal strategies for general price-quality decision models of new products with learning production costs, European Journal of Operational Research, **93**, pp. 476–489.

Chapter 14

Forecasting for Supply Chain and Portfolio Management

Katta G. Murty
Department of Industrial and Operations Engineering
University of Michigan
Ann Arbor, MI 48109-2117, USA
e-mail:murty@umich.edu

Abstract

Material imbalances at some companies have been traced to the procedures they use for forecasting demand based on the usual normality assumption. In this paper we discuss a simple and easy to implement nonparametric technique to forecast the demand distribution based on statistical learning, and ordering policies based on it, that are giving satisfactory results at these companies. We also discuss an application of this nonparametric forecasting method to portfolio management.

Key Words: Forecasting demand, updating demand distribution, nonparametric method, overage and underage costs, order quantity determination, newsvendor approach; returns from investment, risk, portfolio management and optimization, statistical learning.

14.1 Introduction

In production planning projects at computer companies (Dell, Sun), filter making companies (Pall), automobile component suppliers (Borg Warner, Federal Mogul), and others, we found that high inventories for some items, and expedited shipments to cover shortages for some others, are common occurrences at some of them. Examination of the materials requirement planning (MRP) systems used for making production and order quantity decisions at these companies has shown that a common cause for these occurrences are the procedures they use for forecasting demand based on the usual normality assumption. This paper discusses the features of a new, simpler nonparametric forecasting method based on statistical learning, and ordering and lot sizing policies based on it, implemented and working satisfactorily at these companies.

We also discuss an application of this nonparametric forecasting method to portfolio management. We then develop a model based on the principles of statistical learning to determine an optimum portfolio WRT (with respect to) a measure of risk that is closer to the common investors perception of risk.

14.2 Costs of High Inventories and Shortages

Models for controlling and replenishing of inventories have the aim of determining order quantities to minimize the sum of total overage costs (costs of excess inventory remaining at the end of the planning period), and underage costs (shortage costs, or costs of having less than the desired amount of stock at the end of the planning period).

In inventory control literature, the total overage (underage) cost is usually assumed to be proportional to the overage (shortage) amount or quantity, to make the analysis easier. But in some of the companies we were told that a piecewise linear (PL) function provides a much closer representation of the true overage and underage costs. In these companies there is a buffer with limited space in which excess inventory at the end of the planning period can be stored and retrieved later at a low cost (i.e., with minimum requirements of manhours needed) per unit. Once this buffer is filled up, any remaining excess quantity has to be held at a location farther away that requires greater number of manhours for storing or retrieval/unit. Similar situation exists for underage cost as a function of the shortage amount. This clearly implies that the overage and underage costs are PL functions of the excess, shortage quantities. Determining optimum order quantities to minimize such unusual overage, underage cost functions is much harder with existing inventory control models using forecasting techniques in current literature.

An advantage of the new forecasting system discussed in this paper is that such unusual overage, underage cost functions can easily be accommodated under it.

14.3 Commonly used Techniques for Forecasting Demand

In almost all inventory management problems in practice the demand during a future planning period is a random variable with an unknown probability distribution, and the models for these problems have the objective of minimizing the sum of expected overage and underage costs. Successful inventory management systems depend heavily on good demand forecasts to provide data for inventory replenishment decisions.

The output of forecasting is usually presented in the literature as the **forecasted demand quantity**, in reality it is an estimate of the expected demand during the planning period. Because of this, the purpose of forecasting is often misunderstood to be that of generating this single number, even though sometimes the standard deviation of demand is also estimated. All commonly used forecasting methods are parametric methods, they usually assume that demand is normally distributed, and update its distribution by updating the parameters of the distribution, the mean μ, and the standard deviation σ. The most commonly used methods for updating the values of the parameters are the method of moving averages, and the exponential smoothing method.

The method of moving averages uses the average of n most recent observations on demand as the forecast for the expected demand for the next period. n is a parameter known as the *order of the moving average method* being used, typically it is between 3 to 6 or larger.

The exponential smoothing method introduced and popularized by [Brown (1959)], is perhaps the most popular method in practice. It takes $\hat{D}_{t+1}$, the forecast of expected demand during next period $t+1$, to be $\alpha x_t + (1-\alpha)\hat{D}_t$, where x_t is the observed demand during current period t, $\hat{D}_t$ is the forecasted expected demand for current period t, and $0 < \alpha \leq 1$ is a *smoothing constant* which is the relative weight placed on the current observed demand. Typically values of α between 0.1 and 0.4 are used, and the value of α is increased whenever the absolute value of the deviation between the forecast and observed demand exceeds a tolerance times the standard deviation. Smaller values of α (like 0.1) yield predicted values of expected demand that have a relatively smooth pattern, whereas higher values of α (like 0.4) lead to predicted values exhibiting significantly greater variation, but doing a better job of tracking the demand series. Thus using larger α makes forecasts more responsive to changes in the demand process, but will result in forecast errors with higher variance.

One disadvantage of both the method of moving averages and the exponential smoothing method is that when there is a definite trend in the demand process (either growing, or falling), the forecasts obtained by them lag behind the trend. Variations of the exponential smoothing method to track trend linear in time in the demand process have been proposed (see [Holt (1957)]), but these have not proved very popular.

There are many more sophisticated methods for forecasting the expected values of random variables, for example the Box-Jenkins ARIMA models (see [Box and Jenkins (1970)] and [Montgomery and Johnson (1976)]), but these methods are not popular for production applications, in which forecasts for many items are required.

14.4 Parametric Methods for Forecasting Demand Distribution

14.4.1 *Using Normal Distribution with updating of Expected Value and Standard Deviation in each Period*

As discussed in the previous section, all forecasting methods in the literature only provide an estimate of the expected demand during the planning period. The optimum order quantity to be computed depends of course on the entire probability distribution of demand, not just its expected value. So, almost everyone assumes that the distribution of demand is the normal distribution because of its convenience. One of the advantages that the normality assumption confers is that the distribution is fully characterized by only two parameters, the mean and the standard deviation, both of whom can be very conveniently updated by the exponential smoothing or the moving average methods.

Let t be the current period; x_r the observed demand, and $\hat{D}_r, \hat{\sigma}_r$ the forecasts (i.e., estimates by whichever method is being used for forecasting) of the expected demand, standard deviation of demand respectively in period r; for $r \leq t$. Then these forecasts for the next period $t+1$ are:

$\hat{D}_{t+1}$ (by method of moving averages of order n)	$= \frac{1}{n}\sum_{r=t-n+1}^{t} x_r$
$\hat{D}_{t+1}$ (by exponential smoothing method with smoothing constant α)	$= \alpha x_t + (1-\alpha)\hat{D}_t$
$\hat{\sigma}_{t+1}$ (by method of moving averages of order n)	$= +\sqrt{(\sum_{r=t-n+1}^{t}(x_r - \hat{D}_{t+1})^2)/n}$

To get $\hat{\sigma}_{t+1}$ by the exponential smoothing method, it is convenient to use the mean absolute deviation (MAD), and use the formula: standard deviation $\sigma \approx (1.25)\text{MAD}$ when the distribution is the normal distribution. Let MAD_t denote the estimate of MAD for current perior t. Then the forecasts obtained by the exponential smoothing method with smoothing parameter α for the next period $t+1$ are:

$$\begin{aligned} \text{MAD}_{t+1} &= \alpha|x_t - \hat{D}_t| + (1-\alpha)\text{MAD}_t \\ \hat{\sigma}_{t+1} &= (1.25)\text{MAD}_{t+1} \end{aligned}$$

Usually $\alpha = 0.1$ is used to ensure stability of the estimates. And the normal distribution with mean $\hat{D}_{t+1}$, and standard deviation $\hat{\sigma}_{t+1}$ is taken as the forecast for the distribution of demand during the next period $t+1$ for making any planning decisions under this procedure.

14.4.2 *Using Normal Distribution with Updating of Expected Value and Standard Deviation only when there is Evidence of Change*

In some applications, the distribution of demand is assumed to be the normal distribution, but estimates of its expected value and standard deviation are left unchanged until there is evidence that their values have changed. Foote [3] discusses several statistical control tests on demand data being generated over time to decide when to reestimate these parameters. Under this scheme, the method of moving averages is commonly used to estimate the expected value and the standard deviation from recent data whenever the control tests indicate that a change may have occurred.

14.4.3 *Using Distributions other than Normal*

In a few special applications in which the expected demand is low (i.e., the item is a slow-moving item) other distributions like the poisson distribution are sometimes used, but by far the most popular distribution for making inventory management decisions is the normal distribution because of its convenience, and because using it has become a common practice historically.

For the normal distribution, the mean is the mode (i.e., the value associated with the highest probability), and the distribution is symmetric around this value. If histograms of observed demand data of an item do not share these properties, it may indicate that the normal distribution is a poor approximation for the actual distribution of demand, in this case order quantities determined using the normality assumption may be far from being optimal.

These days industrial environment is very competitive with new products replacing the old periodically due to rapid advancements in technology. In this dynamic environment, the life cycles of components and end products are becoming shorter. Beginning with the introduction of the product, its life cycle starts with a *growth period* due to gradual market penetration of the product. This is followed by a *stable period* of steady demand. It is then followed by a final *decline period* of steadily declining demand, at the end of which the item disappears from the market. Also, the middle stable period seems to be getting shorter for many major components. Because of this constant rapid change, it is necessary to periodically update demand distributions based on recent data.

The distributions of demand for some components are far from being symmetric around the mean, and the skewness and shapes of their distributions also seem to be changing over time. Using a probability distribution like the normal defined by a mathematical formula involving only a few parameters, it is not possible to capture changes taking place in the shapes of distributions of demand for such components. This is the disadvantage of existing forecasting methods based on an assumed probability distribution. Our conclusions can be erroneous if the true probability distribution of demand is very different from the assumed distribution.

Nonparametric methods use statistical learning, and base their conclusions on knowledge derived directly from data without any unwarranted assumptions. In the next section we discuss a nonparametric method for forecasting the entire demand distribution that uses the classical empirical probability distribution derived from the relative frequency histogram of time series data on demand. It has the advantage of being capable of updating all changes occurring in the probability distribution of demand, including those in the shape of this distribution.

Then in the following section we illustrate how optimal order quantities that optimize piecewise linear and other unusual cost functions discussed in Section 14.2 can be easily computed using these empirical distribution.

14.5 A Nonparametric Method for Updating and Forecasting the Entire Demand Distribution

In supply chain management, the important random variables are daily or weekly (or whatever planning period is being used) demands of various items (raw materials, components, sub-assemblies, finished goods, spare parts, etc.) that companies either buy from suppliers, or sell to their customers. Observed values of these random variables in each period are generated automatically as a time series in the production process, and are usually available in the production data bases of companies. In this section we discuss a simple nonparametric method for updating changes in the probability distributions of these random variables using this data directly.

Empirical Distributions and Probability Density Functions

The concept of the probability distribution of a random variable evolved from the ancient practice of drawing histograms for the observed values of the random variable. The observed range of variation of the random variable is usually divided into a convenient number of value intervals (in practice about 10 to 25) of equal length, and the relative frequency of each interval is defined to be the proportion of observed values of the random variable that lie in that interval. The chart obtained by marking the value intervals on the horizontal axis, and erecting a rectangle on each interval with its height along the vertical axis equal to the relative frequency is known as the relative frequency histogram of the random variable, or its discretized probability distribution. The relative frequency in each value interval I_i is the estimate of the probability p_i that the random variable lies in that interval, see Figure 1 for an example.

Let $I_1, \ldots, I_n$ be the value intervals with $u_1, \ldots, u_n$ as their midpoints, and $p = (p_1, \ldots, p_n)$, the probability vector in the discretized probability distribution of the random variable. Let $\bar{\mu} = \sum_{i=1}^{n} u_i p_i$, $\bar{\sigma} = \sqrt{\sum_{i=1}^{n} p_i (u_i - \bar{\mu})^2}$.

Then $\bar{\mu}, \bar{\sigma}$ are estimates of the expected value μ, standard deviation σ of the random variable respectively.

We will use the phrase **empirical distribution** to denote such a discretized probability distribution of a random variable, obtained either through drawing the histogram, or by updating a previously known discretized probability distribution based on recent data.

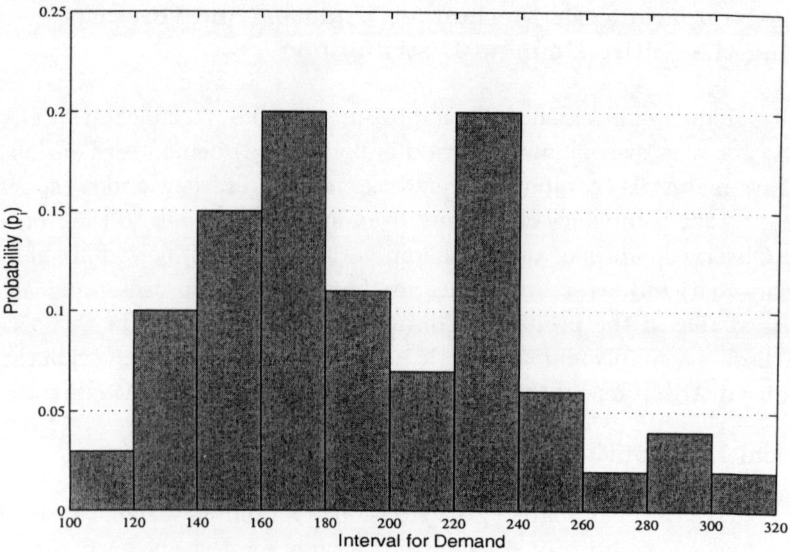

Figure 1: Relative frequency histogram for daily demand for a major component at a PC assembling plant in California.

When mathematicians began studying random variables from the 16th century onwards, they found it convenient to represent the probability distribution of the random variable by the **probability density function** which is the mathematical formula for the curve defined by the upper boundary of the relative frequency histogram in the limit as the length of the value interval is made to approach 0, and the number of observed values of the random variable goes to infinity. So the probability density function provides a mathematical formula for the height along the vertical axis of this curve as a function of the variable represented on the horizontal axis. Because it is a mathematically stated function, the probability density function lends itself much more nicely into mathematical derivations than the somewhat crude relative frequency histogram.

It is rare to see empirical distributions used in decision making models these days. Almost everyone uses mathematically defined density functions characterized by a small number of parameters (typically two or less) to represent probability distributions. In these decision making models, the only freedom we have in incorporating changes, is to change the values of those parameters. This may be inadequate to capture all dynamic changes occurring in the shapes of probability distributions from time to time.

Extending the Exponential Smoothing Method to update the Empirical Probability Distribution of a Random Variable

We will now see that representing the probability distributions of random variables by their empirical distributions gives us unlimited freedom in capturing any type of change including changes in shape, [Murty (2002)].

Let $I_1, \ldots, I_n$ be the value intervals, and $p_1, \ldots, p_n$ the probabilities associated with them in the present empirical distribution of a random variable. In updating this distribution, we have the freedom to change the values of all the p_i, this makes it possible to capture any change in the shape of the distribution.

Changes, if any, will reflect in recent observations on the random variable. Following table gives the present empirical distribution, histogram based on most recent observations on the random variable (for example most recent k observations where k could be about 30), and x_i to denote the probabilities in the updated empirical distribution to be determined.

Value interval	Probability vector in the		
	Present empirical distribution	Recent histogram	Updated empirical distribution (to be estimated)
I_1	p_1	f_1	x_1
$\vdots$	$\vdots$	$\vdots$	$\vdots$
I_n	p_n	f_n	x_n

$f = (f_1, \ldots, f_n)$ represents the estimate of the probability vector in the recent histogram, but it is based on too few observations. $p = (p_1, \ldots, p_n)$ is the probability vector in the empirical distribution at the previous updating. $x = (x_1, \ldots, x_n)$, the updated probability vector, should be obtained by incorporating the changing trend reflected in f into p. In the theory of statistics the most commonly used method for this incorporation is the weighted least squares method [Murty (2002)], which provides the following model (1) to compute x from p and f. In it, β is a weight between 0 and 1, similar to the smoothing constant α in the exponential smoothing method for updating the expected value (like that α there, here β is the relative weight placed on the probability vector from the histogram composed from recent observations).

$$\begin{array}{ll} \text{Minimize} & (1-\beta)\sum_{i=1}^{n}(p_i - x_i)^2 + \beta\sum_{i=1}^{n}(f_i - x_i)^2 \\ \text{subject to} & \sum_{i=1}^{n} x_i = 1 \\ & x_i \geq 0,\ i = 1,\ldots,n \end{array} \tag{1}$$

x is taken as the optimum solution of this convex quadratic program. $\beta =$ 0.1 to 0.4 works well, the reason for choosing this weight for the second term in the objective function to be small is because the vector f is based on only a small number of observations. Since the quadratic model minimizes the weighted sum of squared forecast errors over all value intervals, when used periodically, it has the effect of tracking gradual changes in the probability distribution of the random variable.

The above quadratic program has a unique optimum solution given by the following explicit formula.

$$x = (1-\beta)p + \beta f \tag{2}$$

So we take the updated empirical distribution to be the one with the probability vector given by (2).

The formula (2) for updating the probability vector in the above formula is exactly analogous to the formula for forecasting the expected value of a random variable using the latest observation in exponential smoothing. Hence the above formula can be thought of as the extension of the exponential smoothing method to update the probability vector in the empirical distribution of the random variable.

When there is a significant increase or decrease in the mean value of the random variable, new value intervals may have to be opened up at the left or right end. In this case the probabilities associated with value intervals at the other end may become very close to 0, and these intervals may have to be dropped from further consideration at that time.

This procedure can be used to update the discretized demand distribution either at every ordering point, or periodically at every rth ordering point for some convenient r, using the most recent observations on demand.

14.6 An Application of the Forecasting Method of Section 14.5 for computing Optimal Order Quantities

Given the empirical distribution of demand for the next period, the well known newsvendor model (see [Karlin (1958)], [Nahmias (1993)] and [Silver and Peterson (1979)]) can be used to determine the optimal order quantity

I_i = interval for demand	Probability p_i	u_i = mid-point of interval
100 – 120	0.03	110
120 – 140	0.10	130
140 – 160	0.15	150
160 – 180	0.20	170
180 – 200	0.11	190
200 – 220	0.07	210
220 – 240	0.20	230
240 – 260	0.06	250
260 – 280	0.02	270
280 – 300	0.04	290
300 – 320	0.02	310

for that period that minimizes the sum of expected overage and underage costs very efficiently, numerically. We will illustrate with a numerical example. Let the empirical distribution of demand (in units) for next period be the one given above. The expected value of this distribution $\bar{\mu} = \sum_i u_i p_i =$ 192.6 units, and its standard deviation $\bar{\sigma} = \sqrt{\sum_i (u_i - \bar{\mu})^2 p_i} = 47.4$ units.

Let us denote the ordering quantity for that period, to be determined, by Q, and let d denote the random variable that is the demand during that period. Then

y = overage quantity in this period = amount remaining after the demand is completely fulfilled = $(Q - d)^+ =$ maximum$\{0, Q - d\}$

z = underage quantity during this period = unfulfilled demand during this period = $(Q - d)^- =$ maximum$\{0, d - Q\}$.

Suppose the overage cost $f(y)$, is the following piecewise linear function of y

Overage amount = y	Overage cost $f(y)$ in \$	Slope
$0 \le y \le 30$	$3y$	3
$30 \le y$	$90 + 10(y - 30)$	10.

Suppose the underage cost $g(z)$ in \$, is the fixed cost depending on the amount given below

Underage amount $= y$	Underage cost $g(z)$ in \$
$0 \leq z \leq 10$	50
$10 < z$	150.

To compute $E(Q)$ = the expected sum of overage and underage costs when the order quantity is Q, we assume that the demand value d is equally likely to be anywhere in the interval I_i with probability p_i. This implies for example that the probability that the demand d is in the interval 120 – 125 is = (probability that d lies in the interval $120 - 140)/4 = (0.10)/4 = 0.025$.

Let $Q = 185$. When the demand d lies in the interval $120 - 140$, the overage amount varies from 65 to 45 and the overage cost varies from \$440 to 240 linearly. So the contribution to the expected overage cost from this interval is $0.10(440 + 240)/2$.

Demand lies in the interval $140 - 160$ with probability 0.15. In this interval the overage cost is not linear, but it can be partitioned into two intervals $140 - 155$ (with probability 0.1125), and $155 - 160$ (with probability 0.0375) in each of which the overage cost is linear. In the interval $140 \leq d \leq 155$ the overage cost varies linearly from \$240 to \$90; and in $155 \leq d \leq 160$ the overage cost varies linearly from \$90 to \$75. So, the contribution to the expected overage cost from this interval is \$(0.115 (240 + 90)/2) + (0.0375(90 + 75)/2).

Proceeding this way we see that $E(Q)$ for $Q = 185$ is: \$(0.03 (640+440)/2)+ (0.10(440 + 240)/2) + [(0.115 (240 + 90)/2) + (0.0375(90 + 75)/2)] + (0.20(75 + 15)/2) + [0.0275(15 +0)/2) + 0.055(50) + 0.0275 (150)] + (0.07 + 0.20 + 0.06 + 0.02 + 0.04 + 0.02)150 = \$ 140.87.

In the same way we computed the values of $E(Q)$ for different values of Q spaced 5 units apart, given below.

Q	$E(Q)$
195	178.00
190	162.27
185	143.82
180	139.15
175	130.11
170	124.20
165	120.40
160	121.95
155	122.60
150	124.40
145	139.70

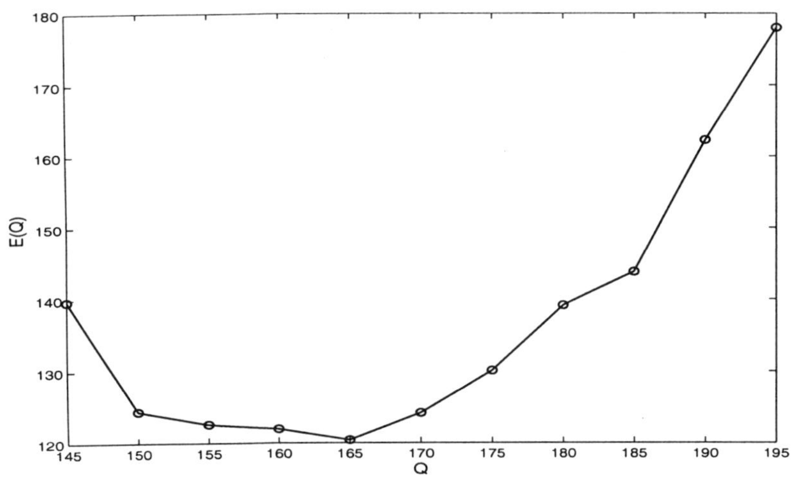

Figure 2: Plot of $E(Q)$ for various values of Q

Figure 2 is a plot of these values of $E(Q)$. Here we computed $E(Q)$ at values of Q which are multiples of 5 units, and it can be seen that $Q = 165$ is the optimum order quantity correct to the nearest multiple of 5. If the optimum is required to greater precision, the above calculation can be carried out for values of Q at integer (or closer) values between 150 to 170 and the best value of Q there chosen as the optimum order quantity.

The optimum value of Q can then be translated into the actual order quantity for the next period by subtracting the expected on-hand inventory at the end of the present period from it.

For each i assuming that demand d is equally likely to be anywhere in the interval I_i with probability p_i, makes the value of $E(Q)$ computed accurate for each Q, However, in many applications people make the simpler assumption that p_i is the probability of demand being equal to u_i, the mid-point of the interval I_i. The values of $E(Q)$ obtained with this assumption will be approximate, particularly when the overage and underage costs are not linear (i.e., when they are piecewise linear etc.); but this assumption makes the computation of $E(Q)$ much simpler, that's why people use this simpler assumption.

14.7 How to incorporate Seasonality in Demand into the Model

The discussion so far dealt with the case when the values of demand in the various periods form a stationary time series. In some applications this series may be seasonal, i.e., it has a pattern that repeats every N periods for some known value of N. The number of periods N, before the pattern begins to repeat is known as the length of the season. In order to use seasonal models, the length of the season must be known.

For example, in the computer industry majority of sales are arranged by sales agents who operate on quarterly sales goals. That's why demand for components in the computer industry, and demand for their own products tends to be seasonal with the quarter of the year as the season. The sales agents usually work much harder in the last month of the quarter to meet their quarterly goals, so demand for products in the computer industry tends to be higher in the third month of each quarter than in the beginning two months. As most of the companies are building to order now-a-days, weekly production levels and demands for components inherit the same kind of seasonality.

At one company in this industry each quarter is divided into three homogeneous intervals. Weeks 1 to 4 of the quarter are slack periods, each of these weeks accounts a fraction of about 0.045 of the total demand in the quarter. Weeks 5 to 8 are medium periods, each of these weeks accounts for a fraction of about 0.074 of the total demand in the quarter. Weeks 9 to 13 are peak periods, each of these weeks accounts for a fraction of about 0.105 of the total demand in the quarter. This fraction of demand in each week of the season is called the *seasonal factor* of that week.

In the same way in the paper industry demand for products exhibits seasonality with each month of the year as the season. Demand for their products in the 2nd fortnight in each month tends to be much higher than in the 1st fortnight.

There are several ways of handling seasonality. One way is for each $i = 1$ to N (= length of the season), consider demand data for the ith period in each season as a time series by itself, and make the decisions for this period in each season using this series based on methods discussed in earlier sections.

Another method that is more popular is based on the assumption that there exists a set of indices c_i, $i = 1$ to N called *seasonal factors* or *seasonal indices* (see [Nahmias (1993)], [Silver and Peterson (1979)]), where c_i represents the demand in the ith period of the season as a fraction of the demand during the whole season (as an example see the seasonal factors given for the computer company described above). Once these seasonal factors are estimated, we divide each observation of demand in the original demand time series by the appropriate seasonal factor to obtain the deseasonalized demand series. The time series of deseasonalized demand amounts still contains all components of information of the original series except for seasonality. Forecasting is carried out using the methods discussed in the earlier sections, on the deseasonalized demand series. Then estimates of the expected demand, standard deviation, and the optimal order quantities obtained for each period must be reseasonalized by multiplying by the appropriate seasonal factor before being used.

14.8 Portfolio Management

One of the most important and very widely studied problems in finance is that of optimizing the return from investments. Everyone in this world from little individual investors to Presidents and CEOs of very large corporations with annual incomes ranging to hundreds of millions of dollars; all the banks, mutual funds, and other financial institutions have great interest in this problem.

There are many different investment opportunities, but the return (also called yield, which may be positive, 0, or negative) from each varies from period to period as a random variable. The area is "data rich" in the sense that the return per unit investment in each investment opportunity in each period in the past is freely available as a time series and can be accessed by anyone.

One important problem in the area is: given a budget B (amount of money available to invest) and a list of investment opportunities 1, ..., n to invest it in; how to optimally divide the budget among the various investment opportunities. Once invested, that investment may be kept for several periods, and the returns from it keep accumulating period to period as long as the investment is kept. So, the important feature is that the return from investment is not in a single installment, but paid out in each period over the life of the investment. In applications, n, the number of investment opportunities under consideration, tends to be large.

Denoting by the decision variable x_i the amount of the budget allocated to investment opportunity i, for $i = 1$, to n; the vector $x = (x_1, \dots, x_n)^T$, called a **portfolio**, is a solution for the problem. The goal of portfolio optimization, is to characterize and find an optimum portfolio.

Let the random variable $P(x)$ denote the total return from portfolio x in a period. For each period in the past we can compute $P(x)$ using x and the available data on returns from individual investment opportunities. So, for any x, $P(x)$ can be generated as a time series. Let $\mu(x) =$ the expected value of the return $P(x)$ in a period from portfolio x.

$P(x)$ varies randomly, this variation is perceived as the **risk** (or **volatility of returns**) associated with portfolio x. Everyone agrees with treating variation in returns from period to period as a risk, but there is not a universal agreement on how to measure this risk. We will use the symbol $r(x)$ to denote this risk as a function of x. The two most important parameters for characterizing an optimum portfolio are:

> $\mu(x) = E(P(x))$, the expected value of the return $P(x)$ in a period, it is a measure of the long term average return per period from portfolio x,
>
> $r(x) =$ a measure of risk associated with portfolio x, a suitable measure is to be selected.

There is universal agreement that an optimum portfolio should maximize $\mu(x)$. In fact some investors select a portfolio x and keep it for a long time. For such investors the period to period variation in the return $P(x)$ may not be that critical, they mainly want to see $\mu(x)$ maximized.

But the majority of investors (particularly large investors like banks, mutual funds etc.) change their portfolio periodically by selling some investments made in earlier periods at current prices, or by investing additional amounts. For these investors the period to period variation in the return is also an important factor to take into consideration. These investors not only want to maximize the long term average return, but would also like to keep the return in every period as high as possible. So, from their perspective, an optimum portfolio should maximize the expected return $\mu(x)$, and minimize the risk $r(x)$; i.e., it should achieve both these objectives simultaneously. So finding an optimum portfolio here is a **multi-objective optimization problem**.

But in multi-objective optimization, there is no concept of "optimality" that has universal acceptance. Also, the two objectives typically conflict with each other; i.e., portfolios that maximize expected return $\mu(x)$ are usually associated with high values for what ever measure $r(x)$ is chosen to represent risk.

Usually the various investment opportunities are partitioned into various sectors by their type (for example utility opportunities, banking opportunities, etc.). Then the decision makers usually impose lower and upper bounds on the amount of the budget that can be invested among investment opportunities in each sector, and may be some other linear constraints also. Suppose the system of all these constraints including the budget constraint is (here e is the column vector of all 1s in R^n)

$$\begin{aligned} &Ax \leq b \\ &e^T x \leq B \\ &x \geq 0 \end{aligned} \tag{3}$$

A portfolio x is said to be a *feasible portfolio* if it satisfies all the constraints in (3). Once a measure $r(x)$ for risk is selected, if $x, \bar{x}$ are two feasible portfolios satisfying: either $\mu(x) > \mu(\bar{x})$ and $r(x) \leq r(\bar{x})$, or $\mu(x) \geq \mu(\bar{x})$ and $r(x) < r(\bar{x})$; then x is said to *dominate* $\bar{x}$, because it is better or the same as $\bar{x}$ WRT (with respect to) both the objective functions in the problem, and strictly better on at least one of the two objectives.

A feasible portfolio $\bar{x}$ is said to be a *nondominated portfolio* or *efficient portfolio* or *pareto optimum portfolio* if there is no other feasible portfolio that dominates it. In multi-objective optimization problems like this one, there is no concept of "optimality" that has universal acceptance, but clearly no investor would like a portfolio that is dominated by another one. So, we should look among efficient Portfolios for a solution to the problem. But usually there are many efficient portfolios, the set of all of them is called the *efficient frontier*.

Mathematicians would consider a multi-objective problem well solved if an algorithm is developed to enumerate the efficient frontier in a computationally efficient way. Here I can mention the entertaining Hollywood movie

A Beautiful Mind based on the life of John Nash who received the Nobel Prize in economics in 1994 for proving that a certain type of two-objective optimization problems always have at least one efficient solution.

In a pair of efficient portfolios, if the 1st is better than the 2nd WRT the average return $\mu(x)$, then the 2nd will be better than the 1st WRT the risk function $r(x)$, so the best portfolio among these two is not defined. Given a feasible portfolio that is not efficient, an efficient portfolio better than it can be found; but there is no universally acceptable criterion for selecting the best among efficient portfolios. The challenge in portfolio optimization is to select a good measure for "risk", and obtain a good portfolio that has satisfactory values for both the objective functions.

Besides portfolio optimization, portfolio management deals with the issues of determining how long an optimum portfolio determined should be kept, and the appropriate tools for tracking its performance while it is kept. Changing the current portfolio and adopting a new one in every period is a very labor-intensive and expensive process, that's why once an optimum portfolio is determined in some period, most investors do not like to change it as long as it is performing upto expectations.

14.9 Variance as a Measure of Risk, and the Markovitz Model for Portfolio Optimization

Most of the work in finance is based on the assumption that the yields in a period from unit investments in the various investment opportunities follow a multivariate normal distribution. Let the vector of expected values in this distribution be $\mu = (\mu_1, \dots, \mu_n)^T$, and let the variance-covariance matrix in it be the symmetric positive definite matrix $\Sigma = (\sigma_{ij})$ of order n.

Then $\mu(x) =$ the expected return from portfolio x in a period is $\mu^T x$, and the variance of this return is $x^T \Sigma x$.

In statistical theory, the variance of a random variable is a well accepted measure of variation of this random variable. Since the "risk" of a portfolio stems from the variation in the returns from it from period to period, Harry Markovitz proposed in 1952 using the variance $x^T \Sigma x$ of returns from the portfolio x as the measure $r(x)$ of risk associated with it under the normality assumption. He suggested the approach of minimizing this risk function subject to the constraint that the expected fractional return per period must be $\geq$ some specified lower bound δ, to define an "optimum portfolio". This leads to the following classical *Markovitz portfolio model* for which he won the 1989 Von Neumann theory prize for contributions to OR of INFORMS, and the 1990 Nobel Prize in economics.

$$\begin{array}{c}\text{Minimize } x^T \Sigma x \\ \text{subject to the feasibility conditions in (3), and } \mu^T x \geq \delta B\end{array} \quad (4)$$

where e is a column vector of all 1's in R^n, and e^T is its transpose. This is a quadratic programming problem, its optimum solution is known as a *minimum variance portfolio*

The minimum variance portfolio does not have universal acceptance as the best portfolio to adopt, since it may not have good practical features. For example, suppose $\delta = 0.07$, and the minimum variance portfolio is a portfolio G with expected return fraction per period of 0.07 and variance of 0.0016. There may be another feasible portfolio H with expected return fraction per period of 0.25 and variance of 0.0018. Portfolio H which is not optimum for this model, yields a higher return than portfolio G with very high probability in every period and is definitely more desirable. Also, as pointed out in [Papahristodoulou and Dotzauer (2004)], many investors and traders as well question whether the variance of the return $x^T \Sigma x$ is an appropriate measure of risk; and many researchers question whether the assumption that the returns from individual investment opportunities follow a multivariate normal distribution is reasonable.

Another problem with this model deals with the computational difficulties in solving it. The distribution of returns may be changing with time, and updating the distribution requires re-computation of the variance-covariance matrix using recent data at frequent intervals, an expensive operation when n is large. Also, the variance-covariance matrix will be fully dense, this makes the model (4) computationally difficult to handle if n is large.

14.10 Other Measures of Risk for Portfolio Optimization

While everyone perceives variation in the returns as an element of risk, no one complains if the variation is taking the returns higher; but they will definitely complain when they begin to decrease. That is, investor's reaction to the two types of variation are highly asymmetric. For this reason the variance of the return used as the measure of risk in the classical Markovitz model is not a fully appropriate measure of risk.

Several other measures of risk of a portfolio have been proposed in the literature. We will use as a measure of risk of a portfolio x, the probability $d(x, \delta)$ that the return $P(x)$ from it in a period is $\leq \delta e^T x$, where δ is a minimum return per unit investment per period demanded by the investor.

A good portfolio should either have as one of its objectives minimizing this risk measure $d(x, \delta)$, or keeping it $\leq$ some specified upper limit γ for it. This measure is closely related to the Value-at-Risk (VaR)measure, and other downside risk measures and the safety-first conditions studied in the literature.

The expected return per period $\mu(x)$ of a portfolio x is a measure of the long term benefit of adopting it; because it measures the average return per period one can expect to get from it if the present distribution of returns continues unchanged. One model that we will consider later in Section 14.12 for defining an optimum portfolio is to maximize $\mu(x)$ subject to the constraint that $d(x, \delta) \leq \gamma$.

14.11 Portfolio Management: Tracking the Distribution of Return from a Portfolio that is kept for a Long Time

The literature in finance has many research publications dealing with models for portfolio optimization, and we will discuss one such model based on statistical learning in the next section. But very few research publications deal with portfolio management, which also deals with tracking the performance of the optimum portfolio determined to check whether it is performing to expectation, and deciding when to change the portfolio. In this section we discuss an application of the simple forecasting method discussed in Section 14.5 to track the performance of the portfolio in current use.

For this, the most important random variables are the per unit return in a period from various investment opportunities. The distributions of these random variables may be changing over time, and unless these changes are incorporated into the decision making process, the selected portfolio may not be a satisfactory one for the next or for any future period.

The distribution of return from a single investment opportunity can be estimated from past data by its discretized probability distribution discussed in Section 14.5. This discretized probability distribution can be updated over time based on recent data by the technique discussed in Section 14.5, we will use the phrase "empirical distribution" to denote the updated distribution.

The updating technique discussed in Section 14.5 is quite convenient for updating the empirical distribution of return from a single investment opportunity, because in this case the probabilities associated with a small number of intervals need to be updated at each updating. But for studying the returns from two or more investment opportunities (2 or more random variables in a dependence relationship), its direct extension becomes unwieldy due to the curse of dimensionality. In the multivariate context, the discretized distribution breaks up the space of the vector of variables into a number of rectangles each with its associated probability. Even when the number of variables in 2, the number of these rectangles is too large, and updating their probabilities becomes impractical.

However we will see that this one-variable technique is itself a useful tool in keeping track of portfolios that are kept for long periods of time.

Suppose an investor likes to keep her/his portfolio $\bar{x}$ unchanged as long as it is performing to his/her expectations. The value of $P(\bar{x})$ in each period can be computed directly from $\bar{x}$ and the available data on the returns from the various investment opportunities, and generated as a time series. Using it, the distribution of $P(\bar{x})$ can be updated over time as explained in Section 14.5. If the distribution of $P(\bar{x})$ is estimated and maintained in the form of an empirical distribution, the expected value of return from the current empirical distribution, is an estimate of the current expected return from portfolio $\bar{x}$. Also, since the empirical distribution is a discretized distribution, an estimate of the risk measure $d(\bar{x}, \delta)$ = probability that the return is $\leq \delta e^T \bar{x}$ in it can be computed very easily. From estimates of expected return, and $d(\bar{x}, \delta)$, the two measures for evaluating a portfolio, the investor can judge whether to continue to keep the portfolio $\bar{x}$, or look for a better portfolio to change to.

14.12 A Model based on Statistical Learning to find an Optimum Portfolio

Let $\bar{x}$ denote the current portfolio in use.

Under the assumption that the returns from various investment opportunities follow a multivariate normal distribution, the measure of risk $d(x, \delta)$ for any portfolio x is a nonlinear function, and the problem of maximizing the expected return μx subject to the constraint that $d(x, \delta) \leq \gamma$ is a complex problem. Even if the optimum solution of this problem can be determined, since the actual distribution of the returns vector is unknown, it is not clear how good the performance of the resulting portfolio derived from the normality assumption will be in reality.

In statistical learning, instead of making assumptions about the distribution of the returns vector, we base our decisions on knowledge derived from actual data. We now develop an MIP (mixed integer programming) model for finding an optimum portfolio based on estimates of relevant quantities obtained from actual data over the most recent m periods, for some selected m. The first model ignores the transaction costs of moving from the current portfolio $\bar{x}$ to the optimum portfolio. Let

e_{ij} = actual return from unit investment in the i-th period from the j-th investment opportunity, $i = 1$ to m, $j = 1$ to n

$E = (e_{ij})$, an $m \times n$ matrix of data on actual returns

$E_{i.} = (e_{i1}, \ldots, e_{in})$, the i-th row vector of E.

The risk condition $d(x, \delta) \leq \gamma$ translates to the requirement that the constraint $E_{i.}x \geq \delta e^T x$ must hold for at least t periods i, where t = ceiling of $((1-\gamma)m)$, the smallest integer $\geq ((1-\gamma)m)$. Define the binary variables $z_1, \ldots, z_m$, where $z_i = 0$ if $E_{i.}x \geq \delta e^T x$, 1 otherwise. In terms of these binary variables, the model for finding an optimum portfolio is (5) to (10), here $L > 0$ is a positive number such that $-L$ is a lower bound for each $E_{i.}x - \delta e^T x$.

$$\text{Maximize } \frac{1}{m} \sum_{i=1}^{m} \sum_{j=1}^{n} e_{ij} x_j \tag{5}$$

$$\text{subject to } \sum_{j=1}^{n} x_j \leq B \tag{6}$$

$$Ax \leq b \tag{7}$$

$$E_{i.}x - \delta e^T x + L z_i \geq 0, \; i = 1, \ldots, m \tag{8}$$

$$\sum_{I=1}^{m} z_i \leq m - t \tag{9}$$

$$x_j \geq 0, z_i \in \{0, 1\}, \text{ for all } i, \tag{10}$$

Transaction costs for selling existing investments, or acquiring additional investments can also be taken into account in the model. Assuming that the transaction costs are linear, suppose c_j^+, c_j^- are the costs of acquiring additional unit investment, selling unit investment respectively in investment opportunity j, for $j = 1$ to n. Then the transaction cost for moving from current portfolio $\bar{x}$ to portfolio x is $\sum_{j=1}^{n}[c_j^+(x_j - \bar{x}_j)^+ + c_j^-(x_j - \bar{x}_j)^-]$, where $(x_j - \bar{x}_j)^+ = \text{Maximum}\{x_j - \bar{x}_j, 0\}$ = additional investment in opportunity j acquired, and $(x_j - \bar{x}_j)^- = \text{Maximum}\{\bar{x}_j - x_j, 0\}$ = investment in opportunity j sold.

Assuming that the transaction cost coefficients c_j^+, c_j^- are all positive, the model for maximizing average return per period $-$ transaction costs is to: Maximize $\frac{1}{m}\sum_{i=1}^{m}\sum_{j=1}^{n} e_{ij}x_j - \sum_{j=1}^{n}(u_j^+c_j^+ + u_j^-c_j^-)$ subject to constraints (6) to (10) and $x_j - \bar{x}_j = u_j^+ - u_j^-$ and $u_j^+, u_j^- \geq 0$ for $j = 1$ to n.

The number of binary variables in either model is m, the number of recent periods considered in the model. Since the distribution of returns may be changing over time, we will not make m too large anyway; so these models can be solved within reasonable time with existing software systems for MIP. For example, if the period is a week, and weekly return data over the most recent 6-month (26 week) period is used to construct the model, it will have only 26 binary variables, and so is quite easy to solve with software tools available in the market.

Solving the same model with different values of δ, γ generates different portfolios which can be compared with each other and the best among them selected for implementation.

The matrix E of returns is expected to be fully dense. So, when n, the number of investment opportunities considered, is large, the LP relaxations of these models will be dense and may turn out to be hard to solve with existing methods based on matrix inversion operations. New descent methods for LP not based on matrix inversion operations discussed in [Murty (2006)] have the potential to solve such models efficiently when n is large.

Bibliography

References on Forecasting and Supply Chain Issues

Box, G. E. P. and Jenkins, G. M. (1970). *Time Series Analysis, Forecasting and Control*, (Holden Day, San Francisco).

Brown, R. G. (1959). *Statistical Forecasting for Inventory Control*, (McGraw-Hill, NY).

Foote, B. L. (1995). On the Implementation of a Control-Based Forecasting System for Aircraft Spare Parts Procurement, *IIE Transactions* **27**, pp. 210-216.

Holt, C. C. (1957). *Forecasting Seasonal and Trends by Exponentially Weighted Moving Averages*, ONR Memo no. **52**.

Karlin, S. (1958). Optimal Inventory Policy for the Arrow-Harris-Marschak Dynamic Model, Ch. 9 in *Studies in the Mathematical Theory of Inventory and Production*, K. J. Arrow, S. Karlin, and H. Scarf (eds.), Stanford University Press.

Montgomory, D. C. and Johnson, L. A. (1976). *Forecasting and Time Series Analysis*, (McGraw-Hill, St. Louis).

Murty, K. G. (2002).*Histogram, an Ancient Tool and the Art of Forecasting*, Technical report, IOE Dept., University of Michigan, Ann Arbor,

Nahmias, S. (1993).*Production and Operations Analysis*, 2nd ed., (Irwin, Boston).

Silver, E. A. and Peterson, R. (1979). *Decision Systems for Inventory Management and Production Planning*, 2nd. ed., (Wiley, NY).

References on Portfolio Optimization

Andersson, F., Mausser, H., Rosen, D. and Uryasev, S. (2001). Credit Risk Optimization with Conditional Value-at-Risk Criterion, *Mathematical Programming B* **89**, pp. 273-292.

Artzner, P., Delban, F., Eber, J. -M. and Heath, D. (1999). Coherent Measures of Risk, *Math. Finance*, **9**, pp. 203-227.

Benati, S. and Rizzi, R. (January 2007). A Mixed Integer Linear Programming Formulation of the Optimal Mean/Value-at-Risk Portfolio Problem, *European Journal of Operastional Research*, **176** 1, pp. 423-434.

Cheklov, A., Uryasev, S. and Zabarankin, M.(2002). Drawdown Measure in Portfolio Optimization, *International Journal of Theoretical and Applied Finance*, **26**, 7 pp. 1443-1471.

Folmer, H. and Schied, A. (2004). *Stochastic Finance.*

Jansen, D. W., Koedijk, K. G. and de Vries, C. G. (2000).Portfolio Selection with Limited Downside Risk, *Journal of Empirical Finance*, **7**, pp. 247-269.

Konno, H. and Yamazaki, H. (1991). Mean-Absolute Deviation Portfolio Optimization Model and Its Application to Tokyo Stock Market, *Management Science*, **37**, pp. 519-531.

Markovitz, H. (1952). Portfolio Selection, *Journal of Finance*, **7**, pp. 77-91.

Murty, K. G. (2006). A New Practically Effficient Interior Point Method for LP, *Algorithmic Operations Research*, 1 3-19; paper can be seen at the website: http://journals.hil.unb.ca/index.php/AOR/index.

Papahristodoulou, C. and Dotzauer, E. (2004). Optimal Portfolios Using Linear Programming Models, *Journal of the Operational Research Society*, **55**, pp. 1169-1177.

Perold, A. (1984). Large-Scale Portfolio Optimization, *Management Science*, **30**, pp. 1143-1160.

Rockafellar, R. T. and Uryasev, S. P. (2000). Optimization of Conditional Value-at-Risk", *J. of Risk*, **2**, pp. 21-42.

Roy, A. D. (1952). Safety-First and the Holding of Assets", *Econometrica*, **20**, pp. 431-449.

Speranza,M. G. (1993). Linear Programming Models for Portfolio Optimization", *Finance*, **14**, pp. 107-123.

Stone, B. K. (1973 September). A Linear Programming Formulation of the General Portfolio Selection Problem, *J. Financ. Quant. Anal.*, pp. 621-636.

Young, M. R. (1998). A minimax-Portfolio Selection Rule with Linear Programming Solution, *Management Science*, **44**, pp. 673-683.

Chapter 15

Variational Analysis in Bilevel Programming

S. Dempe
T. U. Freiberg, Germany
J. Dutta
Indian Institute of Technology Kanpur, India
B. S. Mordukhovich[1]
Wayne State University, USA

Dedicated to the memory of Professor S. R. Mohan

Abstract

The paper is devoted to applications of advanced tools of modern variational analysis and generalized differentiation to problems of optimistic bilevel programming. In this way, new necessary optimality conditions are derived for two major classes of bilevel programs: those with partially convex and with fully convex lower-level problems. We provide detailed discussions of the results obtained and their relationships with known results in this area.

Key Words: Bilevel programming, variational analysis, fully convex lower-level problems, partially convex lower-level problems

15.1 Introduction

In this paper we intend to discuss the interplay of variational analysis and bilevel programming. The term *Variational Analysis* is of quite recent origin, and most probably the monograph by [Rockafellar and Wets (1998)]

[1]Research of this author was partly supported by the US National Science Foundation under grants DMS-0304989 and DMS-0603846 and by the Australian Research Council under grant DP-0451168

had led the popularization of the term. In modern optimization, set-valued maps play a major role. Their role shot into prominence with the advent of nonsmooth analysis and nonsmooth optimization, since the role of the derivative in modern optimization is taken over by set-valued maps known as subdifferentials. The solution set map of a parametric optimization problem is another important example of a set-valued map appearing in optimization theory. This map plays a very fundamental role in bilevel programming; see, e.g., [Dempe S. (2002)]. Further, an important role in optimization is now played by derivatives and coderivatives of set-valued maps. For more details see [Rockafellar and Wets (1998)] and the very recent two-volume monograph by [Mordukhovich (2006a,b)].

On the other hand, bilevel programming grew out of the now classical Stackelberg games (see [von Stackelberg (1934)]) where a leader and a follower interact so that both can achieve their targeted objectives. In the language of optimization this can be framed as a two-level optimization problem as follows:

$$\min_{x} F(x,y) \quad \text{subject to} \quad x \in X, \;\; y \in S(x),$$

where $F : \mathbb{R}^n \times \mathbb{R}^m \to \mathbb{R}$, $X \subseteq \mathbb{R}^n$, and where $S : \mathbb{R}^n \rightrightarrows \mathbb{R}^m$ is the solution set mapping to the lower-level problem:

$$\min_{y} f(x,y) \quad \text{subject to } y \in K(x),$$

where $f : \mathbb{R}^n \times \mathbb{R}^m \to \mathbb{R}$ and $K(x)$ is a closed set for each x. We denote the above optimization problem by (BP). So the idea is that the upper-level decision maker, or the leader, chooses a decision vector x and passes it onto the lower-level decision maker, or the follower, who then—based on the leader's choice x—minimizes his/her objective function and returns the solution y to the leader who then uses it to minimizes his/her objective function.

If for each x the lower-level problem has a unique solution, then the problem (BP) is well defined. However, if there are multiple solutions to the lower-level problem for a given x, then the upper-level objective becomes a set-valued map. In order to overcome this difficulty, two different solution concepts have been defined in the literature. These are namely the optimistic solution and the pessimistic solution.

For the optimistic case one first defines the function

$$\phi_0(x) := \inf_{y}\{F(x,y) : y \in S(x)\}.$$

Then the optimistic problem is:

$$\min \phi_0(x) \quad \text{subject to} \quad x \in X. \tag{15.1}$$

Thus a pair of points $(\bar{x}, \bar{y})$ is said to be an optimistic solution to the bilevel problem (BP) if $\phi_0(\bar{x}) = F(\bar{x}, \bar{y})$ and $\bar{x}$ is the optimal solution (local or global) to (15.1). On the other hand, in the pessimistic case we define the function

$$\phi_p(x) := \sup_y \{F(x,y) : y \in S(x)\}$$

and formulate the pessimistic problem as follows:

$$\min \phi_p(x) \quad \text{subject to} \quad x \in X.$$

In this paper we concentrate on the optimistic bilevel programming problem. An important situation where an optimistic bilevel formulation can be used is, e.g., that between a supplier and a store owner of some commodities. Since both want to do well in their businesses, the supplier will always give his/her best output to the store owner who in turn would like to do his/her best in the business. In some sense, both would like to minimize their loss or rather maximize their profit and thus act in the optimistic pattern. It is clear that in this example the store owner is the upper-level decision maker and the supplier is the lower-level decision maker. Thus in the study of supply chain management the optimistic bilevel problem can indeed play a fundamental role.

As it has been seen in [Dutta and Dempe (2006)] and [Dempe, Dutta and Mordukhovich (to appear)], in studying the optimistic formulation of the bilevel programming problem it is useful to concentrate on the following problem (BPO):

$$\min_{x,y} F(x,y) \quad \text{subject to} \quad x \in X, (x,y) \in \text{gph } S$$

If we consider global optimal solutions, then (BPO) is equivalent to the optimistic formulation of the bilevel problem (BP). This relationship is slightly more subtle when we consider local optimistic solutions. If the solution set map is uniformly bounded around the optimistic solution of the problem (BP), then the optimistic solution is a local minimum for problem (BPO). The converse however need not be true. Hence we will concentrate our efforts to analyze the local optimal points of problem (BPO).

A major bottleneck in developing necessary optimality conditions for bilevel programs is that most of the standard constraint qualifications (like, e.g., the Mangasarian-Fromovitz constraint qualification or the Abadie constraint qualification) are never satisfied for bilevel programs; see, e. g., [Scheel H. and Scholtes S. (2000)]. This problem comes to light when the lower-level problem is replaced by its corresponding Karush-Kuhn-Tucker

(KKT) conditions. This approach of replacing the lower-level problem by KKT conditions seems to be rather adequate if the lower-level problem is convex in the variable y and satisfies some regularity conditions; see [Dutta and Dempe (2006)] for more detailed discussions. The presence of the complementarity slackness condition actually brings forth this violation of constraint qualifications; see, e.g., [Dempe S. (2002)]. Thus various approaches have been used to develop necessary optimality conditions in bilevel programming. The reader may consult the book by [Dempe S. (2002)] and the references therein for various necessary optimality conditions in bilevel programming. Let us mention that the approach in [Dempe S. (2002)] requires an explicit representation of the feasible set of the lower-level problems via equality and inequality constraints. [Dutta and Dempe (2006)] consider the case when the lower-level feasible sets are not explicitly expressed via functional constraints but are convex sets depending on the parameter x, and the lower-level objective function is convex in y for each x. In this setting, for smooth functions F and no constraint situation $X = \mathbb{R}^n$, necessary optimality conditions are expressed as

$$0 \in \nabla F(\bar{x}, \bar{y}) + N_{\text{gph}S}(\bar{x}, \bar{y}),$$

where $(\bar{x}, \bar{y})$ is a locally optimal solution of (BPO) and $N_{\text{gph}S}(\bar{x}, \bar{y})$ is the basic/Mordukhovich normal cone to the graph of the solution set map S at the point $(\bar{x}, \bar{y})$; see Section 15.2. We can now shift our attention to variational analysis, since in order to develop necessary optimality conditions, we need to focus on calculating the basic normal cone in the above expression when the lower-level feasible set is explicitly defined, and also to see under what qualification conditions such a computation is possible. Thus the approach in [Dutta and Dempe (2006)] brings forth the fundamental role that variational analysis plays in bilevel programming. Our aim here is to present the state-of-the-art on the role of variational analysis in bilevel programming.

This paper is planned as follows. In Section 15.2 we present some basic tools and facts from variational analysis, which are widely used in the sequel. In Section 15.3, which is one of the main sections of this paper, we aim to study bilevel programming problems with partially convex lower-level problems. The computation of the coderivative of the solution set map plays a major role in the analysis of the optimality conditions. This has been shown in [Dutta and Dempe (2006)], where results of coderivative computations from [Levy and Mordukhovich (2004)] have been used. We begin Section 15.3 with the explicit computation of the normal cone to the graph of a set-valued map defined as a solution set to a certain generalized variational inequality. Using this, we derive necessary optimality conditions for bilevel programs when the lower-level problem is partially convex, the feasible set does not depend on x, and $X = \mathbb{R}^n$. Then we move on to the case where X still equals $\mathbb{R}^n$ while the feasible set of the lower-level problem depends on x. At the end of this section we consider the general optimistic bilevel programming problem (BPO), where X is a proper subset of $\mathbb{R}^n$ and the lower-level feasible set depends on x. We provide examples where the qualification condition used hold and where they do not hold. It happens that the qualification conditions of Section 15.3 do not hold when the lower-level problem is linear. That leads us to consider the notion of partial calmness due to [Ye and Zhu (1995)]. Then we move to Section 15.4, where we study the case in bilevel programming when the lower-level problem is fully convex, which covers the case where the lower-level problem is linear. We derive necessary optimality conditions, which improve those in Section 15.3, at least for the fully convex lower-level problem.

15.2 Tools from Variational Analysis

In this section we briefly describe the basic tools of variational analysis needed in the sequel. We start with the variational geometry of constraint sets and describe various conic approximations associated with them.

Let us begin with the notion of the *regular normal cone* or the *Fréchet normal cone* at a point $\bar{x} \in C$, where C is a subset of $\mathbb{R}^n$. A vector $v \in \mathbb{R}^n$ is called a regular normal to C at $\bar{x}$ if

$$\langle v, x - \bar{x} \rangle \leq o(\|x - \bar{x}\|),$$

where $\lim_{x \to \bar{x}} \dfrac{o(\|x - \bar{x}\|)}{\|x - \bar{x}\|} = 0$. The collection of all regular normals to C at $\bar{x}$ is a cone denoted by $\hat{N}_C(\bar{x})$.

It is easy to show that if C is a convex set, the regular normal cone reduces to the standard normal cone of convex analysis (see, e.g., [Rockafellar (1970)]). Though this definition of the regular normal cone might look as a natural generalization of the normal cone from the convex case to the nonconvex case, there are some serious pitfalls. One of the major drawbacks is that at points on the boundary of the set C the regular normal cone may just reduce to the trivial cone containing only the zero vector. To overcome this, a limiting procedure is employed, which leads us to the more robust notion of the *basic normal cone*.

A vector $\bar{v} \in \mathbb{R}^n$ is an element of the basic normal cone $N_C(\bar{x})$ to the set C at $\bar{x} \in C$ if there exist sequences $\{x_k\}$ with $x_k \in C$ and $x_k \to \bar{x}$ as well as $\{v_k\}$ with $v_k \to \bar{v}$ and $v_k \in \hat{N}_C(x_k)$. In a more compact form this is written as

$$N_C(\bar{x}) := \limsup_{x \to \bar{x}} \hat{N}_C(\bar{x}), \quad (x \in C),$$

in terms of the so-called Painlevé-Kuratowski upper/outer limit. It is important to note that the basic normal cone is closed but need not be a convex set. Further, when the set C is convex, it reduces to the classical normal cone of convex analysis. Further it is important to note that if $\bar{x}$ is an interior point of C then $N_C(\bar{x}) = \{0\}$. Further it is clear that $\hat{N}_C(\bar{x}) \subseteq N_C(\bar{x})$.

Another concept, which is important for our study, is the notion of *normal regularity* of a set at a given point. The set C is said to be normally regular at $\bar{x} \in C$ if $\hat{N}_C(\bar{x}) = N_C(\bar{x})$.

Associated with the notion of the regular normal cone is the notion of the regular/Fréchet subdifferential of a function. Since in this study our functions are locally Lipschitz, we describe the regular subdifferential only for locally Lipschitz functions.

Let $f : \mathbb{R}^n \to \mathbb{R}$ be a locally Lipschitz function, and let $\bar{x} \in \mathbb{R}^n$ be given. The regular subdifferential $\hat{\partial} f(\bar{x})$ of the function f at $\bar{x}$ is given by

$$\hat{\partial} f(\bar{x}) := \{v \in \mathbb{R}^n : (v, -1) \in \hat{N}_{\text{epi} f}(\bar{x}, f(\bar{x}))\},$$

where epi f denotes the epigraph of f. The regular subdifferential also has a major drawback in the sense that there are points crucial, e.g., for optimization, where this subdifferential becomes empty. These are precisely the points where the regular normal cone to the epigraph of the function f reduces to the trivial cone containing only the zero element.

This trouble with the regular subdifferential is overcome by passing to the limit in order to obtain a more robust object called the *basic subdifferential*, which is given by

$$\partial f(\bar{x}) := \limsup_{x \to \bar{x}} \hat{\partial} f(x)$$

The above expression means that $v \in \partial f(\bar{x})$ if there exist sequences $\{v_k\}$ and $\{x_k\}$ with $x_k \in C$ such that $v_k \to v$ and $x_k \to \bar{x}$ with $v_k \in \hat{\partial} f(x_k)$. Knowing the fact that every basic normal can be realized as the limit of regular normals, we have the equivalent representation of the basic subdifferential:

$$\partial f(\bar{x}) = \{v \in \mathbb{R}^n : (v, -1) \in N_{\text{epi} f}(\bar{x}, f(\bar{x}))\}.$$

The basic normal cone and the basic subdifferential were first introduced by [Mordukhovich (1976)] in 1976. For more details see [Rockafellar and Wets (1998)] or the recent monographs of [Mordukhovich (2006a)], [Mordukhovich (2006b)].

Set-valued maps arise naturally in optimization, and it is very important to look at their differential properties. A significant concept in this direction is the notion of coderivative by (see, e.g., his book [Mordukhovich (2006a)]). Given a set-valued map $F : \mathbb{R}^n \rightrightarrows \mathbb{R}^m$ and a point $(\bar{x}, \bar{y}) \in \text{gph}\, F$, the coderivative of F at $(\bar{x}, \bar{y})$ is a set-valued map $D^*F(\bar{x}, \bar{y}) : \mathbb{R}^m \rightrightarrows \mathbb{R}^n$ defined by

$$D^*F(\bar{x}, \bar{y})(y^*) := \{x^* \in \mathbb{R}^n : (x^*, -y^*) \in N_{\text{gph}F}(\bar{x}, \bar{y})\}.$$

We now consider the following optimization problem (P):

$$\min f_0(x) \quad \text{subject to} \quad F(x) \in U, \quad x \in X, \tag{15.2}$$

where $f_0 : \mathbb{R}^n \to \mathbb{R}$ and $F : \mathbb{R}^n \to \mathbb{R}^m$ are smooth functions, $U \subseteq \mathbb{R}^m$, and $X \subseteq \mathbb{R}^n$. The necessary optimality condition for (P) formulated in the next theorem can be found in [Rockafellar and Wets (1998)] and [Mordukhovich (2006a)].

Theorem 15.1. *Consider problem (P) from (15.2), and let $\bar{x}$ be a local minimum to (P). Assume that the following qualification condition (Q) holds at $\bar{x}$:*

$$y \in N_U(F(\bar{x})) \quad \text{with} \quad 0 \in \nabla F(\bar{x})^T y + N_X(\bar{x}) \text{ implies that } y = 0.$$

Then there exists $\bar{y} \in N_U(F(\bar{x}))$ such that

$$0 \in \nabla f_0(\bar{x}) + \nabla F(\bar{x})^T y + N_X(\bar{x}).$$

Using this result, we can compute the normal cone to the feasible set C, which is explicitly given in the above theorem by

$$C = \{x \in X : F(x) \in U\}. \tag{15.3}$$

However, the explicit computation of the normal cone can be done under certain qualification conditions, and we present the full result in the next theorem.

Theorem 15.2. *Consider the set C given by (15.3), where $F : \mathbb{R}^n \to \mathbb{R}^m$ is a smooth function and X is a closed set. Assume that the qualification condition (Q) of Theorem 15.1 holds at $\bar{x}$. Then one has*

$$N_C(\bar{x}) \subset \bigcup \left\{\nabla F(\bar{x})^T y + N_X(\bar{x}) : y \in N_U(F(\bar{x}))\right\}.$$

Furthermore, if the set X is normally regular at $\bar{x}$ and the set U is normally regular at $F(\bar{x})$, then equality holds in the above expression.

The two theorems presented in this section play a fundamental role in the next section. We show there how to use these theorems to derive necessary optimality conditions for bilevel programs with partially convex lower-level problems.

15.3 Partially Convex Lower-level Problems

In this section we consider partially convex lower-level problems in the bilevel programs (BPO) of our study. By a partially convex lower-level problem we mean that $y \mapsto f(x, y)$ is convex in y for each $x \in X$ and the set $K(x)$ is convex for each x. For simplicity of the presentation we assume the upper-level objective function to be smooth, i.e., with its data to be continuously differentiable. Furthermore, we assume that the lower-level objective function is twice continuously differentiable.
Our first step is to provide an explicit computation of the basic normal cone to the graph of a set-valued map defined as a solution set of a generalized variational inequality. This will play a fundamental role in the subsequent study, since—as we have discussed in Section 15.1—deriving necessary optimality condition for optimistic bilevel programming is based on computing the normal cone to the solution set of the lower-level problem. Let us begin with considering a set-valued map $S : \mathbb{R}^n \rightrightarrows \mathbb{R}^m$ given by

$$S(x) = \{y \in \mathbb{R}^m : 0 \in G(x, y) + M(x, y)\},$$

where $G : \mathbb{R}^n \times \mathbb{R}^m \to \mathbb{R}^d$ is a smooth single-valued map and $M : \mathbb{R}^n \times \mathbb{R}^m \rightrightarrows \mathbb{R}^d$ is a set-valued map of closed graph. We first concern a more simpler version, where the set-valued map does not depend on x, i.e., $M(x, y) = M(y)$. Thus we concentrate on calculating the coderivative of the set-valued map $S : \mathbb{R}^n \rightrightarrows \mathbb{R}^m$ defined above. This is done through the following result.

Theorem 15.3. *Consider* $S : \mathbb{R}^n \rightrightarrows \mathbb{R}^m$ *given by*

$$S(x) = \{y \in \mathbb{R}^m : 0 \in G(x, y) + M(y)\},$$

where $G : \mathbb{R}^n \times \mathbb{R}^m \to \mathbb{R}^d$ *is smooth map and* $M : \mathbb{R}^m \rightrightarrows \mathbb{R}^d$ *is closed-graph. Taking* $(\bar{x}, \bar{y}) \in gph\ S$, *impose the qualification condition:*

$$-\nabla_x G(\bar{x}, \bar{y})^T z = 0 \quad \text{and} \quad w - \nabla_y G(\bar{x}, \bar{y})^T z = 0$$

with $(w, z) \in N_{gph\ M}(\bar{y}, -G(\bar{x}, \bar{y}))$ *implies that* $w = 0$ *and* $z = 0$. *Then one has*

$$N_{gph\ S}(\bar{x}, \bar{y}) \subseteq \{(x^*, y^*) \in \mathbb{R}^n \times \mathbb{R}^m : x^* = -\nabla_x G(\bar{x}, \bar{y})^T \bar{z},$$
$$y^* = \bar{w} - \nabla_y G(\bar{x}, \bar{y})^T \bar{z}, (\bar{w}, \bar{z}) \in N_{gph\ M}(\bar{y}, -G(\bar{x}, \bar{y}))\}.$$

Equality holds in the above expression if any of the following two additional conditions are satisfied:

i) The graphical set gph M is normally regular at $(\bar{y}, -G(\bar{x}, \bar{y}))$.
ii) The matrix $\nabla_x G(\bar{x}, \bar{y})$ is of full row rank.

Proof. To begin with, let us observe that the inclusion $y \in S(x)$ implies that $-G(x, y) \in M(y)$. Then set

$$H(x, y) := (y, -G(x, y))^T.$$

Thus we can equivalently rewrite $S(x)$ as

$$S(x) = \{y \in \mathbb{R}^m : H(x, y) \in \text{gph}M\},$$

which means that

$$\text{gph}S = \{(x, y) : H(x, y) \in \text{gph}M\}.$$

Observe that the qualification condition imposed in the theorem can be equivalently written as

$$\left[\nabla H(\bar{x}, \bar{y})^T (w, z) = 0, \quad (w, z) \in N_{\text{gph } M}(F(\bar{x}, \bar{y}))\right] \Longrightarrow \left[w = 0, \quad z = 0\right],$$

where $\nabla H(\bar{x}, \bar{y})$ stands for the Jacobian of H at $(\bar{x}, \bar{y})$. It is easy to see that

$$\nabla H(\bar{x}, \bar{y}) = \begin{pmatrix} 0 & I \\ -\nabla_x G(\bar{x}, \bar{y}) & -\nabla_y G(\bar{x}, \bar{y}) \end{pmatrix}.$$

Thus we have

$$\nabla H(\bar{x}, \bar{y})^T = \begin{pmatrix} 0 & -\nabla_x G(\bar{x}, \bar{y})^T \\ I & -\nabla_y G(\bar{x}, \bar{y})^T \end{pmatrix}.$$

The above observation allows us to apply Theorem 15.2 and conclude that

$$\begin{aligned} N_{\text{gph } S}(\bar{x}, \bar{y}) \subseteq \{(x^*, y^*) \in \mathbb{R}^m \times \mathbb{R}^n : (x^*, y^*) = \nabla H(\bar{x}, \bar{y})^T (\bar{w}, \bar{z}), \\ (\bar{w}, \bar{z}) \in N_{\text{gph}M}(H(\bar{x}, \bar{y}))\}. \end{aligned}$$

This immediately gives

$$\begin{aligned} N_{\text{gph } S}(\bar{x}, \bar{y}) \subseteq \{(x^*, y^*) \in \mathbb{R}^n \times \mathbb{R}^m : x^* = -\nabla_x G(\bar{x}, \bar{y})^T \bar{z}, \\ y^* = \bar{w} - \nabla_y G(\bar{x}, \bar{y})^T \bar{z}, (\bar{w}, \bar{z}) \in N_{\text{gph } M}(\bar{y}, -G(\bar{x}, \bar{y}))\}. \end{aligned}$$

If gph M is normally regular at $(\bar{y}, -G(\bar{x}, \bar{y})$, we conclude from Theorem 15.2 that the equality holds. If furthermore $\nabla_x G(\bar{x}, \bar{y})$ has full row rank, then the qualification condition is automatically satisfied, and the equality follows by application of Exercise 6.7 (page 202) from [Rockafellar and Wets (1998)]. □

Remark 15.1. The above theorem estimates the normal cone to the graph of the set-valued map S defined as the set of solutions to a generalized variational inequality. This estimate naturally allows one to provide an estimation for the coderivative of S. It is not hard to see that

$$D^*S(\bar{x},\bar{y}) \subseteq \{x^* \in \mathbb{R}^n : \exists v^* \in \mathbb{R}^d, x^* = \nabla_x G(\bar{x},\bar{y})^T v^*,$$
$$-y^* = \nabla_y G(\bar{x},\bar{y})^T v^* + D^*M(\bar{y}, -G(\bar{x},\bar{y})(v^*)\}.$$

Of course, this estimate holds under the assumptions of Theorem 15.3, with equality holding under the same conditions as in Theorem 15.3.
Let us note that the conclusions of Theorem 15.3 can be easily extended to the case when the set-valued mapping M depends on both x and y. To proceed, we need to modify the qualification conditions in order to derive the corresponding estimate, which is slightly different from the previous one due to the change in the dependence pattern of M.

Theorem 15.4. *Consider the set-valued map $S : \mathbb{R}^n \rightrightarrows \mathbb{R}^m$ defined by*

$$S(x) = \{y \in \mathbb{R}^m : 0 \in G(x,y) + M(x,y)\},$$

where $G : \mathbb{R}^n \times \mathbb{R}^m \to \mathbb{R}^d$ is smooth and $M : \mathbb{R}^n \to \mathbb{R}^m \rightrightarrows \mathbb{R}^d$ is closed-graph. Given $(\bar{x},\bar{y}) \in gph\ S$, assume the qualification condition:

$$u - \nabla_x G(\bar{x},\bar{y})^T z = 0, \quad \text{and} \quad w - \nabla_y G(\bar{x},\bar{y})^T z = 0$$

with $(u,w,z) \in N_{gph\ M}(\bar{x},\bar{y},-G(\bar{x},\bar{y}))$ implies that $u = 0$, $w = 0$ and $z = 0$. Then one has the inclusion

$$N_{gph\ S}(\bar{x},\bar{y}) \subseteq \{(x^*,y^*) \in \mathbb{R}^n \times \mathbb{R}^m : x^* = \bar{u} - \nabla_x G(\bar{x},\bar{y})^T \bar{z},$$
$$y^* = \bar{w} - \nabla_y G(\bar{x},\bar{y})^T \bar{z}, (\bar{u},\bar{w},\bar{z}) \in N_{gph\ M}(\bar{x},\bar{y},-G(\bar{x},\bar{y}))\}.$$

Equality holds in the above expression if gphM is normally regular at $(\bar{x},\bar{y},-G(\bar{x},\bar{y}))$.

Proof. Let us begin by defining the function $H : \mathbb{R}^n \times \mathbb{R}^m \to \mathbb{R}^{n+m+d}$ given as

$$H(x,y) = (x,y,-G(x,y))^T.$$

Hence we can write

$$S(x) = \{y \in \mathbb{R}^m : H(x,y) \in \text{gph}M\}.$$

Therefore we have

$$\text{gph}S = \{(x,y) \in \mathbb{R}^n \times \mathbb{R}^m : H(x,y) \in \text{gph}M\}.$$

Observe that the qualification condition mentioned in hypothesis of the theorem can be equivalently stated as

$$\left[\nabla H(\bar{x},\bar{y})^T(u,w,z)=0, \quad (u,w,z)\in N_{\text{gph }M}(H(\bar{x},\bar{y}))\right]$$

$$\Longrightarrow \left[u=0, \quad w=0, \quad z=0\right],$$

where the Jacobian $\nabla H(\bar{x},\bar{y})$ of H at $(\bar{x},\bar{y})$ is given as

$$\nabla H(\bar{x},\bar{y})=\begin{pmatrix} I & 0 \\ 0 & I \\ -\nabla_x G(\bar{x},\bar{y}) & -\nabla_y G(\bar{x},\bar{y})\end{pmatrix},$$

and hence

$$\nabla H(\bar{x},\bar{y})^T=\begin{pmatrix} I \; 0 \; -\nabla_x G(\bar{x},\bar{y})^T \\ 0 \; I \; -\nabla_y G(\bar{x},\bar{y})^T\end{pmatrix}.$$

Now we can apply Theorem 15.2 to conclude that

$$N_{\text{gph}S}(\bar{x},\bar{y})\subseteq\{(x^*,y^*)\in\mathbb{R}^n\times\mathbb{R}^m:(x^*,y^*)=\nabla H(\bar{x},\bar{y})^T(\bar{u},\bar{w},\bar{z})$$
$$(\bar{u},\bar{w},\bar{z})\in N_{\text{gph}M}(F(\bar{x},\bar{y}))\}$$

Thus we can conclude that

$$N_{\text{gph}S}(\bar{x},\bar{y})\subseteq\{(x^*,y^*)\in\mathbb{R}^n\times\mathbb{R}^m:x^*=\bar{u}-\nabla_x G(\bar{x},\bar{y})^T\bar{z},$$
$$y^*=\bar{w}-\nabla_y G(\bar{x},\bar{y})^T\bar{z},(\bar{u},\bar{w},\bar{z})\in N_{\text{gph}M}(\bar{y},-G(\bar{x},\bar{y}))\}.$$

Further if gph$M$ is regular at $(\bar{x},\bar{x},-G(\bar{x},\bar{y}))$ then equality follows from Theorem 15.2. This proves the result. □

Theorem 15.3 allows us to derive necessary optimality conditions for bilevel programs (BPO) with partially convex lower-level problems. It is important to observe that, since the lower-level problem is partially convex, we can equivalently represent S as

$$S(x)=\{y\in\mathbb{R}^m:0\in\nabla_y f(x,y)+N_K(y)\}.$$

It is convenient in what follows to define $N_K(y)$ for all $y\in\mathbb{R}^m$ extending it to $y\notin K$ by $N_K(y)=\emptyset$.

Theorem 15.5. *Consider problem (BPO) with $X=\mathbb{R}^n$ and $K(x)=K$ for all x. Let $(\bar{x},\bar{y})\in gphS$ be a local optimal solution to (BPO), and let the following qualification condition hold:*

$$-(\nabla^2_{xy}f(\bar{x},\bar{y}))^T z=0, \quad w-(\nabla^2_{yy}f(\bar{x},\bar{y}))^T z=0$$

with $(w,z)\in N_{gphN_K}(\bar{y},-\nabla_y f(\bar{x},\bar{y}))$ implies that $w=0$ and $z=0$. Then there exists $(\bar{w},\bar{z})\in N_{gphN_K}(\bar{y},-\nabla_y f(\bar{x},\bar{y}))$ such that

i) $\nabla_x F(\bar{x},\bar{y}) = (\nabla^2_{xy} f(\bar{x},\bar{y}))^T \bar{z}$,
ii) $-\nabla_y F(\bar{x},\bar{y}) = \bar{w} - (\nabla^2_{yy} f(\bar{x},\bar{y}))^T \bar{z}$.

Proof. Since $(\bar{x},\bar{y})$ is local optimal to (BP) with $X\mathbb{R}^n$, we have

$$0 \in \nabla F(\bar{x},\bar{y}) + N_{\text{gphs}}(\bar{x},\bar{y}),$$

which implies that

$$-(\nabla_x F(\bar{x},\bar{y}), \nabla_y F(\bar{x},\bar{y})) \in N_{\text{gphs}}(\bar{x},\bar{y}). \tag{15.4}$$

Now setting $G(x,y) = \nabla_y f(x,y)$ and $N_K = M$, we see that the qualification condition in the theorem is the same as in Theorem 15.3. Thus applying Theorem 15.3, we get the inclusion

$$N_{\text{gph } S}(\bar{x},\bar{y}) \subseteq \{(x^*,y^*) \in \mathbb{R}^n \times \mathbb{R}^m : x^* = -\nabla^2_{xy} f(\bar{x},\bar{y})^T \bar{z},$$
$$y^* = \bar{w} - \nabla^2_{yy} f(\bar{x},\bar{y})^T \bar{z}, (\bar{w},\bar{z}) \in N_{\text{gph } N_K}(\bar{y}, -\nabla_y f(\bar{x},\bar{y}))\}.$$

Now combining the above estimate with (15.4), we arrive at the desired result. □

Remark 15.2. Note that the above theorem is also derived in [Dutta and Dempe (2006)] by using Theorem 3.1 from [Outrata (2000)]. Here we give a direct proof of this result, focusing more on structural and computational issues. Observe further that the qualification condition in the above theorem holds true if we assume that $\nabla^2_{xy} f(\bar{x},\bar{y})$ is of full row rank. To illustrate this, consider the function $f(x,y) := \langle y, Ax\rangle$, where $(x,y) \in \mathbb{R}^n \times \mathbb{R}^m$ and A is a $m \times n$ matrix of full row rank m. We have $\nabla^2_{xy} f(x,y) = A$, and hence the qualification condition of the above theorem is clearly satisfied.

Next we turn to the case where the feasible set of the lower-level problem need not to remain constant for each x, assuming nevertheless that $X = \mathbb{R}^n$. In this case, the solution set to (BPO) is given by

$$S(x) = \{y \in \mathbb{R}^m : 0 \in \nabla_y f(\bar{x},\bar{y}) + N_{K(x)}(y)\}.$$

Setting $N_K(x,y) := N_{K(x)}(y)$ if $y \in K(x)$ and $N_K(x,y) := \emptyset$ otherwise, we rewrite $S(x)$ as

$$S(x) = \{y \in \mathbb{R}^m : 0 \in \nabla_y f(\bar{x},\bar{y}) + N_K(x,y)\}.$$

Theorem 15.6. *Consider problem (BPO), where $X = \mathbb{R}^n$ and the feasible set to the lower-level problem varies with each x. Let $(\bar{x},\bar{y}) \in gphS$ be a local optimal solution to (BPO), and let the following qualification condition hold:*

$$u - \nabla^2_{xy} f(\bar{x},\bar{y})^T z = 0, \quad w - \nabla^2_{yy} f(\bar{x},\bar{y})^T z = 0$$

with $(u,w,z) \in N_{gphN_K}(\bar{x},\bar{y},-\nabla_y f(\bar{x},\bar{y}))$ implies that $u = 0, w = 0, z0$. Then there exists $(\bar{u},\bar{w},\bar{z}) \in N_{gphN_K}(\bar{x},\bar{y},-\nabla_y f(\bar{x},\bar{y}))$ such that

i) $-\nabla_x F(\bar{x},\bar{y}) = \bar{u} - \nabla^2_{xy} f(\bar{x},\bar{y})^T \bar{z}$,
ii) $-\nabla_y F(\bar{x},\bar{y}) = \bar{w} - \nabla^2_{yy} f(\bar{x},\bar{y})^T \bar{z}$.

Proof. Since $(\bar{x},\bar{y})$ is a local optimal solution to (BPO), we have

$$-(\nabla_x F(\bar{x},\bar{y}), \nabla_y F(\bar{x},\bar{y})) \in N_{\text{gph}S}(\bar{x},\bar{y}).$$

Now the result follows from Theorem 15.4 by setting $G(x,y) := \nabla_y f(x,y)$ and $N_K := M$. □

Remark 15.3. Note that the problem of Theorem 15.6 was studied in [Dutta and Dempe (2006)][Theorem 4.1], while our approach here is different.

The next most relevant question is about necessary optimality conditions for the constrained case $x \in X$ in (BPO). Here is the result in this case.

Theorem 15.7. *Let $(\bar{x},\bar{y}) \in gphS$ be a local optimal solution to (BPO), and let the following qualification condition be satisfied:*

$$u - \nabla^2_{xy} f(\bar{x},\bar{y})^T z + \gamma = 0, \quad w - \nabla^2_{yy} f(\bar{x},\bar{y})^T z = 0$$

with $(u,w,z) \in N_{gphN_K}(\bar{x},\bar{y},-\nabla_y f(\bar{x},\bar{y}))$ and $\gamma \in N_X(\bar{x})$ implies the equalities $u = 0, w = 0, z = 0$.
Then there are $(\bar{u},\bar{w},\bar{z}) \in N_{gphN_K}(\bar{x},\bar{y},-\nabla_y f(\bar{x},\bar{y})$ and $\bar{\gamma} \in N_X(\bar{x})$ such that

i) $-\nabla_x F(\bar{x},\bar{y}) = \bar{u} - \nabla^2_{xy} f(\bar{x},\bar{y})^T \bar{z} + \bar{\gamma}$,
ii) $-\nabla_y F(\bar{x},\bar{y}) = \bar{w} - \nabla^2_{yy} f(\bar{x},\bar{y})^T \bar{z}$.

Proof. Observe that problem (BPO) can be equivalently rewritten as

$$\min_{x,y} F(x,y), \quad \text{subject to } (x,y) \in C,$$

where the set C is given by

$$C = \{(x,y) \in X \times \mathbb{R}^m : H(x,y) \in \text{gph}N_K\}$$

$$\text{with} \quad H(x,y) = (x,y,-\nabla_y f(x,y))^T.$$

It is well known that $N_{X\times\mathbb{R}^m}(\bar{x},\bar{y}) = N_X(\bar{x}) \times N_{\mathbb{R}^m}(\bar{y})$, and thus

$$N_{X\times\mathbb{R}^m}(\bar{x},\bar{y}) = \{(\gamma,0) : \gamma \in N_X(\bar{x})\}.$$

Therefore, the qualification condition of the theorem is equivalent to

$$\Big[0 \in \nabla H(\bar{x},\bar{y})^T q + N_{X\times\mathbb{R}^m}(\bar{x},\bar{y}),$$

$$q = (u,w,z) \in N_{\text{gph}N_K}(\bar{x},\bar{y},-\nabla_y f(\bar{x},\bar{y}))\Big]$$

$$\Longrightarrow \big[u = 0, w = 0, z = 0\big].$$

Observe further that

$$\nabla H(\bar{x},\bar{y})^T = \begin{pmatrix} I \; 0 \; -\nabla^2_{xy} f(\bar{x},\bar{y})^T \\ 0 \; I \; -\nabla^2_{yy} f(\bar{x},\bar{y})^T \end{pmatrix}.$$

Thus the qualification condition of this theorem reduces to the qualification condition of Theorem 15.1, and the result follows. □

Remark 15.4. We would like to note that in [Dutta and Dempe (2006)] the optimistic bilevel programming problem with partially convex lower-level problems was not considered in its full generality as it is done in the above Theorem 15.7.

It is time to present an illustrative example for the reader's convenience.

Example 15.1. Consider the optimistic bilevel programming problem in a two-dimensional setting:

$$\min_{x,y} (x-1)^2 + y^2 \quad \text{subject to} \quad x > 0, y \in S(x),$$

where S denotes the solution set mapping to the following lower-level problem:

$$\min_{y} x^2 y \quad \text{subject to} \quad y \geq 0.$$

Observe that $S(x) = \{0\}$ for all $x > 0$, and that the only solution to the above optimistic bilevel programming problem is $(1,0)$. It is clear that $\nabla^2_{xy} f(1,0) = 2$. Let us check that the qualification condition of Theorem 15.7 is satisfied. To proceed, observe that the lower-level feasible set is $[0,+\infty)$, which is thus a convex set independent of x. Note that the vector u actually does not appear in the qualification condition of Theorem 15.7. Hence we may just set $u = 0$ throughout in this particular case. Since $X = \mathbb{R}_+$, we get $N_X(1) = \{0\}$, which easily yields $z = 0$ and $w = 0$. It is easy to check furthermore that the necessary condition of the theorem holds with $\bar{w} = 0$ and $\bar{z} = 0$. Observe finally that $\bar{y} = 0$, since $N_X(1) = \{0\}$.

One of our primary goals of this section is to highlight the fact that necessary optimality conditions for optimistic bilevel programs with partially convex lower-level problems can be basically deduced from Theorem 15.1 and Theorem 15.2, which are indeed fundamental results in optimization theory. An interesting fact that emerges here is that the second-order partial derivatives of the lower-level objective function naturally appear in the first-order optimality conditions for this class of bilevel programs. Another observation that emerges here is that the qualification conditions in the above results do not work if the lower-level problem is a linear optimization problem. This issue is addressed in the next section, where we discuss the property of partial calmness that automatically holds when the lower-level problem is linear. Further, in the next section we approach necessary optimality conditions in bilevel programs by using the idea of optimal value functions.

15.4 Fully Convex Lower-level Problems

Now we investigate problem (BPO) under the assumption that both the lower-level and the upper-level objective functions are convex with respect to both x and y, and that gph K is also a convex set. Denote the optimal value function of the lower-level problem by

$$\varphi(x) := \min_y \{f(x,y) : y \in K(x)\}.$$

Then problem (BPO) is equivalent to the following problem (VPO):

$$\min_{x,y} F(x,y) \text{ subject to } f(x,y) \leq \varphi(x),\ \ y \in K(x),\ \ x \in X.$$

Usual constraint qualifications as, e.g., the Mangasarian-Fromowitz one (in its nondifferentiable version) are not satisfied at each feasible point of (VPO); see [Ye and Zhu (1995)].
Following [Ye and Zhu (1995)], we say that problem (VPO) is *partially calm* at a given point $(\overline{x}, \overline{y})$ if there is a constant $M > 0$ and an open neighborhood D of the triple $(\overline{x}, \overline{y}, 0)$ such that for each feasible point $(x, y, u) \in D$ of the problem

$$\min_{x,y} F(x,y) \ \text{ subject to } f(x,y) - \varphi(x) + u = 0,\ \ y \in K(x),\ \ x \in X$$

we have the relation

$$F(x,y) - F(\overline{x}, \overline{y}) + M|u| \geq 0.$$

By [Ye and Zhu (1995)], partial calmness is satisfied for problem (VPO) if, in particular, all optimal solutions to the lower-level problem are weak sharp minima in the sense of [Burke and Ferris (1993)]: for fixed $\overline{x}$ there exists $\alpha > 0$ such that

$$f(\overline{x}, y) \geq f(\overline{x}, \overline{y}) + \alpha \text{ dist}(y, S(\overline{x})),$$

whenever $y \in K(\overline{x})$, where dist$(y, S(\overline{x}))$ denotes the Euclidean distance of a point y to the set $S(\overline{x})$ and where $\overline{y} \in S(\overline{x})$. It has been shown in [Burke

and Ferris (1993)] that optimal solutions to linear programming problems are weak sharp minima (cf. also [Mangasarian and Meyer (1979)]) whenever the problem has an optimal solution. Also, optimal solutions to quadratic programming problems are weak sharp minima provided that a certain relatively weak assumption is satisfied; see [Burke and Ferris (1993)], [Ye and Zhu (1995)]. Note that the assumption of partial calmness can be replaced by other assumptions; see [Ye (2006)] for more discussions.
The main feature of partial calmness is the validity of an exact penalty function approach to problem (VPO):

Theorem 15.8. *([Ye and Zhu (1995)][Proposition 3.2]) Let $(\overline{x}, \overline{y})$ be a local optimal solution to (VPO). Then, problem (VPO) is partially calm at $(\overline{x}, \overline{y})$ if and only if there exists $\lambda > 0$ such that $(\overline{x}, \overline{y})$ is a local optimal solution to the problem*

$$\min_{x,y} F(x,y) + \lambda(f(x,y) - \varphi(x)) \text{ subject to } y \in K(x),\ x \in X. \qquad (15.5)$$

This is a significant tool in the proof of the next theorem, where the symbols ∂, ∂_x, ∂_y denote, respectively, the subdifferential, the partial subdifferential with respect to x and to y of convex functions in the sense of convex analysis.

Theorem 15.9. *Consider problem (VPO) under the assumptions that:*

i) $K(x) = \{y : g(x,y) \leq 0\}$, $X = \mathbb{R}^n$, $g : \mathbb{R}^n \times \mathbb{R}^m \to \mathbb{R}^p$;

ii) all functions F, f, g_i are convex on $\mathbb{R}^n \times \mathbb{R}^m$, $i = 1, \ldots, p$;

iii) the point $(\overline{x}, \overline{y})$ is a local optimal solution, problem (VPO) is partially calm at $(\overline{x}, \overline{y})$, there exists a compact set C such that $\{(x,y) : g(x,y) \leq 0\} \subseteq C$, and there is a point $(\widehat{x}, \widehat{y})$ with $g_i(\widehat{x}, \widehat{y}) < 0$, $i = 1, \ldots, p$.

Then there exist $\lambda > 0$, λ_i, μ_i, and a point $\widetilde{y} \in S(\overline{x})$ such that the following conditions are satisfied:

$$0 \in \partial_x F(\overline{x}, \overline{y}) + \lambda(\partial_x f(\overline{x}, \overline{y}) - \partial_x f(\overline{x}, \widetilde{y})) + \sum_{i=1}^{p}(\mu_i \partial_x g_i(\overline{x}, \overline{y}) - \lambda\lambda_i \partial_x g_i(\overline{x}, \widetilde{y})),$$

$$0 \in \partial_y F(\overline{x}, \overline{y}) + \lambda \partial_y f(\overline{x}, \overline{y}) + \sum_{i=1}^{p} \mu_i \partial_y g_i(\overline{x}, \overline{y}),$$

$$0 \in \partial_y f(\overline{x}, \widetilde{y}) + \sum_{i=1}^{p} \lambda_i \partial_y g_i(\overline{x}, \widetilde{y}),$$

$$\lambda_i \geq 0,\ \lambda_i g_i(\overline{x}, \widetilde{y}) = 0,\ i = 1, \ldots, p,$$

$$\mu_i \geq 0,\ \mu_i g_i(\overline{x}, \overline{y}) = 0,\ i = 1, \ldots, p.$$

Proof. By our assumptions on the lower-level problem, the optimal value function $\varphi(\cdot)$ is convex, and hence it is locally Lipschitzian; see [Rockafellar (1970)]. Thus (VPO) is a problem of Lipschitzian programming. By partial calmness, the local optimal solution $(\overline{x}, \overline{y})$ is also a local optimal solution to the Lipschitz optimization problem (15.5) for some $\lambda > 0$. Applying to this problem the generalized multiplier rule from [Mordukhovich (2006b)][Theorem 3.21 (iii)] together with the calculus rules for the basic subdifferential from [Mordukhovich (2006a)][Theorem 2.23 (c)], we obtain the existence of multipliers $(\lambda_0, \mu_1, \ldots, \mu_p)$ such that $\lambda_0 \geq 0$ and

$$\mu_i \geq 0, \ \mu_i g_i(\overline{x}, \overline{y}) = 0, \ i = 1, \ldots, p, \tag{15.6}$$

$$0 \in \lambda_0 \partial F(\overline{x}, \overline{y}) + \lambda_0 \lambda (\partial f(\overline{x}, \overline{y}) - \partial_x \varphi(\overline{x}) \times \{0\}) + \sum_{i=1}^{p} \mu_i \partial g_i(\overline{x}, \overline{y}).$$

Observe that we have in fact $\lambda_0 > 0$, i.e., we can set $\lambda_0 = 1$ by the Slater-type qualification conditions assumed in the theorem. Using the important relationship between partial and full subdifferentials in convex analysis

$$\partial \theta(x, y) \subseteq \partial_x \theta(x, y) \times \partial_y \theta(x, y),$$

we obtain the inclusions

$$0 \in \partial F_x(\overline{x}, \overline{y}) + \lambda (\partial_x f(\overline{x}, \overline{y}) - \partial_x \varphi(\overline{x})) + \sum_{i=1}^{p} \mu_i \partial_x g_i(\overline{x}, \overline{y}), \tag{15.7}$$

$$0 \in \partial_y F(\overline{x}, \overline{y}) + \lambda \partial_y f(\overline{x}, \overline{y}) + \sum_{i=1}^{p} \mu_i \partial_y g_i(\overline{x}, \overline{y}). \tag{15.8}$$

By the symmetry property

$$\partial(-\varphi)(\overline{x}) \subseteq -\partial \varphi(\overline{x})$$

and the estimate

$$\partial \varphi(\overline{x}) \subseteq \bigcup_{y \in S(\overline{x})} \ \bigcup_{\lambda \in \Lambda(\overline{x}, y)} \{\partial_x f(\overline{x}, y) + \sum_{i=1}^{p} \lambda_i \partial_x g_i(\overline{x}, y)\}$$

given, e.g., in [Mordukhovich, Nam and Yen (to appear)] with

$$\begin{aligned} \Lambda(\overline{x}, y) = \{\lambda_i \geq 0 : \lambda_i g_i(\overline{x}, y) = 0, \ i = 1, \ldots, p, \\ 0 \in \partial_y f(\overline{x}, y) + \sum_{i=1}^{p} \partial_y g_i(\overline{x}, y)\}, \end{aligned} \tag{15.9}$$

we transform (15.7) to

$$\begin{aligned} 0 \in \partial F_x(\overline{x}, \overline{y}) + \lambda (\partial_x f(\overline{x}, \overline{y}) - (\partial_x f(\overline{x}, \widetilde{y}) \\ + \sum_{i=1}^{p} \lambda_i \partial_x g_i(\overline{x}, \widetilde{y}))) + \sum_{i=1}^{p} \mu_i \partial_x g_i(\overline{x}, \overline{y}) \end{aligned} \tag{15.10}$$

for some $\widetilde{y} \in S(\overline{x})$ and $(\lambda_1, \ldots, \lambda_p) \in \Lambda(\overline{x}, \widetilde{y})$. Conditions (15.10), (15.8), (15.9), (15.6) together with $(\lambda_1, \ldots, \lambda_p) \in \Lambda(\overline{x}, \widetilde{y})$ are the desired necessary conditions, which thus completes the proof the theorem. □

Corollary 15.1. *If the compactness assumption of the theorem is replaced by the inner semicontinuity assumption that for each point $(\widehat{x}, \widehat{y}) \in gph\, S$ and any sequence $\{x^k\}$ with $S(x^k) \neq \emptyset$ converging to $\widehat{x}$ there is a sequence $\{y^k\}$ with $y^k \in S(x^k)$ converging to $\widehat{y}$, then we obtain by [Mordukhovich, Nam and Yen (to appear)][Corollary 4] the inclusion*

$$\partial\varphi(\overline{x}) \subseteq \bigcup_{\lambda \in \Lambda(\overline{x},y)} \{\partial_x f(\overline{x}, y) + \sum_{i=1}^{p} \lambda_i \partial_x g_i(\overline{x}, y)\}. \tag{15.11}$$

Replacing the formula for the subdifferential of φ at $\overline{x}$ in the above proof, we can take $\widetilde{y} = \overline{y}$ in the assertion of the theorem. If, moreover, the functions f, g_i, $i = 1, \ldots, p$, are continuously differentiable, the following necessary optimality conditions result from Theorem 15.9:

There exists $\lambda > 0, \lambda_i$, μ_i, $i = 1, \ldots, p$, satisfying

$$0 \in \partial_x F(\overline{x}, \overline{y}) + \sum_{i=1}^{p} (\mu_i - \lambda\lambda_i)\nabla_x g_i(\overline{x}, \overline{y}),$$

$$0 \in \partial_y F(\overline{x}, \overline{y}) + \lambda\nabla_y f(\overline{x}, \overline{y}) + \sum_{i=1}^{p} \mu_i \nabla_y g_i(\overline{x}, \overline{y}),$$

$$0 \in \nabla_y f(\overline{x}, \overline{y}) + \sum_{i=1}^{p} \lambda_i \nabla_y g_i(\overline{x}, \overline{y}),$$

$$\lambda_i \geq 0,\ \lambda_i g_i(\overline{x}, \overline{y}) = 0,\ i = 1, \ldots, p,$$

$$\mu_i \geq 0,\ \mu_i g_i(\overline{x}, \overline{y}) = 0,\ i = 1, \ldots, p.$$

For a related result, obtained using different assumptions and a different method, we refer to [Ye (2006)][Theorem 4.1].

Optimal solutions to linear programming problems are weak sharp as shown by [Burke and Ferris (1993)]. Moreover, the solution set map to linear programming problems of the type

$$\min c^\top y, \text{ subject to } Ay \leq x$$

with right-hand side perturbations x is lower semicontinuous by [Bank, Guddat, Klatte, Kummer and Tammer (1982)][Theorem 4.3.5] and hence also inner semicontinuous. This allows us to deduce the following simple necessary optimality conditions.

Corollary 15.2. *Consider the bilevel linear programming problem (VOP) with*

$$\varphi(x) = \min_y \{c^\top y : Ay \leq x\}$$

and $X = \mathbb{R}^n$. *Assume for simplicity that* F *is continuously differentiable. If* $(\overline{x}, \overline{y})$ *is a local optimal solution of this problem, then there exist multipliers* $\lambda > 0, \mu \geq 0, \beta \geq 0$ *such that*

$$\begin{aligned} \nabla_x F(\overline{x}, \overline{y}) - (\mu - \lambda\beta) &= 0, \\ \nabla_y F(\overline{x}, \overline{y}) + \lambda c + \mu^\top A &= 0, \\ c + \beta^\top A &= 0, \\ \beta \geq 0, \ \beta^\top (A\overline{y} - \overline{x}) &= 0, \\ \mu \geq 0, \ \mu^\top (A\overline{y} - \overline{x}) &= 0. \end{aligned}$$

Remark 15.5. We would like to conclude the paper by making some brief comments on the usefulness of the optimality conditions studied in this paper from the computational viewpoint. One of the special features of the optimality conditions studied in Section 15.3 is the presence of second-order partial derivatives in the representation of the first-order optimality conditions. This makes the conditions obtained computationally expensive. Further, in Section 15.3 one would observe that the Lagrange multipliers associated with the optimality conditions themselves belong to a set, which is the normal cone to the graph of the normal cone map associated with the feasible set of the lower-level problem. This set is rather difficult to compute; see, e.g., the detailed discussion on this issue in [Dutta and Dempe (2006)]. However, the optimality conditions in Section 15.3 clearly outline the geometric structure associated with bilevel programming involving partially convex lower-level problems.

On the other hand, the optimality conditions studied in Section 15.4 seem to be more easily tractable from the computational viewpoint. Observe that in Section 15.4 the lower-level problem is fully convex and thus more amenable to computation. Further, the optimality conditions derived in this section do not have any second-order partial derivatives and thus much for simple to tackle. Moreover, in the case of smooth lower-level problems these optimality conditions do not even depend on the partial derivatives of the lower-level objective function with respect to the upper-level variable x as shown in Corollary 15.1. However, an important requirement that needs to be satisfied for the optimality conditions in Section 15.4 to work is that of *partial calmness.* To this end, it has been shown in [Dempe, Dutta and Mordukhovich (to appear)] that there can be a large class of optimistic bilevel programs, specially those with quadratic convex lower-level problems, where the partial calmness requirement holds. While on the other hand, there could be a large class of problems, where the partial calmness may fail. The power of the results of Section 15.4 is fully manifested in the case where the lower-level problem is linear. Since partial calmness is automatically satisfied in the case of linear lower-level problems, the optimality conditions laid out in Corollary 15.2 can be used to design algorithms to solve bilevel programs with linear lower-level problems. This can be considered as an area of future research. It is important to keep in mind that even if the lower-level problem is linear or fully convex, the overall problem is a highly nonconvex and nonsmooth. Thus in this paper we attempt to develop optimality conditions, which bring out the nonsmooth geometric structures associated with bilevel problems as well those, which are more suited for computation.

Acknowledgements

The authors are grateful to the anonymous referees whose suggestions have improved the presentation of this paper.

Bibliography

Bank B., Guddat J., Klatte D., Kummer B. and Tammer K. (1982). *Non-Linear Parametric Optimization*, (Akademie-Verlag, Berlin).

Burke J. V. and Ferris M. C. (1993). Weak sharp minima in mathematical programming. *SIAM Journal on Control and Optimization* **31**, pp. 1340–1359.

Dempe S. (2002). *Foundations of Bilevel Programming*, (Kluwer Academie Publishers, Dordrecht).

Dempe S., Dutta J. and Mordukhovich B. S. (to appear). New necessary optimality conditions in optimistic bilevel programming, *Optimization*.

Dutta J. and Dempe S. (2006). Bilevel programming with convex lower level problems. In S. Dempe and V. Kalashnikov, editors, *Optimization with Multivalued Mappings: Theory, Applications and Algorithms*,(Springer Science+Business Media, LLC).

Levy A. B. and Mordukhovich B. S. (2004). Coderivatives in parametric optimization, *Math. Program. Ser. A*, **99**, pp.311-327.

Mangasarian O. L.and Meyer R. R. (1979). Nonlinear perturbations of linear programs, *SIAM Journal on Control and Optimization*, **17**, pp. 745-752.

Mordukhovich B. S. (1976). Maximum principle in problems of time optimal control with nonsmooth constraints, *J. Appl. Math. Mech.*, **40**, pp. 960–969.

Mordukhovich B. S. (2006). *Variational Analysis and Generalized Differentiation, Vol. 1: Basic Theory*, (Springer Verlag, Berlin).

Mordukhovich B. S. (2006). *Variational Analysis and Generalized Differentiation, Vol. 2: Applications*, (Springer Verlag, Berlin).

Mordukhovich B. S., Nam N. M. and Yen N. D. (to appear). Subgradients of marginal functions in parametric mathematical programming, *Mathematical Programming*.

Outrata J. V. (2000). A generalized mathematical program with equilibrium constraints, *SIAM Journal on Control and Optimization*, **38**, pp. 1623–1638.

Rockafellar R. T. and Wets R. J.-B. (1998). *Variational Analysis*,(Springer Verlag, Berlin).

Rockafellar R. T. (1970). *Convex Analysis*, (Princeton University Press, Princeton).

Scheel H. and Scholtes S. (2000). Mathematical programs with equilibrium constraints: stationarity, optimality, and sensitivity, *Mathematics of Operations Research*, **25**, pp. 1–22.

von Stackelberg H. (1934). *Marktform und Gleichgewicht, (Engl. transl.: The Theory of the Market Economy*, 1952, Oxford University Press), Springer-Verlag, Berlin.

Ye J. J. (2006). Constraint qualifications and KKT conditions for bilevel programming problems, *Mathematics of Operations Research*, **31**, pp. 811–824.

Ye J. J. and Zhu D. L. (1995). Optimality conditions for bilevel programming problems, *Optimization*, (with correction in *Optimization* **39**, pp. 361–366, 1997), **33**, pp. 9–27.

Chapter 16

Game Engineering

Robert J. Aumann[1]
Nobel Laureate
Einstein Institute of Mathematics
The Hebrew University of Jerusalem
Israel

Abstract
This talk was delivered at the International Symposium on Mathematical Programming for Decision Making: Theory and Applications organized during January 10-11, 2007 at Indian Statistical Institute, Delhi Centre as part of the platinum jubilee celebrations of the Indian Statistical Institute.

Key Words: Two-person zero-sum game, auctions, arbitration, traffic

Ladies and Gentlemen, I am very glad to be here at this Jubilee celebration of this very very distinguished Institute. This is something which has been from my early childhood in my consciousness and I am very glad to be here to help celebrate this very significant event.

In its beginning, one might say in the formative years when Game Theory was starting out with von Neumann and Morgenstern, there was a very close relationship between Game Theory and Mathematical Programming and indeed Linear Programming. As you know the Duality Theorem of Linear Programming is equivalent, one might say, closely related, to the Minimax Theorem for two-person zero-sum games. In the ensuing decades, two-person zero-sum games lost some of their centrality in applications. It became apparent that most games that one encountered were

[1]The editors thank Professor Arunava Sen, Indian Statistical Institute, Delhi Centre and Professor T. E. S. Raghavan, University of Illinois at Chicago, USA for their help to prepare this text version from the video recorded talk by Professor Robert J. Aumann.

not two-person, zero-sum. However although they are not pre-eminent by themselves in applications, two-person zero-sum games are very important in the mathematical foundation of Game Theory to this day. They sort of form a corner stone of Game Theory, literally in the sense of a corner stone. In other words, two-person zero-sum games are not in the center of Game Theory but are in the corner. But just as a corner stone is very important for a building, so the theory of two-person zero-sum games is very important for Game Theory as a whole.

This morning I would like to speak about a subject which I call Game Engineering. One might say that there are two kinds of applications of Game Theory. One is the kind of application where you get insights into an interactive situation. What I mean by insight is that you understand sort of what is going on. An example is my Nobel Lecture that I gave in Stockholm over a year ago. It is called "War and Peace". The meat of this topic is the theory of repeated games and how one can apply this theory to understand why people go to war and what can be done to bring about peace. This is a matter of insight or understanding. People ask "Why there are strikes"? Eventually after a lot of pain and suffering on the part of the employer, on the part of the employees and on the part of the public when some settlement is reached people ask why is it that the settlement could not have been reached in the first place. There are good reasons for this which I will not discuss this morning because it belongs to that part of the applications of Game Theory which are matters of insight and understanding but are not mechanical matters. This morning I would like to discuss Game Engineering. In Game Engineering, Game Theory tells you precisely what to do. I will put some of the topics on the screen:

(1) Auctions
(2) Arbitration
(3) Traffic
(4) Elections
(5) Job Machine
(6) Cut and Choose

Auctions:

You know that there was a Nobel Prize for auctions. William Vickrey got it for a brilliant idea, a very simple idea by the way. To do something important does not mean that you have to do a very complicated thing, a very deep thing. In fact very often the important ideas are simple ones.

One of the most important ideas in modern Mathematics (you must remember that from my vantage point, modern Mathematics means Mathematics which is 150 years old!) that really covers all of modern Mathematics is Cantor's diagonal method. Now Cantor's diagonal method is a very simple idea. Many mathematicians now call it trivial but it covers all of mathematics and has led to things like Gödel's Impossibility Theorem and so on and its basic idea is really very simple. William Vickrey got the Nobel prize in 1996 for a very simple idea called second-price auctions. I will explain it very briefly. In a closed bid auction everybody writes down on a piece of paper what he is willing to pay for the object being auctioned of. Then the one who wrote the highest price pays the highest price and gets the object. William Vickrey said that that was the wrong way to do it. The right way is that everybody writes down the price on a piece of paper and then the person who writes the highest price gets the object just like before but he doesn't pay what he wrote, he pays the second highest price. This seems crazy because the auctioneer appears to be throwing away money by charging the second-highest price. But it isn't crazy. The reason is that if you must pay the highest price you will be careful about what you write down and you will not write down what you are willing to pay. You will write what you think the second highest price is going to be but everybody will do that. So the price will be depressed whereas if you only have to pay the second highest price, what you write down doesn't matter. So you will write exactly what you are willing to pay. You won't write more than what you are willing to pay because you may be stuck with the object and the second highest price could be more than what you are willing to pay. Therefore the maximum that you are willing to pay that is what you are more motivated to write down in the second- price auction and the receipts are much higher. It is a very simple idea and Vickrey got the Nobel Prize for it. As many of you know he did not live long enough to receive the Prize. The Prize was announced in early October. Three days after the announcement was made, he died of a heart attack. The Nobel Prize is not given to dead people but once the announcement has been made, they give it. Paul Milgrom who is one of the big people in auction theory received it in his name. By the way, I think that the Nobel Prize is a very nice thing and I do recommend it, but I don't think one should take it too seriously. I think Vickery took it very seriously.

Let me just point out two things in this example. First of all, what it builds on is incentives. Game Theory is all about incentives - incentives in interactive situations. In Game Engineering you take these incentives and you build them into very precisely defined systems like that of second-price auctions. You say "do this, this, this etc". It is not fuzzy, it is very precise. I will give another example.

Arbitration:

Here is another thing that sounds crazy but it really works. Suppose there is a labor dispute between employees and an employer. They take an arbitrator and each side presents their case to him. They come up eventually with their final demands. The arbitrator listens to these arguments and usually will do some kind of compromise between their demands. He'll listen, he'll be impressed by the arguments, he'll say these people have a point, those people have a point and arrive at a compromise. And that is good, correct? No, wrong! Why is that bad? Because the employer and the employee are motivated to overstate their demands. They know that the arbitrator will listen to both sides and will compromise. Let's say the employer is willing to give 80 and the union is willing to take 90. The employer says that I will not give more than 50 and the arbitrator will have to take the compromise down. The union says we are not going to take less than 120 and that will take the compromise up. Both sides are bloating their demands, giving wrong figures, throwing light on the wrong part of the problem and so on. The arbitrator gets mixed up and instead of compromising between 80 and 90, he compromises between 50 and 120 and who knows where will it end up. The arbitrator is getting much less information because the system encourages both sides to give false or misleading information. There is another method and it sounds crazy. This method is called final offer arbitration and works in the following way. Each side presents its final offer and the arbitrator must choose one of the two final offers. He is not allowed to compromise. Compromise sounds great and it brings about World Peace. Right? In this case it is wrong. Final offer arbitration works much better because both sides are motivated to do the exact opposite of what they did before. They want now to appear reasonable. They want to convince the arbitrator that they are reasonable and not crazy to make him choose their side. Both sides may reason as follows. "This is a reasonable demand. I am going out of my way towards the other side. So choose me". So the sides will be close together. The employer will say "I will give 80", the union will say "We want to take 90". The arbitrator has to choose one of these offers. It could be 80 or 90 (it can't be 85 because compromise is not allowed) but the difference between 80 and 90 is not that big and this will not matter much.

We have another example where the basic intuition at first sight is the exact opposite of the Game Engineering way to do it. The intuition in auctions is of course, you take the highest price and not the second price. And in arbitration, the intuition is to compromise. But in both cases, Game Engineering tells us that these may be the wrong way to do things.

Traffic:

My final example is going to be about traffic. Let me tell you a little story about traffic. This relates to a wonderful institution which was mentioned here before and that is the State University of New York at Stony Brook. Everybody knows about 9-11 but there was an attempt to blow up the World Trade Center several years before 9-11. I was in Stony Brook at that time - I had a part time job there. In Stony Brook, the Jewish community is fairly small and I prefer to spend the Sabbath in New York City. So I was going to New York City (which is 50 miles away from the University) on Friday afternoon. The attempt to blow up the World Trade Center happened on Friday morning. Now I don't read newspapers or watch television. So I went on Friday afternoon to pick up my valise before driving to the City. My landlord said "Bob, are you going to the City? I said "Yeah, sure" and he said "You can't do that". I said, "Why, not"? He said, "You didn't hear about the World Trade Center? The City is in chaos. You can't get in. Traffic is snarled. Bridges are closed. People are advised very strongly to stay away from Manhattan." I didn't know how to spend the Sabbath in Stony Brook - I hadn't prepared any food, didn't know where I would say my prayers and so on. And then I said, "Let's try and get in and see what happens. Maybe I'll have to turn back but lets see". Ladies and Gentlemen, I have never been in the City faster. Everybody was warned by the TV and the radio to stay in. So Bob Aumann goes in. That's Game Theory, that's Game Engineering. You know there is GPS (Global Positioning System) in certain places. The GPS tells you the shortest way to go from one place to another. What they don't take into account is traffic conditions. This can be done with modern technology. There is no problem in principle. The interesting part is the game theoretic part of it because you want to tell people not only where the traffic is now but where it is going to be as a result of where you tell them to go. If you say that there is a traffic jam at a particular place nobody will go there and by the time you get there will be no traffic jam there. So telling people where to go is a non-trivial and important problem. This isn't something fuzzy about how to bring about World Peace and why people strike. This is something very definite and it is an Engineering problem.

Let me close with another example of Game Engineering. Let me tell you a tale of two cities.

There are two cities A and B with a mountain range between them. Both the cities are in valleys. One valley is south of the mountain range and the other, north of the mountain range (Figure-1). You can go from one city to the other by either driving either along a super highway AX and then going by a winding, slow road over the mountain range XB or you can first take another slow winding road over the mountain range, AY and then take another super highway YB. The super highway drive takes 3 hours and the slow way over the mountain range takes 5 hours. There is just one equilibrium here. That equilibrium says that half the traffic will go one way i.e. AX and XB and half the traffic will go the other way, i.e. AY and YB.

Why is this the only equilibrium? As soon as more than half the traffic takes one of the routes, the extra traffic increase the time taken on that route for everyone especially on the slow mountain road. There will be less traffic on the other route and so the people will take other route. So there is just one equilibrium $(\frac{1}{2}, \frac{1}{2})$. The number of hours taken in equilibrium is $3 + 5 = 8$.

This took place in a country with an extremely dynamic forward thinking Department of Highways. Since people were wasting lot of time on these slow winding roads, they decided to build a very expensive tunnel XY which takes $\frac{1}{2}$ hour to cross(Figure 2).

They estimated that people would now take $6\frac{1}{2}$ hours to get from A to B (3 hours on the two super highways AX and YB and $\frac{1}{2}$ hour on the tunnel XY) and thus save $1\frac{1}{2}$ hours. Did that happen? No! Once the tunnel was built everyone started taking the route AX, XY and YB. Earlier each of the super highways was bearing half of the traffic and now they were bearing all of it. So now it took 4 hours instead of 3 to take them. Therefore the total time taken now was $8\frac{1}{2}$ hours whereas it took 8 hours before. So one person decided to go the old way but that now took 4 hours on super highway and 5 hours over the slow mountain road, i.e. 9 hours altogether! So he went back to the $8\frac{1}{2}$ route and the person who built the tunnel was given an early pension. Again this is Game Engineering.

Don't be too hasty to build new highways! This is a theoretical example called the Brass's Paradox. However there is nothing pathological about it. In fact something like this was observed in Amsterdam a few years ago. Because of some construction they had to close some of the roads and as a result traffic in the whole city got better. Thank you very much.

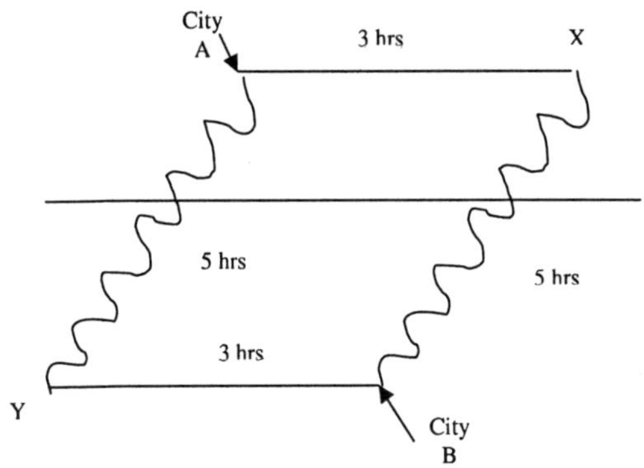

Figure – 1

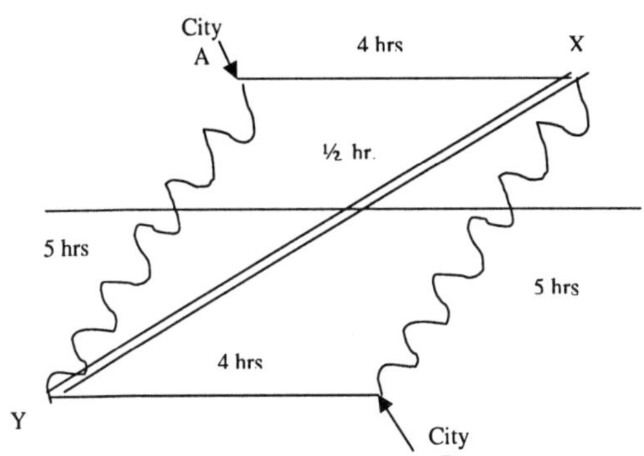

Figure – 2

Chapter 17

Games of Connectivity

Pradeep Dubey
Center for Game Theory in Economics
Stony Brook University, USA and
Cowles Foundation, Yale University, USA
Rahul Garg
IBM India Research Lab
New Delhi, India

Abstract

We consider a communications network in which users transmit beneficial information to each other at a cost. We pinpoint conditions under which the induced cooperative game is supermodular (convex). Our analysis is in a lattice-theoretic framework, which is at once simple and able to encompass a wide variety of seemingly disparate models.

Key Words: Information lattice, multicast/unicast transmission, cooperative games, Shapley value, convex/supermodular games.

17.1 Introduction

The Shapley value [Shapley (1953)] constitutes a scheme for the fair division of the benefits in a cooperative game. Unfortunately it is not always "stable" in that some coalitions may have incentive to break away because they can obtain more on their own than what the Shapley value awards them. In other words, the Shapley value can fail to be in the core of the game.

In a seminal paper [Shapley (1971)], Shapley identified the class of "convex" games in which the Shapley value is not only in the core, but is the "center of gravity" of the core. These are games that exhibit increasing returns to cooperation: the marginal contribution of a player to a coalition goes up as the coalition is enhanced. (An equivalent condition – see [Shapley (1971)] – is that the game be a supermodular function on the lattice of coalitions.)

In this paper we pinpoint conditions under which certain games of connectivity are convex. Players in our model are located at the vertices of a communications network and can stand to gain a lot by sharing disparate bits of information that they initially hold. Indeed information is more amenable to sharing than standard commodities. Commodities are typically lost to the person who gives them away. Information in contrast has "the quality of mercy", blessing him that gives and him that takes, since the giver retains all his information even as he sends it out[1]. Nevertheless it is not automatic that all information will be shared. This is because, though costlessly duplicable, information may be costly to transmit (e.g., on account of setup costs of links in the communications network). Any coalition must do a careful cost-benefit analysis, choosing that pattern of information transmission which maximizes the total net benefit to its members.

It should be pointed out that our model takes its cue from, and includes as a special case, the multicast transmission games presented in [Feigenbaum, Papadimitriou and Shenker (2004)] (and recalled in section 17.2.1 below). There, too, the Shapley value was examined though the focus was on its computation and incentive-compatibility (more precisely, on deriv-

---

[1]This "blessedness" of information hinges critically on the fact that we are in a *cooperative* game. In a noncooperative and strategic setting, it can happen that enhancing player i's information can hurt him (as well as others), even though i gets endowed with new additional strategies by virtue of his information. The reason is essentially as follows. The change in i's information is common knowledge, i.e., the others don't get more information but know that i has got it, i knows that they know, etc. This creates a new game in the minds of all the players. Now it may well be that this new game has a wholly different Nash Equilibrium (NE) than before, in which i's opponents have altered their strategies. If it turns out that, in the face these altered strategies, i's old and new strategies all yield lower payoffs to i than what he was getting at the old NE, he will be hurt.

In contrast, such a phenomenon is impossible in the context of a cooperative game. For if some members of a coalition get more strategies (on account of enhanced information), the coalition can choose to ignore these new strategies – precisely because its members are acting collectively – and always get its old payoff. New strategies will only be brought into play if they increase the coalition's payoff (i.e., the sum of the payoffs of its members).

We thank an anonymous referee for spurring us to make this clarification.

ing a group strategy-proof mechanism from it). Our analysis shows, as was said, that the Shapley value is not only fair, but also stable, in all the games of [Feigenbaum, Papadimitriou and Shenker (2004)], further bolstering its plausability as a solution concept there.

An important feature of our approach is that we formulate information in terms of a *lattice*. This leads to a framework that is at once universal and simple. We can encompass a wide variety of seemingly different models, involving unicast and multicast modes of transmission, setup and variable costs in the communications network, and information that comes in various guises (from finite dimensional vectors, to partitions of a set, to layered encoding). The lattice framework makes for a remarkably transparent analysis in all cases.

The paper is organized as follows. In Section 17.2 we present some motivating examples, starting with the model in [Feigenbaum, Papadimitriou and Shenker (2004)]. The abstract lattice-theoretic framework is presented in in Section 17.3. In Section 17.4 we establish our main result which states that games of connectivity are convex. Section 17.5 points out a monotonicity property of optimal transmissions. Finally, in Section 17.6, we show how to fit the examples into our lattice-theoretic framework; and we also examine the tightness of our assumptions and indicate some generalizations of the model.

17.2 Examples

We present a series of examples of information transmission in a network, all of which yield supermodular games, as we shall see in Sections 17.4 and 17.6.

17.2.1 *Multicast Transmission*

First let us recall the game presented in [Feigenbaum, Papadimitriou and Shenker (2004)]. There is a finite tree Γ with a sender δ located at its root and and a distinct receiver at each leaf (terminal vertex). Any receiver α can get information from δ if α is connected to δ using the edges of Γ. The tree Γ is viewed as a digital network which carries a public broadcast by δ, and it is assumed that information flowing into any vertex of the tree can be costlessly duplicated and sent out (multicast) on any subset of the outgoing edges. But the edges of Γ do have setup costs associated to them.

Offsetting these costs are benefits $B(\alpha)$ to α when he receives information from δ.

A cooperative game is induced on the player-set N of receivers in a natural manner. Any coalition $S \subset N$ can use an arbitrary subtree Γ' of Γ at the cost $C(\Gamma')$ of all the edges of Γ'. The benefit S derives from Γ' is $B(S, \Gamma') = \sum_\alpha B(\alpha)$, where the summation runs over all α in S which are connected to δ via Γ'. Thus the "worth" $w(S)$ of coalition S (i.e., the most S can guarantee to itself) is obtained by maximizing the net benefit $B(S, \Gamma') - C(\Gamma')$ over all possible subtrees Γ'.

There can be several senders located at different vertices of the tree, each with its own distinctive information to transmit. Moreover not all senders need be "dummies" as in [Feigenbaum, Papadimitriou and Shenker (2004)]. Some of them could be bona fide players in the game with the power to withhold their information. One could also imagine them to have different transmission trees, possibly with significant overlap.

In spite of these complications, the game remains supermodular and so the Shapley value continues to be centrally located in the core (but its computation may no longer be as felicitous as in [Feigenbaum, Papadimitriou and Shenker (2004)]).

17.2.2 *Unicast Transmission*

Imagine a set of users connected to each other through a hierarchical network (as in telephony). Again suppose they are located on the leaves of a tree Γ with other vertices acting as relays. But the communication is private rather than public, and the users transmit information to each other on a one-to-one basis.

The user at leaf α can choose the amount of information $\tau_{\alpha\beta} \in [0, m], m > 0$, to be sent to β. The total benefit derived at β is $\sum_\alpha B_{\alpha\beta}(\tau_{\alpha\beta})$, where $B_{\alpha\beta}$ is an arbitrary non-decreasing function. As before, it costs to use the tree. Each edge now has not only a setup cost, but also an arbitrary non-decreasing variable cost for every α–to-β flow on it. (The variable costs here add across flows, but the setup cost is invariant of them.)

This unicast scenario also gives rise to a cooperative game in an obvious way. Any coalition S chooses $\tau = \{\tau_{\alpha\beta} : \alpha \in S, \beta \in S\}$, and a subforest of Γ to carry τ, so as to maximize the net benefit.

It turns out that this game is also supermodular.

17.2.3 *Transmission of Layered Information*

We turn to a situation where information is encoded or organized in layers (e.g., as in a video transport system, see [Wang and Zhu (1998)]). To be precise, suppose layer L_i consists of " information bricks" numbered by integers $m_{i-1}+1, m_{i-1}+2, \ldots, m_i$. The bricks in $L = \cup_{i=1}^{k} L_i$ are, however, distributed arbitrarily among the n players located at the vertices of a communication tree Γ, with no duplication. So, denoting by Σ_α the set of bricks held at vertex α, we have $\Sigma_\alpha \cap \Sigma_\beta = \phi$ if $\alpha \neq \beta$. Players wish to receive bricks in order to build a "knowledge pyramid", but they cannot construct layer L_i unless all previous layers $L_1, L_2, \ldots, L_{i-1}$ are in place. Of course, since these bricks are not standard commodities but signify information, no player loses any of his own bricks by sending them to others. The player at vertex α may transmit any subset $Q_e \subset \Sigma_\alpha$ on any edge e emanating from α. Then for any edge e' that follows from e, he can send $Q_{e'} \subset Q_e$, and so on. In short he can contemplate multicast transmission on Γ with α as the root.

There is a set-up cost for every edge e as earlier, and additional flow costs $C_{e,\alpha}(x)$ for $x \in \Sigma_\alpha$.

Benefits accrue as follows. Denoting by $Q_{\beta\alpha} \subset \Sigma_\beta$ the subset of bricks that α receives from β, the benefit to α is $f_\alpha(n)$, where

$$n = \max\{j : L_i \subset \Sigma_\alpha \cup (\cup_\beta Q_{\beta\alpha}) \forall i \leq j\}$$

and $f(n)$ is an arbitrary non-decreasing function.

The idea here, as was said, is that information is organized in pyramidical form. Information of layer L_i is not usable unless all layers $L_1, L_2, \ldots, L_i$ are complete.

The cooperative game, arising in this setup, is once again supermodular.

17.2.4 *Transmission of Information Partitions*

As before, Γ is a tree with players located at its vertices. Let $Q = \{1, 2, \ldots, k\}$ be the set of states of nature, and let $\{Q_\alpha : \alpha \in V\}$ be a partition of Q. (Here V denotes the set of vertices of Γ and Q_α is understood to be the empty set if no player is located at α.) Further let P_α be a partition of Q_α. The interpretation is that $\{P_\alpha, Q \backslash Q_\alpha\}$ is the private information initially held by the player at vertex α. Notice that private information is disjoint across players, i.e., each player is in the dark about states that other players can distinguish.

For simplicity every player α has a state-contingent endowment $(a_1(\alpha), \dots, a_k(\alpha))$ of a single non-tradeable resource (such as his skill), to be used as input in his individual production. He must, of course, use the same input in states that he cannot distinguish. But since expected profit of any player depends on his state-contingent vector of inputs, there are inherent gains from sharing information. The precise model is as follows.

Each player can transmit its information partition (or any coarsening thereof) to other vertices prior to the production stage. If the player at vertex α winds up with the partition P of Q, his profit (via production) is

$$\max f_\alpha(x_1, x_2, \dots, x_k)$$

$$\begin{aligned} \text{Subject to: } x_i &\le a_i(\alpha) \\ x_i &\ge 0 \\ \text{and } i \sim_P j \Rightarrow x_i &= x_j \end{aligned}$$

where $i \sim_P j$ means that i and j are in the same cell of the partition P. We assume that the production function f_α is supermodular on R^k_+, i.e., (assuming differentiability):

$$\frac{\partial}{\partial x_i}\frac{\partial f_\alpha}{\partial x_j} \ge 0$$

for all i, j and α. In other words the inputs $x_1, x_2, \dots, x_k$ are weakly complementary: if α increases his input in some state, this does not diminish his marginal productivity in any state.

When a coalition S forms, its members can transmit information to each other through any subforest of Γ after paying the setup costs, and then they can pool their profits.

This, too, induces a game that is supermodular.

17.2.5 *General Network with Controlled Edges*

Let G be an arbitrary undirected graph with edge set E and vertex set V. For each vertex $\alpha \in V$, let $\Gamma(\alpha) \subset G$ be a tree rooted at α on which α is constrained to transmit its information. Further suppose that edges of G are subject to the control of coalitions.

Thus when a coalition S forms, each $\alpha \in S$ has access to only those edges in $\Gamma(\alpha)$ whose controllers are contained in S.

In this setup, players who are neither senders nor receivers of information, may nevertheless have a vital role to play in the game on account of their control of edges (such as cable operators or monopoly network providers).

All of our preceding examples can be embedded in this larger framework. The games induced will still be supermodular.

17.3 The Abstract Model

We build an abstract lattice-theoretic model of information and its transmission, which unifies the above (and more) examples and makes for a particularly transparent analysis.

17.3.1 *The Communications Network*

Let $G = (V, E)$ be a graph where V is a finite set of vertices and E is a set of undirected edges.

For every $\alpha \in V$ there is a tree $\Gamma(\alpha) \equiv (V(\alpha), E(\alpha)) \subset G$, rooted at α, that can be used by α to transmit its information to other vertices.

17.3.2 *Information*

Information is modeled as a lattice $\mathcal{L}$ with $\geq$ denoting the partial order and $\vee, \wedge$ the join and the meet operators[2]. We assume that $0 \equiv \wedge\{x : x \in \mathcal{L}\}$ exists in $\mathcal{L}$ and that that $\wedge$ distributes over $\vee$, i.e.,

$$x \wedge (y \vee z) = (x \wedge y) \vee (x \wedge z)$$

for all $x, y, z \in \mathcal{L}$. This property holds in a variety of contexts and is well-known (see [Birkhoff (1977)]).

The canonical examples we have in mind is that $\mathcal{L}$ is the power set of a finite set with $\geq$ corresponding to the set-theoretic notion of $\supset$; or that $\mathcal{L}$ is the set of all partitions of a finite set with $\geq$ corresponding to refinement; or that $\mathcal{L}$ is a closed interval of the real line with $\geq$ corresponding to the standard order; or that $\mathcal{L}$ is the product lattice of finitely many such lattices. In all of these cases 0 exists in $\mathcal{L}$ and the distributive property holds.

Any vertex $\alpha \in V$ can transmit information from a sub-lattice $\mathcal{L}(\alpha)$ of $\mathcal{L}$. A key assumption we make is that the information held at different

[2]Recall (see e.g. [Birkhoff (1977)]) that for and x and y in $\mathcal{L}$, there exists a greatest lower bound w.r.t. $\geq$ (denoted $x \wedge y$) and a least upper bound (denoted $x \vee y$).

vertices is disjoint, i.e.,

$$x \in \mathcal{L}(\alpha), y \in \mathcal{L}(\beta), \alpha \neq \beta \Rightarrow x \wedge y = 0$$

We also assume that each vertex can opt to send no information, i.e., $0 \in \mathcal{L}(\alpha)$ for all $\alpha \in V$.

17.3.3 *Location of Players and Public Facilities*

Let $N = \{1, 2, \ldots, n\}$ be the set of players. There is an additional dummy player, labeled $n+1$, used to model public facilities available to all players in N. Denote $\tilde{N} = N \cup \{n+1\}$.

Each vertex is occupied by a player[3] as specified by a *location map*

$$\eta : V \rightarrow \tilde{N}$$

where $\eta(\alpha)$ denotes the player (possibly, dummy) at vertex α. Let $V(S)$ represent the set of all the vertices occupied by players in $S \cup \{n+1\}$ i.e.,

$$V(S) = \{\alpha \in V : \eta(\alpha) \in S \cup \{n+1\}\}$$

17.3.4 *Control of Edges*

Edges are controlled by coalitions of players in accordance with a *control map*

$$\kappa : E \rightarrow 2^N$$

where $\kappa(e)$ denotes the coalition that controls[4] the use of edge e. (If $\kappa(e) = \phi$, then e is accessible to everyone.)

17.3.5 *The Transmission of Information*

Each vertex α can transmit information $x \in \mathcal{L}(\alpha)$ to other vertices on its tree $\Gamma(\alpha) \equiv (V(\alpha), E(\alpha))$. Concatenating across vertices, the total transmission may be viewed as a map $\tau : E \times V \rightarrow \mathcal{L}$ with the interpretation that $\tau(e, \alpha)$ is the information transmitted by the vertex α on the edge e. Some natural conditions must be imposed on this map τ. Any vertex α can send information only out of $\mathcal{L}(\alpha)$ i.e.,

$$\tau(e, \alpha) \in \mathcal{L}(\alpha) \tag{17.1}$$

[3] The case where several players occupy a vertex is included in our set-up (see remark 3 in Section 17.6).

[4] A natural case: if $e = (\alpha, \beta)$, then $\kappa(e) = (\eta(\alpha) \cup \eta(\beta)) \cap N$.

for all $\alpha \in V$ and $e \in E(\alpha)$. Moreover, no vertex α can send any (except null) information on edges outside its tree i.e.,

$$\tau(e, \alpha) = 0 \text{ if } e \notin E(\alpha) \tag{17.2}$$

for all $\alpha \in V$ and $e \in E$. Finally, the join of all the information of α that flows out of a vertex must be no more than the information of α that arrives at it, i.e.,

$$\tau(e, \alpha) \geq \vee\{\tau(e', \alpha) : e' \in F(e, \alpha)\} \tag{17.3}$$

for all $\alpha \in V$ and $e \in E(\alpha)$, where $F(e, \alpha)$ denotes the set of immediate offspring edges of e in the tree $\Gamma(\alpha)$.

Let $\mathcal{T}$ denote the set of all possible transmissions, i.e.,

$$\mathcal{T} = \{\tau : E \times V \to \mathcal{L} : \tau \text{ satisfies (17.1), (17.2) and (17.3)}\}$$

The set $\mathcal{T}$ itself forms a lattice under the natural definitions: $\tau \geq \tau'$ if $\tau(e, \alpha) \geq \tau'(e, \alpha)$ for all e, α; $(\tau \vee \tau')(e, \alpha) = \tau(e, \alpha) \vee \tau'(e, \alpha)$ for all e, α; $(\tau \wedge \tau')(e, \alpha) = \tau(e, \alpha) \wedge \tau'(e, \alpha)$ for all e, α.

For any coalition $S \subset N$, define the subset $\mathcal{T}(S) \subset \mathcal{T}$ of *transmissions feasible for* S as follows:

$$\mathcal{T}(S) = \{\tau \in \mathcal{T} : \text{ for any } e \text{ and } \alpha,\ \tau(e, \alpha) > 0 \Rightarrow \kappa(e) \subset S \text{ and } \alpha \in S \cup \{n+1\}\}$$

In other words, only members of S or public vertices can transmit information in $\mathcal{T}(S)$; and only the edges under the control of S may be used.

17.3.6 *The Reception of Information*

A transmission $\tau \in \mathcal{T}$ induces a reception $\sigma(\tau, \alpha) \in \mathcal{L}$ at every vertex $\alpha \in V$ as follows:

$$\sigma(\tau, \alpha) = (x^*(\alpha)) \vee (\vee\{\tau(e(\beta, \alpha), \beta) : \beta \in V \backslash \{\alpha\} \text{ and } \alpha \in \Gamma(\beta)\})$$

where $e(\beta, \alpha)$ is the edge coming into α from β in $\Gamma(\beta)$ and $x^*(\alpha) \equiv \vee\{x : x \in \mathcal{L}(\alpha)\}$.

Here $x^*(\alpha)$ represents the maximum information in $\mathcal{L}(\alpha)$. Since α can costlessly receive its own information, and since information is valuable, we suppose that α always "sends" $x^*(\alpha)$ to itself. The total reception at α is obtained by joining $x^*(\alpha)$ with the bits of information $\tau(e(\beta, \alpha), \beta)$ sent to α by other vertices β.

17.3.7 *The Cost of a Transmission*

The cost of transmitting information (originating at different vertices) on any edge is given by[5] $c_e : \mathcal{L}^V \to R_+$, where $c_e((x(\alpha))_{\alpha \in V}) \equiv$ the cost of the flow $(x(\alpha))_{\alpha \in V}$ on e. We postulate that c_e is *submodular* on $\mathcal{L}^V$, i.e.,

$$c_e(x \vee y) + c_e(x \wedge y) \leq c_e(x) + c_e(y)$$

for all $e \in E$ and $x, y \in \mathcal{L}^V$. Such costs can arise in several ways. For instance, suppose there is a set-up cost $f(e)$ for e, and a further set-up cost $f(e, \alpha)$ for every vertex α that uses e, i.e.,

$$c_e((x(\alpha))_{\alpha \in V}) = \begin{cases} 0, \text{ if } x(\alpha) = 0 \text{ for all } \alpha \\ f(e) + \sum_{x:x(\alpha)>0} f(e, \alpha), \text{ otherwise} \end{cases}$$

It is evident that this cost function is submodular, and that it remains so if we add variable costs $\sum_{\alpha \in V} g_\alpha(x(\alpha))$ provided each $g_\alpha : \mathcal{L} \to R_+$ is itself submodular (i.e., evinces economy of scale).

The *cost of transmission* $\tau \in \mathcal{T}$ is the sum of the costs incurred on all the edges, i.e.,

$$C(\tau) = \sum_{e \in E} c_e((\tau(e, \alpha))_{\alpha \in V})$$

It is easy to verify that C is submodular on $\mathcal{T}$, i.e.,

$$C(\tau) + C(\tau') \geq C(\tau \vee \tau') + C(\tau \wedge \tau') \tag{17.4}$$

17.3.8 *The Benefit from a Transmission*

For every vertex $\beta \in V$, there is a benefit function $B_\beta : \mathcal{L} \to R_+$, where $B_\beta(x)$ represents the benefit to β from receiving information $x \in \mathcal{L}$. We assume that B_β is supermodular and non-decreasing for all $\beta \in V$ i.e.,

$$B_\beta(x \vee y) + B_\beta(x \wedge y) \geq B_\beta(x) + B_\beta(y)$$

and

$$x \geq y \Rightarrow B_\beta(x) \geq B_\beta(y)$$

The benefit to a coalition $S \subset N$ from transmission $\tau \in \mathcal{T}$ is given by

$$B(S, \tau) = \sum_{\beta \in V(S)} B_\beta(\sigma(\tau, \beta))$$

It is again easy to verify that B is supermodular on $\mathcal{T}$ (with S fixed). But the supermodularity of B and the submodularity of C do not immediately lead to the supermodularity of the game w defined in the next section.

[5]Note that $\mathcal{L}^V$ is a finite product of $\mathcal{L}$ with itself (V times) and is a product lattice.

17.4 The Connectivity Game

We consider the cooperative game that arises from the communications network. A non-empty coalition $S \subset N$ can choose any $\tau \in \mathcal{T}(S)$ to transmit information between its members or to receive information from public vertices. The coalition obtains total benefit $B(S, \tau)$ but at a cost $C(\tau)$. The maximum net benefit that S can guarantee is therefore given by

$$w(S) = \max_{\tau \in \mathcal{T}(S)} B(S, \tau) - C(\tau)$$

(with $w(\phi)$ understood to be 0). We call w the *connectivity game.*

Recall that a game $w : 2^N \to R$ is called *supermodular* (or, as in [Shapley (1971)], convex) if w is supermodular on the lattice 2^N, i.e.,

$$w(S \cup T) + w(S \cap T) \geq w(S) + w(T)$$

for all $S \subset N$ and $T \subset N$. Our main result is:

Theorem 17.1. *The connectivity game w is supermodular.*

For the proof, see the Appendix.

17.5 The Growing Transmissions Property

It is worth noting that optimal transmissions grow with the coalitions in the sense made precise by Theorem 17.2 below.

Theorem 17.2. *Let $S \subset T \subset N$ and let $\tau_1 \in \mathcal{T}(S)$ be an optimal transmission for S. Then there exists an optimal transmission $\tau \in \mathcal{T}(T)$ for T such that $\tau \geq \tau_1$.*

For the proof, see the Appendix.

17.6 Remarks

Remark 1 (Embedding the examples) We briefly indicate how to fit our examples (from Section 17.2) into the abstract model.

For Section 17.2.1, take $\Gamma(\alpha) = \Gamma$ rooted at α, $\kappa(e) = \phi$ for all e, $\mathcal{L}(\delta) = \{0, 1\}$, $\mathcal{L}(\alpha) = \{0\}$ for all $\alpha \neq \delta$, $\mathcal{L} =$ the cross product of all these lattices, $B_\delta = 0$, $B_\alpha(0) = 0$ and $B_\alpha(1) = B(\alpha)$ for all $\alpha \neq \delta$. Finally the cost of an edge is its setup cost if there is a non-zero transmission on it and zero otherwise.

For Section 17.2.2, let $\mathcal{L}(\alpha) = [0, m]^V$, each of whose elements specifies the information sent by α to all the other vertices. The lattice operations $\vee$ and $\wedge$ are obtained by taking component-wise maximum and minimum. $\mathcal{L}$ as usual is the cross product of all the $\mathcal{L}(\alpha)$. The cost functions are obvious. The rest of the construction is as before. (Notice that despite the fact that the components of the benefit and cost functions have no supermodularity or concavity assumptions on them, the benefit/cost functions are supermodular/submodular in our lattice framework. This follows from the fact that they are additive over their components and that super or sub-modularity is no constraint on a function of one variable.)

For the example in Section 17.2.3, take $\mathcal{L}(\alpha)$ to be the totally ordered set $\{0\} \cup \Sigma_\alpha$, and $\mathcal{L}$ to be the cross product. We leave it to the reader to verify that the benefit function is supermodular.

Finally, for the example in Section 17.2.4, take $\mathcal{L}(\alpha)$ to be the lattice of all partitions of Q which are coarser than $\{P_\alpha, Q \backslash Q_\alpha\}$. The supermodularity of the benefit functions follows from that of $f_\alpha, \alpha \in V$.

Remark 2 (Acyclicity) Cycles in the transmissions network $\Gamma(\alpha)$ can cause our result to break down. Consider the network in Figure 17.1 in which players 1, 2, 3, 4, each have access to the whole graph, with costs as shown and with $\epsilon < 1$.

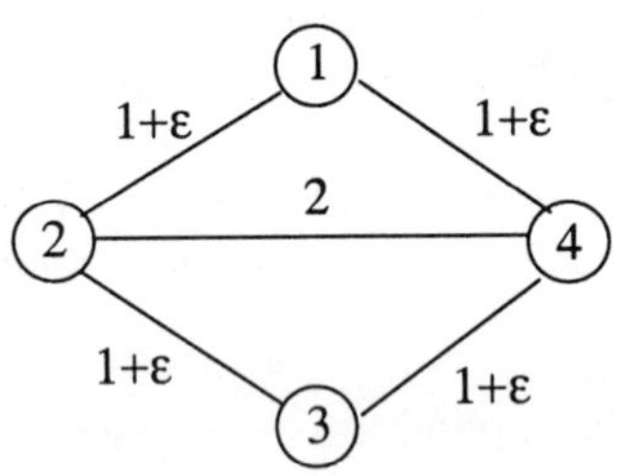

Fig. 17.1 Cycles in the communications network

Further suppose that 1, 2, 3 each derive benefit $B > 2(1+\epsilon)$ from being connected to 4. Then it is clear that

$$\begin{aligned} w(2,4) &= B - 2 \\ w(2,3,4) &= 2B - 2(1+\epsilon) \\ w(1,2,4) &= 2B - 2(1+\epsilon) \\ w(1,2,3,4) &= 3B - 3(1+\epsilon) \end{aligned}$$

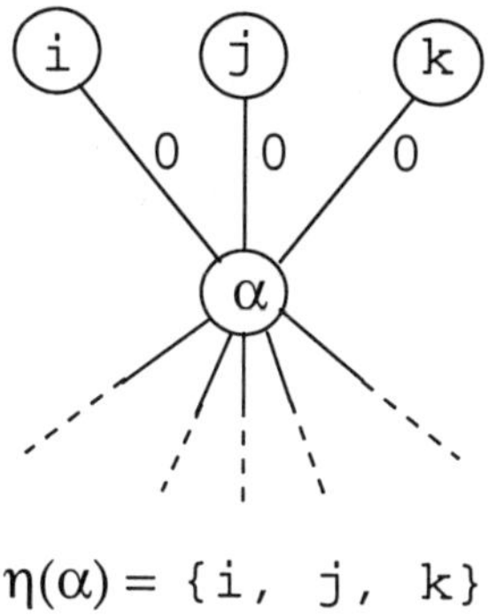

Fig. 17.2 Modeling multiple players at a vertex

But then
$w(1,2,3,4) + w(2,4) = 4B - 5 - 3\epsilon < 4B - 4 - 2\epsilon = w(1,2,4) + w(2,3,4)$
showing that w is not supermodular.

Remark 3 (Multiple players at a vertex) Our model allows for many players to be located at the same vertex α. Indeed, by creating a new vertex for each player present at α, and joining these with zero-cost edges to α, we create an expanded graph which fits our model (see Figure 17.2).

Remark 4 (Control of vertices) Our model also permits coalitions to control vertices by the graph expansion shown in Figure 17.3. Every edge incident at α is intercepted with a zero-cost edge controlled by the coalition controlling α.

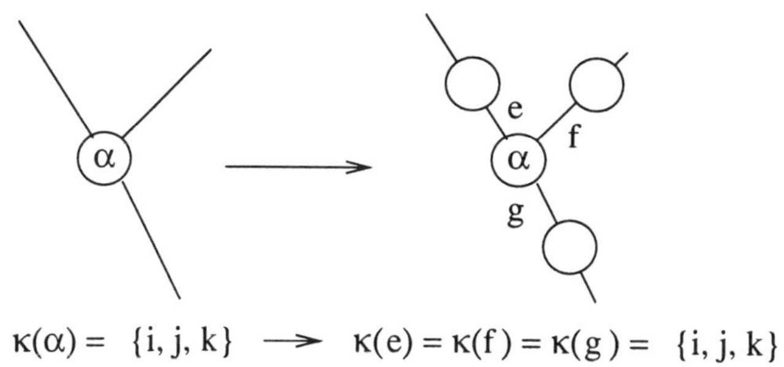

Fig. 17.3 Modeling control of vertices

Remark 5 (Veto players) A more general control of edges by veto players renders our results invalid. Consider a player set $\{1,2,3\}$ and suppose that there is common tree available to everyone, which consists of just one zero-cost edge connecting player 1 to a public vertex. The edge can be sanctioned by player 1 (the veto player), in conjunction with any player in $\{2,3\}$. The only benefit B is obtained by player 1 when he gets connected to the public vertex. In this game $w(1) = 0$ and $w(1,2) = w(1,3) = w(1,2,3) = B$. Hence $w(1,2,3) + w(1) = B < 2B = w(1,2) + w(1,3)$, showing that w is not supermodular.

Remark 6 (Dropping distributivity) In the special case where $\mathcal{L}$ is the cross product of the lattices $\mathcal{L}(\alpha)$ over $\alpha \in V$, our results hold without postulating that $\wedge$ distributes over $\vee$. In this case, the analogue of (17.7) for $\wedge$ holds trivially (and this was the only step that required distributivity).

But in general distributivity is indispensable.

Remark 7 (Enhancement of information) So far we have taken information to be fixed a priori. But it could well happen that the information of an agent gets enhanced by virtue of the information he receives from others. He can turn around and send his enhanced information back to them, enhancing theirs', and so on. Even in this setting, under suitable hypotheses, the induced cooperative game is well-defined (i.e., the enhancement sequence converges) and is supermodular, as we shall show in a sequel paper.

17.7 Appendix

17.7.1 *Proof of Theorem 17.1*

We first establish some lemmas.

Lemma 17.1. *Let $S \subset N$, $T \subset N$, $\tau \in \mathcal{T}(S)$ and $\tau' \in \mathcal{T}(T)$. Then $\tau \vee \tau' \in \mathcal{T}(S \cup T)$ and $\tau \wedge \tau' \in \mathcal{T}(S \cap T)$.*

Proof. Since τ and τ' are in $\mathcal{T}$, $\tau(e,\alpha)$ and $\tau'(e,\alpha)$ are in $\mathcal{L}(\alpha)$. Since $\mathcal{L}(\alpha)$ is a lattice, $(\tau\vee\tau')(e,\alpha) \equiv \tau(e,\alpha)\vee\tau'(e,\alpha) \in \mathcal{L}(\alpha)$ and $(\tau\wedge\tau')(e,\alpha) \equiv \tau(e,\alpha) \wedge \tau'(e,\alpha) \in \mathcal{L}(\alpha)$.

Next, if $e \notin E(\alpha)$, then $\tau(e,\alpha) = \tau'(e,\alpha) = 0$ and therefore $(\tau \vee \tau')(e,\alpha) = 0$ and $(\tau \wedge \tau')(e,\alpha) = 0$ as well.

Finally, since

$$\tau(e,\alpha) \geq \vee\{\tau(e',\alpha) : e' \in F(e,\alpha)\} \tag{17.5}$$
$$\tau'(e,\alpha) \geq \vee\{\tau'(e',\alpha) : e' \in F(e,\alpha)\} \tag{17.6}$$

we have

$$\begin{aligned}(\tau \vee \tau')(e,\alpha) &= \tau(e,\alpha) \vee \tau'(e,\alpha)\\ &\geq \vee\{(\tau \vee \tau')(e',\alpha) : e' \in F(e,\alpha)\}\end{aligned}$$

from the associativity of $\vee$ and the fact that $x \geq x'$ and $y \geq y'$ implies $x \vee y \geq x' \vee y'$. This shows that $\tau \vee \tau' \in \mathcal{T}$. Also from (17.5) and (17.6)

$$\begin{aligned}\tau(e,\alpha) \wedge \tau'(e,\alpha) &\geq (\vee\{\tau(e',\alpha) : e' \in F(e,\alpha)\}) \wedge (\vee\{\tau'(e',\alpha) : e' \in F(e,\alpha)\})\\ &\geq \vee\{(\tau(e',\alpha) \wedge \tau'(e',\alpha)) : e' \in F(e,\alpha)\}\end{aligned}$$

The first inequality follows from the fact that $x \geq x'$ and $y \geq y'$ implies $x \wedge y \geq x' \wedge y'$; the second from the fact $(x \vee y) \wedge z \geq (x \wedge z) \vee (y \wedge z)$ and the commutativity and associativity of $\vee, \wedge$. This proves that $\tau \wedge \tau' \in \mathcal{T}$.

To check that $\tau \vee \tau' \in \mathcal{T}(S \cup T)$, observe that, for any e and α

$$\begin{aligned}&\tau(e,\alpha) \vee \tau'(e,\alpha) > 0\\ &\Rightarrow \tau(e,\alpha) > 0 \text{ or } \tau'(e,\alpha) > 0\\ &\Rightarrow \kappa(e) \subset S \text{ or } \kappa(e) \subset T\\ &\Rightarrow \kappa(e) \subset S \cup T\end{aligned}$$

To check that $\tau \wedge \tau' \in \mathcal{T}(S \cap T)$, observe that

$$\begin{aligned}&\tau(e,\alpha) \wedge \tau'(e,\alpha) > 0\\ &\Rightarrow \tau(e,\alpha) > 0 \text{ and } \tau'(e,\alpha) > 0\\ &\Rightarrow \kappa(e) \subset S \text{ and } \kappa(e) \subset T\\ &\Rightarrow \kappa(e) \subset S \cap T\end{aligned}$$

□

Lemma 17.2. *For $S \subset N$, $T \subset N$, $\tau \in \mathcal{T}$ and $\tau' \in \mathcal{T}$,*

$$B(S,\tau_1) + B(T,\tau_2) \leq B(S \cup T, \tau_1 \vee \tau_2) + B(S \cap T, \tau_1 \wedge \tau_2)$$

Proof. From the definition of σ and the associativity of $\vee$ it is immediate that

$$\sigma(\tau \vee \tau', \alpha) = \sigma(\tau,\alpha) \vee \sigma(\tau',\alpha) \tag{17.7}$$

The analogous result holds for $\wedge$ only when the lattice $\mathcal{L}$ is distributive and the sub-lattices $\mathcal{L}(\alpha)$ are disjoint across $\alpha \in V$, as we now check.

Let α and β be two distinct vertices. Denote by $\rho(\tau,\alpha,\beta)$ the information that α receives from β in the transmission τ, i.e.,

$$\rho(\tau,\alpha,\beta) = \begin{cases} \tau(e(\beta,\alpha),\beta), & \text{if } \alpha \in \Gamma(\beta) \\ 0, & \text{otherwise} \end{cases}$$

where, recall $e(\beta,\alpha)$ is the edge coming into α from β in the tree $\Gamma(\beta)$. Then,

$$\sigma(\tau,\alpha) = x^*(\alpha) \vee (\vee\{\rho(\tau,\alpha,\beta) : \beta \in V\backslash\{\alpha\}\})$$

So,

$$\begin{aligned}\sigma(\tau,\alpha) \wedge \sigma(\tau',\alpha) = &(x^*(\alpha) \vee (\vee\{\rho(\tau,\alpha,\beta) : \beta \in V\backslash\{\alpha\}\})) \\ &\wedge(x^*(\alpha) \vee (\vee\{\rho(\tau',\alpha,\beta) : \beta \in V\backslash\{\alpha\}\}))\end{aligned}$$

By the distributivity of $\wedge$ over $\vee$, and the commutativity and associativity of $\wedge$ and $\vee$, the right hand side of the above equation simplifies to

$$x^*(\alpha) \vee (\vee\{\rho(\tau,\alpha,\beta) \wedge \rho(\tau',\alpha,\beta') : \beta \in V\backslash\{\alpha\}, \beta' \in V\backslash\{\alpha\}\})$$

Since the sub-lattices $\mathcal{L}(\beta)$ and $\mathcal{L}(\beta')$ are disjoint when $\beta \neq \beta'$ all the cross-terms in the above expression disappear, reducing it to

$$x^*(\alpha) \vee (\vee\{\rho(\tau,\alpha,\beta) \wedge \rho(\tau',\alpha,\beta) : \beta \in V\backslash\{\alpha\}\})$$

which obviously equals

$$x^*(\alpha) \vee (\vee\{\rho(\tau \wedge \tau',\alpha,\beta) : \beta \in V\backslash\{\alpha\}\})$$

proving that

$$\sigma(\tau \wedge \tau',\alpha) = \sigma(\tau,\alpha) \wedge \sigma(\tau',\alpha) \tag{17.8}$$

From the definition of the benefit function B,

$$B(S,\tau) + B(T,\tau') = \sum_{\beta\in V(S)} B_\beta(\sigma(\tau,\beta)) + \sum_{\beta\in V(T)} B_\beta(\sigma(\tau',\beta))$$

By rearranging terms we get

$$\begin{aligned}B(S,\tau) + B(T,\tau') = &\sum_{\beta\in V(S)\backslash V(T)} B_\beta(\sigma(\tau,\beta)) + \sum_{\beta\in V(T)\backslash V(S)} B_\beta(\sigma(\tau',\beta)) \\ &+ \sum_{\beta\in V(S)\cap V(T)} (B_\beta(\sigma(\tau,\beta)) + B_\beta(\sigma(\tau',\beta)))\end{aligned} \tag{17.9}$$

From (17.7), (17.8) and the supermodularity of B_β we have

$$\begin{aligned}B_\beta(\sigma(\tau,\beta)) + B_\beta(\sigma(\tau',\beta)) &\leq B_\beta(\sigma(\tau,\beta) \vee \sigma(\tau',\beta)) + B_\beta(\sigma(\tau,\beta) \wedge \sigma(\tau',\beta)) \\ &= B_\beta(\sigma(\tau \vee \tau')) + B_\beta(\sigma(\tau \wedge \tau'))\end{aligned}$$

Therefore (17.9) becomes

$$
\begin{aligned}
& B(S,\tau) + B(T,\tau') \\
&\quad \le \sum_{\beta\in V(S)\setminus V(T)} B_\beta(\sigma(\tau,\beta)) + \sum_{\beta\in V(T)\setminus V(S)} B_\beta(\sigma(\tau',\beta)) \\
&\quad + \sum_{\beta\in V(S)\cap V(T)} (B_\beta(\sigma(\tau\vee\tau',\beta)) + B_\beta(\sigma(\tau\wedge\tau',\beta))) \\
&\quad \le \sum_{\beta\in V(S)\setminus V(T)} B_\beta(\sigma(\tau\vee\tau'),\beta) + \sum_{\beta\in V(T)\setminus V(S)} B_\beta(\sigma(\tau\vee\tau',\beta)) \\
&\quad + \sum_{\beta\in V(S)\cap V(T)} B_\beta(\sigma(\tau\vee\tau',\beta)) + \sum_{\beta\in V(S)\cap V(T)} B_\beta(\sigma(\tau\wedge\tau',\beta)) \\
&\quad = B(S\cup T,\tau\vee\tau') + B(S\cap T,\tau\wedge\tau')
\end{aligned}
$$

(The last inequality follows from the fact that B_β is a non-decreasing function on $\mathcal{L}$ for all $\beta \in V$). □

Completion of the Proof Let S and T be arbitrary coalitions of N. Let τ_1^*, τ_2^* be optimal transmissions for coalitions S, T respectively, i.e.,

$$w(S) = B(S,\tau_1^*) - C(\tau_1^*)$$

$$w(T) = B(T,\tau_2^*) - C(\tau_2^*).$$

From Lemma 17.2 and the fact that C is submodular (see (17.4)), we have

$$
\begin{aligned}
w(S) + w(T) \le\; & B(S\cup T,\tau_1^*\vee\tau_2^*) - C(\tau_1^*\vee\tau_2^*) \\
& + B(S\cap T,\tau_1^*\wedge\tau_2^*) - C(\tau_1^*\wedge\tau_2^*) \qquad (17.10)
\end{aligned}
$$

Since τ_1^* is an optimal transmission for coalition S, $\tau_1^* \in \mathcal{T}(S)$. Similarly $\tau_2^* \in \mathcal{T}(T)$. By Lemma 17.1, $\tau_1^* \vee \tau_2^* \in \mathcal{T}(S\cup T)$ and $\tau_1^* \wedge \tau_2^* \in \mathcal{T}(S\cap T)$. But then,

$$w(S\cup T) \ge B(S\cup T,\tau_1^*\vee\tau_2^*) - C(\tau_1^*\vee\tau_2^*) \qquad (17.11)$$

$$w(S\cap T) \ge B(S\cap T,\tau_1^*\wedge\tau_2^*) - C(\tau_1^*\wedge\tau_2^*) \qquad (17.12)$$

Inequalities (17.10), (17.11) and (17.12) give

$$w(S) + w(T) \le w(S\cup T) + w(S\cap T)$$

showing that the game w is convex.

17.7.2 Proof of Theorem 17.2

Proof. Let τ_2 be an optimal transmission of T. Denote $\tau' \equiv \tau_1 \wedge \tau_2$ and $\tau \equiv \tau_1 \vee \tau_2$. By Lemma 17.1 and the fact that $S \subset T$, we have $\tau' \in \mathcal{T}(S)$ and $\tau \in \mathcal{T}(T)$.

The optimality of τ_1 for S implies

$$B(S, \tau_1) - B(S, \tau') \geq C(\tau_1) - C(\tau')$$

By the submodularity of C we have

$$C(\tau_1) - C(\tau') \geq C(\tau) - C(\tau_2)$$

From Lemma 17.2 we also have

$$B(T, \tau) - B(T, \tau_2) \geq B(S, \tau_1) - B(S, \tau')$$

The above three inequalities imply

$$B(T, \tau) - B(T, \tau_2) \geq C(\tau) - C(\tau_2)$$
$$\Rightarrow B(T, \tau) - C(\tau) \geq B(T, \tau_2) - C(\tau_2)$$

Since τ_2 is an optimal transmission for T, the above inequality shows that τ is also optimal for T. But $\tau \equiv \tau_1 \vee \tau_2 \geq \tau_1$, proving the theorem. □

Bibliography

Birkhoff G. (1977). Lattice Theory, *Encyclopaedia Brittanica, Macropaedia*, Encyclopaedia Brittanica Inc., Helen Hemingway Benton, London, **1**, pp. 519–524.

Feigenbaum J., Papadimitriou C. and Shenker S. (2004). Sharing the Cost of Multicast Transmissions, *ACM Symposium on Theory of Computing (STOC'00)*, (Portland, Oregon, USA).

Shapley L. (1953). A value for n-person games in *Contributions to the Theory of Games* ed. by Kuhn H. W.and Tucker A. W., *Annals of Mathematics Studies*, **2 (28)**, pp. 307–317

Shapley L. (1971). Cores of convex games, *International Journal of Game Theory*, **1**, pp. 11–26.

Wang Y. and Zhu Q. (1998). Error control and concealment for video communications: A review, *Proceedings of IEEE, special issue on Multimedia Signal Processing*, **86 (5)** pp. 974-997.

Chapter 18

A Robust Feedback Nash Equilibrium in a Climate Change Policy Game

Magnus Hennlock[1]
Department of Economics and Statistics, Gothenburg University
P. O. Box 640, S-405 30 Gothenburg, Sweden
e-mail: Magnus.Hennlock@economics.gu.se

Abstract

A robust feedback Nash equilibrium is defined and solved analytically in a differential climate model with N regions based on an approach of IPCC 2001 scientific report for calculating radiative forcing due to anthropogenic CO_2 emissions. In addition, uncertainty is introduced by perturbing the climate change dynamics such that future radiative forcing and global mean temperature will have unknown outcomes *and* probability distributions. There are n asymmetric investors, each investing in a portfolio containing N regional capital stocks used in production that generates CO_2 emissions. In each region there is one policy maker, acting as a regional social planner, that chooses regionally optimal abatement policies. Dynamic maximin decision criteria are applied for the policy makers in a robust feedback Nash equilibrium for N policy makers' abatement strategies and n investors' investment strategies.

Key Words: Closed-loop equilibrium, subgame perfect, robust control, feedback Nash equilibrium, uncertainty aversion, minimax decision criteria, climate change

18.1 Introduction

Climate change policy is an example of a decision-making process that is subject to fundamental uncertainties concerning the underlying scientific

[1]Research supported by Adlerbertska Forskningsstiftelsen.

information available. Policy makers' decision to take or not take measures today are based on scientists' projections, usually generated by climate models, such as Global Circulation Models (GCM) evaluated for different emissions scenarios. Previously, there has been a dispute among scientists whether the increase in mean global temperature during the last century is caused by the increase in anthropogenic GHG gases or not. In the 1995 report, the IPCC stated that 'the evidence suggests a discernible human influence on global climate'. The IPCC 2001 report concluded that 'most of the observed warming over the last fifty years is likely to have been due to the increase in greenhouse gas concentrations'. The projections of future climate impacts and sea level raise are subject to great uncertainty and the projections, which may span over several hundreds years, differ between models as they are highly dependent upon uncertain model parameters such as subgrid-scale diffusion coefficients, precipitation and evaporation fluxes [Harvey (2000)]. Moreover, there may be thresholds leading to abrupt sudden changes which cannot be simulated and anticipated by the models such as reorganizations of the oceanic circulation, abrupt disappearance of the Arctic sea ice and abrupt increase in climate sensitivity. Due to different model assumptions, there also exists a variety of climate models generating different results, predicting an increase in global surface mean temperature within the range 1.5 - 7.5 K. The IPCC Climate Change 2001, The Scientific Basis Report (p. 745) mentions mainly four sources of uncertainties

(1) Uncertainties in converting emissions to atmospheric concentrations.
(2) Uncertainties in converting concentrations to radiative forcing.
(3) Uncertainties in modeling the climate response to a given forcing.
(4) Uncertainties in converting model response into inputs for impact studies.

The fundamental uncertainty concerns not only future outcomes but also future probability distributions. True or inferred probability distributions are not available from samples, and hence, expected utility theory may not be a proper tool to analyze optimal climate change policies. In GCM simulations, where current data and knowledge are used to build models that predict events 100-200 years into the future, the probability distributions are usually results from scientists' ad hoc assumptions and guesswork, which differ among scientists and models.

In the literature on the theory of decision-making it is common to distinguish between risk and uncertainty. The former refers to a process where the actual outcome is unknown but probability distribution is known or can be estimated from samples. However, already Knight (1921) suggested that for many choices, the assumption of known probability distributions is too strong. Keynes (1921), in his treatise on probability, put forward the question whether we should be indifferent between two scenarios that have equal probabilities, but one of them is based on greater knowledge. Savage (1954) argued that we should, while Ellsberg (1961) showed in an experiment that we tend not to do so. A person that is facing two uncertain lotteries with the same (subjective) probability to success, but with less information provided in the second lottery, tends to prefer the first lottery where more information is available. Having Ellsberg's paradox in mind, [Gilboa and Schmeidler (1989)] formulated a maximin decision criterion, by weakening Savage's Sure-Thing Principle, to explain the result from the Ellsberg experiment. In plain words, the decision-maker is suggested to maximize expected utility under the belief that the worst case scenario will happen in the future (a maximin decision criterion). The maximin decision criterion has been applied before in static models by e.g. Chichilnisky (2000) and Bretteville (2002) with the general result that it leads to an increase in abatement effort.

This paper takes the maximin decision criterion further into dynamic modeling and performs an analysis of the criterion in a dynamic climate model with stock effects. The first problem to encounter is that [Gilboa and Schmeidler (1989)] is based on static decision making and their axioms are not sufficient for dynamic models. For example, they do not state how the decision-maker's beliefs are affected by new information (which could increase or decrease scientific uncertainty) during the play of the game. In this paper we suggest that a rational decision-maker updates her beliefs to new information due to scientific progresses by a rule derived from backward induction in addition to the maximin decision criterion.[2] This is in accordance with the IPCC 1995 report stating 'the challenge is not to find the best policy for the next 100 years, but to select a prudent strategy and to adjust it over time in the light of new information'. We develop the single-player model in [Hennlock (2006)], presenting a robust abatement

[2] [Gilboa and Schmeidler (1989)] view uncertainty aversion as a minimization of the set of probability measures while [Hansen, Sargent, Turmuhambetova and Williams (2001)] set a robust control problem and let its perturbations be interpreted as multiple priors of the max-min expected utility theory.

policy in a dynamic climate change model, to a game with N uncertainty averse policy makers and n investors and solves for a robust feedback Nash equilibrium in regional abatement policies and investments. The paper is formal and presents the analytical solution and the coefficients of the players' value functions and provides an introductory analysis. Further analyzes and simulations are left to future studies.

The following sections are organized in the following way. In the section 18.2, the climate models and climate change impacts are presented. Section 18.3 presents players and payoff functions and the optimal strategies which is followed by an concluding comments and summary in 18.4 and 18.5 respectively.

18.2 The Climate Models

There are $j = 1, 2, \ldots, N$ regions, each endowed with $i = 1, 2, \ldots, n$ physical capital stocks k_{ij} and a regional natural capital stock x_j (e.g agriculture, water resources or ecosystem etc.). The production process using k_{ij} generates anthropogenic CO_2 net emission flow $E_{ijt} - \eta_{ij}q_{ijt}$ in period t where $\eta_{ij}q_{ijt}$ is abatement with q_{ijt} being abatement effort undertaken in production process i in region j and $\eta_{ij} > 0$ an industry-specific efficiency parameter. In equation (18.1), the sum of net emissions flows $E_{ijt} - \eta_{ij}q_{ijt}$ at time $t \in [0, \infty)$ from production processes $i = 1, 2, \ldots, n$ in regions $j = 1, 2, \ldots, N$ accumulates to the global atmospheric concentration CO_2 stock M_t measured in ppm. $\xi_j > 0$ is the regional marginal atmospheric retention ratio and Ω the rate of assimilation.

$$dM = \left[\sum_{m=1}^{n}\sum_{k=1}^{N}\xi_j(E_{mkt} - \eta_{mk}q_{mkt}) - \Omega M_t\right] dt \tag{18.1}$$

$$dF = \alpha(\nu + \gamma M_t/M_0)dt + \beta\sigma\sqrt{M_t}[h_t dt + d\hat{B}] - \beta\sqrt{M_0}dt \tag{18.2}$$

$$dT = \lambda dF \tag{18.3}$$

$$\alpha = 4.841 \tag{18.4}$$

$$\beta = 0.0906 \tag{18.5}$$

$$\lambda = 0.5 \quad K/Wm^{-2} \tag{18.6}$$

The equation system (18.2) - (18.6) shows the atmospheric concentration rate M_t influence on radiative forcing dF (measured in Wm^{-2}) and global mean temperature T_t according to IPCC Climate Change 2001, The scientific basis, referring to [Shi (1992)] for calculating radiative forcing due to CO_2 where M_0 is the 1990 CO_2 concentration rate.[3] A change in radiative forcing affects the energy budget of the climate system and hence the global mean surface temperature through the relationship (18.3).

[Visser et al. (2000)] suggest that uncertainty in radiative forcing models is the greatest contributor to uncertainty in climate change predictions. Hence, we introduce scientific uncertainty about the climate model that goes beyond a stochastic process of radiative forcing by perturbing the model. The second term in (18.2) is multiplied by $h_t dt + d\hat{B}$ where dB is determined by the process

$$B_t = \hat{B}_t + \int_0^t h_s ds \tag{18.7}$$

and $d\hat{B}$ is the increment of the Wiener process $\hat{B}t$ on the probability space (Ξ, Φ, G) with variance $\sigma^2 \geq 0$. $\{\hat{B}_t : t \geq 0\}$ and $\{h_t : t \geq 0\}$ is a progressively measurable drift distortion, implying that the probability distribution itself is distorted such that the probability measure G is replaced be another unknown probability measure Q on the space (Ξ, Φ, Q). The drift term h_t represents different projections of future radiative forcing, and hence, global mean surface temperature T_t by (18.3). The projection of h_t is unknown as well as its probability distribution. Hence, (18.2) is interpreted as a set of radiative forcing models, one model for each h_t, with the restriction that the set of models is bounded by the constraint $h_t^2 \leq \Theta^2$.

18.2.1 *Climate Change Impacts*

The damages of the changes in global mean surface temperature are discussed in the IPCC Climate Change 2001, The scientific basis report. An overview of benefits and costs and further references are suggested by e.g. [Tol (2002a)] and [Tol (2002b)]. The considered impacts are often on natural capitals such as agriculture, forestry, water resources, sea level (loss of dry- and wetland), increased consumption of energy resources (heating and cooling), but also health (diseases and human heat and cold stress). Most research has been conducted on the effects of sea level rise e.g. [Titus and Narayan (1991)]. In this model we also focus on damages on natural capital losses that derive from a global mean temperature deviation $T_t - T_0$.

[3] For analytical tractability $\ln(M_t/M_0)$ in [Shi (1992)] is replaced by the linear approximation $\nu + \gamma M_t$ with parameters set to $\nu = -0.89179$ and $\gamma = 0.8974$.

All regions face the same global radiative forcing and climate sensitivity, resulting in a change in global mean surface temperature as for example in [Nordhaus and Yang (1996)], while benefits and costs may differ across regions $j = 1, 2, \ldots, N$.

Regional physical capital accumulation in (18.8) follows the structure of [Merton (1975)] and [Yeung (1995)] with a Cobb-Douglas investment function and depreciation rate $\delta_{ij} > 0$ where I_{ijt} is the investment by investor i in region j in period t

$$dk_{ij} = \left[I_{ijt}^{1/2} k_{ijt}^{1/2} - \delta_{ij} k_{ijt} \right] dt \tag{18.8}$$

$$dx_j = \left[r_j \left(1 - \frac{\sqrt{x_{jt}}}{K_j} \right) \sqrt{x_{jt}} - \frac{\Psi_j (T_t - T_0)}{\sqrt{x_{jt}}} x_{jt} \right] dt \tag{18.9}$$

$$i = 1, 2, \ldots, n \quad j = 1, 2, \ldots, N \tag{18.10}$$

The equations of motion of x_{jt} in (18.9) are adopted from [Hennlock (2005)] and consist of a modified natural growth function with intrinsic growth $r_j > 0$ and regional carrying capacity $\bar{x}_{jt} = K_j^2$. The loss of x_{jt} due to a deviation in global mean temperature rise $T_t - T_0$, where T_0 is the 1990 mean temperature level, is determined by a non-linear endogenous decay rate, given by the ratio $(T_t - T_0)\sqrt{x_{jt}}$, suggesting that the damage from a given mean temperature deviation accelerates as the stock x_{jt} decreases. (18.1) - (18.10) define the dynamic system with $2 + M(1 + n)$ state variables. The introduction of the unknown variable h_t in (18.2) implies that the dynamics of the system corresponds the set Θ^2 of radiative forcing models.

18.3 Players and Payoffs

There are two types of players in the model, investors $i = 1, 2, \ldots, n$ who are investing money in regional physical capital stocks k_{ij} located in regions $j = 1, 2, \ldots, N$. The investors are not physically tied to any specific region but allocate investments internationally between their capital stocks k_{ij} in regions $j = 1, 2, \ldots, N$. In each region j, there is a policy-maker, acting as a regional social planner and taking into account socio-economic interests in region j when enforcing *regionally* optimal emission reductions in period t for each regional industry i located in region j. The game is solved for the feedback Nash equilibrium, in which policy-makers and investors act independently of each other, possibly using asymmetric discount rates, benefit and cost structures subject to the dynamic models defined by the system (18.1)-(18.10).

18.3.1 *Investor* $i \in [1, n]$

Each investor $i \in [1, n]$ solves a stochastic optimization problem and allocates his total investment $\sum_{k=1}^{N} I_{ikt}$ in period t between his production processes located in all regions $j \in [1, N]$ for the production of a good y that is sold on the world market at unit price. Thus total industry production in region j is $y_{jt} = \sum_{m=1}^{n} \phi_{mj} k_{mjt}^{1/2}$, where $\phi_{ij} > 0$ is a industry-specific technology parameter. Production generates regional emissions flow $E_{jt} = \sum_{m=1}^{n} \varphi_{mj} y_{mjt}$ where $\varphi_{mj} > 0$ is an industry-specific pollution parameter. Profit-maximization by each investor i is achieved by allocating his investment (and thereby production activity) between the regions $j \in [1, N]$. The expected payoff of investor i, where ϵ is the expectation operator, is

$$\max_{u_{ikt}} \epsilon \int_0^{\infty} \sum_{k=1}^{N} \left\{ p(1 - u_{ikt}) y_{ikt} - \frac{c_{ik} (q_{ikt})^2}{E_{ikt}} \right\} e^{-\rho_i t} dt \tag{18.11}$$

Investor i seeks the optimal cash dividend from each regional business $j \in [1, N]$ in each period by controlling the share $u_{ijt} \in [0, 1]$ of revenue that is reinvested in his regional capital stocks k_{ij}. By investing I_{ijt} investor i contributes to total industry output y_{jt} in region j, which is $y_{jt} = \sum_{m=1}^{n} \phi_{mj} k_{mjt}^{1/2}$. The amount reinvested I_{ijt} in period t by investor i is the remainder $I_{ijt} = u_{ijt} y_{ijt}$. Investor i's discount rate is $\rho_i > 0$.

The last term in (18.11) is firm i's abatement cost, which is quadratic in regional abatement activity q_{ijt} due to capacity constraints as more local abatement effort q_{ijt} is employed. Abatement cost is decreasing in E_{ijt}, suggesting that it requires more expensive techniques as E_{ijt} becomes smaller. $c_{ij} > 0$ is an abatement cost parameter in production i in region j. The total emissions flow from region j is $E_{jt} = \sum_{m=1}^{n} E_{mjt} = \sum_{m=1}^{n} \varphi_{mj} y_{mjt}$. The industry-specific level of q_{ijt}^{o} is set by policy maker j and is taken as given when investor i seeks optimal investment u_{ikt}. We consider the following game rule:

Definition 1. Regional Policy Enforcement Policy maker j can enforce $\eta_{ij} q_{ijt}^{o}$ in region j such that every investor $i \in [1, n]$ investing in regional capital k_{ij} in region j has to take regional abatement command $q_{ijt}^{o} \geq 0$ as given when choosing u_{ijt}^{*} in a robust feedback Nash equilibrium.

Investor i maximizes the payoff function (18.11) subject to the dynamic system (18.1) to (18.10) and $q_{ijt}^{o} \geq \forall i \in [1, n]$ and $\forall j \in [1, N]$ in definition 1. Investor i's stochastic optimal problem is written as:

$$\max_{u_{ikt}} \epsilon \int_0^\infty \sum_{k=1}^{N} \left\{ (1-u_{ikt})y_{ikt} - \frac{c_{ik}(q^o_{ikt})^2}{E_{ikt}} \right\} e^{-\rho_i t} dt \tag{18.12}$$

subject to

$$dk_{ij} = \left[I_{ijt}^{1/2} k_{ijt}^{1/2} - \delta_{ij} k_{ijt} \right] dt \tag{18.13}$$

$$dx_j = \left[r_j \left(1 - \frac{\sqrt{x_{jt}}}{K_j} \right) \sqrt{x_{jt}} - \frac{\Psi_j(T_t - T_0)}{\sqrt{x_{jt}}} x_{jt} \right] dt \tag{18.14}$$

$$dM = \left[\sum_{m=1}^{n} \sum_{k=1}^{N} \xi_j (E_{mkt} - \eta_{mk} q_{mkt}) - \Omega M_t \right] dt \tag{18.15}$$

$$dF = \alpha(\nu + \gamma M_t / M_0)dt + \beta\sigma\sqrt{M_t}[h_{jt}dt + d\hat{B}] - \beta\sqrt{M_0}dt \tag{18.16}$$

$$dT = \lambda dF \tag{18.17}$$

$$i = 1, 2, \ldots, n \quad j = 1, 2, \ldots, N \tag{18.18}$$

Definition 2. *If there exist n value functions $V_i(\boldsymbol{k}, \boldsymbol{x}, M, T, t)$ where*

$$\boldsymbol{k} = (k_{11}, k_{12}, \ldots, k_{1N}, k_{21}, k_{22}, \ldots, k_{2N}, k_{n1}, k_{n2}, \ldots, k_{nN}) \tag{18.19}$$

and $\boldsymbol{x} = (x_1, x_2, \ldots, x_N)$ that satisfy

$$V_i(\boldsymbol{k}, \boldsymbol{x}, M, T, t) = \tag{18.20}$$

$$\epsilon \int_0^\infty \sum_{k=1}^{N} \left\{ (1-u^*_{ikt})y_{ikt} - \frac{c_{ik}(q^o_{ikt})^2}{E_{ikt}} \right\} e^{-\rho_i t} dt$$

$$\geq \epsilon \int_0^\infty \sum_{k=1}^{N} \left\{ (1-u_{ikt})y_{ikt} - \frac{c_{ik}(q^o_{ikt})^2}{E_{ikt}} \right\} e^{-\rho_i t} dt$$

for strategies $u^*_{ijt}(k_j, t) \subseteq R^1 \quad \forall i \in [1, n]$ and $\forall j \in [1, N]$ which satisfy the state equations,

$$dk_{ij} = \left[I_{ijt}^{1/2} k_{ijt}^{1/2} - \delta_{ij} k_{ijt} \right] dt \tag{18.21}$$

$$dx_j = \left[r_j \left(1 - \frac{\sqrt{x_{jt}}}{K_j} \right) \sqrt{x_{jt}} - \frac{\Psi_j(T_t - T_0)}{\sqrt{x_{jt}}} x_{jt} \right] dt \tag{18.22}$$

$$dM = \left[\sum_{m=1}^{n} \sum_{k=1}^{N} \xi_j (E_{mk} - \eta_{mk} q^o_{mkt}) - \Omega M_t \right] dt \tag{18.23}$$

$$dF = \alpha(\nu + \gamma M_t/M_0)dt + \beta\sigma\sqrt{M_t}[h_{jt}dt + d\hat{B}] - \beta\sqrt{M_0}dt \quad (18.24)$$

$$dT = \lambda dF \quad (18.25)$$

$$i = 1, 2, \ldots, n \quad j = 1, 2, \ldots, N \quad (18.26)$$

The feedback Nash controls strategies

$$\Gamma^*_{ijt} = \{u^*_{ijt}(k_{ijt})\} \quad \forall i \in [1, n] \quad \forall j \in [1, N] \quad (18.27)$$

provide a feedback Nash equilibrium solution of the game defined in (18.12) to (18.18) given (18.37) to (18.43) [Basar and Olsder (1999)].

The value functions in definition 2 satisfy the partial differential equation system (18.28) - (18.29). Using (18.12) - (18.18) and definition 1 and 2 yield the dynamic programming problem [Fleming and Richel (1975)] of investor i

$$-\frac{\partial V_i}{\partial t} = \max_{u_{ikt}} \sum_{k=1}^{N} \left\{ (1 - u_{ikt})y_{ikt} - \frac{c_{ik}(q^o_{ikt})^2}{E_{ikt}} \right\} e^{-\rho_i t} \quad (18.28)$$

$$+ \sum_{m=1}^{n} \sum_{k=1}^{N} \frac{\partial V_i}{\partial k_{mk}} \left[I^{1/2}_{mkt} k^{1/2}_{mkt} - \delta_{mk} k_{mkt} \right]$$

$$+ \sum_{k=1}^{N} \frac{\partial V_i}{\partial x_k} \left[r_k \left(1 - \frac{\sqrt{x_{kt}}}{K_k} \right) \sqrt{x_{kt}} - \frac{\Psi_k(T_t - T_0)}{\sqrt{x_{kt}}} x_{kt} \right]$$

$$+ \frac{\partial V_i}{\partial M} \left[\sum_{m=1}^{n} \sum_{k=1}^{N} \xi_j (E_{mkt} - \eta_{mk} q^o_{mkt}) - \Omega M_t \right]$$

$$+ \frac{\partial V_i}{\partial T} \lambda[\alpha(\nu + \gamma M_t/M_0) + \beta\sigma\sqrt{M_t}h_t - \beta\sqrt{M_0}] + \frac{1}{2}\frac{\partial^2 V_i}{\partial T^2}\sigma^2 M_t$$

$$i = 1, 2, \ldots, n \quad j = 1, 2, \ldots, N \quad (18.29)$$

The feedback Nash controls strategies

$$\Gamma^*_{ijt} = \{u^*_{ij}(k_{ijt})\} \quad \forall i \in [1, n] \quad \text{and} \quad \forall j \in [1, N] \quad (18.30)$$

are given by maximizing the partial differential equations (18.28) with respect to (18.30) for n players and solving for the feedback Nash control variables.

$$u^*_{ijt} = \left(\frac{1}{2} \frac{\partial V_i}{\partial k_{ij}} \right)^2 \frac{k^{1/2}_{ijt}}{\phi_{ij} e^{-2\rho_i t}} \quad \forall i \in [1, n] \quad \text{and} \quad \forall j \in [1, N] \quad (18.31)$$

The partials derivatives in (18.31) are investor i's expected feedback Nash shadow price of k_{ij}. In order to identify shadow price paths, n value functions $V_i(\boldsymbol{k}, \boldsymbol{x}, M, T, t)$ that satisfy definition 2 and the partial differential equation system formed by (18.28) - (18.29) must be identified.

Proposition 1.

$$V_i(\boldsymbol{k}, \boldsymbol{x}, M, T, t) = \left(\sum_{m=1}^{n} \sum_{k=1}^{N} a_{imk} k_{mk}^{\frac{1}{2}} + \sum_{k=1}^{N} b_{ik} x_k^{\frac{1}{2}} + d_i M + g_i T + m_i \right) e^{-\rho_i t} \tag{18.32}$$

The value functions $\forall i \in [1, n]$ satisfy definition 2 and the partial differential equation system formed by system (18.28) - (18.29).

Proof: Appendix A.1.

Substituting the feedback Nash shadow prices and costs into (18.31) yields the feedback strategies in terms of parameter values, where the values of the undetermined coefficients $(\boldsymbol{a}_{iij}, \boldsymbol{b}_{ij}, d_i, g_i, m_i)$ for all investors $i \in [1, n]$ and regions $j \in [1, N]$ are determined in Appendix A.1.

$$u_{ijt}^{*} = \left(\frac{a_{ij}}{4} \right) \frac{1}{\phi_{ij} k_{ijt}^{1/2}} \in [0, 1] \quad \forall i \in [1, n] \quad \text{and} \quad \forall j \in [1, N] \tag{18.33}$$

Investor i's feedback Nash investment rate u_{ijt}^{*} is decreasing in k_{ijt}, implying that the share of revenue used for investment is large when k_{ijt} is low during business start up. As k_{ijt} grows, investor i will increase cash dividend and reduce the share of revenue reinvested in his capital k_{ij} located in region j. From Appendix A.1 follows that a greater investor's discount rate ρ_i and capital deprecation rate δ_{ij} of k_{ijt} in region j, the greater is the share of revenue that the investor withdraw as cash dividend in each period and the lower is the share used for reinvestment in k_{ijt}.

18.3.2 *Policy Maker $j \in [1, N]$*

The policy makers $j = 1, 2, \ldots, N$ faces a region-specific loss of natural capital x_{jt} due to climate change $T_t - T_0$ and seek to find optimal abatement commands $\eta_{ij} q_{ijt}^{o}$, for the industries within region j while the investors $i = 1, 2, \ldots, n$ choose regional investments u_{ijt}^{*}.[4] In the Nash equilibrium,

[4]In the case of auctioned permits it is straightforward to show that the investor's shadow price of k_{jt} would fall for given levels of k_{jt} leading to a reduction in investment.

each policy maker $j \in [1, N]$ seeks optimal abatement commands $\eta_{ij} q^o_{ijt}$ to each industry i in region j given that the remaining $N-1$ policy makers individually seek the optimal $\eta_{ik} q^o_{ikt}$ $\forall k \neq j$ and that every investor $i \in [1, n]$ individually seek optimal u^*_{ijt}. The expected payoff of policy maker $j \in [1, N]$ is

$$\epsilon \int_0^\infty \sum_{m=1}^n \left\{ \omega_j y_{mjt} + \psi_j x_{jt}^{1/2} - \frac{c_{mj} q^2_{mjt}}{E_{mjt}} \right\} e^{-\rho_j t} dt + \theta_j R(Q) \quad (18.34)$$

The first term is social benefit of employment that is assumed to be proportional to total regional production $y_{mjt} = \phi_{mj} k^{1/2}_{mjt}$ where the parameter $\omega_j > 0$. The second term is the benefit from the regional natural capital x_{jt} in region $j \in [1, N]$ with the parameter $\psi_j > 0$. The last term within the brackets is the industry-specific abatement cost function. Policy maker j's discount rate is $\rho_j > 0$.

Following [Hansen, Sargent, Turmuhambetova and Williams (2001)], policy maker j's payoff function can be written as (18.34) in a multiplier robust problem where $1/\theta_j \geq 0$ denotes the policy maker's preference for robustness when $\{h_s\}$ in (18.7) is unknown. The actual evolution of h_s will change the future probability distribution of B_t having probability measure Q relative to the distribution of $\hat{B}_t$ having measure G. The Kullback-Leibler distance between Q and G is

$$R(Q) = \int_0^\infty \epsilon_Q \left(\frac{|h_s|^2}{2} \right) e^{-\rho_j t} ds \quad (18.35)$$

As long as $R(Q) < \Theta_j$ in (18.34) is finite

$$Q \left\{ \int_0^t |h_s|^2 ds < \infty \right\} = 1 \quad (18.36)$$

which has the property that Q is locally continuous with respect to G, implying that G and Q cannot be distinguished with finite data, and hence, modeling a situation with a decision maker that cannot know the future probability distribution when using current data.

Every policy maker $j \in [1, N]$ seeks for a robust industry-specific cost-efficient amount of abatement $\eta_{ij} q^o_{ijt}$ for each industry i given the Nash investment decisions $I^*_{ijt} \equiv u^*_{ijt} \phi_{ij} k^{1/2}_{ijt}$ by investor $i = 1, 2, \ldots, n$:

$$\max_{q_{jt}} \min_{h_{jt}} \epsilon \int_0^\infty \sum_{m=1}^n \left\{ \omega_j y_{mjt} + \psi_j x_{jt}^{1/2} - \frac{c_{mj} q^2_{mjt}}{E_{mjt}} + \frac{\theta_j h_t^2}{2} \right\} e^{-\rho_j t} dt \quad (18.37)$$

subject to

$$dk_{ij} = \left[(I^*_{ijt})^{1/2} k_{ijt}^{1/2} - \delta_{ij} k_{ijt}\right] dt \tag{18.38}$$

$$dx_j = \left[r_j \left(1 - \frac{\sqrt{x_{jt}}}{K_j}\right) \sqrt{x_{jt}} - \frac{\Psi_j(T_t - T_0)}{\sqrt{x_{jt}}} x_{jt}\right] dt \tag{18.39}$$

$$dM = \left[\sum_{m=1}^{n} \sum_{k=1}^{N} \xi_j (E_{mkt} - \eta_{mk} q_{mkt}) - \Omega M_t\right] dt \tag{18.40}$$

$$dF = \lambda\alpha(\nu + \gamma M_t/M_0)dt + \beta\sigma\sqrt{M_t}[h_{jt}dt + d\hat{B}] - \beta\sqrt{M_0}dt \tag{18.41}$$

$$dT = \lambda dF \tag{18.42}$$

$$i = 1, 2, \ldots, n \quad j = 1, 2, \ldots, N \tag{18.43}$$

Definition 3. Robust Feedback Nash Equilibrium *If there exist N value functions $W_j(\boldsymbol{k}, \boldsymbol{x}, M, T, t)$ where*

$$\boldsymbol{k} = (k_{11}, k_{12}, \ldots, k_{1N}, k_{21}, k_{22}, \ldots, k_{2N}, k_{n1}, k_{n2}, \ldots, k_{nN}) \tag{18.44}$$

and $\boldsymbol{x} = (x_1, x_2, \ldots, x_N)$ that satisfy

$$W_j(\boldsymbol{k}, \boldsymbol{x}, M, T, t) = \tag{18.45}$$

$$\epsilon \int_0^\infty \sum_{m=1}^{n} \left\{\omega_j y_{mjt} + \psi_j x_{jt}^{1/2} - \frac{c_{mj}(q^o_{mjt})^2}{E_{mjt}} + \frac{\theta_j (h^o_{jt})^2}{2}\right\} e^{-\rho_j t} dt$$

$$\geq \epsilon \int_0^\infty \sum_{m=1}^{n} \left\{\omega_j y_{mjt} + \psi_j x_{jt}^{1/2} - \frac{c_{mj} q^2_{mjt}}{E_{mjt}} + \frac{\theta_j h^2_{jt}}{2}\right\} e^{-\rho_j t} dt$$

for strategies $q^o_{jt}(k_j, t) \subseteq R^1$ and $h^o_{jt}(M, t) \subseteq R^1$ given that $h^o_{jt}(M, t) \equiv$ $\arg\min W_j(\boldsymbol{k}, \boldsymbol{x}, \boldsymbol{L}, M, T, t)$ $\forall j \in N$ and which satisfy the state equations,

$$dk_{ij} = \left[(I^*_{ijt})^{1/2} k_{ijt}^{1/2} - \delta_{ij} k_{ijt}\right] dt \tag{18.46}$$

$$dx_j = \left[r_j \left(1 - \frac{\sqrt{x_{jt}}}{K_j}\right) \sqrt{x_{jt}} - \frac{\Psi_j(T_t - T_0)}{\sqrt{x_{jt}}} x_{jt}\right] dt \tag{18.47}$$

$$dM = \left[\sum_{m=1}^{n} \sum_{k=1}^{N} \xi_j (E_{mkt} - \eta_{mk} q^o_{mkt}) - \Omega M_t\right] dt \tag{18.48}$$

$$dF = \alpha(\nu + \gamma M_t/M_0)dt + \beta\sigma\sqrt{M_t}[h_t dt + d\hat{B}] - \beta\sqrt{M_0}dt \tag{18.49}$$

$$dT = \lambda dF \tag{18.50}$$

$$i = 1, 2, \ldots, n \quad j = 1, 2, \ldots, N \tag{18.51}$$

The feedback Nash controls strategies

$$\Gamma^o_{ijt} = \{q^o_{ij}(k_{ij}), h^o_j(k_j)\} \quad \forall i \in [1, n] \quad \forall j \in [1, N] \tag{18.52}$$

provide a robust feedback Nash equilibrium solution of the game defined in (18.37) to (18.43) given (18.12) to (18.18)[Basar and Olsder (1999)].

The value functions in definition 3 satisfy the partial differential equation system (18.53) - (18.54). Using (18.12) - (18.18) and (18.37) - (18.43) and definition 2 and 3 yield the Isaacs-Bellman-Fleming equation [Fleming and Richel (1975)] of policy maker j:

$$-\frac{\partial W_j}{\partial t} = \tag{18.53}$$

$$\max_{q_{jt}} \min_{h_{jt}} \sum_{m=1}^{n} \left\{ \omega_j y_{mjt} + \psi_j x_{jt}^{1/2} - \frac{c_{mj} q_{mjt}^2}{E_{mjt}} + \frac{\theta_j h_{jt}^2}{2} \right\} e^{-\rho_j t}$$
$$+ \sum_{m=1}^{n} \sum_{k=1}^{N} \frac{\partial W_j}{\partial k_{mk}} \left[(I_{mkt}^*)^{1/2} k_{mkt}^{1/2} - \delta_{mk} k_{mkt} \right]$$
$$+ \sum_{k=1}^{N} \frac{\partial W_j}{\partial x_k} \left[r_k \left(1 - \frac{\sqrt{x_{kt}}}{K_k} \right) \sqrt{x_{kt}} - \frac{\Psi_k (T_t - T_0)}{\sqrt{x_{kt}}} x_{kt} \right]$$
$$+ \frac{\partial W_j}{\partial M} \left[\sum_{m=1}^{n} \sum_{k=1}^{N} \xi_j (E_{mkt} - \eta_{mk} q_{mkt}) - \Omega M_t \right]$$
$$+ \frac{\partial W_j}{\partial T} \lambda [\alpha(\nu + \gamma M_t / M_0) + \beta \sigma \sqrt{M_t} h_{jt} - \beta \sqrt{M_0}]$$
$$+ \frac{1}{2} \frac{\partial^2 W_j}{\partial T^2} \sigma^2 M_t$$

$$i = 1, 2, \ldots, n \quad j = 1, 2, \ldots, N \tag{18.54}$$

The robust feedback Nash controls strategies

$$\Gamma_{ijt}^o = \{ q_{ij}^o(k_{ij}), h_j^o(M_t) \} \quad \forall i \in [1, n] \quad \forall j \in [1, N] \tag{18.55}$$

are given by maximizing the partial differential equations (18.53) with respect to (18.55) for the N policymakers and solving for the robust control variables.

$$q_{ijt}^o = -\frac{\partial W_j}{\partial M} \frac{\xi_j \eta_{ij}}{2 c_{ij} e^{-\rho_j t}} E_{ijt} \tag{18.56}$$

$$h_{jt}^o = -\frac{\partial W_j}{\partial T} \frac{\lambda \beta \sigma}{\theta_j e^{-\rho_j t}} M_t^{1/2} \tag{18.57}$$

$$i = 1, 2, \ldots, n \quad j = 1, 2, \ldots, N \tag{18.58}$$

The partials derivatives in (18.56) and (18.57) are policy makers j's expected Nash shadow price shadow cost of concentration rate CO_2 and global mean temperature, respectively. The optimal abatement command $\eta_{ij} q_{ijt}^o$

is proportional to policy maker j's shadow cost of pollution and E_{ijt}. As expected, the optimal abatement effort is $q^o_{ijt} \geq 0$ for all t since $\partial W_j / \partial M_t \leq 0$ $\forall t$.

The robust feedback Nash strategies in definition 3 are subgame perfect *strategies* and will therefore be responses to the evolution of state variables. These properties of the players' controls, make the robust Nash abatement controls credible and efficient threats in every subgame starting at $t < \infty$, given the policy makers' preferences for robustness.

The N value functions $W_j(\boldsymbol{k}, \boldsymbol{x}, M, T, t)$ that satisfy definition 3 and the partial differential equation system formed by (18.53) - (18.54) must be identified in order to identify shadow prices.

Proposition 2.

$$W_j(\boldsymbol{k}, \boldsymbol{x}, M, T, t) = \tag{18.59}$$

$$\left(\sum_{m=1}^{n} \sum_{k=1}^{N} a_{jmk} k_{mk}^{\frac{1}{2}} + \sum_{k=1}^{N} b_{jk} x_k^{\frac{1}{2}} + d_j M + g_j T + m_j \right) e^{-\rho_j t}$$

The value functions $\forall j \in [1, N]$ satisfy definition 3 and the partial differential equation system formed by system (18.53) - (18.54).

Proof: Appendix A.2.
Substituting the robust Nash equilibrium shadow costs into (18.56) - (18.57) yields the robust feedback Nash equilibrium strategies in terms of parameter values. The values of undetermined coefficients $(\boldsymbol{a}_{jjj}, \boldsymbol{b}_{jj}, d_j, g_j, m_j)$ for all $j \in [1, N]$ are derived in Appendix A.2.

$$q^o_{ijt} = -d_j \frac{\xi_j \eta_{ij}}{2 c_{ij} e^{-\rho t}} E_{ijt} \geq 0 \tag{18.60}$$

$$h^o_{jt} = -g_j \frac{\lambda \beta \sigma}{\theta_j} M_t^{1/2} \geq 0 \tag{18.61}$$

$$i = 1, 2, \ldots, n \quad j = 1, 2, \ldots, N \tag{18.62}$$

18.3.3 *Game Formulations*

Several game formulations are possible. In the previous sections, every investor and every policy makers optimize individually solving for a robust feedback Nash equilibrium involving investors' Nash investment strategies u^*_{ijt} and policy makers' robust Nash strategies in abatement commands

$q_{ijt}^o \geq 0$.[5] Another interesting formulation is to let policy makers $j \in [1, N]$ cooperate or form coalitions while the n investors still play Nash investment strategies. This and other cooperative structures are left for further studies.

18.4 Concluding Comments

We have defined an analytical solution to a robust feedback Nash equilibrium with $n + N$ asymmetric players with non-linear feedback (Markov) Nash strategies based on the IPCC 2001 climate model assumptions about radiative forcing, adding perturbation, making it impossible for policy makers to infer correct future probability distribution about climate variables using current data. Policy makers are then assumed to play robust strategies. The advantage of an analytical solution is that it allows for a deeper understanding than numerical simulations. However, to explore all dimensions of this analytical asymmetric solution would be too extensive in this paper and is left to forthcoming studies. Some clarifying features and conclusions though are mentioned here.

Within each region j there is a second-best solution between investors $i = 1, 2, \ldots, n$ and policy maker j. The policy maker j's social benefit of employment in region j in (18.34) is not embedded in investor i's Nash shadow price of k_{ij}. This contributes to make Nash investment controls u_{ijt}^* too small compared to a *regional* Pareto optimal investment control u_{ijt}^o. This effect is counteracted by the fact that investor i does not consider the negative effects on natural capital x_{jt} in region j, which make investment in region j too big compared to the *regional* Pareto optimal level as investor i's Nash shadow price of capital k_{ij} is too high compared to the *regional* Pareto optimal shadow price of k_{ij}. This externality is internalized to a second-best solution by the enforcement statement in definition 1. Another dimension of externality derives from the asymmetry between the uncertainty neutral investors and the (more or less) uncertainty averse policy makers. An uncertainty neutral investor i has a greater expected Nash shadow price of capital k_{ij} than an uncertainty averse policy maker *ceteris paribus*. Applying the enforcement statement in definition 1, internalizes this effect to a second-best solution at the *regional* level as the policy maker sets optimal abatement commands taking into account her uncertainty aversion. Finally, moving across borders, investor i's invest-

[5] Technically, it is straightforward to also let investors be uncertainty averse in this equilibrium.

ment u^*_{ijt} in region j, as well as policy maker j's abatement command q^o_{ijt} in region j, do not take into account the negative effects on foreign natural capital in regions $k \in [1, N], k \neq j$. This conveys to an externality between policy makers as globally Pareto optimal choices are unenforced.

Proposition 3. *If investor i's discount rate ρ_i increases, then i's expected Nash shadow prices fall for all k_{ij} $\forall j \in [1, N]$, resulting in an increase in i's feedback Nash cash dividend flows and corresponding decrease in i's feedback Nash investment rate strategies u^*_{ijt} at given k_{ij} levels in all regions $j \in [1, N]$.*

Proof: (18.33) and Appendix A.1.

Proposition 4. *If production efficiency ϕ_{ij} of k_{ij} in region j increases and/or capital depreciation rate δ_{ij} in region j decreases, then investor i's expected Nash shadow price of k_{ij} increases, resulting in an increase in i's feedback Nash investment rate strategy u^*_{ijt} at given k_{ij} levels in region $j \in [1, N]$.*

Proof: (18.33) and Appendix A.1.

Proposition 5. *If abatement cost c_{ij} in region j decreases, then investor i's expected Nash shadow price of k_{ij} decreases, resulting in an decrease in i's feedback Nash investment rate strategy u^*_{ijt} at given k_{ij} levels in region $j \in [1, N]$.*

Proof: (18.33) and Appendix A.1.

Proposition 6. *An uncertainty averse ($\theta_j << \infty$) policy maker j faces a greater expected Nash shadow cost of M compared to an uncertainty neutral policy maker ($\theta_j \rightarrow \infty$), using the expected utility decision criterion in the feedback Nash equilibrium. As $\theta_j \rightarrow 0$, policy maker j's expected Nash shadow cost of M_t increases toward infinity and also becomes highly sensitive (quadratic) to radiative forcing dF and climate sensitivity parameters λ and parameters α and β in the climate model, resulting in significant increases in robust feedback Nash abatement command strategies q^o_{ijt} $\forall \in [1, n]$.*

Proof: (18.60) and Appendix A.2.

Proposition 7. *The abatement cost burden due to $q^o_{ijt} \geq 0$ lowers investor i's expected Nash shadow price of capital k_{ij} in region j for given k_{ij} levels and hence feedback Nash investment rate u^*_{ijt} falls by (18.33).*

Proof: (18.33) and Appendix A.1.

Proposition 8. *If marginal regional atmospheric retention ratio ξ_j and/or pollution parameter φ_{ij} increases, then policy maker j increases robust Nash abatement command strategies q^o_{ijt} at given k_{ij} levels.*

Proof: (18.60) and Appendix A.2.

Proposition 9. *If the climate sensitivity parameter λ increases ceteris paribus, then all policy makers $j \in [1, N]$ increase robust feedback Nash abatement commands q^o_{ijt} at given k_{ij} levels, each policy maker j with a magnitude that corresponds to their specific benefits, costs, parameters and capital endowments.*

Proof: (18.60) and Appendix A.2.

Proposition 10. *If social utility parameter ψ_j of natural capital x_{jt} increases, then policy maker j's expected Nash shadow cost of concentration rate M_t increases ceteris paribus, which increases robust feedback Nash abatement command strategies q^o_{ijt} at given k_{ij} levels.*

Proof: (18.60) and Appendix A.2.

Proposition 11. *If abatement efficiency η_{ij} increases, then policy maker $j \in [1, N]$ increases robust feedback Nash abatement command strategies q^o_{ijt} at given k_{ij} levels.*

Proof: (18.60) and Appendix A.2.

Another conclusive result that is strengthened by the introduction of preferences for robustness among policy makers is the possibility of a positive shadow price of foreign polluting physical capital k_{ik} in region j, which according to (18.81) in Appendix A.2 occurs when $2c_{ik} + d_k\xi_k\eta_{ik} < 0$. This has conclusive results when introducing transfer to the model as it may induce policy makers in all regions $j \neq k$ to give transfers to region $k \neq j$ that *increases CO_2 generating capital in region k also in a feedback Nash equilibrium.*

Proposition 12. *If $2c_{ik} + d_k\xi_k\eta_{ik} < 0$ holds in region $k \neq j$ then policy makers in all other regions $j \neq k$ face a positive expected Nash shadow cost $\partial W_j/\partial k_{ijk}$ of physical capital stock k_{ik} where $i = 1, 2, \ldots, n$ in region k.*

Proof: (18.81) in Appendix A.2.

It follows from (18.81) that increased robustness of the policy maker's preferences in region k increases this effect. The intuition is that a robust policy maker k, provided that climate model parameters λ, α, β and σ, are sufficiently high, eventually will use the future increase in income from a greater current k_{ikt} to reduce future net emissions more than if current k_{ikt} was smaller. Using (18.81) - (18.86) it follows that this situation occurs more likely if policy maker k has a strong preference for robustness, is non-myopic and hosts a region k endowed with a natural capital stock with big natural carrying capacity $\bar{x}_k$, that faces great damage Ψ_k from climate change $T_t - T_0$ and has attracted industries with low abatement costs c_{ik} relative to the size of physical parameter values λ, α, β and σ that increase climate sensitivity and radiative forcing for given CO_2 concentration rates in the climate model.

18.5 Summary

The paper has presented an analytically tractable feedback Nash equilibrium with corresponding robust nonlinear feedback Nash strategies and $n + N$ asymmetric players subject to IPCC 2001 radiative forcing model assumptions. Finding analytically tractable solutions to non-linear feedback Nash equilibria beyond linear and quadratic games with many asymmetric players is extremely rare. The analytical solution in this paper opens up for analyzes of different game formulations with asymmetric players using the radiative forcing simple expressions suggested by IPCC Climate Change 2001, The scientific basis, referring to [Shi (1992)] for calculating radiative forcing due to CO_2. Further sensitivity analysis and simulations of the dynamics are left to future studies.

We have shown that a robust policy maker faces a greater expected robust feedback Nash shadow cost of atmospheric CO_2 compared to the feedback Nash expected utility decision criterion. Moreover, the stronger the preference for robustness the policy maker's expected feedback Nash shadow cost becomes highly sensitive to scientific parameters of radiative forcing and climate sensitivity in the climate model. The greater robust feedback Nash shadow cost of CO_2 rate, the greater are robust feedback Nash abatement command levels compared to the levels in the expected utility decision criterion under feedback Nash conjectures. The greater abatement command levels by uncertainty averse policy makers results in a fall in investors' expected Nash shadow price of physical capital. Hence, investors' feedback Nash responses are to decrease reinvestments (by increasing current cash dividends) in regions with uncertainty-averse policy makers. As expected, physical capital tends to move away from regions which host policy makers with stronger preference for robustness in abatement policy.

On the other hand, the introduction of robustness strengthens the possibility that a policy maker in region $j \neq k$, also in a feedback Nash equilibrium, may face a positive shadow price of foreign polluting physical capitals k_{ik} in region $k \neq j$ which would induce policy maker j to give transfer that increases physical capital in region k. This effect is strengthened also in a robust Nash equilibrium where policy makers have strong preferences for robustness, are non-myopic and host regions endowed with a natural capital stock with big natural carrying capacity and have industries with low abatement costs relative to the magnitude of physical parameter values that increases radiative forcing and climate sensitivity in the climate model.

18.6 Appendix

18.6.1 *Appendix A.1 Proof of Proposition 1*

Substituting (18.33) into the partial differential equations (18.28) for all $i \in n$ forms the n indirect Isaacs-Bellman-Fleming equations of investors $i = 1, 2, \ldots, n$. The coefficients of the indirect values functions in proposition 1 are then determined by the block recursive equation system

$$\rho_i a_{iij} = \phi_{ij} - \frac{(d_j \xi_j \eta_{ij})^2 \varphi_{ij} \phi_{ij}}{4c_j} - \frac{a_{iij}}{2} \delta_{ij} + d_i \xi_j \varphi_{ij} \phi_{ij} + d_i \xi_j \eta_{ij} \frac{d_j \xi_j \eta_{ij} \varphi_{ij} \phi_{ij}}{2c_{ij}} \tag{18.63}$$

$$\rho_i a_{imj} = -\frac{a_{imj}}{2} \delta_{mj} + d_i \xi_j \varphi_{mj} \phi_{mj} + d_i \xi_j \eta_{mj} \frac{d_j \xi_j \eta_{mj} \varphi_{mj} \phi_{mj}}{2c_{mj}} \quad \forall m \neq i \tag{18.64}$$

$$\rho_i b_{ij} = -\frac{b_{ij} \rho_i}{2K_j} \tag{18.65}$$

$$\rho_i d_i = \frac{g_i \lambda \alpha \gamma}{M_0} - d_i \Omega - g_i \lambda \beta \sigma \frac{g_j \lambda \beta \sigma}{\theta_j} \tag{18.66}$$

$$\rho_i g_i = -\sum_{j=1}^{N} b_{ij} \Phi_j \tag{18.67}$$

$$a_{iij} = \frac{\phi_i \left(1 - d_i \xi_k \varphi_{ij} + \frac{(\xi_k \eta_{ij})^2 \varphi_{ij} \phi_{ij}}{c_{ij}} \left(\frac{d_i d_k}{2} - \frac{(-d_k)^2}{4}\right)\right)}{\rho_i + \delta_{ij}/2} \tag{18.68}$$

$$a_{imj} = d_i \xi_k \varphi_{mj} \phi_{mj} \left(1 + \frac{d_k \xi_k \eta_{mj}^2}{2c_{mj}}\right) \quad \forall m \neq i \tag{18.69}$$

$$b_{ij} = 0 \tag{18.70}$$

$$d_i = \frac{g_i \lambda \left(\frac{\alpha\gamma}{M_0} - \frac{g_j \lambda (\beta\sigma)^2}{\theta_j}\right)}{\rho_i + \Omega} \tag{18.71}$$

$$g_i = -\sum_{j=1}^{N} \frac{b_{ij} \Phi_j}{2\rho_i} \tag{18.72}$$

$$i = 1, 2, \ldots, n \quad j = 1, 2, \ldots, N \quad m \in [1, n] \tag{18.73}$$

Since $b_{ij} = 0$, then $a_{imj} = 0$, $d_i = 0$ and $g_i = 0$, which further simplify (18.68). The coefficient m_i in proposition 1 is uniquely determined by the coefficients in (18.68) - (18.73) and (18.80) - (18.86).

18.6.2 *Appendix A.2 Proof of Proposition 2*

Substituting (18.60) - (18.61) into the partial differential equations (18.53) for all $j \in H$ forms the N indirect Isaacs-Bellman-Fleming equations of policy makers $j = 1, 2, \ldots, N$. The coefficients of the indirect values functions in proposition 2 are then

$$\rho_j a_{jij} = \omega_j \phi_{ij} + \frac{(d_j \xi_j \eta_{ij})^2 \varphi_{ij} \phi_{jj}}{2c_{ij}} - \frac{a_{jij} \delta_{ij}}{2} + d_j \xi_j \varphi_{ij} \phi_{ij} \tag{18.74}$$

$$\rho_j a_{jik} = -\frac{a_{jik} \delta_{ik}}{2} + d_j \xi_k \varphi_{ik} \phi_{ik} \left(1 + \frac{d_k \xi_k \eta_{ik}^2}{2c_{ik}}\right) \quad k \neq j \tag{18.75}$$

$$\rho_j b_{jj} = \psi_j - \frac{b_{jj} r_j}{2K_j} \tag{18.76}$$

$$\rho_j b_{jk} = -\frac{b_{jk} r_k}{2K_k} \quad k \neq j \tag{18.77}$$

$$\rho_j d_j = -\frac{(g_j \lambda \beta \sigma)^2}{2\theta_j} + \frac{g_j \lambda \alpha \gamma}{M_0} - d_j \Omega \tag{18.78}$$

$$\rho_j g_j = -\sum_{k=1}^{N} \frac{b_{jk} r_k}{2} \Phi_k \quad \forall k \tag{18.79}$$

$$a_{jij} = \frac{\phi_{ij}}{\rho_j + \delta_{ij}/2} \left(\omega_j + \frac{(d_j \xi_j \eta_{ij})^2 \varphi_{ij}}{2c_{ij}} + d_j \xi_j \varphi_{ij} \right) \tag{18.80}$$

$$a_{jik} = \frac{d_j \xi_k \varphi_{ik} \phi_{ik}}{\rho_j + \delta_{ik}/2} \left(1 + \frac{d_k \xi_k \eta_{ik}}{2c_{ik}} \right) \quad \forall k \neq j \tag{18.81}$$

$$b_{jj} = \frac{\psi_j u}{\rho_j + \frac{r_j}{2K_j}} \tag{18.82}$$

$$b_{jk} = 0 \quad k \neq j \tag{18.83}$$

$$d_j = \frac{\frac{g_j \lambda \alpha \gamma}{M_0} - \frac{(-g_j \lambda \beta \sigma)^2}{2\theta_j}}{\rho_j + \Omega} \tag{18.84}$$

$$g_j = -\frac{\sum_{k=1}^{N} b_{jk} \Phi_k}{\rho_k} \quad \forall k \tag{18.85}$$

$$j = 1, 2, \ldots, N \quad i = 1, 2, \ldots, n \quad k \in [1, N] \tag{18.86}$$

The undetermined coefficients in appendices A.1 and A.2 are uniquely defined, and hence, this corresponding feedback Nash equilibrium is unique. The coefficient m_j in proposition 2 is uniquely determined by the coefficients in (18.68) - (18.73) and (18.80) - (18.86).

Bibliography

Basar, T. and Olsder, G. (1999). *Dynamic Noncooperative Game Theory*, second edn. (Philadelpiha SIAM).

Fleming, W. and Richel, R. (1975). *Deterministic and Stochastic Optimal Control* (Springer Verlag).

Gilboa, I. and Schmeidler, D. (1989). Maxmin expected utility with non-unique prior, *Journal of Mathematical Economics* **18**, pp. 141–153.

Hansen, L., Sargent, T., Turmuhambetova, G. and Williams, N. (2001). *Robustness and uncertainty aversion*, Tech. rep., http://home.uchicago.edu/lhansen/uncert12.pdf.

Harvey, D. (2000). *Global Warming - the Hard Science* (Pearson Education Limited, Essex).

Hennlock, M. (2005). A differential game on the management of natural capital subject to emissions from industry production, *Swiss Journal of Economics and Statistics* **141**, pp. 411–436.

Hennlock, M. (2006). A robust abatement policy in a climate change policy model, Working paper, Department of Economic and Statistics, Gothenburg University, Gothenburg.

Houghton, J., Ding, Y., Griggs, D., Noguer, M., van der Linden, P., Dai, X. and Johnson, C. (2001). *Climate change 2001: The scientific basis*, Tech. rep., Intergovernmental Panel on Climate Change (IPCC).

Merton, R. (1975). An asymptotic theory of growth under uncertainty, *Review of Economic Studies* **42**, pp. 375–393.

Nordhaus, W. and Yang, Z. (1996). A regional dynamic general-equilibrium model of alternative climate-change strategies, *The American Economic Review* **86**, 4, pp. 741–765.

Shi, G. (1992). Radiative forcing and greenhouse effect due to the atmospheric trace gases, *Science in China* **35**, pp. 217–229.

Titus, J. and Narayan, V. (1991). *Probability distribution of future sea level rise: A monte carlo analysis based on the ipcc assumptions*, Tech. rep., USEPA and the Bruce Company, Washington, D.C.

Tol, R. (2002a). Estimates of the damage costs of climate change part i. benchmark estimates, *Environmental and Resource Economics* **21**, pp. 47–73.

Tol, R. (2002b). Estimates of the damage costs of climate change part ii. dynamic estimates, *Environmental and Resource Economics* **21**, pp. 135–160.

Visser, H., Folkert, R., Hoekstra, J. and deWolff, J. J (2000). Identifying key sources of uncertainty in climate change projections, *Climate Change* **45**, pp. 421–457.

Yeung, D. (1995). *Pollution-induced business cycles: A game theoretical analysis*, in C. Carraro and J. Filar (eds.), Control and Game-Theoretic Models of Environment (Birkhauser).

Chapter 19

De Facto Delegation and Proposer Rules

Haruo Imai
Kyoto University, Japan
Katsuhiko Yonezaki
Kyoto University, Japan

Abstract

We consider multi-person bargaining problem where players interests are correlated. In particular, we investigate the limit outcomes of the stationary subgame perfect equilibrium outcomes of the sequential bargaining game with a coalition under two different bargaining protocols, and correlation of interests are found within each coalition. By limit, we mean the case where the interval between the two consecutive offers vanishes. The result shows that an endogenous delegation occurs in each coalition to its toughest member. The outcome exhibits a sharp distinction that under the fixed order rule, the size of coalition does not matter, while under the predetermined proposer rule, it matters. This result extends the finding for two sided problems.

Key Words: Bargaining problem, delegation, toughness, coalition, bargaining protocol, sequential bargaining game

19.1 Introduction

A pure n-persoin bargaining problem like the one for splitting-a-dollar has a structure that everybody is an enemy of another. Naturally, the typical solution like the n-person extension of the Nash bargining solution is determined to balance the power of each player. As a result, if a characteristic of one player changes, then the solution changes in most cases.

In reality, it is not rare that some people's intersts are bound together due to some contracts, law, or customs, indicating that people's interests are aligned. An extreme example is the case of wage bargianing where each worker receives the same wage. On the other side, owners of the firm may receive a fixed share from the pool of funds left to the owners as a whole. Another example may be the case where the issue is working condition which may affect each worker differently but the direction of the change in individual welfare must be the sam in the case of labor-management bargaining. Likewise, some particular change in the policy in the case of negotiation among parties forming a coalition government may have similar effect to some member parties. If this sort of correlated intersts among participants is present, a change in the bargained outcome affects members of the group with aligned interests, i.e. the coalition, in the same direction, and so the structure of bargaining changes from the general case where such an alignment of interets is absent. One possibility is that the bargaining position of one player may become irrelevant in the course of bargaining process, because of a presence of a "tougher" player with the same direction of interests, and whenever the softer player rejects the proposal, so does the tougher player. In this paer, we like to examine the bargaining outcome when there is such an extreme alignment of interests among subgroups of the participants of the negotiation, and in particular, to show the way some player's characterstic may become irrelvant in determining the outcome.

To express this extreme alignment of interests among players in a standard splitting-a-dollar problem, we employ the following setups. Players $\{1, 2, ..., n\}$ are partitioned into M coalitions $\{I_1, I_2, ..., I_M\} = \Pi$ (such that for any m, $I_m \neq \varnothing$and for any $m \neq m'$, $I_m \cap I_{m'} = \varnothing$). Π is called a coalition structure. What is bargained over is the division of a dollar among M coalitions, i.e. $(x_1, x_2, ..., x_M)$ with $x_m \geqq 0$ and $\sum x_m = 1$. Given $x = (x_1, x_2, ..., x_M)$, each player i's payoff is determined by an "indirect" utility function $u^i(x_m)$ where $i \in I_m$, and u^i is continuous, concave, and increasing in x_m. This reduced form fits to the situation described above, although not always a derivation yields the desired property like the concavity. We also assume that u^i represents a period payoff for a discouting representation of time preference of a player i.

We adopt a standard sequential bargaining game and investigate the stationary subgame perfect equilibrium outcomes. Since participants' payoffs are correlated, endogenous delegation takes place naturally. There is a strand of literatures discussing delegation in negotiation, but they essentially investigate the situation where the way to choose a representative (either from inside the group or the outside the group) is exogenously given (e.g. Cai (2000), Burtraw (1993), and also see references therein). Here following the setup discussed in Imai and Salonen (2000), we show that a perfect correlation among group members makes one member a de facto representative. The chosen one is the "toughest" player of that group, where the measure of toughness is the reciprocal of the boldness proposed in Aumann and Kurz (1977) (also see Roth (1989) and Burgos, Grant, and Kajii (2002)).

In the sequential bargaining game among more than two players, there is a choice among several alternative rules in selecting a new proposer in the event of a rejection of a standing offer. Recently, in the literature on legislative bargaining (over coalition formation), attention is paid on a potential difference created by the two rules called fixed order (FO) rule and random proposer (RP) rule (cf. Montero (1999), and Merlo (1997)). The former refers to the rule under which the player who rejected the current offer is entitled to make the next proposal (which could be one natural extension of Rubinstein (1982) model to n-person baragaining) , whereas the latter implies a random selection of a proposer according to a certain probability distribution (which is attributed to Binmore (1987) in pure baragaining and Okada (1996) for coalition formation with equal probability, and Baron and Ferejhon (1989) for legislative bargaining. (A mixture of or some variation of these rules to accomodate th epolitical reality is also examined (cf, Montero and Vidal-Puga (2006)). Here we replace FO with another rule under which the next proposer is predetermined and independent of the identity of the rejecting player. Admittedly the naming is a bit confusing but we call this rule the predetermined proposer (PP) rule. This rule is a natural extension inthe pure bargaining literature (in fact the early and basic attempt by Shaked reported in Sutton (1986) and elucidated in Osborne and Rubinstein (1990) utilizes this rule), and can be seen as a counterpart of RP in terms of its independence on the identity of the rejecting player. We compare these two rules and the results stand in a stark contrast. In the main section we describe the model and result with respect to PP, and in the discussion we describe the result under FO which is an easy adaptation of the analysis under PP.

19.2 Model and Result

Let $I = \{1, ..., N\}$ be the set of players, where $N > 3$. The coalition structure is $\Pi = \{I_1, ..., I_m, ..., I_M\}$. The set of agreements is Δ^M (=

$\{(x_1, x_2, ..., x_M) \in R_+^M : \sum x_m = 1\})$ with x as a generic element. The preference of a player i is given by a utility function such that an agreement at t periods later yields utility level $\delta^{t-1}u_i(x_m), i \in I_m$, where $\delta \in (0,1)$ and u_i is a continuous, non negative and concave function of x. Also we assume that it is strictly increasing and the right derivatives of u_i's at 0 are finite. A perpetual disagreement yields the utility level 0.

We consider a bargaining game in which offers are made alternately and an offer consists of a particular value of (x_m). This process is one extension of two person bargaining process considered by Binmore and Rubinstein and the n-person process examined by Shaked and many others is of this type.

The particular process of bargaining we consider is as follows – given an order on I, say $(1, 2, \ldots, n)$, in the first period, player 1 offers x^1, and then players $2, 3, \ldots, n$ in that order replies if s/he rejects or accepts that offer. If all players accept, then $x^1 = (x_m^1)_{m=1}^M$ realized. If some player rejects, the process moves to the second period, and under (PP) player 2 makes an offer x^2, and players $3, 4, \ldots, n, 1$ replies in that order. If all players agree on x^2, then (x_m^2) realizes in this period, while if some player rejects, then the process moves to the third period in which player 3 makes an offer. The process continues unless an agreement is reached.

We consider the limit of stationary subgame perfect equilibria of the above game as the length of a period converges to 0 (or $\delta \to 1$). The steps to derive the result are essentially the same as those for the two person bargaining game. The solution obtained is the modified n-person Nash bargaining solution.

In order to define the solution for general cases, it is convenient to define the "toughness" of player. It turns out that here "toughness" means the degree to which each player can withstand delay. For each $i \in S_m$ and $x_m \in [0,1]$, define $W_i(x_m)$ to be $x_m \in [0,1]$ such that $u_i(x'_m) = \delta u_i(x_m)$ if there is such x'_m, and otherwise to be 0. Note that $u_i(W_i(x_m))$ is the least utility level feasible via the current agreement which is greater than or equal to the utility level in the case of the agreement at x_m in the next stage. Also note that W_i is continuous and nondecreasing.

Given x_m, $W_i(x_m)$ provides one way to measure "toughness" of each player. In the bargaining situation under consideration, the consent of every player must be obtained, and hence what matters is the "toughness" of the "toughest" player in each coalition I_m. Define $W_m(x_m)$ to be $\max_{i \in I_m} W_i(x_m)$. W_m are also continuous and nondecreasing.

The outcomes of a bargaining naturally depend on the order of players in making an offer. As usual, if the length of the time interval between periods shrinks, in the limit this dependence of the outcome on the ordering is expected to disappear. However the measure of "toughness" of players raised above becomes inconvenient in the limiting procedure as $W_i(x_m)$ tends to x_m for any i. Instead, one may utilize the rate at which $W_i(x_m)$ tends to x_m as δ tends to 1, i.e. the limit value of $\frac{x_m - W_i(x_m)}{1-\delta}$ for $i \in I_m$. This rate is 0 at $x_m = 0$. In other cases, as $\frac{x_m - W_i(x_m)}{1-\delta} = \frac{(x_m - W_i(x_m))u_i(x_m)}{u_i(x_m) - u_i(W_i(x_m))}$ holds, in the limit, this rate converges to $\frac{u_i(x_m)}{u_i'(x_m)}$ if u_i is differentiable. Naturally, the lower this rate $\left(\frac{u_i(x_m)}{u_i'(x_m)}\right)$ is, the "tougher" a player may be called (which is nothing but the reciprocal of the boldness).

Given the assumption of differentiable utility functions, for the n-person case, we can proceed similarly. Denote by $\Omega_i(x_m)$ the measure of toughness of i, $\frac{u_i(x_m)}{u_i'(x_m)}$ for $i \in S_m$. Then define $\Omega_m(x_m)$ to be the toughness of the toughest player $\min_{i \in S_m} \Omega_i(x_m)$. Then define $\Omega(0)$ to be any value y between $\lim_{x_m \to 1} \Omega_m(x_m)$ and $+\infty$.

The modified n-person Nash solution outcome is then defined to be the x_m at which $n_m \Omega_m(x_m) = n_{m'} \Omega_{m'}(x_{m'})$ for any m, m' holds where number $|S_m| = n_m$.

If one wishes to characterize this solution as a maximizer of some function like the Nash product, one may consider the logarithm of the utility functions. Define $T_m(x_m) = A - \int_{x_m}^1 \frac{1}{\Omega_m(x_m)} dx$ for $x_m > 0$, and $T_m(x_m) = B - \int_0^{x_m} \frac{1}{\Omega_m(x)} dx$ for $x_m < 1$ where $A = \log u_i(1)$ for some $i \in S_m$ with $\Omega_i(1) = \Omega_m(1)$. The solution we consider is characterized as the maximizer of $\sum_m n_m T_m(x_m)$, or equivalently $\Pi \exp\{n_m T_m(x_m)\}$

In general where differentiability is not assumed, one takes the gradient correspondences g_i where $i \in S_m$, $z = g_i(x_0)$ for $x_0 \in [0,1]$ if $z \geq 0$ and $z(x - x_0) + u_i(x_0) \geq u_i(x)$ holds for any $x \in [0,1]$. Note that by definition, g_i is upper semicontinuous (with $g_i(0)$ not necessarily compact) and g_i is nonincreasing in the sense that given $x < x'$, for any $i = g_i(x)$ and any $z' = g_i(x_0')$, $z \leq z'$ holds. Note that $\sup g_i(0) = \infty$.

Then define correspondence Ω_i by $\Omega_i(x_m) = \frac{u_i(x_m)}{g_i(x_m)}$ for each $x_m \in [0,1]$ and for each $i \in S_m$. Finally define a correspondence Ω_m by $z = \Omega_m(0)$ if $z \leq \min_{i \in S_m} \sup \Omega_i(0)$,and $z = \Omega_m(x_m)$ for $x \in (0,1)$ if $z = \Omega_i(x_m)$ for some $i \in S_m$ and $z \leq \lim_{x_m' \downarrow x_m} \min_{i \in S_m} \inf \Omega_i(x_m')$. Ω_m is upper semicontinuous and increasing with $0 = \Omega_m(0)$ and $\sup \Omega_m(1) = \infty$.

Based on these correspondences, the solution and the surrogate utility functions are defined exactly in the same manner as above, with a suitable selection necessary in evaluating the integrals.

The result of this paper is —

Proposition 19.1. *There is a unique stationary subgame perfect equilibrium outcome, and as δ tends to 1, this outcome converges to the modified Nash solution.*

This result indicates the modification made to the Nash bargaining solution due to the extreme correlatoin of interests among players. Also one can see that the outcome is affected by the choice of the rules of bargaining process.

19.3 Proof

Equilibrium

First we characterize a stationary subgame perfect equilibrium offers, (x^{*i}).

$$x^{*i} = \begin{pmatrix} W_1(x_1^{*i+1}) \\ \vdots \\ W_{m(i)-1}(x_{m(i)-1}^{*i+1}) \\ 1 - \sum_{m \neq m(i)} W_m(x_m^{*i+1}) \\ W_{m(i)+1}(x_{m(i)+1}^{*i+1}) \\ \vdots \\ W_M(x_M^{*i+1}) \end{pmatrix} \tag{19.1}$$

for all i, where $m(i)$ indicates the coalition i belongs to, i.e. $i \in I_{m(i)}$. If there is a solution to the above system of equations, then they will comprise a stationary subgame perfect equilibria, so that each period, proposer i proposes x^{*i} which will be accepted.

In the sequel, what matters is the equilibrium offer each player makes for his/her own, and so one may summarize the above conditions by

$$(x^{*i}_{m(i)}) = \begin{pmatrix} x^{*1}_{m(1)} \\ \vdots \\ x^{*i}_{m(i)} \\ \vdots \\ x^{*M}_{m(M)} \end{pmatrix} = \begin{pmatrix} 1 - \sum_{m \neq m(1)} W_m^{j(1,m)}(x_m^{*1+j(1,m)}) \\ \vdots \\ 1 - \sum_{m \neq m(i)} W_m^{j(i,m)}(x_m^{*i+j(i.m)}) \\ \vdots \\ 1 - \sum_{m \neq m(M)} W_m^{j(M,m)}(x_m^{*M+j(M,m)}) \end{pmatrix} \tag{19.2}$$

where $j(i,m)$ is the number of periods necessary to reach the chance where a member of I_m makes an offer after the period with player i's offer for the first time, and $i + j(i,m)$ stands for the player $i' = (i + j(i,m)) \bmod N$.

Contraction Mappings

First we note that W_i's are contraction mappings. If $i \in S_m$ for $x < x'$ with $W_i(x_m) > 0$ we have $\frac{(u_i(x') - u_i(x))}{x' - x} \leq \frac{(u_i(W_i(x')) - u_i(W_i(x)))}{W_i(x') - W_i(x)}$. If $u_i(W_i(x)) = \delta u_i(x)$ and $u_i(W_i(x')) = \delta u_i(x')$ hold, then the above inequality yields $W_i(x') - W_i(x) = \delta(x' - x)$. If $W_i(x_m) = 0$ and $W_i(x'_m) > 0$, $0 < \delta(u_i(W_i(x')) - u_i(W_i(x)))/(u_i(x') - u_i(x)) < 1$ holds and so we still obtain $W_i(x') - W_i(x) \leq \delta(x' - x)$ trivially holds. Thus W_i is contracting mapping.

Then, for $x < x'$, $W_m(x') - W_m(x) \geq W_i(x') - W_i(x)$ for i with $W_m(x') = W_i(x')$ and so W_m is also a contraction mapping.

Fixed Point

Our solution is given by the fixed point of composite of N mappings F_i from Δ^M to Δ^M such that

$$F_i(x) = \begin{pmatrix} W_i(x_1) \\ \vdots \\ W_{m(i)-1}(x_{m(i)-1}) \\ 1 - \sum_{m \neq m'} W_m(x_m) \\ W_{m(i)+1}(x_{m(i)+1}) \\ \vdots \\ W_M(x_M) \end{pmatrix} \tag{19.3}$$

we verify that F_i is a contraction mapping with respect to the absolute sum norm. To this end, we take $x, x' \in \Delta^M$ and show that $\|x - x'\| > \|F_i(x) - F_i(x')\|$. Given i, x and $x' \in \Delta^M$, let $Q_+ =$

$\{m \neq m(i) : x_m - x'_m \geqq 0\}$ and $Q_- = \{m \neq m(i) : x_m - x'_m < 0\}$. Without loss of generality, we may assume that $\sum_{m \neq m(i)} (x_m - x'_m) \geqq 0$.

We wish to show that

$$\begin{aligned}\|x - x'\| &= \sum_{m \neq m(i)} |x_m - x'_m| + \left| \sum_{m \neq m(i)} (x_m - x'_m) \right| = 2\sum_{Q_+} |x_m - x'_m| \\ &> \sum_{m \neq m(i)} |W_m(x_m) - W_m(x'_m)| + \left| \sum_{m \neq m(i)} (W_m(x'_m) - W_m(x_m)) \right| \\ &= \|F_m(x) - F_m(x')\|.\end{aligned}$$

First, if $\sum_{m \neq m(i)} (W_m(x'_m) - W_m(x_m)) \leqq 0$, then $\|F_m(x) - F_m(x')\| = 2\sum_{Q_+} |W_m(x_m) - W_m(x'_m)| < 2\sum_{Q_+} |x_m - x'_m| = \|x - x'\|$.

Next, if $\sum_{m \neq m(i)} (W_m(x'_m) - W_m(x_m)) > 0$, then $\|F_m(x') - F_m(x)\| = 2\sum_{Q_-} |W_m(x'_m) - W_m(x_m)| < 2\sum_{Q_-} |x_m - x'_m| = \|x - x'\|$.

Thus F_i is a contraction mapping and so is $\widetilde{F} = F_1 \circ F_2 \circ \cdots \circ F_N$, which possesses a unique fixed point.

Convergence

For each δ, W_i and W_m are defined as above. Let M^δ be the upper bound on $1 - W_i(1)$ for all i, given δ. Let $\iota(\delta, m, x_m)$ be the player whose number is the earliest among those $n \in \arg\max W_i(x_m)$. Then define $\Phi_m(\delta, m, x_m)$ to be $(1-\delta)/(x_m - W_m(x_m))$ and the surrogate utility function $u_m\delta$ for I_m be $\exp(\int \Phi_m(\delta, m, x_m) + u(\iota(\delta, m, 0))$

First, we claim that given $\varepsilon > 0$, there is $\delta' < 1$, such that for any $\delta > \delta'$, m, j and i, $\left|x^i_m - x^j_m\right| < N\varepsilon$. This can be shown that by choosing δ such that $M^\delta < N\varepsilon$, then from the definition of the SSPE ((1)), one sees that if $i+1$ does not belong to I_m, then $x^i_m - x^{i+1}_m < \varepsilon$: if $i+1$ belongs to I_m, then $x^{i+1}_m - x^i_m < (N-1)\varepsilon$: furthermore, they have to come back to the same value after N periods. Then, it follows from the continuity of u_i's, that given $\varepsilon > 0$ there is $\delta < 1$ such that $\left|u_m(x^i_m) - u_m(x^j_m)\right| < \varepsilon$ holds for any $\delta > \delta'$, m, j and i. Further, one can choose δ' so that $1 - \delta'^N < \varepsilon$. Then from (2), one sees that $\left|\Pi u_m(x^i_m)^{n_m} - \Pi u_m(x^j_m)^{n_m}\right| < \varepsilon$ holds for any i, j. Thus all offers are close to each other although there are some distinct offers, and their modified Nash products are close to each other, implying that they are close to the maximizer. And indeed as δ tends toward 1, offers converge to the maximizers.

19.4 FO rule and Discussion

Under FO rule, everything remains the same except for the effect of the size of a coalition. I.e. the bargaining game essentially that among representatives from each coalition. As a consequence, the resulting solution is the maximizer of the M person Nash product: Πu_m.

The difference the proposer rules creates was extensively discussed in the literature concerning political negotiation. There the reality of the rule was investigated and they focused upon random proposer rule and FO. Here, we considered another rule where regardless of the identity of the player who rejected the standing proposal, next proppser is determined by the rule. This rule may not have much counterpart in reality. However, the result it produces is the same as the one by random proposer rule. In fact, this rule can be considered as one representation of the random proposer rule under certainty. .

In the n-person bargaining literature, this rule has been used by several authors as well as FO. One reason why the difference has not been paid much attention was that the result does not differ much. Here, by the coalitional constraint, interests of members are perfectly correlated, which creates the distinction as in Imai and Salonen (2000) as well as in Montero (1999).

One return from this exercise is the examination of the effect of coalition formation prior to negotiation, although one needs to reexamine concavity assumtion for such adaptation. If the protocol of the bargaining stage is that of fixed order, then joining coalition does not give an extra merit to the toughest player, and two identically tough players also do not find it advantageous to merge, and hence coalitions would not form (which corresponds to Harsanyi's joint bargaining paradox; Harsanyi (1977)). By contrast, if the protocol is that of predetermined proposer or to that effect random proposer, then there could be a merit from coalition formation, because softer player can borrow the bargaining power of the toughest player in the coalition, while the toughest player can obtain the leverage through the size of a coalition. This shows that there may be no agreement among the members as to which rule to prevail before the coalition formation stage, and status quo may continue to reign. (The multitude of the outcomes corresponding to different rules may have counterparts in the cooperative analysis carried out by Chae and Heidhues (2004).)

19.5 Conclusion

We investigated the stationary subgame perfect equilibrium outcome of the sequential bargaining game with a coalition structure in the limit under two different bargaining protocols, where there is a perfect correlation of interests among the members of each coalition. The result shows an endogenous delegation occurs in each coalition to its "toughest member". The outcome exhibits a sharp distinction that under the fixed order rule, the size of coalition does not matter, while under the predetermined proposer rule, it matters.

Acknowledgements

Authors appreciate the referee's comment gratefully. The first author wish to acknowledge the financial support by Grant-in-Aid for Scientific Research (C) of Japanese Ministry of Education (#16530116).

Bibliography

Aumann, R. J. and M. Kurz. (1977). Power and Taxes, *Econometrica* **45**, pp. 1137–1161.

Baron, D. and J. Ferejohn (1989). Bargaining in legislatures, *American Political Science Review*, **83**, pp. 1181–1206.

Binmore, K. (1987). Perfect Equilibria in Bargaining Models, *The Economicsof Bargaining*, ed by K. Binmore and P. Dasgupta, pp. 77–105.

Burgos, A., S. Grant, and A. Kajii, (2002). Bargaining and boldness, *Games and Economic Behavior*, **38**, pp. 28–51.

Burtraw, D. (1993). Bargaining with Noisy Delegation, *RAND Journal of Economics*, **24**, pp. 40–57.

Cai, H. (2000). Bargaining on Behalf of a Constituency. *Journal of Economic Theory*, **92**, pp. 234–273.

Chae, S. and P. Heidhues, (2004). A Group Bargaining Solution, *Mathematical Social Sciences*, **48**, pp. 37–53.

Harsanyi, J., (1977). Rational Behavior and Bargaining Equilibrium in *Games and Social Situations*, (Cambridge Univ. Press, Cambridge).

Imai, H. and H. Salonen, (2000). Representative Bargaining Solution for Two-Sided Bargaining Problems, *Mathematical Social Sciences*, **39**, pp. 349–365

Merlo, A., (1997). Bargaining over governments in a stochastic environment, *Journal of Political Economy*, **105**, pp. 101–131.

Montero, M. (1999). Coalition Formation in Games with Externalities, mimeo., (Tilburg University).

Montero M. and J. Vidal-Puga. Demand Commitment in Legislative Bargaining, mimeo, (University of Nottingham).

Okada, A. (1996). A Noncooperative Coalitional Bargaining Game with Random Proposers, *Games and Economic Behavior*, **16**, pp. 97–108.

Osborne, M. J. and Rubinstein A. (1990). Bargaining and Markets, (Academic Press, San Diego, CA).

Roth, A. (1989). Risk aversion and the relationship between Nash's solution and subgame perfect equilibrium of sequential bargaining, *Journal of Risk and Uncertainty*, **2**, pp. 353–365

Rubinstein, A. (1982). Perfect Equilibrium in a Bargaining Model, *Econometrica*, **50**, pp. 97–109.

Sutton, J. (1986). Non-Cooperative Bargaining Theory: An Introduction, *The Review of Economic Studies*, **53**, pp. 709–724.

Chapter 20

The Bargaining Set in Effectivity Function

Dawidson Razafimahatolotra[1]
Panthéon-Sorbonne Economie.
University Paris 1 Panthéon Sorbonne
e-mail: razafimahatolotra@univ-paris1.fr or razafimahatolotra@yahoo.ca

Abstract

This paper investigates stability properties of effectivity functions. The Bargaining Set in Effectivity Function generalizes the concept of cycles and connects it with the well known stability notion of bargaining sets.
At first, we propose to study relations between cycles and implement a class of effectivity functions for which theses cycles are equivalent. The part two of this work will be devoted to analyze the stability of the Bargaining Sets and gives relations between them. Bargaining sets are the Zhou's, the Mass-Colell's and the Aumann Davis Maschler's bargaining sets.

Key Words: Effectivity function, cycle, stability, core, bargaining sets.

20.1 Introduction

This paper investigates stability properties of effectivity functions. Effectivity functions were first introduced by Moulin and Peleg (1982, Journal of Mathematical Economics) as a way of describing effectiveness of coalitions in game forms. Keiding (1985, International Journal of Game Theory) provided a necessary and sufficient condition for an effectivity function to be

[1]Dawidson Razafimahatolotra is a teacher at the University of Antananarivo, a member of the project DELICOM (Agence Nationale de la Recherche, JC-JC05) and CES (Centre d'Economie de la Sorbonne) at the University of Paris 1 Pantheon Sorbonne. Postal address: Maison des Sciences Économiques, 5e étage, 106 - 112 boulevard de l'Hpital 75013 Paris - France.

core stable. He introduced the concept of cycles of effectivity function and showed that an effectivity function is stable if and only if it is acyclic.The Bargaining Set in effectivity function generalizes the concept of cycles and connects it with the well known stability notion of bargaining sets.

Given a set of alternatives A and a set of players N, an effectivity function E is a map that assigns to each coalition $S \subset N$ a family of subsets of A. The interpretation is: if the set B is assigned by the effectivity function to coalition S, then S can force the outcome to be in B and if $C \subsetneq B$ where $C \notin E(E)$, then S is not to be able to precise that social state is members of C. Players are endowed with preference ordering. A profile is a vector of preferences of members of N

A society can choice a social state a at a profile u for some rule $\mathcal{R}$ if members of N can select $a \in A$ where agents $(i \in N)$ accepts the rule $\mathcal{R}$.

An effectivity function is called stable (either in terms the core, the Mass-Colell's, Aumann Devis Maschler's bargaining set or Zhou's bargaining set) if for all profile u, N can choice at least an alternative with the rule of the core etc.

The notion of cycle was discussed at first by Condorcet where he shows that social choice can be empty with majority rule. The definitions of cycles are generalized in Abdou & Keiding where they essentially capture the following simple idea. Suppose S_1 and S_2 are two coalitions such that $S_1 \cap S_2 = \emptyset$ and $S_1 \cup S_2 = N$, and B_1 and B_2 are two sets of alternatives satisfying $B_1 \cap B_2 = \emptyset$ and $B_1 \cup B_2 = A$. The effectivity function is that S_k is effective for B_k. To show the emptiness of social choice in theses rules we considers the following profile. Each agent $i \in S_1$ prefers any element of B_1 to those in B_2, that is for all $x_k \in B_k$, $k = 1, 2$; $u^i(x_1) > u^i(x_2)$. Similarly all agents in S_2 prefers every alternatives in B_2 to every alternatives in B_1. Our effectivity function E, allows S_1 to block B_2 (i.e social state do not belong to B_2) and S_2 to block B_1 then E is said to have a cycle(of order 2). One can check that under such a profile, Zhou's bargaining set of E will be empty. The argument here is very simple. Suppose to the contrary $a \in B_1$ (if $a \in B2$, the argument is similar) is in the bargaining set. Then S_2 will object this proposal via B_2. However, this objection does not have a valid counter objection because a counter objecting coalition, by definition, must contain some members from S_1 as well as S_2. Clearly, for such a coalition, there is no alternative which can make 1 both the members from S_1 (compared to a) and members from S_2 (compared to B_2) better off. Therefore a does not belong to the bargaining set.

Bargaining sets can be defined as soon as in a priori structure of coalitions or without structure of coalitions. In this paper, we focalize our work only in the case where no structure of coalition is given. So, coalition formation is not an object of this paper. Our intention is to analyze different formulations of the bargaining set, but preferences are a linear order on the set of alternatives and set of alternatives is finite, some of theses definitions are equivalent. So, we propose to analyze only three definitions: Zhou's, Mass-Colell's and Aumann Devis Maschler's bargaining sets.

At first, we propose to study relations between cycles and implement a class of effectivity functions where theses cycles are equivalent. The second part is devoted to analyze stabilities of the Bargaining Sets and gives relations between them.

20.2 Definitions and Notations

The set of players N and the set of alternatives A are finite. For any set X, $|X|$ represent the cardinal of X. We denote $\mathcal{X} = \{B \subset X, B \neq \emptyset\}$ (or $\mathcal{P}(X)$ if $\mathcal{X}$ can not be read) the set of non empty subset of N and $\mathcal{X}^* = \mathcal{X} \cup \{\emptyset\}$. If $D \in \mathcal{P}(X)$ then $D^c = X \setminus D$ and $D^+ = \{C \supset D, C \subset X\}$. A partition of a set X is a family of sets $(D_k)_{k \in I}$ where $\cup_{k \in I} D_k = X$, and $D_k \cap D_l = \emptyset, \forall k \neq l$. We denote $\mathfrak{p}(X)$ the set of partition of X.

An *effectivity function* is a correspondence $E\ \mathcal{N} \to \mathcal{P}(\mathcal{A})$ satisfying

$$B \in E(N), \forall B \in \mathcal{A} \text{ and } A \in E(S), \forall S \in \mathcal{N}$$

E is *regular* if $[S, T \in \mathcal{N}, B \in E(S) \text{ and } B' \in E(T)] \Rightarrow$ either $B \cap B' \neq \emptyset$ or $S \cap T \neq \emptyset$ and *maximal* if $[S \in \mathcal{N}, B \notin E(S)] \Rightarrow B^c \in E(S^c)$. E is *superadditive* if $[S_k \in \mathcal{N}, S_1 \cap S_2 = \emptyset \text{ and } B_k \in E(S_k)] \Rightarrow B_1 \cap B_2 \in E(S_1 \cup S_2)$. E is *monotonic* if $[B \in E(S), C \supset B \text{ or } T \supset S] \Rightarrow C \in E(T)$.

For $s = 1 \dots |N|$, we denote $A^s = A \otimes \cdots \otimes A$ (s-times) and $\mathcal{L}(A)$ the set of linear order on A. A preference is a member of $\mathcal{L}(A)$ and a profile is a vector $u = (u^i)_{i \in N}$. For $B \subset A$, $u^i(B) = \min_{b \in B} u^i(b)$. If $B, C \subset A$ and $S \subset N$, the notation $u^S(B) \gg u^S(C)$ means $u^i(B) > u^i(C), \forall i \in S$, the notation $u^S(B) > u^S(C)$ means $u^i(B) \geq u^i(C), \forall i \in S$ and $u^i(B) > u^i(C)$ for at least $i \in S$, and we write $u^S(B) \geq u^S(C)$ if $u^i(B) \geq u^i(C), \forall i \in S$[2].

Let $x \in A$. An *objection* against x is a pair (S, B) s.t $B \in E(S)$ and

$$u^S(B) > u^S(x)$$

[2]If $C = \{c\}$ or $B = \{b\}, b \notin C$, then $u^S(B) \gg u^S(C)$ and $u^S(B) > u^S(C)$ are equivalents. In this case, we retain the notation $u^S(B) > u^S(C)$. So, the only version of objection against $x \in A$ is $u^S(B) > u^S(C)$

A *Zhou's* ($\mathfrak{m}_z$) *counter objection* against the objection (S, B) is a pair (T, C) s.t

$$T \cap S \neq \emptyset, T \setminus S \neq \emptyset, S \setminus T \neq \emptyset$$

$$\left(u^{S\cap T}(C), u^{T\cap S^c}(C)\right) \geq \left(u^{S\cap T}(B), u^{T\cap S^c}(x)\right)$$

A *Mass-Colell's* ($\mathfrak{m}_m$) *counter-objection* against the objection (S, B) is a pair (T, C) s.t

$$\left(u^{S\cap T}(C), u^{T\cap S^c}(C)\right) > \left(u^{S\cap T}(B), u^{T\cap S^c}(x)\right)$$

An objection is $\mathfrak{m}_\alpha$ justified, $\alpha \in \{z, m\}$, if there is no $\mathfrak{m}_\alpha-$ counter objection against it.

The Aumann-Davis-Maschler's (ADM) bargaining set is defined as follow

An objection of k against x player j at $x \in A$ is a pair (S, B) s.t $B \in E(S)$, $k \notin S \ni j$ and

$$u^S(B) > u^S(x)$$

A counter objection of j against k is a pair (T, C) s.t $C \in E(T)$, $j \notin T \ni k$ and

$$\left(u^{T\cap S}(C), u^{(T\setminus S)}(C)\right) \geq \left(u^{T\cap S}(B), u^{(T\setminus S)}(x)\right)$$

The *core*, the *Zhou's bargaining set*, the *Mass-Colell's bargaining set* and the ADM's bargaining set of an effectivity function E at the profile u are

$$C(N, u) = \{x/ \text{ there is no objection against } x\},$$
$$\mathcal{M}_z(E, u) = \{x \in A, \text{ no objection against } x \text{ is } \mathfrak{m}_z\text{-justified}\},$$
$$\mathcal{M}_m(E, u) = \{x \in A, \text{ no objection against } x \text{ is } \mathfrak{m}_m\text{-justified}\}$$
$$\mathcal{M}_1(E, u) = \{x \in A, \text{ no player has a justified objection against any other playe[r]}$$

An effectivity function E is $\mathfrak{c}$-stable (resp. $\mathfrak{m}_\alpha, \alpha \in \{z, m, 1\}$-stable) if for every profile u, $\mathcal{C}(E, u) \neq \emptyset$ (resp. $\mathcal{M}_\alpha(E, u) \neq \emptyset, \alpha \in \{z, m, 1\}$)

Notational Conventions : In this paper we assume that $|N| = n, |A| = m \geq 2$.

The integer $2 \leq r \leq \min(n, m)$ denote especially an order of a cycle. In this case I or $I_r = \{1, \ldots, r\}$ is a set of integers modulo r, and $I_r^* = I_r \setminus \{r\}$

The set of indexes $J \subset I$ is s.t $\cap_{k\in J} S_k \neq \emptyset$ where (S_k) is a family of coalitions and the set of indexes L is for any subset of I_r

In general, a family pairwise disjoints of N is denoted by $(T_k)_{k\leq r}$ and the one for A is denoted by $(C_k)_{k\leq r}$. The notation (S_k, B_k) is s.t $B_k \in E(S_k)$

If D is a finite set and (X_d) a family of sets, $X_D = \prod_{d\in D} X_d$.

20.3 Some Analysis of Cycles

For a r-tuple of coalition $(S_1,..,S_r) \in \mathcal{N}^r$, we note $\mathcal{J} = \{J \subset I_r, \cap_{k\in J} S_k \neq \emptyset\}$ and $\mathcal{J}_k = \{J \in \mathcal{J}/J \ni k\}$. A $\mathcal{J}$- selection is a map $\sigma : \mathcal{J} \longrightarrow \{1,..,r\}$ such that $\sigma(J) \in J$.

Definition 20.1. A cycle of order r is a family $(S_1,..,S_r,\ B_1,..,B_r,\ C_1,..,C_r)$ satisfying $B_k \in E(S_k)$, $(C_k)_k \in \mathfrak{p}(A)$, $B_k \cap C_k = \emptyset$ and for every $J \in \mathcal{J}$ there is $k_J \in J$ s.t $B_{k_J} \cap C_l = \emptyset\ \forall l \in J$.

Definition 20.2. A 2r-tuples $(S_1,..,S_r, B_1,..B_r)$, $B_k \in E(S_k)$ is balanced if for every $\mathcal{J}$- selection σ we have $\bigcap_{k=1\ldots r} \left(\bigcup_{J\in\mathcal{J}_k} B_{\sigma(J)} \right) \neq \emptyset$. An effectivity function E is *balanced* if every family $(S_1,..,S_r, B_1,..,B_r)$ satisfying $B_k \in E(S_k)$ is balanced.

A *circular interval of length* $p \in I_r^*$ *starting at* $k \in I_r$ is a $L_k(p) = \{k,\ldots,k+p\} \subset I_r$. A family $\mathcal{L}_p \subset \mathcal{P}(I_r)$ is circular of length p if every $L \in \mathcal{L}_p$ is a circular interval of length p.

Definition 20.3. An effectivity function E is *circular of order* r if for some $(T_1,\ldots,T_r,C_1,\ldots,C_r)$ where $T_k \cap T_l = \emptyset$ and $C_k \cap C_l = \emptyset \forall k \neq l$, there is a circular family of length $1 \leq p \leq r-1$ s.t $C_{L^c} \in E(T_L), \forall L \in \mathcal{L}_p$. It is equivalent to $C_L \in E(T_{L^c}), \forall L \in \mathcal{L}_{r-p}$

Remark 20.1. An *upper cycle* of length r is a family $(T_k, B_k)_{k\in I_r}$ s.t $T_k \cap T_l = \emptyset, \forall k \neq l \in I_r$, $B_k \in E(T_k)$ and $\cap_{k\in I_r} B_k = \emptyset$. A *lower cycle* of length r is a family $(S_k, C_k)_{k\in I_r}$ s.t $C_k \cap C_l = \emptyset, \forall k \neq l \in I_r$, $C_k \in E(S_k)$ and $\cap_{k\in I_r} S_k = \emptyset$. [Abdou & Keiding or Vannucci]

Remark 20.2. If E is monotonic, E has a lower cycle if E is circular with $p = 1$ and E has an upper cycle if E is circular with $p = r-1$

Theorem 20.1. *An effectivity function E is acyclic if and only if E is balanced.*

Proof. Let E be a cyclic effectivity function and $(S_1,..,S_r,\ B_1,..,B_r,\ C_1,..,C_r)$, $B_k \in E(S_k)$ be a cycle. Let σ be s.t for $k \in I_r$, $\sigma^{-1}(k) = \{J \in \mathcal{J}_k/B_k \cap C_l = \emptyset, \forall l \in J\}$ and take $J_k \in \sigma^{-1}(k)$. Denote $I_k = \{l \in I_r, B_l \cap C_s = \emptyset \forall s \in J, J \in \mathcal{J}_k\}$. We have

$$\bigcup_{J\in\mathcal{J}_k} B_{\sigma(J)} = \bigcup_{l\in I_k} B_l$$

$$\bigcap_{k\in I_r}\left(\bigcup_{J\in\mathcal{J}_k} B_{\sigma(J)}\right)=\bigcap_{k\in I_r}\left(\bigcup_{l\in I_k} B_l\right)$$
$$\subseteq\bigcap_{k\in I_r}\left(\bigcup_{l\in I_k} C_{J_l^c}\right)$$
$$\left(\bigcap_{k\in I_r}\left(\bigcup_{J\in\mathcal{J}_k} B_{\sigma(J)}\right)\right)^c\supseteq\bigcup_{k\in I_r}\left(\bigcap_{l\in I_k} C_{J_l}\right)$$
$$\supseteq\bigcup_{k\in I_r} C_k$$

Conversely Suppose that $\exists(S_1,..,S_r,B_1,..,B_r)$ and a $\mathcal{J}$-selection σ s.t $\bigcap_{k=1..r}\bigcup_{J\in\mathcal{J}_k} B_{\sigma(J)}=\emptyset$ and put $C_k=\left(\bigcup_{J\in\mathcal{J}_k} B_{\sigma(J)}\right)^c$.

We claim that $(S_1,..,S_r,B_1,..,B_r,C_1,..,C_r)$ is a cycle. In fact,

(1) $\bigcup_{k=1..r} C_k=A$

(2) $\forall J\in\mathcal{J},\exists k=\sigma(J)\ B_{\sigma(J)}\cap C_l=\emptyset,\ \forall l\in J$

Proposition 20.1. *Let E be a circular effectivity function of order r. Then, E has a cycle.*

Proof. Let $(T_1,\ldots,T_r,C_1,\ldots,C_r)\in\mathfrak{p}(N\otimes A)$ and let $\mathcal{L}_p$ a circular family of length p. We prove that E is not balanced. Define $S_k=T_{L_k}$ and $B_k=C_{L_k^c}$ where L_k is the circular interval of length p starting at k. A subset of I_r, $J=\{k_1,\ldots,k_s\}\in\mathcal{J}$ if $\cap_{j=1}^s L_{k_j}\neq\emptyset$ and we define $\sigma(J)=k_1$.

If $\alpha\in I_r$, denote $J^1,\ldots,J^\nu$ the members of $\mathcal{J}_\alpha$ and k_1^s the smallest element of $J^s, s=1\ldots\nu$. We have that $\alpha\in\cap_{s=1}^\nu C_{L_{k_1^s}}$. So

$$\bigcap_{\alpha=1}^{r}\left(\bigcup_{J\in\mathcal{J}_\alpha} B_{\sigma(J)}\right)=\bigcap_{\alpha=1}^{r}\left(\bigcup_{J\in\mathcal{J}_\alpha} C_{L_{k_1^s}}\right)$$
$$\bigcup_{\alpha=1}^{r}\left(\bigcap_{J\in\mathcal{J}_\alpha} C_{L_{k_1^s}}\right)\supset\bigcup_{\alpha=1}^{r} C_\alpha=A$$

The Theorem 20.1 achieve the proof. □

Proposition 20.2. *If E has a cycle of order $r\leq 3$, then E has either an upper or a lower cycle of order $\rho\leq 3$.*

The proposition is trivial for $r=2$. Then, let $r\geq 3$ and $(S_k,B_k,C_k)_{k=1\ldots r}$ be a cycle of order r. For simplification, we denote

$$\mathcal{Q}_k=\{L\subset I_r/|L|=k\}$$

We need the two following lemma to prove this proposition

Lemma 20.1. *Let E be a cyclic effectivity function of order r. If for some $L \in \mathcal{Q}_{r-1}$ we have that $\cap_{k\in L} S_k \neq \emptyset$, then E has a lower cycle of order $\rho \leq r$.*

Proof. Let $(S_k, B_k, C_k)_{k\in I_r}$ be a cycle of order r s.t $S_1 \cap \cdots \cap S_{r-1} \neq \emptyset$. We obtain $B_1 \subset C_r, \ldots, B_k \subset C_r \cup \cdots \cup C_{k-1}, \ldots, B_{r-1} \subset C_r \cup \cdots \cup C_{r-2}$[3] and suppose that E has no lower cycle of order $\rho \leq r$.

By $B_1 \subset C_r$, we would have $S_1 \cap S_r \neq \emptyset$. Hence, $B_r \subset C_2 \cup \cdots \cup C_{r-1}$. i.e $B_2 \cap B_r = \emptyset$. Also, $S_2 \cap S_r \neq \emptyset$. Then either $B_2 \subset C_1$ or $B_r \subset C_3 \cup \cdots \cup C_{r-1}$. If $S_2 \cap S_1 \cap S_r = \emptyset$, then $B_r \subset C_3 \cup \cdots \cup C_{r-1}$. If not $(S_k, B_k)_{k\in\{2,1,r\}}$ is a lower cycle. If $S_2 \cap S_1 \cap S_r \neq \emptyset$, by the definition of cycle with the selection of indexes $2, 1, r$, we have $B_r \subset C_3 \cup \cdots \cup C_{r-1}$. I.e, necessary $B_r \subset C_3 \cup \cdots \cup C_{r-1}$.

Now, suppose that $B_k \subset C_r \cup \cdots \cup C_{k-1}$ implies $B_r \subset C_{k+1} \cup \cdots \cup C_{r-1}$, and prove that the assertion is true for $k + 1$.

As $B_{k+1} \subset C_r \cup \cdots \cup C_k$ then $B_r \cap B_{k+1} = \emptyset$, it gives $S_r \cap S_{k+1} \neq \emptyset$. Hence either $B_{k+1} \subset C_1 \cup \cdots \cup C_k$ or $B_r \subset C_{k+2} \cup \cdots \cup C_{r-1}$. In the second case, we achieve the proof.

Suppose that $B_{k+1} \subset C_1 \cup \cdots \cup C_k$. If $S_{k+1} \cap S_1 \cap S_r = \emptyset$, $(S_k, B_k)_{k\in\{k+1,1,r\}}$ is a lower cycle. If $S_{k+1} \cap S_1 \cap S_r \neq \emptyset$ the existence of cycle with the selection of indexes $k+1, r, 1$ gives either $B_{k+1} \subset C_2 \cup \cdots \cup C_k$ or $B_r \subset C_{k+2} \cup \cdots \cup C_{r-1}$[4]. In the second case the recurrence is hold, and in the first: We suppose again by recurrence that $B_{k+1} \subset C_{l-1} \cup \cdots \cup C_k$ implies either $B_{k+1} \subset C_l \cup \cdots \cup C_k$ or $B_r \subset C_{k+2} \cup \cdots \cup C_{r-1}$ and we prove that the assertion is true for l. In fact, suppose that $B_{k+1} \subset C_l \cup \cdots \cup C_k$. Put $J_l = \{s/B_l \cap C_s \neq \emptyset\}$ and $l_1 = \min\{s, s \in J_l\}$[5], ..., $J_{l_k} = \{s/B_{l_k} \cap C_s \neq \emptyset\} \subset C_r \cup \cdots \cup C_{l_k - 1}$ and $l_{k+1} = \min\{s/s \in J_{l_k}\}$[6]. Let p be the number s.t $l_p = r$, that exist because at least $B_1 \subset C_r$.

So, we have a family as represented in the following figure

[3] For this assertion, observe that for each J satisfying $\cap_{k\in J} S_k \neq \emptyset$, there is $k_J \in J$ s.t $B_k \cap C_l = \emptyset, \forall l \in J$. We begin by $J_1 = \{1, \ldots, r-1\}$ and after $J_2 = \{2, \ldots, r-1\}$ until $J_{r-1} = \{r-1\}$. We can suppose after reorder the set of indexes that $k_{J_1} = 1$.
In this proof, and in some cases, the definition of [Abdou & Keiding] of a cycle is adequate: If $S_{k_1} \cap S_{k_2} \neq \emptyset, \ldots, S_{k_{s-1}} \cap S_{k_s} \neq \emptyset$ and $B_{k_1} \cap C_{k_2} \neq \emptyset, \ldots, B_{k_s - 1} \cap C_{k_s} \neq \emptyset$, we obtain $B_{k_s} \cap C_{k_1} = \emptyset$, for all selection of indexes $\{k_1, \ldots, k_s\} \subset J$ s.t $\cap_J S_k \neq \emptyset$

[4] We have that B_{k+1}, B_r, B_1 are pairwise disjoints, then if $S_{k+1} \cap S_r \cap S_1 = \emptyset$ we obtain a lower cycle of order 3. So, we can suppose $S_{k+1} \cap S_r \cap S_1 \neq \emptyset$, and then we can use the existence of cycle

[5] The minimization on L is defined as follow $r < 1 < \cdots < l < \cdots < k$

[6] $1 \leq l_1 < \cdots < l_{k+1} < l_k < \cdots < l_p$

$$\begin{array}{ccccc} S_{k+1} & S_l & \dots & S_{l_p} & S_r \\ C_l \cup \dots \cup C_k & C_{l_1} \cup \dots \cup C_{l-1} & \dots & C_r & C_{k+1} \cup \dots \cup C_{r-1} \\ C_{k+1} & C_l & \dots & C_{l_{p-1}} & C_r \end{array}$$

The family $B_{k+1}, B_l, \dots, B_{l_p}, B_r$ is composed of a pairwise disjoint of sets. Hence, to avoid a lower cycle of length $p+2$, we would have $S_{k+1} \cap S_l \cap \dots \cap S_r \neq \emptyset$. Also, by the definition of cycle on the selection of indexes $\{k+1, l, l_1, \dots, l_p, r\}$, we obtain either $B_{k+1} \subset C_{l+1} \cup \dots \cup C_k$ or $B_r \subset C_{k+2} \cup \dots \cup C_{r-1}$. That achieve the recurrence on l.

Consequently, we have always $B_r \subset C_{k+2} \cup \dots \cup C_{r-1}$. That achieve the recurrence on k.

This conclusion gives either $B_r = \emptyset$ or, by a discursive recurrence, $B_k \subset C_{k-1}, \forall k \in I_r$ i.e to a lower cycle of order r. □

Lemma 20.2. *Let E be a cyclic effectivity function of order r. If for some $L \in \mathcal{Q}_{r-1}$ we have that $S_k \cap S_l = \emptyset$ for all $k \neq l \in L$, then E has an upper cycle of order $\rho \leq r$.*

Proof. Let $(S_k, B_k, C_k)_{k \in I_r}$ be a cycle of order r, $L = \{1, \dots, r-1\} \in \mathcal{Q}_{r-1}$[7] s.t $S_k \cap S_l = \emptyset \; \forall k \neq l \in L$, and suppose that E has no upper cycle of order $\rho \leq r$.

Put

$$K_r = \{\alpha \in I_r, S_r \cap S_\alpha \neq \emptyset\}$$

We can proof that either $K_r = \{r\}$ or $K_r = I_r$.

By the definition of L we have $\emptyset \neq \cap_{k \in L} B_k \subset C_r$. i.e $B_k \cap C_r \neq \emptyset$ $\forall k \in L$.

So, if $l \in K_r$ we have $B_r \cap C_l = \emptyset$ and

$$B_r \subset \bigcup_{k \in K_r^c} C_k$$

If $k \in K_r^c$ i.e $S_k \cap S_r = \emptyset$, then by the definition of L and the absence of upper cycle

$$B_r \cap \left(\bigcap_{k \in K_r^c} B_k \right) \neq \emptyset$$

As

$$\left(\bigcup_{k \in K_r^c} C_k \right) \cap \left(\bigcap_{k \in K_r^c} B_k \right) = \bigcup_{k \in K_r^c} \left(C_k \cap \left(\bigcap_{k \in K_r^c} B_k \right) \right) = \emptyset$$

[7]Without loosing the generality, we can suppose $L = \{1, \dots, r-1\}$

Then if $K_r \ni k \neq r$ and $K_r^c \neq \emptyset$, we obtain $B_r = \emptyset$
Knowing that $\cap_{k\in I_r} B_k = \emptyset$[8], then the absence of upper cycle gives $S_r \cap S_k \neq \emptyset$ for at least $k \neq r \in I_r$ i.e $K_r = I_r$. Hence,

$$S_r \cap S_k \neq \emptyset, \forall k \in I_r$$

In conclusion, the absence of upper cycle is not possible. □

Proof of the Proposition : A consequence of Lemma 20.1 and Lemma 20.2.

Proposition 20.3. *There is a monotonic effectivity function E and cyclic of order 4, but E is not circular.*

Proof. Let $N = \{1, \ldots, 5\}$, $A = \{x_1, \ldots, x_4\}$ and E the effectivity function defined by $E(S_1) = \{x_2, x_3\}^+, E(S_2) = \{x_3, x_4\}^+, E(S_3) = \{x_1, x_4\}^+, E(S_4) = \{x_1, x_2\}^+$ where $S_1 = \{1,2\}, S_2 = \{2,3\}, S_3 = \{4,5\}, S_4 = \{1,3,5\}$. For $T \supset S_k, k \in \{1,2,3,4\}$, then $E(T) = E(S)$ and for T do not contain S_k, $E(T) = \{A\}$.
It is easy to verify that E is not circular but $(S_k, B_k, C_k)_{k\in I_4}$ where $B_k = \min\{B | B \in E(S_k)\}$ and $C_1 = \{x_4\}$, $C_3 = \{x_1\}$, $C_3 = \{x_2\}$, $C_4 = \{x_3\}$ is a cycle of order 4. □

Proposition 20.4. *Let E be a superadditive and cyclic effectivity function of order $r < 5$, then E has a lower or an upper cycle of order $\rho < 5$.*

Proof. For $r = 3$, see the Proposition 20.2. For $r = 4$, see annex. □

Proposition 20.5. *There is a monotonic and superadditive effectivity function E and cyclic of order $r \geq 5$, but E is not circular.*

Proof. Let $N = \{1, \ldots, 7\}$ and $A = \{x_1, \ldots, x_5\}$, and consider the effectivity function E defined as:

$$\begin{cases} E(S_1) = \{x_2, x_3, x_4\}^+, E(S_2) = \{x_3\}^+, E(S_3) = \{x_4, x_5\}^+, \\ E(S_4) = \{x_5, x_2\}^+ \text{ and } E(S_5) = \{x_1\}^+ \text{ where } S_1 = \{1,2,3,4\}, \\ S_2 = \{1,3,4,5,7\}, S_3 = \{1,2,5,6\}, S_4 = \{3,6,7\} \text{ and } S_5 = \{4,5,6,7\}. \\ \text{If } S \in \{S | S \supset S_k, k \in I_5\}, \text{ then } E(T) = E(S_k). \text{ Other wise } E(T) = \{A\}. \end{cases}$$

We claim that E is not circular but have a cycle of order 5.

Remark 20.3. If E is a monotonic $r-$circular effectivity function and if $r < \min\{m, n\}$, then E is $(r+1)-$ circular.

[8]Because the family (C_k) form a partition of A

First, E is not circular.

Suppose the contrary and let $(T_k)_{k\in I_5} \in \mathfrak{p}(N)$ and put $C_k = \{x_k\}$. For $p \in I_5^r$, denote $B_k(p) = C_k \cup \cdots \cup C_{k+p-1}$ and $S_k(p) = T_{k+p} \cup \cdots \cup T_{k-1}$. The cardinal of $(T_k), \lambda \in \{\pi(1,1,1,1,3), \pi(1,1,1,2,2) | \ \pi$ is a permutation$\}$, and every coalition S effective for some $B \subsetneq A$ is $S = S_4$ or s.t $|S| \geq 4$.
If $p = 4$, we need at least one $S_k(4)$ s.t $|S_k(p)| = 1$, and
If $p = 3$, we need at least two $S_k(3)$ s.t $|S_k(3)| \leq 3$. Then $p \leq 2$.
If $p = 1$, we need five districts coalitions $S_k(1)$ effective for one alternative. It is in opposition to the definition of E, where only superset of S_2 or S_5 is able to force social state to be member of a singleton.
If $p = 2$ and $\lambda \in \{\pi(1,1,1,1,3), \pi \in \mathcal{S}_5\}$, we need a leat two $S_k(2)$ s.t $|S_k(2)| \leq 3$. It is in contradiction with the definition of E.
If $p = 2$ and $\lambda \in \{\pi(1,1,1,2,2), \pi \in \mathcal{S}_5\}$, then[9]

$$(S_1(2), S_2(2), S_3(2), S_4(2), S_5(2)) = \pi\left((S_4, S_u, S_2, \bar{S}_v, S_w)\right)$$

Where $u \neq v \neq w \in \{1,3,5\}$ and $\bar{S}_v \supsetneq S_v$ s.t $\bar{S}_v \neq S_2$
And

$$\forall k \in I_5, S_k \text{ or } \bar{S}_k \text{ is able for } B, |B| = 2 \quad (*)$$

In view of S_1, (*) is not possible.

In conclusion, there is no p s.t $\forall k \in I_5; B_k(p) \in E(S_k(p))$ i.e E is not 5-circular.

Second, $(S_k, B_k, C_k)_{k\leq 5}$ where $C_k = \{x_k\}$ is a cycle of order 5.
So, define $H_i = \{l \in I | S_l \ni i\}$ and $(J_k)_k$ the family of maxima of $(H_i)_{i\in N}$. We have that

$$H_1 = \{1,2,3\}, H_2 = \{1,3\}, H_3 = \{1,2\}, H_4 = \{1,2,5\}, H_5 = \{2,3,5\},$$

$H_6 = \{3,4,5\}$, $H_7 = \{2,4,5\}$, and the following structure shows that E has a cycle of length 5

$$\begin{aligned}
J_1 &= \{1,2,3\} \text{ and } B_3 \subset C_4 \cup C_5; B_2 \subset C_3 \cup C_4 \cup C_5 \\
J_2 &= \{1,2,5\} \text{ and } B_2 \subset C_3 \cup C_4; B_1 \subset C_2 \cup C_3 \cup C_4 \\
J_3 &= \{2,3,5\} \text{ and } B_5 \subset C_1 \cup C_4; B_3 \subset C_1 \cup C_4 \cup C_5 \\
J_4 &= \{3,4,5\} \text{ and } B_5 \subset C_1 \cup C_2; B_4 \subset C_1 \cup C_2 \cup C_5 \\
J_5 &= \{2,4,5\} \text{ and } B_5 \subset C_1 \cup C_3; B_2 \subset C_1 \cup C_3 \cup C_5
\end{aligned}$$

□

Remark 20.4. We can prove that E is not balanced. In fact, we have $\mathcal{J}_1 = \{1k, 123, 125\}$, $\mathcal{J}_2 = \{2k, 123, 125, 235, 245\}$, $\mathcal{J}_3 = \{4k, 123, 235, 345\}$,

[9] $\forall k \neq l \in I_5, \forall p \in I_5 \setminus \{r\}; S_k(p) \setminus S_l(p) \neq \emptyset$ and $S_l(p) \setminus S_k(p) \neq \emptyset$

$\mathcal{J}_4 = \{4k, 345, 245\}$, $\mathcal{J}_5 = \{5k, 125, 235, 345, 245\}$, where the k in $\mathcal{J}_l$ is that $k \in I \setminus \{l\}$. Let σ be the $\mathcal{J}-$ selection defined as $\sigma(J) = k_J$ where $B_{k_J} \cap C_l = \emptyset, \forall l \in J$. Then, if $\Sigma_k = \{\sigma(J) | J \in \mathcal{J}_k\}$, we have that

$$\Sigma_1 = \{1,2,3,4\}, \Sigma_2 = \{2,3,5\}, \Sigma_3 = \{3,4,5\}, \Sigma_4 = \{2,4,5\}, \Sigma_5 = \{1,2,5\}$$

and

$$\bigcap_{k\in I}\left(\bigcup_{J\in\mathcal{J}_k} B_{\sigma(J)}\right) = \bigcap_{k\in I}\left(\bigcup_{l\in\Sigma_k} B_l\right) = \emptyset$$

Proposition 20.6. *Let E be an effectivity function. Then,*
lower or upper cycle $\Rightarrow$ circular $\Rightarrow$ cycle, and
cycle $\not\Rightarrow$ circular $\not\Rightarrow$ lower or upper cycle.

Proof. Upper or lower cycle $\Rightarrow$ circular is trivial, and circular $\Rightarrow$ cycle is by the Proposition 20.2.

These implications are strict :

Cycle $\not\Rightarrow$ circular is by the Proposition 20.5.

Circular $\not\Rightarrow$ upper or lower cycle is shown in the following example.

Let $N = \{1,2,3,4,5\}$, $A = \{x_1, x_2, x_3, x_4, x_5\}$ and E is s.t $E(12) = x_1x_2x_3^+$, $E(23) = x_2x_3x_4^+$, $E(34) = x_3x_4x_5^+$, $E(45) = x_4x_5x_1^+$, $E(51) = x_5x_1x_2^+$, and if $T \in \{S \subset N | S \supset \{i, (i+1) | i = 1..5 \text{mod } [5]\}\}$, then $E(T) = E(S)$, in the other case, $E(T) = \{\{A\}\}$.

$\forall S \subset N, B \in E(S)$ we have $|B| \geq 3$. consequently, if $S_1, S_2 \subset N, B_k \in E(S_k)$ $B_1 \cap B_2 \neq \emptyset$ i.e E have neither a lower nor an upper cycle.

If $C_k = \{x_k\}$, $T_k = \{k+4\}$ and $J_k = \{k, k+1, k+2\}$ then $C_{J_k} \in E(T_{J_k^c})$ i.e E is circular of order 5 with $p = 3$. □

In the following, we focused our intention on the minimal order r s.t E has a cycle or circular of order r. Denote $\sigma(S) = \min\{r, \text{E is cyclic of order r}\}$ and $\sigma(E) = +\infty$ if E is acyclic. In the same way $\nu(E) = \min\{r, \text{E is circular of order} r\}$ and $\nu(E) = +\infty$ if E is not circular.

Proposition 20.7. *Let E be a monotonic and maximal effectivity function. Then*

$$\text{either } \sigma(E) = \nu(E) \leq 3 \text{ or } \sigma(E) = \nu(E) = +\infty$$

Proof. If E is a maximal effectivity function, then either $\sigma(E) \leq 3$ or $\sigma E = +\infty$[Abdou] If $\sigma(E) = 3$, then E has either a lower or an upper cycle. As E is monotonic, then E has a lower or an upper cycles of order 3 if and only if E is circular. □

We finish this section with class of effectivity function where non circularity and acyclicity are equivalents.

An effectivity function E is additive if for some measures λ and ν on A and N

$$B \in E(S) \text{ if and only if } \lambda(B) + \nu(S) > 1$$

An effectivity function E is simple if

$$\text{either } E(S) = \{A\} \text{ or } E(S) = \mathcal{N} \text{ for all } S \subset N$$

E is anonymous if

$$E(S \cup \{i\}) = E(S \cup \{j\}) \text{ for all } S, S \cap \{i,j\} = \emptyset \text{ and for all } i,j \in N$$

E is neutral if for all $B, C \subset A$ s.t $|B| = |C|$

$$B \in E(S) \text{ if and only if } C \in E(S) \text{ for all } S \subset N$$

Proposition 20.8. *Let E be a simple effectivity function. Then, E is circular if and only if E has a cycle.*

Proposition 20.9. *Let E be a monotonic, anonymous and neutral effectivity function. Then, E is additive if and only if E is non circular.*

Proof. Every additive effectivity function is acyclic [Abdou& Keiding] then non circular[Proposition 20.1]. The following proof concern the reciprocal.

Let E be a non additive effectivity function. Then, $\forall\lambda$ a measure on A and ν a measure on N, there is $B \in E(S)$ s.t $\lambda(B) + \nu(S) \leq 1$. (*)

Let λ and ν s.t

$$\lambda(x) = \frac{1}{m}, \forall x \in A \text{ and } \nu(i) = \frac{1}{n}, \forall i \in N$$

Put $r = \min(n, m)$. We can suppose $m \geq n$ and then $r = n$. By (*), take $B \in E(S)$ s.t $\lambda(B) + \nu(S) \leq 1$ and choice $(C)_{k \in I_r} \in \mathfrak{p}(A)$ s.t for some $J \subset I_r$

$$B \subset \left(\bigcup_{k \in J} C_k\right) \text{ and } B \subset \left(\bigcup_{k \in J'} D_k\right) \Rightarrow |J'| \leq |J|, \forall (D)_{k \in I_r} \in \mathfrak{p}(A), \forall J' \subset I_r$$

Then

$$\lambda\left(\bigcup_{k \in J} C_k\right) \leq \lambda\left(\bigcup_{k \in J'} C_k\right), \forall J' \subset I_r \text{ and } |J'| = |J| \qquad (2*)$$

Rearranging the indexation of (C_k), we can suppose $J = \{1, \ldots, p\}$ and take $(T_k)_{k \in I_r} \in \mathfrak{p}(N)^{10}$ s.t $S \subset \cup_{k \notin J} T_k$. Put $J_k = \{k, \ldots, k+p\}$, then by (2*), the neutrality, the anonymously and the monotonicity

$$\bigcup_{l \in J_k} C_l \in E\left(\bigcup_{l \in J_k} T_l\right), \forall k \in I_r$$

□

In the following denote $\mathcal{C}$, $\mathcal{M}_z$, $\mathcal{M}_m$ and $\mathcal{M}_1$ be the set of $\mathfrak{c}$, $\mathfrak{m}_z$, $\mathfrak{m}_m-$, $\mathfrak{m}_1$ stable effectivity functions. Denote $\mathcal{E}$ the class of monotonic effectivity functions and $\mathcal{S}$ the class of superadditive effectivity functions. Finally, put $\mathcal{F}$ the class of effectivity function s.t $T \cap S = \emptyset, S, T \in \mathcal{N} \Rightarrow B \cup C = A, \forall B \in E(S), C \in E(T)$

20.4 Stability and Cycles

Theorem 20.2. *Let E be an effectivity function. E is $\mathfrak{c}-$ stable if and only if E is acyclic.*

Proof. Abdou & Keiding was given a proof of this theorem. We propose here a short proof in coherence with our notation and construction.

Let $E \notin \mathcal{C}$ and u a profile s.t $\mathcal{C}(E, u) = \emptyset$.
Take $x_1 \in A$ and (S_1, B_1) s.t $B_1 \in E(S_1)$ and $u^{S_1}(B_1) > u^{S_1}(x_1)$, and define

$$C_1 = \left\{x, u^{S_1}(B_1) > u^{S_1}(x_1)\right\}$$

Now, define the sequence C_k as follow

Take $x_k \in A \setminus (\cup_{l \leq k-1} C_l)$, $B_k \in E(S_k)$ s.t $u^{S_k}(B_k) > u^{S_k}(x_k)$ and

$$C_k = \left\{x \in A | u^{S_k}(B_k) > u^{S_k}(x)\right\}$$

Let r be the minimal integer satisfying $\cup_{k \leq r} C_k = A$, and put $I_r = \{1, \ldots, r\}$
If E is acyclic then, for some $J \subset I_r$, $\cap_J S_k \neq \emptyset$ and for all $k \in J$, there is $l \in J$ satisfying $B_k \cap C_l \neq \emptyset$. In this case,

$$\exists k_1, \ldots, k_s \in J \text{ s.t } B_{k_1} \cap C_{k_2} \neq \emptyset, \ldots, B_{k_s} \cap B_{k_l} \neq \emptyset, l \leq s$$

It implies that for some $i \in \cap_{k \in J} S_k$,

$$u^i(B_{k_l}) > \cdots > u^i(B_{k_s}) > u^i(B_{k_l})$$

[10] T_k is a singleton for every $k \in I_r$

A contradiction

Reciprocally: Let u be the profile: For $i \notin \cup_{k \in I_r} S_k$, u^i is an arbitrary element of $\mathcal{L}(A)$. For $i \in \cap_{k \in J} S_k$, $J = \{k_1, \ldots, k_s\}$ s.t $B_{k_l} \cap C_{k_j} = \emptyset, \forall s \geq j \geq l \geq 1$, u^i is s.t

$$u^i(B_{k_l}) > u^i(C_{k_l}) > \cdots > u^i(C_{k_s})$$

The preference $u_i \in \mathcal{L}(A)$ is well defined $\forall i \in N$, and

$$\forall i \in S_k : u^i(B_k) > u^i(C_k)$$

Knowing that (C_k) is a partition of A, then for all $x \in A$, there is $1 \leq k \leq r$ s.t (S_k, B_k) is an objection against a. □

20.4.1 *Comparison of the Bargaining Sets*

Let (S, B) an objection against x. If (T, C) is a $\mathfrak{m}_z$-counter objection against (S, B), then $S \setminus T, T \setminus S, S \cap T \neq \emptyset$ and

$$(u^{T \cap S}(C), u^{T \setminus S}(C)) \geq (u^{T \cap S}(B), u^{T \setminus S}(x)) \quad (*)$$

By $S \cap T \neq \emptyset$, we have $x \notin C$. Also, (*) is equivalent to

$$(u^{T \cap S}(C), u^{T \setminus S}(C)) > (u^{T \cap S}(B), u^{T \setminus S}(x)) \quad (2*)$$

If (T, C) is a $\mathfrak{m}_m$-counter objection against (S, B), then

$$(u^{T \cap S}(C), u^{T \setminus S}(C)) > (u^{T \cap S}(B), u^{T \setminus S}(x)) \, (3*)$$

By (2*) and (3*) $\mathfrak{m}_x, x \in \{z, m\}$ counter objection against (S, B) is an objection against a, and every $\mathfrak{m}_z-$ counter objection is a $\mathfrak{m}_m$ counter objection.

If (T, C) is a counter objection by $k \in T$ against $l \in S \setminus T$ at x, then

$$(u^{T \cap S}(C), u^{T \setminus S}(C)) \geq (u^{T \cap S}(B), u^{T \setminus S}(x)) \quad (4*)$$

If $S \cap T \neq \emptyset$, $x \notin C$: Every $\mathfrak{m}_z$-counter objection is a $\mathfrak{m}_1-$ counter objection[11].

Proposition 20.10.

$$\mathcal{C} \subset \mathcal{M}_z \subsetneq \mathcal{M}_m \subsetneq \mathcal{M}_1$$

[11] If $S \cap T = \emptyset$, it is possible that $x \in C$

Proof. As in the above, inclusions are naturals.

$\mathcal{M}_m \subsetneq \mathcal{M}_1$

Let $N = \{1,2\}$, $A = \{x,y\}$ and E be the effectivity function $E(1) = x^+, E(2) = y^+$. The only profile at which $\mathcal{C}(E,u) = \emptyset$ is s.t $u^1(x) > u^1(y)$ and $u^2(y) > u^2(x)$. Because $u^2(y) \geq u^2(y)$ and $u^1(x) \geq u^1(x)$, then $\mathcal{M}_1(E,u) = \{x,y\}$ i.e $E \in \mathcal{M}_1$.

However, at this profile $\mathcal{M}_m(E,u) = \emptyset$ i.e $E \in \mathcal{M}_m \setminus \mathcal{M}_1$

$\mathcal{M}_z \subsetneq \mathcal{M}_m$

Let $N = \{1,\ldots,m+1\}$, $A = \{x_1,\ldots,x_m\}$ and E the effectivity function s.t every $\{k\}, k \in N$ can block one alternative and every S, $2 \leq |S| \leq n-1$ can block nothing. The effectivity E has an upper cycle, then $E \notin \mathcal{C}$.

Let u a profile at which the core of E is empty. Denote x_{l_k} the worse alternative for k. We have that $|N| = |A| + 1$, then the emptiness of the core of E gives $x_{l_\alpha} = x_{l_\beta}$ for some $\alpha \neq \beta \in N$ and $x_{l_p} \neq x_{l_q}$, $\forall p, q \neq \alpha$. Coarsely $\mathcal{M}_z(E,u) = \emptyset$ but $\mathcal{M}_m(E,u) \neq \emptyset$.

So, $E \in \mathcal{M}_m \setminus \mathcal{M}_z$. □

Remark 20.5.

$$\mathcal{E} \cap \mathcal{S} \cap \mathcal{M}_m \subsetneq \mathcal{E} \cap \mathcal{S} \cap \mathcal{M}_1$$

The effectivity function in the first example above is monotonic and super-additive.

Remark 20.6.

$$\mathcal{F} \cap \mathcal{M}_z = \mathcal{F} \cap \mathcal{M}_m$$

If (T,C) is a $\mathfrak{m}_m-$counter objection against (S,B), then $a \notin B \cup C$. So $S \cap T \neq \emptyset$. However

$$\mathcal{F} \cap \mathcal{M}_m \subsetneq \mathcal{F} \cap \mathcal{M}_1$$

20.4.2 *The Zhou's Bargaining Set*

Proposition 20.11. *Let $E \in \mathcal{E} \cap \mathcal{M}_z$. Then, E is not circular.*

Proof. Suppose the contrary and let E be a circular effectivity function. Then, there is a $(T_k, C_k)_{k \in I_r}$, $(T_k) \in \mathfrak{p}(N)$ and $(C_k) \in \mathfrak{p}(A)$, and $p \in I_r^*$ s.t $\forall L \in \mathcal{L}_p : C_L \in E(T_{L^c})$ Now, define a profile $u = \left(u^i\right)$

$$\begin{cases} \forall k \in I_r\ \forall i,j \in T_k : u^i = u^j, \\ \forall i \in T_k\ u^i(C_k) < \cdots < u^i(C_{k-1}), \\ \forall i \in N \text{ argmax } u^i_{c \in C_k}(c) = c_k. \end{cases}$$

We claim that $\mathcal{M}_z(E,u) = \emptyset$

Let $x \in A$. Then for some $k \in I_r$, $x \in C_{k-1}$. If $x \neq c_{k-1}$, the pair $(N, \{c_k\})$ is a $\mathfrak{m}_z$-justified objection against x. Hence $x \notin \mathcal{M}_z(E,u)$.

Now, we shall prove that $c_{k-1} \notin \mathcal{M}(E,u)$.

Set

$$\rho = \min\left\{\beta | \bigcup_{l=k}^{k+\beta-1} C_l \in E\left(\bigcup_{l=k+\beta}^{k-1} T_l\right)\right\}$$

By the circularity of E; $1 \leq \rho \leq p-1$.

Put $B = C_k \cup \ldots \cup C_{k+\rho-1}$ and $S = T_{k+\rho} \cup \ldots \cup T_{k-1}$. By the definition of ρ, the pair (S,B) is an objection against c_{k-1}, which will be $\mathfrak{m}_z-$ justified

Suppose the contrary and assume that (T,C) is a $\mathfrak{m}_z-$counter objection against (S,B). Then $C \in E(T)$ and

$$\forall i \in S \cap T \quad u^i(C) \geq u^i(B) \quad (*)$$

$$\forall i \in T \setminus S \quad u^i(C) \geq u^i(c_{k-1})$$

The set $T \setminus S \subset T_k \cdots \cup T_{k+\rho-1}$. Then, if $\theta \in [0,\ldots,\rho-1]$ is the first index satisfying $(T \setminus S) \cap T_{k+\theta} \neq \emptyset$, we have that

$$\begin{cases} u^i(C_{k+\theta}) < \cdots < u^i(C_{k-1}) \\ u^i(C_{k-1}) < \cdots < u^i(C_{k+\theta-1}) \end{cases}$$

and then, $C \subset C_{k-1} \cup \cdots \cup C_{k+\theta-1}$

This inclusion, in view of (*) implies that $C \cap C_{k-1} \neq \emptyset$.

In conclusion

$$C \subset C_k \cup \ldots \cup C_{k+\theta-1} \qquad (**)$$

As θ is the first index s.t $T \cap T_{k+\theta} \neq \emptyset$, then by the definition of S, we get

$$T \setminus S \subset T_{k+\theta} \cup \ldots \cup T_{k+\rho-1}$$

$$T \cap S \subset T_{k+\rho} \cup \ldots \cup T_{k-1}$$

Consequently,

$$T \subset T_{k+\theta} \cup \ldots \cup T_{k+\rho+1} \cup \ldots \cup T_{k-1}$$

So, by (**) and the monotonicity of E

$$\theta \in \left\{\beta | \bigcup_{l=k}^{k+\beta-1} C_l \in E\left(\bigcup_{l=k+\beta}^{k-1} T_l\right)\right\}$$

It contradicts the definition of ρ, because $1 \leq \theta \leq \rho - 1$ i.e $\theta < \rho$. □

Corollary 20.1. *Let E be a maximal effectivity function. Then,*

$$E \in \mathcal{E} \cap \mathcal{M}_z \Leftrightarrow E \in \mathcal{E} \cap \mathcal{C}$$

It is a natural implication of the Proposition 20.7.

Corollary 20.2. *Let E be a monotonic, anonymous and neutral effectivity function. Then,*

$$E \in \mathcal{M}_z \Leftrightarrow E \in \mathcal{C}$$

Corollary 20.3. *Let $E \in \mathcal{S} \cap \mathcal{M}_z$. If E is circular, then the order of the circularity $r \geq 5$*

It is a consequence of the Proposition 20.4 and the Proposition 20.7.

Proposition 20.12.

Let E be a monotonic and super additive effectivity function. Then the non circularity is not sufficient for the $\mathfrak{m}_z-$ stability.

Proof. Let E be the effectivity function of the Proposition 20.5, and let u be the profile

1	2	3	4	5	6	7
x_5	x_5	x_5	x_4	x_1	x_1	x_1
x_4	x_4	x_3	x_3	x_4	x_5	x_5
x_3	x_2	x_2	x_2	x_5	x_2	x_3
x_2	x_3	x_4	x_1	x_3	x_4	x_2
x_1	x_1	x_1	x_5	x_2	x_3	x_4

We claim that $\mathcal{M}_z(E,u) = \emptyset$
Denote $B_k = \min\{B|B \in E(S_k)\}$. We observe that at u, (S_k, B_k) is an objection against x_k and not x_l, $\forall k \in I_5, l \in I_5 \setminus \{k\}$.
Let $x_{k_0} \in A$, then the pair (S_{k_0}, B_{k_0}) is a justified objection against x_{k_0}.
So, suppose that (T, C) is a $\mathfrak{m}_z$counter objection against (S_{k_0}, B_{k_0}). i.e

$$\begin{cases} \forall i \in S \cap T & u^i(C) \geq u^i(B_{k_0}) \\ \forall i \in T \setminus S & u^i(C) \geq u^i(x_{k_0}) \\ \text{And then} & u^T(C) > u^T(x_{k_0}) \end{cases}$$

By the definition of E, we have

$$T \supset S_k \text{ and } C \supset B_k \text{ for some } k \in I_5, l \neq k_0$$

Therefore, (T, C) is an objection against x_k and against x_{k_0}. In this case, $(S_k, B_k)_{k \in I \setminus \{k_0\}}$ is a cycle of order 4. By the Proposition 20.4, E is circular. In contradiction with the Proposition 20.5. □

Proposition 20.13. *If E is a non monotonic effectivity function, then the non circularity is not necessary for the $\mathfrak{m}_z$−stablity.*

Proof. Let $N = \{1,2,3\} \cup \{\alpha_1, \ldots, \alpha_5\}$ and $A = \{x_1, x_2, x_3\}$, and consider the effectivity function E defined as follow : Denote

$$\mathcal{Q}_k = \{\{k, k+1\}, \{\alpha_p, k, k+1\}, \{\alpha_1, \alpha_p, k, k+1\} \,|p \in \{1, \ldots, 5\}\}$$

$$\begin{cases} \forall k \in I_3 \forall S \in \mathcal{Q}_k, E(S) = \{x_{k+2}\}^+ \\ \forall S \notin \mathcal{Q}_k, k \in I_3, E(S) = \{A\}. \end{cases}$$

First, the effectivity function E is 3-circular.

If $T_1 = \{\alpha_1, 1\}$, $T_2 = \{\alpha_p, 2\}$, $T_3 = \{3\}$ and $C_k = \{x_k\}$, we have that

$$C_k \in E(T_{k+1} \cup T_{k+2}), \forall k \in I_3$$

Second, we claim that $E \in \mathcal{M}_z$

Because $\mathcal{C} \subset \mathcal{M}_z$, then it is sufficient to prove that $\mathcal{M}_z(E,u) \neq \emptyset$ for every u s.t $\mathcal{C}(E,u) = \emptyset$. By the circularity of E, $\mathcal{C}(E,u) = \emptyset$ for some profile u.

So, let u be a profile at which the core of E is empty.

The objections against x_k are of the form $(\{\alpha_p, k, s\}, x_{\bar{s}})$ or of the form $(\{\alpha 1, \alpha_p, k, s\}, x_{\bar{s}})$ where $s \neq \bar{s} \in \{k, k+1\}$.

Then, preferences of k, s satisfy

$$\begin{cases} u^k(x_{\bar{s}}) > u^k(x_k) \\ u^s(x_{\bar{s}}) > u^k(x_k) \end{cases}$$

The objections against x_s are of the form $(\{\alpha_q, s, l\}, x_{\bar{l}})$ or of the form $(\{\alpha_1, \alpha_q, s, l\}, x_{\bar{l}})$ where $l \neq \bar{l} \in \{k, \bar{s}\}$.

If $l = k$, preferences of k, s satisfy

$$\begin{cases} u^k(x_{\bar{s}}) > u^k(x_s) \\ u^s(x_{\bar{s}}) > u^k(x_s) \end{cases}$$

In this case, $x_{\bar{s}}$ is at the top for k and s. So, an objection against $x_{\bar{s}}$ do not contain either k nor s. It is not possible by the definition of E, and then $l = \bar{s}$.

Finally, in the same argument, the objections against $x_{\bar{s}}$ are of the form $(\{\alpha_t, \bar{s}, k\}, x_k)$ or of the form $(\{\alpha_1, \alpha_t, \bar{s}, k\}, x_k)$

In conclusion, preferences of $1, 2, 3$ are

1	2	3	or	1	2	3
x_3	x_1	x_2		x_2	x_3	x_1
x_2	x_3	x_1		x_3	x_1	x_2
x_1	x_2	x_3		x_1	x_2	x_3

By the definition of E, there is $\alpha_{p_0} \in \{\alpha_1, \ldots, \alpha_5\}$ s.t α_{p_0} do not participate for the emptiness of $\mathcal{C}(E, u)$.
Suppose that x_k is the worse alternative for α_{p_0}. Then, the objections against x_k are $(\{\alpha_p, k, s\}, x_{\bar{s}})$ and $(\{\alpha_{p_0}, k, s\}, x_{\bar{s}})$, eventually one or both two coalitions are with α_1, for some $p \neq p_0$. Coarsely, if one of them is an objection, the other is a $\mathfrak{m}_z-$ counter objection. i.e $x_k \in \mathcal{M}_z(E, u)$. □

Corollary 20.4.

$$\mathcal{C} \subsetneq \mathcal{M}_z{}^{12}$$

20.4.3 *The Mass-Colell's Bargaining Set*

Remark 20.7. If (T, C) is a $\mathfrak{m}_m-$ but not $\mathfrak{m}_z-$ counter objection against an objection (S, B), we have $S \subset T$, $T \subset S$ or $S \cap T = \emptyset$.

For $S \in \mathcal{P}(N)$, denote $\mathcal{B}_S = \min\{B | B \in E(S)\}$.

Lemma 20.3. *Let E be a regular $\mathfrak{m}_z-$ unstable effectivity function. Then, there is a profile v satisfying $\mathcal{M}_z(E, v) = \emptyset$ and if (T, C) is a $\mathfrak{m}_m-$ counter objection against a $\mathfrak{m}_z-$justified objection (S, B) at v, we have $S \cap T = \emptyset$*

Proof. First, let $x \in A$ and (S, B) is a $\mathfrak{m}_z-$justified objection against x s.t $B \in \mathcal{B}_S$ and if (S', B') is a $\mathfrak{m}_z-$justified objection against x, then $S' \not\supset S$. Let (T, C) be a $\mathfrak{m}_m-$ counter objection against (S, B)

If $T \supset S$, then

$$u^T(C) > \left(u^S(B), u^{T \setminus S}(x)\right)$$

Consequently, if (T', C') is a $\mathfrak{m}_z-$ counter objection against (T, C), we have

$$\begin{aligned} u^{T'}(C') &> \left(u^{T' \cap S}(C), u^{T' \cap (T \setminus S)}(C), u^{T' \setminus T}(x)\right) \\ &> \left(u^{T' \cap S}(B), u^{T' \cap (T \setminus S)}(x), u^{T' \setminus T}(x)\right) \\ &> \left(u^{T' \cap S}(B), u^{T' \setminus S}(x)\right) \end{aligned}$$

As (S, B) is $\mathfrak{m}_z-$justified objection against x, then (T, C) is a $\mathfrak{m}_z-$justified objection against x. In contradiction with the definition of (S, B).

So, we can suppose that: *A pair (T, C), a $\mathfrak{m}_m-$ but not a $\mathfrak{m}_z-$ counter objection against an objection (S, B) satisfy either $S \cap T = \emptyset$ or $T \subset S$*

[12]We think that $\mathcal{E} \cap \mathcal{M}_z = \mathcal{E} \cap \mathcal{C}$

Second, let $x_1 \in A$ and (S_1, B_1)[13] be a $\mathfrak{m}_z$−justified objection against x_1 satisfying: There is a $\mathfrak{m}_m$−counter objection (T, C) against (S_1, B_1) and $T \subset S_1$.
Take

$$T_1 \in \min\{T \subset S_1 | (T, C) \text{ is a } \mathfrak{m}_m - \text{counter objection against } (S_1, B_1)\}$$

Put

$$J_1 = S_1 \setminus T_1$$

And define a profile u_1 as follow

$$\begin{cases} u_1^i = u^i, \text{ for } i \notin J_1 \\ \forall i \in J_1; u_1^i(x_1) = \max_{x \in A} u_1^i(x) \text{ and} \\ u_1^i(x) > u_1^i(y) \Leftrightarrow u^i(x) > u^i(y), \forall x, y \neq x_1 \end{cases}$$

Now, let $x_2 \neq x_1 \in A$ and (S_2, B_2) be a $\mathfrak{m}_z$−justified objection against x_2 at u.
If (T_2, C_2) is a $\mathfrak{m}_z$− counter objection against (S_2, B_2)[14] or a $\mathfrak{m}_m$− counter objection against (S_2, B_2) s.t $T_2 \subset S_2$ at u_1, then

$$\begin{array}{lllclll} u_1^{T_2}(C_2) & \geq & \left(u_1^{T_2 \cap S_2}(B_2), u_1^{T_2 \setminus S_2}(x_2)\right) & \text{or} & u_1^{T_2}(C_2) & > & u_1^{T_2}(B_2) \\ u^{T_2}(C_2) & \not\geq & \left(u^{T_2 \cap S_2}(B_2), u^{T_2 \setminus S_2}(x_2)\right) & & u^{T_2}(C_2) & \not> & u^{T_2}(B_2) \end{array}$$

Each of these system implies $T_2 \cap J_1 \neq \emptyset$, $x_1 \notin B_2$ and

$$C_2 = \{x_1\}$$

Put

$$J_2 = J_1 \cap T_2$$

and define a profile v as follow

$$\begin{cases} v^i = u^i, \text{ for } i \notin J_1 \\ v^i = u_1^i, \text{ for } i \in J_1 \setminus J_2 \\ \forall i \in J_2; v^i(x_2) > v^i(x_1) > v^i(x) \forall x \neq x_1, x_2 \text{ and} \\ v^i(x) > v^i(y) \Leftrightarrow u^i(x) > u^i(y), \forall x, y \neq x_k, k = 1, 2. \end{cases}$$

We claim that v satisfy the desired condition of the lemma.
First case: the assertion is true for (T_1, C_1) against x_1

[13]We suppose that a pair satisfying this condition is unique. We can prove this assertion by a recursive argument.
[14]Without loosing the generality, we suppose that a pair satisfying this condition is unique.

If (T, C) is $\mathfrak{m}_z-$ counter objection against (T_1, C_1) at v, then

$$v^T(C) \geq \left(v^{T\cap T_1}(C_1), v^{T\setminus T_1}(x_1)\right)$$

- If $T \cap J_1 = \emptyset$, we have $T \cap T_1 = T \cap S_1$ and $T \setminus T_1 = T \setminus S_1$. I.e

$$v^T(C) \geq \left(v^{T\cap S_1}(B_1), v^{T\setminus S_1}(x_1)\right)$$

In contradiction with (S_1, B_1) is a $\mathfrak{m}_z-$ justified objection against x_1.
- If $T \cap (J_1 \setminus J_2) \neq \emptyset$, then there is $i \in T$ s.t x_1 is at the top for i i.e $v^T(C) > v^T(x_1)$ is not possible.
- If $T_\cap J_2 \neq \emptyset$, we have $C = \{x_2\}$. Then by the regularity of E, $T \cap T_2 \neq \emptyset$. So, for some $i \in T \cap T_2$, we have $u^i(x_2) > u^i(x_1)$ and $u^i(x_1) > u^i(x_2)$. A contradiction, i.e $T \cap J_2 = \emptyset$.

Consequently, (T_1, C_1) is a $\mathfrak{m}_z-$justified objection against x_1.

By the minimality of T_1, we can not have a $\mathfrak{m}_m-$ counter objection (T, C) against (T_1, C_1) s.t $T \not\subseteq T_1$.

Second case: The assertion is true for (S_2, B_2) against x_2.

The pair (S_2, B_2) is a $\mathfrak{m}_z-$ justified objection against x_2 at v.
In fact, (S_2, B_2) is $\mathfrak{m}_z-$ justified objection against x_2 at u_1, and $x_1 \notin C, B_2$. Then

$$v^T(x) > v^T(y) \Leftrightarrow u^T(x) > u^T(y), \forall x, y \in C \cup B_2$$

The choice of T_2 and J_2 gives that if (T, C) is a $\mathfrak{m}_m-$ counter objection against (S_2, B_2) at v then $T \subsetneq S_2$.

Third case: the assertion is true for any (S, B) against x.
If (T, C) is a $\mathfrak{m}_z-$ counter objection against (S, B), or a $\mathfrak{m}_m-$counter objection s.t $T \subset S$, at v but not neither at u nor at u_1, we have $T \cap J_2 \neq \emptyset$ and $C \subset \{x_1, x_2\}$. If $B \cap \{x_1, x_2\} = \emptyset$, T, C is a counter-objection at u_1, then $x_1 \in B(x_2$ is at the top). Consequently $C = \{x_2\}$ and by the regularity $T \cap T_2 \neq \emptyset$. Again, if $i \in T \cap T_2$, then $u^i(x_2) > u^i(x_1)$ and $u^i(x_1)$ $> u^i(x_2)$. □

Theorem 20.3.

$$\mathcal{S} \cap \mathcal{M}_z = \mathcal{S} \cap \mathcal{M}_m$$

Proof. Every super additive effectivity function is regular, then by the precedent lemma, we can suppose that every (T, C), a $\mathfrak{m}_m-$ counter objection against (S, B), a $\mathfrak{m}_z-$ justified objection against x is s.t $S \cap T = \emptyset$

Let $E \in \mathcal{M}_z \setminus \mathcal{M}_m$ and u be a profile s.t $\mathcal{M}_z(E, u) = \emptyset$. Suppose that there is a pair (S, B), a $\mathfrak{m}_z-$ but not a $\mathfrak{m}_m-$ justified objection against an

alternative x at u, and denote $\mathcal{O} = \{(T,C)|(T,C)$is a $\mathfrak{m}_m-$ but not a $\mathfrak{m}_z-$ counter objection against$(S,B)\}$.
$B_u \in \min\{D \subset B \cap C | D \in E(Q)$ for some $Q \in \mathcal{P}(N), C \in E(T)$ and $(T,C) \in \mathcal{O}\}$.

$$S_u = \max\left\{Q \in \mathcal{P}(N) | B_u \in E(Q) \text{ and } u^Q(B_u) > u^Q(x_u)\right\}$$

We can prove that (S_u, B_u) is a justified $\mathfrak{m}_m-$ objection against x_u.
Let (T,C) be a $\mathfrak{m}_m-$ counter objection against (S_u, B_u), then

$$u^T(C) > \left(u^{T\cap S_u}(B_u), u^{T\setminus S_u}(x_u)\right)$$

As $S_u \cap T \supset S \cap T$, $(S_u \setminus T) \supset (S \setminus T)$ and $u^i(C) > u^i(x_u), \forall i \in T \supset (T \setminus S)$, then

$$u^T(C) > \left(u^{T\cap S}(B_u), u^{T\setminus S}(x_u)\right)$$

If $T \cap S \neq \emptyset$, then the pair (T,C) is a $\mathfrak{m}_z-$counter objection against (S,B). A contradiction. Necessarily, $S \cap T = \emptyset$. By the super additivity of E, we have that $C \cap B \neq \emptyset$ and $C \cap B \in E(S \cup T)$
If (T,C) is a $\mathfrak{m}_m-$ but not $\mathfrak{m}_z-$ counter objection against (S_u, B_u), then $S_u \cap T = \emptyset$. By the super additivity, $B_u \cap C \in E(S_u \cup T)$. It is in contradiction with the definition of B_u and S_u. □

Theorem 20.4.

$$\mathcal{E} \cap \mathcal{M}_z = \mathcal{E} \cap \mathcal{M}_m$$

Remark 20.8. We know by the Proposition 20.11 that every $E \in \mathcal{E} \cap \mathcal{M}_z$ is non circular, then regular. So, the theorem do not change if we add regular on the condition i.e the Lemma 20.3 is valid for this theorem.

Lemma 20.4. *Let $E \notin \mathcal{M}_z$ be a monotonic effectivity function and u a profile s.t $\mathcal{M}_z(E,u) = \emptyset$. If $\mathcal{M}_m(E,u) \neq \emptyset$, then there is a profile v s.t $\mathcal{M}_m(E,v) \subsetneq \mathcal{M}_m(E,u)$.*

Proof. Let $E \in \mathcal{M}_m \setminus \mathcal{M}_z$ and u a profile s.t $\mathcal{M}_z(E,u) = \emptyset$.
Set

$$A_0 = \mathcal{M}_m(E,u) \text{ and } A_1 = A \setminus A_0$$

Take $x_0 \in A_0$ and let (S_0, B_0) be a $\mathfrak{m}_z-$ but not $\mathfrak{m}_m-$ justified objection against (S_0, B_0). Consider $(T_k^0, C_k^0)_{k=1\ldots s_0}$ the list of $\mathfrak{m}_m-$ counter objection against (S_0, B_0). Then, by the precedent lemma

$$T_k^0 \cap S_0 = \emptyset, \forall k \in \{1, \ldots, s_0\}$$

Set

$$J_0 = \bigcup_{k=1}^{s_0} T_k^0$$

and define the profile

$$\begin{cases} u_1^i = u_0^i, \forall i \notin J_0. \\ \forall i \in J_0 : u_1^i(x_0) = \max_{x\in A} u_1^i(x) \text{ and} \\ u_1^i(x) > u_1^i(y) \Leftrightarrow u_0^i(x) > u_0^i(y), \forall x, y \neq x_0 \end{cases}$$

If $\mathcal{M}_m(E, u_1) \cap A_1 \neq \emptyset$, we note $\{x_1, \ldots, x_\alpha\} = \mathcal{M}_m(E, u_1) \cap A_1$.

Let $(S_l, B_l)_{l=1\ldots\alpha}$ be a $\mathfrak{m}_m$−justified objection against x_l at u_0, and consider $(T_k^l, C_k^l)_{k=1\ldots s_l}$ the list of $\mathfrak{m}_m$− counter objection against (S_l, B_l) at u_1. Then,

$$\begin{cases} u_1^{T_k^l}(C_k^l) > \left(u_1^{T_k^l\cap S_l}(B_1), u_1^{T_k^l\setminus S_l}(x_l)\right) \\ u_0^{T_k^l}(C_k^l) \not> \left(u_0^{T_k^l\cap S_l}(B_l), u_0^{T_k^l\setminus S_l}(x_l)\right) \end{cases} \qquad (*)$$

As $u_1^i = u_0^i, \forall i \notin J_0$, we should have $T_k^l \cap J_0 \neq \emptyset, \forall l, k$.
So, take $i \in T_k^l \cap J_0$ s.t

$$\begin{cases} u_1^i(C_k^l) \geq u_1^i(x) \\ u_0^i(C_k^l) < u_0^i(x) \end{cases} \quad \text{or} \quad \begin{cases} u_1^i(C_k^l) > u_1^i(x) \\ u_0^i(C_k^l) \leq u_0^i(x) \end{cases} \qquad (2*)$$

Where $x = \text{argmin } u_1^i(B_l)$ if $i \in T_k^l \cap S_1$ and $x = x_l$ if $i \in T_k^l \setminus S_l$.

By (2*), if argmin $u_1^i(C_k^l) = b \neq x_0$ we have $u_1^i(y) > u_1^i(x) \Leftrightarrow u_0^i(y) > u_0^i(x)$, and then inequalities (2*) lead to a contradiction.

In conclusion,

$$\text{argmin } u_1^i(C_k^l) = x_0$$

As $i \in J_0$ i.e x_0 is at the top for i at u_1, we obtain

$$C_k^l = \{x_0\}, \forall k = 1\ldots s_l, \forall l = 1\ldots\alpha \qquad (3*)$$

By the (*) and (3*), put

$$J_1 = J_0 \setminus \{i | i \text{ satisfy } (2*)\}$$

We have that

$$\emptyset \neq J_1 \subsetneq J_0\text{[15]}$$

Define u_2 the profile

$$\begin{cases} u_2^i = u_0^i, \forall i \notin J_1. \\ u_2^i = u_1^i, \forall i \in J_1 \end{cases}$$

[15] If $T_k^l \subset J_0, \forall l, k$, the pair $(\{i | x_0 \text{ is at the top for } i \text{ at } u_1\}, \{x_0\})$ is a $\mathfrak{m}_m$−justified objection against (S_l, B_l) at u_1. i.e $\mathcal{M}_m(E, u_1) \subset \mathcal{M}_m(E, u_0)$

We claim that

$$\mathcal{M}_m(E,u_2) \subsetneq \mathcal{M}_m(E,u)$$

First, $x_0 \notin \mathcal{M}(E,u_2)$.
We prove at first that $x_0 \notin \mathcal{M}_m(E,u_1)$.
If (T,C) is a $\mathfrak{m}_m-$ counter objection against (S_0,B_0) at u_1, then either

$$\begin{cases} u_1^T(C) > \left(u_1^{T\cap S_0}(B_0), u_1^{T\setminus S_0}(x_0)\right) \\ u_0^T(C) \ngeq \left(u_0^{T\cap S_0}(B_0), u_0^{T\setminus S_0}(x_0)\right) \end{cases}$$

Or

$$T \cap S_0 = \emptyset \text{ and } u_1^T(C) > u_1^T(x_0)$$

In the first case, we have $T \cap J_0 \neq \emptyset$ and then $C = \{x_0\}$. A contradiction. In the second case, $T = T_k^0 \subset J_0$ for some $k \in \{1, \dots s_0\}$ i.e x_0 is at the top for every $i \in T$ i.e we cannot have $u_1^T(C) > u_1^T(x_0)$ for some $C \subset A$.

Now, if (T,C) is a $\mathfrak{m}_m-$ counter objection against (S_0,B_0) at u_2, then

$$\begin{cases} u_2^T(C) > \left(u_2^{T\cap S_0}(B_0), u_2^{T\setminus S_0}(x_0)\right) \\ u_1^T(C) \ngtr \left(u_1^{T\cap S_0}(B_0), u_1^{T\setminus S_0}(x_0)\right) \end{cases} \qquad (4*)$$

So, to maintain $u_2^T(C) > \left(u_2^{T\cap S_0}(B_0), u_2^{T\setminus S_0}(x_0)\right)$, necessarily $T \cap J_1 = \emptyset$. As $u_2^i = u_1^i, \forall i \in J_1 \cup J_0^c$, then by (4*), $T \cap (J_0 \setminus J_1) \neq \emptyset$.
Let $i \in T \cap (J_0 \setminus J_1) \neq \emptyset$, *i.e* $u_2^i = u_0^i$, s.t

$$\begin{cases} u_0^i(C) \geq u_0^i(x) \\ u_1^i(C) < u_1^i(x) \end{cases} \quad \text{or} \quad \begin{cases} u_0^i(C) > u_0^i(x) \\ u_1^i(C) \leq u_1^i(x) \end{cases}$$

Where $x = \text{argmin } u_0^i(B_0)$ if $i \in T \cap S_0$ and $x = x_0$ if $i \in T \setminus S_0$.
Again, as in (2*), we have $C = \{x_0\}$, that is impossible. i.e we conclude that

$$x_0 \notin \mathcal{M}(E,u_2)$$

Second, $\mathcal{M}_m(E,u_2) \subset A_0$
Let $x \in \mathcal{M}_m(E,u_2) \setminus \mathcal{M}_m(E,u)$. If $x \in \mathcal{M}_m(E,u_1) \cap A_1$ i.e $x = x_l$ for some $l \in \{1,\dots,\alpha\}$, let $(R_k,D_k)_{k=1\dots s_r}$ the list of $\mathfrak{m}_m-$counter objection against x_l at u_2. By (*) and the definition of J_1 i.e $u_2^i = u_1^i, \forall i \in (J_0 \setminus J_1)^c$, we have $R_k \neq T_{k'}^l, \forall k' \in \{1,\dots,s_l\}$. Then,

$$\begin{cases} u_2^{R_k}(D_k) > \left(u_2^{R_k\cap S_l}(B_l), u_2^{R_k\setminus S_l}(x_l)\right) \\ u_1^{R_k}(D_k) \ngtr \left(u_1^{R_k\cap S_l}(B_l), u_1^{R_k\setminus S_l}(x_l)\right) \\ u_0^{R_k}(D_k) \ngtr \left(u_0^{R_k\cap S_l}(B_l), u_0^{R_k\setminus S_l}(x_l)\right) \end{cases} \qquad (5*)$$

As $u_2^i = u_1^i, \forall i \in (J_0 \setminus J_1)^c$ and $u_2^i = u_0^i, \forall i \in J_1^c$, by the first and the second, and the first and third relation of (5*), we can choice $i \in (J_0 \setminus J_1) \cap R_k$ and $j \in J_1 \cap R_k$ s.t

$$\begin{cases} u_0^i(D_k) \geq u_0^i(x) \\ u_1^i(D_k) < u_1^i(x) \end{cases} \quad \text{or} \quad \begin{cases} u_0^i(D_k) > u_0^i(x) \\ u_1^i(D_k) \leq u_1^i(x) \end{cases} \quad (6*)$$

and

$$\begin{cases} u_1^j(D_k) \geq u_1^j(y) \\ u_0^j(D_k) < u_0^j(y) \end{cases} \quad \text{or} \quad \begin{cases} u_1^j(D_k) > u_1^j(y) \\ u_0^j(D_k) \leq u_0^j(y) \end{cases} \quad (7*)$$

Where $x = \text{argmin } u_1^i(B_l)$ if $i \in R_k \cap S_l$ and $x = x_l$ if $i \in R_k \setminus S_0$, and $y = \text{argmin } u_0^j(B_l)$ if $j \in R_k \cap S_l$ and $y = x_l$ if $i \in R_k \setminus S_0$
Again, in the same argument to (2*), we have by (6*) or (7*) that $D_k = \{x_0\}$. If $x = x_l$ or $y = x_l$, we can see easily that (6*) and (7*) are incompatible. So, we suppose

$$x = \text{argmin } u_1^i(B_l) \neq y = \text{argmin } u_0^j(B_l)$$

In this case, (6*) and (7*) give:

$$\begin{cases} u_0^i(x_0) \geq u_0^i(B_l) \\ u_0^j(x_0) < (\text{or } \leq) u_0^j(B_l) \end{cases} \quad \text{or} \quad \begin{cases} u_0^i(x_0) > u_0^i(B_l) \\ u_0^j(x_0) \leq (\text{or } <) u_0^j(B_l) \end{cases}$$

and

$$\begin{cases} u_1^j(x_0) \geq u_1^j(B_l) \\ u_1^i(x_0) < (\text{or } \leq) u_1^i(B_l) \end{cases} \quad \text{or} \quad \begin{cases} u_1^j(x_0) > u_1^j(B_l) \\ u_1^i(x_0) \leq (\text{or } <) u_1^i(B_l) \end{cases}$$

As $i \in J_0$ i.e x_0 is at the top for i at u_1, then the second equation of 3.11 gives $B_l = \{x_0\}$. Finally, $D_k = B_l = \{x_0\}$ i.e (R_k, D_k) can not be a $\mathfrak{m}_m$−counter objection against (S_l, B_l).

Therefore,

$$\mathcal{M}_m(E, u) \subsetneq \mathcal{M}_m(E, u_2)$$

□

Proof of the Theorem

Let $E \in \mathcal{M}_m \backslash \mathcal{M}_z$, u a profile s.t $\mathcal{M}_z(E, u) = \emptyset$. Choice $(u_p)_{p \geq 0}$ a sequence of profiles s.t $u_0 = u$ and $\mathcal{M}_m(E, u_{p-1}) \subsetneq \mathcal{M}_m(E, u_p)$.
As $\mathcal{M}_m(E, u_0) \subset A$, A is a finite set. Then, there is p_{max} s.t $\mathcal{M}_m(E, u_{max}) = \emptyset$. □

Concluding Remarks

The initial aim of this paper is to gives necessary and sufficient conditions for the stability of a bargaining set. We was considered this problem as formulation of cycles in term of partition of $N \otimes A$.

Results in this paper prove the non relevance of stability of the ADM's bargaining set as acyclicity. For the two others, which are equivalent in monotonicity, bargaining sets stability and the core stability are equivalent in several class of effectivity function : Maximal effectivity function, simple effectivity function, symmetric and neutral effectivity function. Moreover, we think that monotonicity is sufficient to get this equivalence. In this case, the formulation of cycle in term of partition of $N \otimes A$ is just to simplify the comprehensiveness of the acyclicity.

20.5 Annexe

Proof of the Proposition 20.7 with $r = 4$

Note that the set of indexation $I = \{1, \ldots, 4\}$ is a set of integer modulo 4.

$$\mathcal{Q}_3 = \{\{1,2,3\}, \{1,2,4\}, \{1,3,4\}, \{2,3,4\}\}$$

If $J \in \mathcal{Q}_3$, then $\cap_{k \in J} S_k = \emptyset$ and $S_k \cap S_l = \emptyset$ for some $k \neq l \in J$. Define

$$\mathcal{R} = \{\{k,l\} \,|S_k \cap S_l \neq \emptyset; k \neq l \in J \in \mathcal{Q}_3\}$$

After counting, the set $\mathcal{R}$ is a super set of a set of the form $\{\{k,l\}, \{p,q\}\}$ where k, l, p, q are different, or of the form $\{\{k, k+1\}, \{k, k+2\}, \{p,q\}\}$ where $p, q \in \{k+1, k+2, y \neq k\}$[16].
In the following, the notation $k_1 \ldots k_s$ represent $\{k_1, \ldots, k_s\}$

First case : $\mathcal{R}$ contain a set of the form $\{kl, pq\}$; k, l, p, q are different
The relation $S_k \cap S_l \neq \emptyset$ gives

$$B_x \subset C_p \cup C_q \text{ for some } x \in \{k,l\} \quad (1)$$

and $S_p \cap S_q \neq \emptyset$ implies

$$B_y \subset C_k \cup C_l \text{ for some } y \in \{p,q\} \quad (2)$$

Denote $\bar{x} = \{k,l\} \setminus \{x\}$ and $\bar{y} = \{p,q\} \setminus \{y\}$. By (1) and (2): $S_x \cap S_y \neq \emptyset$ and then

$$B_y \subset C_{\bar{x}} \text{ or } B_x \subset C_{\bar{y}}$$

[16]A set like $\{\{1,2\}, \{3,4\}, \{2,3\}\}$ is not considered in this case because it is a super set of $\{\{1,2\}, \{3,4\}\}$. Every set with 4 elements is considered in this class

Suppose that

$$B_y \subset C_{\bar{x}}{}^{17} \qquad (2')$$

Then $S_{\bar{x}} \cap S_y \neq \emptyset$. Hence,

$$B_{\bar{x}} \subset C_x \cup C_{\bar{y}} \qquad (3)$$

If $S_x \cap S_{\bar{y}} = \emptyset$, then by the super additivity

$$C_y \supset B_x \cap B_{\bar{y}} \in E(S_x \cup S_{\bar{y}}) \qquad (4)$$

Knowing that $(S_{\bar{y}} \cup S_x) \cap S_{\bar{x}} \cap S_y = \emptyset$ whatever $S_{\bar{x}} \cap S_y \neq \emptyset$, we obtain $S_{\bar{y}} \cup S_x \neq N$. So, by (2'), (3) and (4)

$(S_{\bar{y}} \cup S_x, S_{\bar{x}}, S_y, B_x \cap B_{\bar{y}}, B_{\bar{x}}, B_y)$ is a lower cycle

Therefore, $S_x \cap S_{\bar{y}} \neq \emptyset$ i.e $B_x \subset C_y$ or $B_{\bar{y}} \subset C_{\bar{x}} \cup C_y$. If $B_x \subset C_y$, then by (2') and (3) $B_x, B_{\bar{x}}, B_y$ is a partition of A. If $B_{\bar{y}} \subset C_{\bar{x}} \cup C_y$, then $S_{\bar{x}} \cap S_{\bar{y}} \neq \emptyset$. In the first case, $(S_x, S_{\bar{x}}, S_y, B_x, B_{\bar{x}}, B_y)$ is a lower cycle and in the second that is $(S_{\bar{x}}, S_y, S_{\bar{y}}, B_{\bar{x}}, B_y, B_{\bar{y}})$. This conclusion achieves the proof for the first case.

Lemma 20.5. *Let E be a superadditive and cyclic affectivity function of order 4. If $\mathcal{R}$ contain a set of the form $\{kk+1, kk+2, kk+3\}$ for some $k \in \{1, \ldots, 4\}$, then E has a lower cycle of order 3.*

Proof. Suppose $k = 1$, then

$$B_1 \subset C_3 \cup C_4(\alpha 1) \text{ or } B_2 \subset C_3 \cup C_4(\beta 1)$$

$$B_1 \subset C_2 \cup C_4(\alpha 2) \text{ or } B_3 \subset C_2 \cup C_4(\beta 2)$$

$$B_1 \subset C_2 \cup C_3(\alpha 3) \text{ or } B_4 \subset C_2 \cup C_3(\beta 3)$$

If two of α and one of β[18], then $B_1 \subset C_x$ and $B_x \subset C_y \cup C_{\bar{y}}$ where $x \in \{2,3,4\}$, $y \in \{2,3,4\} \setminus \{x\}$ and $\bar{y} \in \{2,3,4\} \setminus \{x, y\}$

-If $S_y \cap S_x = \emptyset$ and $S_{\bar{y}} \cup S_x = \emptyset$, by super additivity we have that

$$C_y \supset B_y \cap C_x \in E(S_y \cup S_{\bar{x}}) \text{ and } C_y \supset B_{\bar{y}} \cap B_x \in E(S_{\bar{y}} \cup S_x) \ (*)$$

By the hypothesis on elements of $\mathcal{Q}_3$,

$$S_1 \cap (Sy \cup S_x) \cap (S_{\bar{y}} \cap S_x) = (S_1 \cap S_y \cap S_{\bar{y}}) \cup (S_x \cap S_y \cap S_{\bar{y}}) = \emptyset$$

Then by (*), $(S_1, S_y \cup S_x, S_{\bar{y}} \cup S_x, B_1, B_y \cup B_x, B_{\bar{y}} \cup B_x)$ is a lower cycle.

- If $S_y \cap S_x \neq \emptyset$, then $B_x \subset C_{\bar{y}}$ or $B_y \subset C_1 \cup C_{\bar{y}}$. In the first case, $S_{\bar{y}} \cap S_x \neq \emptyset$ and then $B_{\bar{y}} \subset C_y \cup C_1$. Hence, $(S_1, S_x, S_{\bar{y}}, B_1, B_x, B_{\bar{y}})$ is a lower cycle. In the second case, if $S_{\bar{y}} \cap S_x = \emptyset$ we have that $(S_1, S_y, S_{\bar{y}} \cup S_x, B_1, B_y, B_{\bar{y}} \cap B_x)$ is lower cycle. Yet, in the second case, if $S_{\bar{y}} \cap S_x \neq \emptyset$ we have that $(S_1, S_x, S_{\bar{y}}, B_1, B_x, B_{\bar{y}})$ is a lower cycle.

If one α and two β, then $B_1 \subset C_x \cup C_y$, $B_x \subset C_y \cup C_z$ and $B_y \subset C_x \cup C_z$ where $x, y, z \in \{2,3,4\}$[19]

[17] the same argument if $B_x \subset C_{\bar{y}}$

[18] The case three α is impossible.

[19] Note that x and y play a symmetric role

- If $S_x \cap S_y = \emptyset$, $S_x \cap S_z = \emptyset$ and $S_y \cap S_z = \emptyset$, then by the super additivity of E

$$C_z \supset B_x \cap B_y \in E(S_x \cup S_y), C_y \supset B_x \cap B_z \in E(S_x \cup S_z) \text{ and}$$

$$C_x \supset B_y \cap B_z \in E(S_y \cup S_z)$$

We have $(S_x \cup S_y) \cap (S_x \cup S_z) \cap (S_y \cup S_z) = \emptyset$, then

$$(S_x \cup S_y, S_x \cup S_z, S_y \cup S_z, B_x \cap B_y, B_x \cap B_z, B_y \cap B_z) \text{ is a lower cycle}$$

- If $S_x \cap S_y \neq \emptyset$, then $B_x \subset C_z$. Hence, $S_x \cap S_z \neq \emptyset$ i.e $B_z \subset C_1 \cup C_y$. It gives $S_z \cap S_y \neq \emptyset$ i.e $B_y \subset C_x$ or $B_z \subset C_1$. In the first case $(S_x, S_y, S_z, B_x, B_y, B_z)$ is a lower cycle. In the second case $(S_1, S_x, S_z, B_1, B_x, B_z)$ is a lower cycle.
- If $S_x \cap S_y = \emptyset$ and $S_x \cap S_z \neq \emptyset$ then, $C_z \supset B_x \cap B_y \in E(S_x \cup S_y)$ and $B_x \subset C_y$ or $B_z \subset C_1 \cup C_y$. In the first case, $B_x \subset C_y$, we obtain $S_x \cap S_y \neq \emptyset$ cf the above. In the second case, $B_z \subset C_1 \cup C_y$ and then $S_y \cap S_z \neq \emptyset$ i.e $B_y \subset C_x$ or $B_z \subset C_1$. In these two cases, we obtain a lower cycle or order 3.

If three β, then $B_2 \subset C_3 \cup C_4$, $B_3 \cup C_2 \cup C_4$ and $B_4 \subset C_2 \cup C_3$.
- If for some $x, y \in \{2,3,4\}$ we have $S_x \cap S_y = \emptyset$ i.e

$$B_\alpha \subset C_z$$

For some $\alpha \in \{x, y\}$ and $z \in \{2,3,4\} \setminus \{x,y\}$, we obtain $S_z \cap S_\alpha \neq \emptyset$ and then

$$B_z \subset C_{\bar{\alpha}}$$

where $\bar{\alpha} \in \{x,y\} \setminus \{\alpha\}$. Hence $S_z \cap S_{\bar{\alpha}} \neq \emptyset$ i.e

$$B_{\bar{\alpha}} \subset C_\alpha$$

So, $(S_2, S_3, S_4, B_2, B_3, B_4)$ is a lower cycle.
- If for all $x, y \in \{2,3,4\}$; $S_x \cap S_y = \emptyset$, by the super additivity of E we have

$$C_4 \supset B_2 \cap B_3 \in E(S_2 \cup S_3), C_3 \supset B_4 \cap B_2 \in E(S_4 \cup S_2) \text{ and}$$

$$C_2 \supset B_3 \cap B_4 \in E(S_3 \cup S_4)$$

Because $(S_2 \cup S_3) \cap (S_3 \cup S_4) \cap (S_4 \cup S_2) = \emptyset$, then we have a lower cycle. □

- **Second case** : $\mathcal{R}$ contain a set of the form $\{kk+1, kk+2, pq\}$ with $p, q \neq k$ i.e $S_k \cap S_{k+3} = \emptyset$
Suppose that $k = 1$, then

$$B_1 \subset C_3 \cup C_4(\alpha 1) \text{ or } B_2 \subset C_3 \cup C_4(\beta 1)$$

$$B_1 \subset C_2 \cup C_4(\alpha 2) \text{ or } B_3 \subset C_2 \cup C_4(\beta 2)$$

$$B_\alpha \subset C_1 \cup C_\beta(\gamma) \quad \text{for some } \alpha \in \{p, q\}, \beta \notin \{1, p, q\}$$

If two α and γ, then $B_1 \subset C_4$, that leads to a cycle of length 2.

If one α, one β and γ, then

$$B_1 \subset C_x \cup C_4, B_x \subset C_{\bar{x}} \cup C_4 \text{ and } B_\alpha \subset C_1 \cup C_\beta, x \neq \bar{x} \in \{2,3\} \quad (5)$$

Because $S_1 \cap S_4 = \emptyset$, then $\alpha \in \{x, \bar{x}\}$. If $\alpha = 2$, with $C_3 \supset B_1 \cap B_4 \in E(S_1 \cup S_4)$, we obtain a lower cycle $(S_1 \cup S_4, S_3, S_2, B_1 \cap B_4, B_3, B_2)$. If $\alpha = 3$, we have $x \neq 3$. Then by (5), $S_2 \cap S_3 \neq \emptyset$ i.e $B_2 \subset C_4$ or $B_3 \subset C_1$.
- If $B_3 \subset C_1$, $B_2 \subset C_3 \cup C_4$ and by the supper additivity we have: $C_2 \supset B_1 \cap B_4 \in E(S_1 \cup S_4)$, then $(S_1 \cup S_4, S_2, S_3, B_1 \cap B_4, B_2, B_3)$ is a lower cycle.
- If $B_2 \subset C_4$, then $S_2 \cap S_4 \neq \emptyset$ i.e $B_4 \subset C_1 \cup C_3$. So, by $C_2 \supset B_1 \cap B_4 \in E(S_1 \cup S_4)$ we have a lower cycle $(S_1 \cup S_4, S_2, S_4, B_1 \cap B_4, B_2, B_4)$

If two β and γ, then $B_2 \subset C_3 \cup C_4$, $B_3 \subset C_2 \cup C_4$ and $B_\alpha \subset C_1 \cup C_\beta$.
- If $S_2 \cap S_3 \neq \emptyset$, then

$$B_x \subset C_4 \text{ for some } x \in \{2,3\}$$

i.e $S_x \cap S_4 \neq \emptyset$ and then $B_4 \subset C_1 \cup C_{\bar{x}}$. So,

$$C_{\bar{x}} \supset B_1 \cap B_4 \in E(S_1 \cup S_4)$$

If $\bar{x} \in \{2,3\} \setminus \{x\}$, then $B_{\bar{x}} \subset C_x \cup C_4$ i.e $S_{\bar{x}} \cap S_4 \neq \emptyset$. By $S_1 \cap S_4 = \emptyset$, then

$$B_{\bar{x}} \subset C_x$$

Therefore, $(S_1 \cup S_4, S_x, S_{\bar{x}}, B_1 \cap B_4, B_x, B_{\bar{x}})$ is a lower cycle.
- If $S_2 \cap S_3 = \emptyset$, then $\{p, q\} \in \{x, 4\}$, $x \in \{2,3\}$. So, we have

$$C_4 \supset B_2 \cap B_3 \in E(S_2 \cup S_3) \ (6)$$

$$B_x \subset C_{\bar{x}} \ (e1) \text{ or } B_4 \subset C_{\bar{x}} \cup C_1 \ (e2)$$

* If $S_{\bar{x}} \cap S_4 = \emptyset$, knowing that $S_1 \cap S_4 = \emptyset$, then

$$C_x \supset B_{\bar{x}} \cap B_4 \in E(S_{\bar{x}} \cup S_4) \text{ and } C_{\bar{x}} \supset B_1 \cap B_4 \in E(S_1 \cup S_4)$$

In this case $(S_1 \cup S_4, S_{\bar{x}} \cup S_4, S_2 \cup S_3, B_1 \cap B_4, B_{\bar{x}} \cap B_4, B_2 \cap B_3)$ is a lower cycle.
* If $S_{\bar{x}} \cap S_4 \neq \emptyset$, then

$$B_{\bar{x}} \subset C_x \ (e3) \text{ or } B_4 \subset C_x \cup C_1 \ (e4)$$

We have that $e1, e3$ with (6) lead to a lower cycle, $e1, e4$ and (6) with the emptiness of $S_1 \cap S_4$ give a lower cycle. $e2, e3$ and (6) with the emptiness of $S_1 \cap S_4$ give a lower cycle. $e2, e4$ simultaneously is not possible. □

Bibliography

Abdou, J and Keiding H. (1991). *Effectivity Function in Social Choice*, (Dordrecht: Kluwer Academic press).

Abdou, J (1982). Stabilit de la fonction veto, cas du veto maximal, *Mathmatiques et Sciences Humaines*, tome **80** pp. 39–65.

Van Deemen Ad M.A (1997). *Coalition Formation and Social Choice*,Dordrecht: (Kluwer Academic press).

Dutta B. (1984). Effectivity Function and Acceptable Game Form, *Economica*, Vol **52**, 5, pp. 512–526.

Zhou Lin (1994). A New Bargaining Set of an N-Person Game and Endogenous Coalition Formation, *Games and Economic Behavior*, **6**, pp. 512–526.

Mizutani M., Nae Chan Lee, Nishino H. (1994). On the Equivalence of Balancedness and the Stability in Effectivity Function Games, *Journal of Operations Research Society of Japan*, **37**, No.3, pp. 243–250.

Moulin H. (1994). The Strategy of Social Choice, *Advanced Textbooks in Economics*, **18**, (North-Holland, Amsterdam).

Peleg B. (2002). *Complete Characterization of Acceptable Game Forms by Effectivity Functions*, online paper.

Vohra R. (1991). An Existence Theorem For a Bargaining Set, *Journal of Mathematical Economics*, **20**, pp. 19–24.

Vannucci S. , Effectivity Functions, Opportunity Rankings, and Generalized Desirability Relations, *Siena Dept. of Economics Working Paper No. 304.*

Vannucci, S. (2004). A Coalitional Game-Theoretic Model of Stable Government Forms with Umpires. *Siena Dept. of Economics Working Paper No. 437.*

Danilov Vladmir I. and Sotskov Alexander I. (2002). *Social Choice Mechanism*, (Springer).

Chapter 21

Dynamic Oligopoly as a Mixed Large Game – Toy Market

Agnieszka Wiszniewska-Matyszkiel
Institute of Applied Mathematics and Mechanics[1]
Warsaw University
ul. Banacha 2, 02-097, Warsaw, Poland
e-mail: agnese@hydra.mimuw.edu.pl

Abstract

In this paper we consider a game modelling a market consisting of two firms with market power and a continuum of consumers. A specific feature of a market for toys is considered with each firm producing two kinds of distinguishable goods. The problem of finding a Nash equilibrium implies firms' optimal advertising and production plans over time, where the aggregate of demands of consumers may depend on firms' past decisions. Equilibria at this market may have strange properties, like oscillatory production and advertising strategies.

Key Words: Nash equilibrium, dynamic game, large game, duopoly, advertising and production plan

21.1 Introduction

The problem of optimal marketing strategies in oligopolistic markets is such that dynamic games are a natural way to model it. Many papers on dynamic oligopolies concern various models with quantity competition under some assumptions of price dynamics, e.g. sticky prices considered by [Fersthman and Kamien (1987)], or [Cellini and Lambertini (2004, 2007)].

[1]The research partly supported by KBN grant no. 5 H02B 008 20.
Scientific work financed by funds for science in years 2005-2007 (grant no. 1 H02B 016 29).

Another way often used in dynamic approach to modelling marketing strategies is the problem of optimal pricing and advertising. In this field there are many generalizations of Dorfman-Steiner theorem according to which expenses on advertising constitute a constant rate of sales (Dorfman, Steiner [Dorfman and Steiner (1954)] in monopolistic and static version), to dynamic optimization framework (e.g. [Schmalensee (1972)]), or to dynamic games (e.g. [Dockner and Feichtinger (1986)]). There is a vast literature on optimal marketing strategies including advertising, e.g. [Schmalensee (1976)], [Dockner, Feichtinger and Sorger (1985)], [Fruchter (2001)], [De Cesare and Di Liddo (2001)]. Some vast surveys of this subject are in [Sethi (1977)]. There are also papers considering positive external effects of advertising effort of one firm on sales of its opponents, e.g. [Cellini and Lambertini (2003)]. Some reviews of game theoretic models of optimal marketing are in [Jørgensen (1982, 1986)], and [Feichtinger and Jørgensen (1983)]. Usually, the game is played only by the firms, with the space of consumers as a mass described by their aggregate demand function only, not subjects facing some decision-making problem.

There are many possible approaches to model marketing at oligopolistic markets. Most of papers consider non-differentiated goods. Besides, since the full model is very compound, there may be its simplifications: advertisement-price, advertisement-quantity and advertisement-quality models of competition.

In our paper a duopolistic market for toys is modelled as a large game of mixed type: we have two "large", atomic players – firms and a continuum of small, *negligible* players – parents deciding what to purchase as a gift for their children. There are four kinds of highly differentiated goods produced and advertisement-quantity type of competition. The demand side is very compound to reflect well known psychological rules which apply especially to relations between parents and their children.

21.1.1 *Large Games*

The simplest characterization of *large games* is contained in the phrase *games with infinitely many players.* In order to make it possible to evaluate the influence of the players on aggregate variables, a measure is introduced on a σ-field of subsets of the set of players, therefore large games are sometimes referred to as *games with a measure space of players.* However, the notion *games with a measure space of players* encompasses also games with finitely many players, where e.g. the counting measure on the power set may be considered.

Large games illustrate situations where the number of agents is large enough to make a single agent from a subset of the set of players (possibly the whole set) insignificant – *negligible* – when we consider the impact of his action on aggregate variables while joint action of this subset of negligible players is not negligible. This happens in many real situations: at competitive markets, stock exchange, or while we consider emission of greenhouse gases and similar global effects of exploitation of the common global ecosystem.

Although it is possible to construct models with countably many players illustrating the phenomenon of this negligibility, they are very inconvenient to cope with. Therefore simplest examples of large games are so called *games with continuum of players*, where players constitute a nonatomic measure space, usually unit interval with the Lebesgue measure. If, additionally, we consider at least one atomic player, then we call such a game a *mixed large game*.

The first attempts to use models with continuum of players are contained in [Aumann (1964, 1966)] and [Vind (1964)].

Some theoretical works on large games are [Schmeidler (1973)], [Mas-Colell (1984)], [Balder (1995)], [Wieczorek (2004, 2005)], [Wieczorek and Wiszniewska (1999)] and [Wiszniewska-Matyszkiel (2000b)].

Although the general theory of dynamic games with continuum of players is still being developed, there are interesting applications of such games: [Wiszniewska-Matyszkiel (2000a, 2001)] concerning models of exploitation of common ecosystems by large groups of players, [Karatzas, Shubik, Sudderth, (1994)] and [Wiszniewska-Matyszkiel (2003b, 2006)] and [Wiszniewska-Matyszkiel (2005)] analyzing dynamic games with continuum of players modelling financial markets and [Wiszniewska-Matyszkiel (2002)] containing example of a dynamic game modelling presidential elections together with the proceeding campaign. To the best of author's knowledge, there are no studies of mixed large games in the dynamic context.

This paper continues a sequence of the author's papers concerning dynamic games with a continuum of players: [Wiszniewska-Matyszkiel (2002)] and [Wiszniewska-Matyszkiel (2003a)] developing a general theory of such games and [Wiszniewska-Matyszkiel (2002, 2003b)] devoted to a certain class of games with discrete time and continuum of players with special focus on applications.

Introducing a continuum of players instead of a finite number, however large, can change essentially properties of equilibria and the way of calculating them even if the measure of the space of players is preserved in order to make the results comparable. Such comparisons were made by the author in [Wiszniewska-Matyszkiel (2005, 2007)].

21.2 Formulation of the Basic Model

The game considered is played over time set $\mathbb{T}$ equal to $\{0, \dots, T\}$ or to $\{0, \dots, +\infty\}$ (for simplicity, we refer to the latter case as to infinite T). In the game we have *two firms* with market power and *a continuum of parents* buying gifts for their children. Since we do not consider strategic behaviour of children (according to the author's observations they are susceptible to advertisements and always make their parents buy any toy they desire if only this is physically possible and there is a reason for it), we can consider the problem reduced to producers and parents, with choices of children incorporated into their parents payoffs as "promises". This constitutes a space of players consisting of two atoms (we can assign to each firm measure equal to 1) and the unit interval $\mathbb{I}$ with the Lebesgue measure λ.

21.2.1 *Producers*

In our model there are two producers of toys, each of them can produce two goods. Goods produced by each producer are distinguishable, although all four goods are similar – the choice of children will therefore depend mainly on the advertising.

Prices of goods are fixed in the game, identical for all goods and denoted by p. This assumption is taken since we consider goods in the same price range and quantity-advertisement competition is taken into account.

By $Q_{i,j}(t)$ we shall denote the production of good j by producer i at time t.

We shall denote by $c(q_1, q_2)$ the cost of production of q_1 units of good 1 and q_2 units of good 2 by each of the producers. The function is symmetric, strictly increasing with $c(0,0) = 0$. Moreover, assume that $c(q_1 + q_2, 0) < c(q_1, q_2)$ (and therefore, $c(0, q_1 + q_2) < c(q_1, q_2)$) for all $q_1, q_2 > 0$. We do not assume convexity in the general case. On the contrary, we even allow for discountinuity at 0, which is assumed to reflect the cost of switching on the production line for each good. The last assumption about the cost functions is that the function $p \cdot q - c(q, 0)$ is strictly increasing in q on the interval $(0, 2\bar{D}]$ for a constant $\bar{D}$ being a constraint for "primary" demand (to be defined later) and it attains its maximum on the interval $[0, 2\bar{D}]$ at $2\bar{D}$. This holds if, for example, the average cost function is decreasing on the interval $(0, 2\bar{D}]$ and p is greater then the average cost at $2\bar{D}$, which is natural if we consider a market where the production is below its competitive long run equilibrium level, as it is in oligopolies.

By $A_{i,j}(t)$ we shall denote the cost of advertising effort (*advertising effort* for short) of good j by i-th producer at time t. We assume that for all t we have a constraint $A_{i,1}(t) + A_{i,2}(t) \leq \bar{A}_i$ – the firm's fund for advertising.

At each stage producers know the dependence of demand for their products on advertising efforts of both producers (to be defined in the sequel).

Every producer i maximizes the sum – over the time set – of his profits from selling his products minus advertising costs discounted by a discounting function $\Xi^i : \mathbb{T} \to \mathbb{R}_+$, which is positive and nonincreasing. The firms' profits depend also on the demand side of the market, therefore they will be formally defined in the sequel.

21.2.2 *Parents (and Children)*

Children want to get the toy whose advertisement they see most frequently. We assume that it means that at stage t the amount of children *asking for* good j of i-th producer is equal to $D(t) \cdot \frac{A_{i,j}(t)}{A_{1,1}(t)+A_{1,2}(t)+A_{2,1}(t)+A_{2,2}(t)}$ if the denominator is positive, while $\frac{1}{4} \cdot D(t)$ when nothing is advertised. The constants $D(t)$ are given a priori e.g. $D(t)$ is 1 for Christmas and about $\frac{1}{12}$ in any other period – then it denotes the ratio of children who have a reason to get a present: for Christmas or birthday (if, for simplicity, we assume that there are no children born exactly at Christmas and divide the rest of the year into 12 equal parts). Parents always promise to buy the toy that children ask for whenever there is a reason for a gift. A *promise* of parent ω at time t will be denoted by $P^{\omega}(t)$. It is a vector $P^{\omega}_{i,j}(t)$ with coordinates denoting numbers of toys promised – it can take values 0 or 1. The whole profile of promises at time t will be denoted by $P(t)$. Whatever the distribution of promises is, it is Lebesgue-integrable and the aggregate of $P^{\omega}_{i,j}(t)$ equals $D(t) \cdot \frac{A_{i,j}(t)}{A_{1,1}(t)+A_{1,2}(t)+A_{2,1}(t)+A_{2,2}(t)}$ if at least one of the toys is advertised, $\frac{1}{4} \cdot D(t)$ when nothing is advertised. We assume that all $D(t) \leq \bar{D}$ for some constant $\bar{D}$.

Another assumption is that in the case when the promised good is not available, parents buy the other product of the same firm, and if that is not available either they buy good 1 or 2 of the other producer (in this ordering to make the choice fully deterministic). Such a reasoning is quite obvious – the goods are differentiable and products of the same firm (e.g. two sets of Lego blocks or two "hotwheel" cars) can be perceived as more similar than products of different firms (e.g. a set of Lego blocks and a "hotwheel" car). Since this part of player's strategy is fixed, we shall consider only reduced strategies, in which it is not taken into account.

Besides, parents can buy a present they have promised before but were not able to buy because it was not available. However, they buy not more than one extra toy at each stage. The part of strategy of player ω concerning this will be denoted by a function $U^\omega : \mathbb{T} \times \mathbb{N} \to \{0,1\}$, with coordinates $U^\omega_{i,j}(t, x^\omega)$ (equal 0 or 1) denoting the number of units of good j produced by i-th producer bought at time t given vector of unkept promises x^ω of player ω, and not because of new promise. The strategies of all players are such that at most one coordinate of $U^\omega(t, x^\omega)$ is equal to 1, while the remaining coordinates are 0. We additionally assume that for every t and $x : \omega \mapsto x^\omega$ measurable, the profile of parents' decisions $U^\omega(t, x^\omega)$ is Lebesgue-integrable with respect to ω. In order to simplify the notation, we shall use the notation $U(t, X(t))$ for $\{U^\nu_{k,l}(t, X^\nu_{k,l}(t))\}_{\nu\in\mathbb{I},\ k,l\in\{1,2\}}$.

The *unkept promises* of parent ω at time t are described by a vector function $X^\omega(t)$ with coordinates $X^\omega_{i,j}(t)$ denoting the number of promised units of good j produced by i-th producer before time t and not bought. If we want to concentrate only on unkept promises $x = \{x^\omega\}_{\omega\in\mathbb{I}}$, without dependence on time, we call x *state* (unkept promises of all parents are state variables of the game and they constitute its trajectory over time).

The trajectory of unkept promises X evolves as follows:

$X^\omega_{i,j}(0) = 0$ for every $\omega \in \mathbb{I}$, $i, j \in \{1, 2\}$.

for $t \geq 0$ $X^\omega_{i,j}(t+1)$ is equal to

$X^\omega_{i,j}(t) - U^\omega_{i,j}(t, X^\omega_{i,j}(t))$ when both $U^\omega_{i,j}(t, X^\omega_{i,j}(t))$ and $P^\omega_{i,j}(t)$ are available to parent ω,

$X^\omega_{i,j}(t) + P^\omega_{i,j}(t) - U^\omega_{i,j}(t, X^\omega_{i,j}(t))$ when $U^\omega_{i,j}(t, X^\omega_{i,j}(t))$ is available to ω while $P^\omega_{i,j}(t)$ is not available,

$X^\omega_{i,j}(t)$ when $U^\omega_{i,j}(t, X^\omega_{i,j}(t))$ is not available to ω while $P^\omega_{i,j}(t)$ is available,

$X^\omega_{i,j}(t) + P^\omega_{i,j}(t)$ when both $U^\omega_{i,j}(t, X^\omega_{i,j}(t))$ and $P^\omega_{i,j}(t)$ are not available.

We can simplify this compound formula by introducing two additional boolean functions $\Phi^\omega_{i,j}$ and $F^\omega_{i,j}$, describing availability of promised toys and strategies concerning fulfillment of unkept promises:

corresponding to current promises – $\Phi^\omega_{i,j}(\pi)$ (for $\pi = P(t)$) is equal to

0 if player ω has $\pi^{\omega}_{i,j} = 1$ and good (i,j) is not available to him, and 1 otherwise;

and corresponding to current fulfillment of past unkept promises $F^{\omega}_{i,j}(u,\pi,x)$ (for $\pi = P(t)$, $x = X(t)$ and $u = U(t, X(t))$) is equal to 1 if player ω has $U^{\omega}_{i,j}(t, X^{\omega}_{i,j}(t)) = 1$ and good (i,j) is available to him, and 0 otherwise.

The rules of availability are implied by the following lexicographic ordering.

I. New promises are fulfilled first.

1. If the amount of goods produced is less than the amount of goods for the new promises, then

a) first goods with higher fitness are assigned

b) then those for parents with higher $x^{\omega}_{i,j}$

c) and finally larger ω is before smaller with assignment to the lower bound of the interval.

2. After buying toys for new promises, toys for unkept promises are bought in the following ordering:

a) first of parents with $u^{\omega}_{i,j} = 1$ buy those who have higher $x^{\omega}_{i,j}$

b) if this criterion does not decide, then larger ω is before smaller with assignment to the lower bound of the interval.

These rules define fully the functions Φ and F.

Using these two functions we can write the equation defining trajectories as

$X^{\omega}_{i,j}(t+1) = X^{\omega}_{i,j}(t) + \left(1 - \Phi^{\omega}_{i,j}(P(t))\right) - F^{\omega}_{i,j}(U(t,X(t)), P(t), X(t))$.

Each parent ω maximizes the sum of instantaneous utilities discounted by a discounting function $\Sigma^{\omega} : \mathbb{T} \to \mathbb{R}_{+}$, which is positive and non-increasing. After elimination of the fixed part of strategy the instantaneous utility of parent ω at each stage given the state x and profile of fulfilling unkept promises at this stage, denoted by u, is reduced to $-C \cdot \sum_{i,j\in\{1,2\}} \left(\left(x^{\omega}_{i,j} - F^{\omega}_{i,j}(u,\pi,x)\right)^{+}\right)^2 - \sum_{i,j\in\{1,2\}} p \cdot F^{\omega}_{i,j}(u,\pi,x)$, where C is a constant greater than p. The first component of the payoff function expresses the "natural human need for consequence", emphasized by psychologists.

This leads to the payoff of parent ω written as (by a slight abuse of notation)

$$\sum_{t=0}^{T} \Sigma^{\omega}(t) \cdot \Big(-C \cdot \sum_{i,j\in\{1,2\}} \left(\left(X^{\omega}_{i,j}(t) - F^{\omega}_{i,j}(U(t,X(t)), P(t), X(t))\right)^{+}\right)^2 - \sum_{i,j\in\{1,2\}} p \cdot F^{\omega}_{i,j}(U(t,X(t)), P(t), X(t)) \Big)$$

21.2.3 *Market Clearing*

In order to state the market clearing condition we first have to define formally the demand function.

Given the profile of parents' strategies U at time t and state x, the demand $\Delta(t,x)$ is rather compound, since demand on each good can depend on availability of other goods. In the simplest case, in which for all i and j the supply fulfills $Q_{i,j}(t,x) \geq \int_{\mathbb{I}} P^{\omega}_{i,j}(t) + U^{\omega}_{i,j}(t,x^{\omega})$, we have $\Delta^{U,Q,P}_{i,j}(t,x) = \int_{\mathbb{I}} P^{\omega}_{i,j}(t)+U^{\omega}_{i,j}(t,x^{\omega})d\lambda(\omega)$. If for some i,j we have $Q_{i,j}(t) \leq \int_{\mathbb{I}} P^{\omega}_{i,j}(t)+U^{\omega}_{i,j}(t,x^{\omega})d\lambda(\omega)$ while $Q_{1,1}(t)+Q_{1,2}(t) \geq \int_{\mathbb{I}} P^{\omega}_{1,1}(t)+P^{\omega}_{1,2}(t)d\lambda(\omega)$ and $Q_{2,1}(t) + Q_{2,2}(t) \geq \int_{\mathbb{I}} P^{\omega}_{2,1}(t) + P^{\omega}_{2,2}(t)d\lambda(\omega)$, then $\Delta^{U,Q}_{i,j}(t,x) = Q_{i,j}(t)$ and $\Delta^{U,Q,P}_{i,\sim j}(t,x) = \min\left(Q_{i,\sim j}(t), \int_{\mathbb{I}} P^{\omega}_{i,\sim j}(t) + U^{\omega}_{i,\sim j}(t,x^{\omega})d\lambda(\omega) + \left(\int_{\mathbb{I}} P^{\omega}_{i,j}(t)d\lambda(\omega) - Q_{i,j}(t)\right)^{+}\right)$ (where $\sim j$ denotes the choice different from j). In this case for the other firm $\sim i$ we have either the same formulae (if a condition analogous to that for i and j holds) or for both goods k produced by him $\Delta^{U,Q,P}_{\sim i,k}(t,x) = \int_{\mathbb{I}} P^{\omega}_{\sim i,k}(t) + U^{\omega}_{\sim i,k}(t,x^{\omega})d\lambda(\omega)$. The formulae in the case when one of the players produces less than new promises implied by his advertisements cannot be satisfied by both his goods are even more complicated, but their formulation has been defined by the description of players' behaviour in the obvious way.

After defining the market demand we can finally formulate payoffs of firm i in our game given a profile of strategies of the players:
$\sum_{t=0}^{T} \Xi^{i}(t)\cdot \ (p\cdot \ \left(\Delta^{U,Q,P}_{i,1}(t,X(t))+ \Delta^{U,Q,P}_{i,2}(t,X(t))\right) \ -c(Q_{i,1}(t),Q_{i,2}(t)) -A_{i,1}(t) - A_{i,2}(t))$.

Since, apparently, the four values $\int_{\mathbb{I}} P^{\omega}_{i,j}(t)d\lambda(\omega)$ are the only characteristics of $P(t)$ influencing the demand and they are defined by the advertising efforts, instead of $\Delta^{U,Q,P}_{i,j}(t,x)$, we shall write, by a slight abuse of notation, $\Delta^{U,Q,A}_{i,j}(t,x)$.

21.3 Results

As in majority of game theoretic models, we are interested in Nash equilibria. In games with a measure space of players the standard definition of Nash equilibrium has the following form.

Definition 21.1. A profile of strategies is a Nash equilibrium, if for almost every (with respect to the measure) player, his strategy at this profile maximizes his payoff given the strategies of the remaining players.

In our game it means that each firm and all elements of a subset of parents of measure 1 maximize their payoffs given the strategies of the remaining players.

In the general model we can state the following properties of Nash equilibria.

Theorem 21.1. *Assume that $T < +\infty$ or the discounting functions of the parents are such that $\sum_{t=0}^{\infty} t^2 \cdot \Sigma^{\omega}(t)$ is finite for a.e. ω. At every Nash equilibrium for a.e. parent ω, for every time t the strategy of parent ω strategy fulfills $U_{i,j}^{\omega}(t, X^{\omega}(t)) = 1$ for one of the pairs (i,j) such that $X_{i,j}^{\omega}(t) > 0$ and $F_{i,j}^{\omega}(U(t,X(t)), P(t), X(t)) = 1$.*

Proof. The condition $T < +\infty$ or the discounting functions of parents are such that $\sum_{t=0}^{\infty} t \cdot \Sigma^{\omega}(t)$ is finite for a.e. ω guarantees that payoffs of all parents in the game are finite. In the case of finite time horizon it is obvious, while in the infinite horizon for all t we have $X_{i,j}^{\omega}(t) > 0$ and the inequality $X_{i,j}^{\omega}(t+1) \leq X_{i,j}^{\omega} + \bar{D}$, therefore $X_{i,j}^{\omega}(t) \leq \bar{D} \cdot t + X_{i,j}^{\omega}(0)$ and player's accumulated payoff is equal to
$\sum_{t=0}^{\infty}[-C \cdot \sum_{i,j} \left(\left(X_{i,j}^{\omega}(t) - F_{i,j}^{\omega} \left(U(t,X(t)), P(t), X(t) \right) \right)^{+} \right)^{2} -$
$\sum_{i,j} p \cdot F_{i,j}^{\omega} \left(U(t,X(t)), P(t), X(t) \right)] \cdot \Sigma^{\omega}(t)$, whose absolute value is constrained by
$\left| \sum_{t=0}^{\infty} \left(-C \cdot \sum_{i,j} \left(X_{i,j}^{\omega}(t) \right)^{2} - 4p \right) \cdot \Sigma^{\omega}(t) \right| \leq d \cdot \sum_{t=0}^{\infty} t^2 \cdot \Sigma^{\omega}(t)$ for a constant d. Therefore the payoff is finite.

Suppose, conversely, that there exists a set of positive measure for which the condition is not fulfilled. Take any ω from this set. For some time t player ω has $U_{i,j}^{\omega}(t, X^{\omega}(t)) = 0$ for all pairs (i,j) available to him while at least one $X_{i,j}$ is positive. We can increase the payoff of player ω by changing his strategy: if good (i,j) is available to him at time t, then we define $\widetilde{U}_{i,j}^{\omega}(t, X^{\omega}(t)) = 1$ and $\widetilde{U}_{i,j}^{\omega}(t', X^{\omega}(t')) = 0$ for t' being the first of time instants s after t at which $U_{i,j}^{\omega}(s, X^{\omega}(s)) = 1$ (obviously when such an s does exist). For other time instants and x's we define $\widetilde{U}_{i,j}^{\omega}(t, x^{\omega}) = U_{i,j}^{\omega}(t, x^{\omega})$. Note that availability of promises will not change at any time, while the instantaneous payoff is fixed for any time before t, for any time in $\{t, \ldots, t'-1\}$ increases while from t' on it is at least the same. Therefore the payoff of player ω in the game increases. Since this holds for ω in a nonnegligible set, a profile cannot be an equilibrium. □

It may seem that for such a utility function of parents all equilibria should be such that for all t and a.e. ω player's ω strategy $U_{\bar{i},\bar{j}}^{\omega}(t, X^{\omega}(t)) = 1$

for $(\bar{\imath},\bar{\jmath})$ for which $X^{\omega}_{i,j}(t) > 0$ attains its maximum over all (i,j) which are available to player ω (i.e. $F^{\omega}_{i,j}(U(t,X(t)),P(t),X(t)) = 1$). This intuition is false, as we can see in example 21.1.

Example 21.1. "Strange" Nash equilibrium for 4 period game with two dead-seasons and costs of switching production on.

Consider a game with time horizon $T = 3$, demands $D(0) = D(1) = 10$, $D(2) = D(3) = 0$, bounds for advertising efforts $\bar{A}_i = 1$, price $p = 10$, cost function $c(q_1,q_2) = \begin{cases} 0 & \text{if } q_1 = q_2 = 0, \\ 2\bar{c} + q_1 + q_2 & \text{if } q_1, q_2 > 0, \\ \bar{c} + q_1 + q_2 & \text{otherwise.} \end{cases}$ (the constant $\bar{c}$ represents the cost of switching on production of each good) and all discounting functions $\Sigma^{\omega} = \Xi^{i} \equiv 1$.

Proposition 21.1. *If $\bar{c} \le 44$, then any profile fulfilling*
$U^{\omega}_{i,j}(t,x^{\omega}) = \begin{cases} 1 \text{ for one of } (i,j) \text{ such that } x^{\omega}_{i,j} = 1, \\ 0 \text{ for all other } (i,j). \end{cases}$
and $A_{1,1}(0) = A_{2,1}(0) = 1$, $A_{1,2}(0) = A_{2,2}(0) = 0$, $Q_{1,1}(0) = Q_{2,1}(0) = 0$, $Q_{1,2}(0) = Q_{2,2}(0) = 5$,
$A_{1,1}(1) = A_{2,1}(1) = 0$, $A_{1,2}(1) = A_{2,2}(1) = 1$, $Q_{1,1}(1) = Q_{2,1}(1) = 5$, $Q_{1,2}(1) = Q_{2,2}(1) = 0$,
$Q_{1,1}(2) = Q_{2,1}(2) = 0$, $Q_{1,2}(2) = Q_{2,2}(2) = 5$, *and for all* (i,j) $A_{i,j}(2) = A_{i,j}(3) = 0$, $Q_{i,j}(3) = 0$
is an equilibrium.

Proof. Checking for each player that his strategy is a best responses to the others' strategies is a simple calculation. □

Theorem 21.2. *Assume that $T < +\infty$ or the discounting functions of the firms are such that $\sum_{t=N}^{\infty} \Xi^{i}(t) \to 0$ as $N \to \infty$ for $i = 1,2$. At every Nash equilibrium at which parents' strategies are such that for all t and a.e. ω player's ω strategy $U^{\omega}_{\bar{\imath},\bar{\jmath}}(t,X^{\omega}(t)) = 1$ for $(\bar{\imath},\bar{\jmath})$ for which $X^{\omega}_{i,j}(t) > 0$ attains its maximum over all (i,j) which are available to player ω and such that whenever both goods of the same producer are available and maximal while at most one good of the other firm is maximal, then the player chooses one of goods of the former producer, the production of firm i is such that $Q_{i,1}(t) + Q_{i,2}(t) \ge \int_{\mathbb{I}} P^{\omega}_{i,1}(t) + P^{\omega}_{i,2}(t) d\lambda(\omega)$.*

This means that at every such equilibrium joint production of each firm is at every stage at least equal to joint amount of new promises for its products.

In order to prove this theorem, we shall need the following Lemma.

Lemma 21.1. *Assume that $T < +\infty$ or the discounting functions of firm 1 is such that $\sum_{t=N}^{\infty} \Xi^1(t) \to 0$ as $N \to \infty$. Consider a profile of parents' strategies U are such that for all t and a.e. ω player's ω strategy $U^{\omega}_{\bar{\imath},\bar{\jmath}}(t, X^{\omega}(t)) = 1$ for $(\bar{\imath},\bar{\jmath})$ for which $X^{\omega}_{i,j}(t) > 0$ attains its maximum over all (i,j) which are available to player ω and such that whenever both goods of the same producer are available and maximal while at most one good of the other firm is maximal, then the player chooses one of goods of the former producer and any strategy of firm 2. Let us define the value function of firm 1, $W_1 : \mathbb{T} \times \mathbb{X} \to \mathbb{R}_+$ by $W_1(t,x) = \sup_{available\ A_{1,1},A_{1,2},Q_{1,1},Q_{1,2}} \sum_{s=t}^{T} \Xi^1(s)\cdot$ $\left(p\left(\Delta_{1,1}^{U,Q,A}(s,X(s)) + \Delta_{1,2}^{U,Q,A}(s,X(s))\right) - c\left(Q_{1,1}(s),Q_{1,2}(s)\right)\right.$ $\left.-A_{1,1}(s) - A_{1,2}(s)\right)$, where X is a trajectory of parents' unkept promises starting at time t from x and corresponding to the players strategies.*

For every t, $x_{2,1}$, $x_{2,2}$ the function W_1 is nondecreasing in $x_{1,1}$ and $x_{1,2}$.

Proof. (of Lemma 21.1) Note that W_1 is the value function for our problem discounted for the moment 0. Therefore it is natural that the proof is by backwards induction, starting from T (for finite T) and using the Bellman equation.

Finite horizon case.

We start at T. At this moment for all x we have $W_1(T,x)$ $= \sup_{a_1,a_2,q_1,q_2} \Xi^1(T)\cdot\left(p\left(\Delta_{1,1}^{U,Q,A}(T,x)) + \Delta_{1,2}^{U,Q,A}(T,x)\right) - c\left(q_1,q_2\right)\right.$ $\left.-a_1 - a_2\right)$, which is nondecreasing in $x_{1,1}$ and $x_{1,2}$.

Now let us assume, that $W_1(t,x)$ is for all x nondecreasing in $x_{1,1}$ and $x_{1,2}$, and we shall prove the analogous fact about $W_1(t-1,x)$. By the Bellman equation for our problem $W_1(t-1,x) = \sup_{a_1,a_2,q_1,q_2} \Xi^i(T)\cdot$ $\left(p\left(\Delta_{1,1}^{U,Q,A}(T,x)) + \Delta_{1,2}^{U,Q,A}(T,x)\right) - c\left(q_1,q_2\right) - a_1 - a_2\right)$ $+W_1(t, x + (1-\Phi(\pi)) - F(U(t-1,x),\pi,x))$ for some profile of parents promises (at time $t-1$) π such that $\int_{\mathbb{I}} \pi^{\omega}_{1,j} d\lambda(\omega)$ equals $D(t-1)\cdot \frac{a_j}{a_1+a_2+A_{2,1}(t)+A_{2,2}(t)}$ if at least one of the toys is advertised, $\frac{1}{4}\cdot D(t)$ when nothing is advertised. The first component of the sum is a nondecreasing function of $x_{1,1}$ and $x_{1,2}$ whatever strategy firm 1 chooses, since demands are nondecreasing functions of $x_{1,1}$ and $x_{1,2}$. At each stage, if we increase $x_{1,j}$ and we do not reduce availability of goods, then we do not decrease $x + (1-\Phi(\pi)) - F(U(t-1,x),\pi,x)$.

Since $W_1(t, \cdot)$ is nondecreasing, we have a supremum of a sum of two nondecreasing functions of x. Although the assumption that we do not reduce availability of good may lead to strategies that are not optimal, note that for this increased x the optimal strategy yields supremum in which the payoff of firm 1 is at least as large as for the restricted strategy. This means that $W_1(t-1, \cdot)$ is nondecreasing, which ends the proof for finite T.

In the proof for infinite T we obtain W_1 as a limit of value functions for finite T. □

Proof. (of Theorem 21.2)

Let us consider a profile of players' strategies for which the property is not fulfilled. Without loss of generality we assume that for firm 1 at some t we have $Q_{1,1}(t) + Q_{1,2}(t) < \int_{\mathbb{I}} P^{\omega}_{1,1}(t) + P^{\omega}_{1,2}(t) d\lambda(\omega)$. It can happen only for t such that $D(t) > 0$. Given the advertising efforts of both firms, we have $\int_{\mathbb{I}} P^{\omega}_{1,1}(t)+P^{\omega}_{1,2}(t)d\lambda(\omega) = D(t) \cdot \frac{A_{1,1}(t)+A_{1,2}(t)}{A_{1,1}(t)+A_{1,2}(t)+A_{2,1}(t)+A_{2,2}(t)}$. We shall show that firm 1 can increase its payoff by changing its strategy.

We shall consider the following cases, which are not mutually exclusive:

1. Firm 1 advertises only good 1 at time t. In this case we have $Q_{1,1}(t) + Q_{1,2}(t) < \int_{\mathbb{I}} P^{\omega}_{1,1}(t) d\lambda(\omega)$. Whatever $Q_{1,1}(t)$ and $Q_{1,2}(t)$ are, by changing $Q_{1,2}(t)$ to $D(t) \cdot \frac{A_{1,1}(t)}{A_{1,1}(t)+A_{2,1}(t)+A_{2,2}(t)} - Q_{1,1}(t)$ the firm increases its instantaneous payoff at time t without changing future payoffs, since for each parent his $X^{\omega}_{i,j}(t+1)$ remain the same.

2. Firm 1 produces only good 1 at time t and has advertising effort of this good greater than 0. In this case we have $Q_{1,1}(t) < \int_{\mathbb{I}} P^{\omega}_{1,1}(t) + P^{\omega}_{1,2}(t) d\lambda(\omega)$. There is no strategy which for sure increases firm's instantaneous payoff without changing its future payoff, but, by Lemma 21.1, we know that by increasing $X(t+1)$ we guarantee that the future payoff will not decrease.

The firm will increase its instantaneous payoff without decreasing future payoffs (by Lemma 21.1) by e.g. changing $A_{1,1}(t)$ to 0 and increasing $A_{1,2}(t)$ by $A_{1,1}(t)$.

3. Firm 1 produces and advertises both goods at time t but it does not produce good 1 after t. In this case the payoff can be improved (by Lemma 21.1) by e.g. changing $A_{1,1}(t)$ to 0 and increasing $A_{1,2}(t)$ by $A_{1,1}(t)$ and switching to production of good 1 only at time t, in the amount $Q_{1,1}(t) + Q_{1,2}(t)$.

4. Firm 1 produces and advertises both goods at time t and produces both goods after t, and $Q_{1,1}(t) \geq \int_{\mathbb{I}} P^{\omega}_{1,1}(t) d\lambda(\omega)$. In this case we do not

decrease the future payoffs (by Lemma 21.1, since $x^{\omega}_{1,2}$ increases without changing other coordinates of x) and increase the instantaneous payoff at time t by choosing production of good 1 at time t in the amount of $Q_{1,1}(t)+Q_{1,2}(t)$ before change and producing no good 2 at time t (by the condition $c(q_1,q_2) > c(0,q_1+q_2)$).

5. Firm 1 produces and advertises both goods at time t and produces both goods in future, and $Q_{1,j}(t) < \int_{\mathbb{I}} P^{\omega}_{1,j}(t)d\lambda(\omega)$ for both goods. In this case we increase the instantaneous payoff at time t without changing future payoffs by e.g. changing the advertising efforts to $A'_{1,2}(t) < A_{1,2}(t)$ such that $D(t)\cdot\frac{A'_{1,2}(t)}{A'_{1,1}(t)+A'_{1,2}(t)+A_{2,1}(t)+A_{2,2}(t)} = \int_{\mathbb{I}} P^{\omega}_{1,2}(t)d\lambda(\omega) - Q_{1,2}(t)$ and with $A'_{1,1}(t) = A_{1,1} + (A_{1,2}(t) - A'_{1,2}(t))$ (preserving the joint advertising effort), and switching to production of good 2 only with the amount equal to the sum of amounts considered before the change. The instantaneous payoff will increase by the condition $c(q_1,q_2) > c(0,q_1+q_2)$. By Lemma 21.1, future payoff will not decrease.

6. Firm 1 advertises neither of goods at time t at which firm 2 does advertise. In such a situation our condition is fulfilled obligatorily, since $\int_{\mathbb{I}} P^{\omega}_{1,1}(t) + P^{\omega}_{1,2}(t)d\lambda(\omega) = 0$.

7. Neither of the firms advertises at time t and $D(t) > 0$. Such a situation can never happen at equilibrium, since increasing the advertising efforts by firm 1 by an arbitrarily small ε will increase demand for its goods from $\frac{D(t)}{2}$ to $D(t)$ now and it is not going to decrease it in future.

8. $D(t) = 0$. In such a situation our condition is also fulfilled obligatorily, since $\int_{\mathbb{I}} P^{\omega}_{1,1}(t) + P^{\omega}_{1,2}(t)d\lambda(\omega) = 0$.

These cases with their obvious analogues for other i and j describe all possible situations in which the condition is not fulfilled. □

21.3.1 *Case A: Two Periods with Dead-Season*

Now we shall consider a two period model with $D(0) > 0$ and $D(1) = 0$ with cost function convex componet-wise.. We consider the behaviour of one firm, without loss of generality firm 1 at an equilibrium.

Proposition 21.2. *Assume that for some* $0 < a \le \bar{A}_1$ *we have* $p\cdot D(0)\cdot\frac{a}{a+\bar{A}_2} - c\left(D(0)\cdot\frac{a}{a+\bar{A}_2},0\right) - a > p\cdot\frac{1}{2}\cdot D(0) - c\left(\frac{1}{2}\cdot D(0),0\right) > 0$ *and that* $\max_{q_1,q_2\le\bar{D}} p\cdot(q_1+q_2) - c(q_1,q_2)$ *is attained for* $q_1 = q_2 = \bar{D}$. *Every equilibrium has the property that at time* 0 *firm* 1 *advertises only product* j *(arbitrary) while produces only* $\sim j$ *in the amount equal to* $D(0)\cdot$

$\frac{A_{i,j}(0)}{A_{1,1}(0)+A_{1,2}(0)+A_{2,1}(0)+A_{2,2}(0)}$ *and at time* 1 *it produces only product* j*, while all parents who promised good* j *of firm* 1 *buy it at time* 1 *(at time* 0 *they buy only* $\sim j$ *of the same firm).*

Proof. The condition $p \cdot D(0) \cdot \frac{a}{a+A_2} - c\left(D(0) \cdot \frac{a}{a+A_2}, 0\right) - a > p \cdot \frac{1}{2} \cdot D(0) - c\left(\frac{1}{2} \cdot D(0), 0\right)$ implies that at time 0, given any level of advertising effort of firm 2, some situation with positive advertising level is better in one stage game than not advertising at all. Since $pq - c(q, 0)$ is an increasing function of q at $(0, 2\bar{D}]$, the r.h.s. of the inequality is equal to the maximal possible payoff if we do not advertise (whatever firm 2 does).

At time 1 no firm advertises, since advertising at this stage only decreases payoffs.

Now we shall prove that firm 1 advertises only one product at time 0 and produces the other. By Theorem 21.1, demand for good $(1, j)$ at time 1 is equal to $\int_{\mathbb{I}} U^{\omega}_{1,j}(1, X^{\omega}(1))d\lambda(\omega) \leq \bar{D}$. By the second assumption about the cost function, we have $Q_{1,j}(1) = \int_{\mathbb{I}} U^{\omega}_{1,j}(1, X^{\omega}(1))d\lambda(\omega) = \int_{\mathbb{I}} X^{\omega}(1))d\lambda(\omega)$. Given the advertising efforts of firm 2 and $a = A_{1,1}(0) + A_{1,2}(0)$, at time 0 firm i can sell at most $q = D(0) \cdot \frac{a}{a+A_{2,1}(0)+A_{2,2}(0)}$. The maximal profit at stage 0 it will get if it produces only one (arbitrary) good in this amount is equal to $pq - c(q, 0) - a$, by the general assumption of our model: $c(q_1 + q_2, 0) < c(q_1, q_2)$ for all positive q_1 and q_2. Assume that the produced good is good 1. Then for $A_{1,1}(0) = 0$ and $A_{1,2}(0) = a$ the profit of firm 1 at time 1 is also maximal. Since at both stages we maximize profits, the discounted sum is also maximal. □

Such a behaviour, apparently counterintuitive, can often be observed. The best proof from the real life are parents who meet in January in toy shops buying gifts which were not available but advertised in December.

21.3.2 *Case B: Two Periods*

We again consider a two period model, but this time we do not impose so many conditions, only $D(0) > 0$, $D(1) \geq 0$.

Proposition 21.3. *Every equilibrium has the property that at time* 0 *firm* i *advertises exactly one product* j *(arbitrary) while produces only the other one in the amount equal to* $D(0) \cdot \frac{A_{i,j}(0)}{A_{1,1}(0)+A_{1,2}(0)+A_{2,1}(0)+A_{2,2}(0)}$ *and at time* 1 *it produces only the product advertised at the time* 0*, while all parents who promised at time* 0 *any advertised good buy it at time* 1 *(at time* 0 *they buy only* $\sim j$ *of the same firm).*

Proof. The proof is similar. We only have to exclude the case in which one of the firms, without loss of generality 1, does not advertise at time 0. First assume that neither of firms advertises. Then the payoff of firm 1 will increase as a result of changing its advertising effort to some small ε. Now let us assume that the advertising effort of firm 2 is $a > 0$. If firm 1 does not advertise, then firm 2 can increase its payoff by changing its advertising effort to $\frac{a}{2}$: it still has the whole market. Producing only one product at time 1 is a consequence of the condition about the cost function: $c(q_1 + q_2, 0) < c(q_1, q_2)$. □

21.3.3 *Case C: Finitely Many Stages and Negligibility of Low Advertising Efforts*

Now we consider a modification of our model with time horizon $0 < T < +\infty$ with an additional assumption that there exists a minimal effective advertising effort ε for each product and choosing $A_{i,j}(t) < \varepsilon$ has influence on $P(t)$ equivalent to choosing $A_{i,j}(t) = 0$ i.e. $\int_{\mathbb{I}} P^{\omega}_{i,j}(t)d\lambda(\omega)$ is equal to 0 if $A_{i,j}(t) < \varepsilon$ and at least one $A_{k,l}(t) \geq \varepsilon$, to $D(t) \cdot \frac{A_{i,j}(t)}{\sum_{k,l \in \{1,2\}, A_{k,l}(t) > \varepsilon}, A_{k,l}(t)}$ if the denominator is positive, and to $\frac{1}{4}D(t)$ otherwise.

In subsequent results we assume that this ε is quite large compared to profits that can be obtained.

Under some assumptions about the cost functions and bounds for advertising efforts, it is possible to obtain an equilibrium at which one of the firms advertises only at even time instants while its opponent at odd time instants, besides the last stage. Moreover, only one product of each firm is advertised at each stage of advertising t. Each of the firms produces at any stage of advertising t only the non-advertised product and the amount produced is equal to the maximal $D(t)$, while at stage $t+1$ it produces the other product also in amount $D(t)$. Parents buy at each stage in which their promise is positive the non-advertised product of the advertising firm, while at the next stage they fulfill unkept promise for the other good. Formally we state as follows.

Proposition 21.4. *Assume that for all t $D(t) = \bar{D}$, and that for $i = 1, 2$ for all $\varepsilon \leq a \leq \bar{A}_i$ and all $q \leq 2\bar{D}$, $\bar{q} \leq \frac{a}{a+\varepsilon}\bar{D}$ we have $p \cdot \bar{q} - \min_{q_1+q_2=q}\left(c(\bar{q} + q_1, q_2) - c(q_1, q_2)\right) < \frac{a}{2}$ and for all $\{q^i\}_{i=1}^{T}$ such that $q^i \leq 2\bar{D}$ and all $\bar{q} \leq \frac{a}{a+\varepsilon}\bar{D}$ and all sequences $\{\bar{q}^i\}_{i=1}^{T}$ such that $\sum_{i=1}^{T} \bar{q}^i = \bar{q}$ we have $p \cdot \bar{q} - \min_{q_1^i+q_2^i=q^i} \sum_{i=1}^{T}\left(c(\bar{q}^i + q_1^i, q_2^i) - c(q_1^i, q_2^i)\right) < \frac{a}{2}$.*

a) There is no equilibrium such that both firms advertise at the same time instant.

b) If for all t $p \cdot D(t) - c(D(t), 0) > \varepsilon$, then there exists an equilibrium at which $A_{1,1}(t) = A_{1,2}(t) = 0$, $Q_{1,1}(t) = D(t-1)$ and $Q_{1,2}(t) = 0$ for even $t > 0$; $A_{1,1}(t) = \varepsilon$, $A_{1,2}(t) = 0$, $Q_{1,1}(t) = 0$ and $Q_{1,2}(t) = D(t)$ for odd t, while $A_{2,1}(t) = A_{2,2}(t) = 0$, $Q_{2,1}(t) = D(t-1)$ and $Q_{2,2}(t) = 0$ for odd t; $A_{2,1}(t) = \varepsilon$, $A_{2,2}(t) = 0$, $Q_{2,1}(t) = 0$ and $Q_{2,2}(t) = D(t)$ for even t. At this equilibrium the production at each stage is equal to the maximal possible demand. At this equilibrium for a.e. $\omega \in \mathbb{I}$ and for all even $t > 0$ we have $X^{\omega}_{1,1}(t) = 1$ and for all odd t we have $X^{\omega}_{2,1}(t) = 1$ while the remaining $X^{\omega}_{i,j} = 0$. The strategy of player ω is such that $U^{\omega}_{1,1}(t, X^{\omega}(t)) = 1$ for all even $t > 0$ and $U^{\omega}_{2,1}(t, X^{\omega}(t)) = 1$ for all odd t.

Proof.

a) Let us consider a profile at which firm 1 decides to change its strategy and advertise good 1 with advertising effort $a \geq \varepsilon$ at time t at which firm 2 also advertises with the expenditures on advertising equal to at least ε. Before the change firm 1 produced the amounts whose sum was q. Since the expected demand for its goods may increase at this time instant at most by $\bar{q} = \frac{a}{a+\varepsilon}\bar{D}$, which is reached for the maximal advertising effort assuming the minimal advertising effort of firm 2, and at most by the same amount jointly in the future. Since the function $pq - c(q, 0)$ is increasing in the interval $(0, 2\bar{D}]$, the maximum is attained at the maximal production that can be sold at each time instant, which is attained at $q + \bar{q}$. Therefore the firm's income now is going to increase by at most $(p \cdot (\bar{q} + q) - c(\bar{q} + q, 0)) - \min_{q_1+q_2=q} (p \cdot q - c(q_1, q_2)) = p \cdot \bar{q} - (c(\bar{q} + q, 0) - \max_{q_1+q_2=q} (c(q_1, q_2)))$, which is less than $\frac{\varepsilon}{2}$. The joint growth of profits in the future is by the analogous reasoning less than $\frac{\varepsilon}{2}$. Since the discounting function is nonincreasing, the joint discounted increase of payoffs minus increase in the discounted advertising effort is negative.

b) We construct such an equilibrium, starting from some strategies of parents, then best responses of the firms and we end by checking whether the initial profile of parents decisions constitutes an equilibrium with these strategies of the firms.

Since the strategies of players have to constitute the best response only to the strategies used by the remaining players, let us assume, for simplicity of further calculations, that for all ω we define parents strategies as follows $U^{\omega}_{1,2}(t, x^{\omega}) = U^{\omega}_{2,2}(t, x^{\omega}) \equiv 0$ (since good 2 is advertised by neither of firms)

$U^{\omega}_{1,1}(t, x^{\omega}) = 1$ if $x^{\omega}_{1,1} > 0$ and t is even, 0 otherwise
$U^{\omega}_{2,1}(t, x^{\omega}) = 1$ if $x^{\omega}_{2,1} > 0$ and t is odd, 0 otherwise.

Now let us examine the strategy of firm 1.

We start at time T. If T is even, then firm 2 advertises, so, by the assumption, it is better for firm 1 not to advertise. Thus the production of firm 1 is constrained by the demand i.e. $\int_{\mathbb{I}} U^{\omega}_{1,1}(T, X^{\omega}(T)) + U^{\omega}_{1,2}(T, X^{\omega}(T))d\lambda(\omega) = \int_{\mathbb{I}} U^{\omega}_{1,1}(T, X^{\omega}(T))d\lambda(\omega)$, which is equal to the measure of the set of parents whose $X^{\omega}_{1,1}(T) > 0$.

For odd T firm 2 does not advertise and $\int_{\mathbb{I}} U^{\omega}_{1,1}(T, X^{\omega}(T)) + U^{\omega}_{1,2}(T, X^{\omega}(T))d\lambda(\omega) = 0$. Therefore the best strategy of firm 1 is the joint advertising effort ε (possibly only advertising good 1) and, because $c(q_1 + q_2, 0) > c(q_1, q_2)$ for $q_1, q_2 > 0$, production of only one good, possibly 1.

What is the value at time $T-1$ of the best response strategy to the other players' strategies, assuming that at time T we choose strategy as we have just proven to be optimal at time T? If T is even, then at time $T-1$ firm 2 does not advertise, while at time T we choose not to advertise and produce only for unkept promises for our good 1. Therefore the optimal decision is to choose at time $T-1$ advertising effort $A_{1,1}(t) = \varepsilon$, and producing good 2 only with production equal to $\bar{D}$, which results in both optimal profit at time $T-1$ and maximal possible $\int_{\mathbb{I}} U^{\omega}_{1,1}(T, X^{\omega}(T))d\lambda(\omega)$.

Now let us take any $s < t$ and assume that our strategy fulfills $A_{1,1}(t) = A_{1,2}(t) = 0$, $Q_{1,1}(t) = D(t-1)$ and $Q_{1,2}(t) = 0$ for even $0 < t \leq s$; $A_{1,1}(t) = \varepsilon$, $A_{1,2}(t) = 0$, $Q_{1,1}(t) = 0$ and $Q_{1,2}(t) = D(t)$ for odd $t \leq s$. By analogous reasoning we prove, that the result holds for all $t \leq s-1$.

So, by backwards induction, we have proven that the strategy of firm 1 defined by $A_{1,1}(t) = A_{1,2}(t) = 0$, $Q_{1,1}(t) = D(t-1)$ and $Q_{1,2}(t) = 0$ for even $t > 0$; $A_{1,1}(t) = \varepsilon$, $A_{1,2}(t) = 0$, $Q_{1,1}(t) = 0$ and $Q_{1,2}(t) = D(t)$ for odd t is the best response to strategies of the remaining players.

For firm 2 an analogous reasoning applies.

And finally, let us note, that whatever equilibrium strategy profile of parents we consider as the best responses to firms' strategies and behaviour of the other parents, we obtain the result that on a set of ω of measure 1 for all t, $U^{\omega}_{i,1}(t, X^{\omega}(t))$ for i such that $X^{\omega}_{i,1}(t) > 0$. □

Remark 21.1. In the case of the cost function linear with switching-on cost $c(q_1, q_2) = \begin{cases} 0 & \text{if } q_1 = q_2 = 0, \\ 2\bar{c} + b \cdot (q_1 + q_2) & \text{if } q_1, q_2 > 0, \\ \bar{c} + b \cdot (q_1 + q_2) & \text{otherwise.} \end{cases}$ the complicated assumption can be reduced to a simple one: $(p - b) \cdot \bar{D} < \varepsilon$.

What is interesting, such alternating advertising sequences in order to increase efficiency of advertising efforts could be observed not only at toy markets – there were similar observations concerning the market of soft drinks – almost equal division of advertising days between Pepsi and Coca-Cola.

21.3.4 *Case D: Infinite Time Horizon and Negligibility of Low Advertising Efforts*

Let us add an additional assumption like in Case C that there exists a minimal effective advertising effort ε for each product and choosing $A_{i,j}(t) < \varepsilon$ has influence on $P(t)$ equivalent to choosing $A_{i,j}(t) = 0$ i.e. $\int_{\mathbb{I}} P^{\omega}_{i,j}(t) d\lambda(\omega)$ is equal to 0 if $A_{i,j}(t) < \varepsilon$ and at least one $A_{k,l}(t) \geq \varepsilon$, to $D(t) \cdot \frac{A_{i,j}(t)}{\sum_{k,l \in \{1,2\}, A_{k,l}(t) > \varepsilon} A_{k,l}(t)}$ if the denominator is positive, and to $\frac{1}{4} D(t)$ otherwise.

This ε is assumed to be large compared to expected profits.

In this case, under the conditions about the cost functions and bounds for advertising efforts, it is possible to obtain an equilibrium at which one of the firms advertises only at even time instants while it opponent at odd time instants. Moreover, only one product is advertised at each stage. Each of the firms produces at any stage of advertising only the non-advertised product and the amount produced is equal to the maximal $D(t)$, while at stage $t + 1$ it produces the other product also in amount $D(t)$. Parents buy at each stage in which their promise is positive the non-advertised product of the advertising firm, while at the next stage they fulfill unkept promise for the other good.

Proposition 21.5. *Assume that for all* t $D(t) = \bar{D}$, *and that for* $i = 1, 2$ *for all* $\varepsilon \leq a \leq \bar{A}_i$ *and all* $q \leq 2\bar{D}$, $\bar{q} \leq \frac{a}{a+\varepsilon} \bar{D}$ *we have* $p \cdot \bar{q} - (c(\bar{q} + q, 0) - \max_{q_1 + q_2 = q} c(q_1, q_2)) < \frac{a}{2}$ *and for all* $\{q^i\}_{i=1}^{+\infty}$ *such that* $q^i \leq 2\bar{D}$ *and all* $\bar{q} \leq \frac{a}{a+\varepsilon} \bar{D}$ *and all sequences* $\{\bar{q}^i\}_{i=1}^{+\infty}$ *such that* $\sum_{i=1}^{T} \bar{q}^i = \bar{q}$ *we have* $p \cdot \bar{q} - \min_{q^i_1 + q^i_2 = q^i} \sum_{i=1}^{+\infty} \left(c(\bar{q}^i + q^i_1, 0) - c(q^i_1, q^i_2) \right) < \frac{a}{2}$.

a) If the discounting functions of the firms are such that $\sum_{t=N}^{\infty} \Xi^i(t) \to 0$ as $N \to \infty$ for $i = 1, 2$, then there is no equilibrium such that both firms advertise at the same time instant.

b) If for all t $p \cdot D(t) - c(D(t), 0) > \varepsilon$, then there exists an equilibrium at which $A_{1,1}(t) = A_{1,2}(t) = 0$, $Q_{1,1}(t) = D(t-1)$ and $Q_{1,2}(t) = 0$ for even $t > 0$; $A_{1,1}(t) = \varepsilon$, $A_{1,2}(t) = 0$, $Q_{1,1}(t) = 0$ and $Q_{1,2}(t) = D(t)$ for odd t, while $A_{2,1}(t) = A_{2,2}(t) = 0$, $Q_{2,1}(t) = D(t-1)$ and $Q_{2,2}(t) = 0$ for odd t; $A_{2,1}(t) = \varepsilon$, $A_{2,2}(t) = 0$, $Q_{2,1}(t) = 0$ and $Q_{2,2}(t) = D(t)$ for even t. At this equilibrium the production at each stage is equal to the maximal possible demand. At this equilibrium for a.e. $\omega \in \mathbb{I}$ and for all even $t > 0$ we have $X_{1,1}^{\omega}(t) = 1$ and for all odd t we have $X_{2,1}^{\omega}(t) = 1$ while the remaining $X_{i,j}^{\omega} = 0$. The strategy of player ω is such that $U_{1,1}^{\omega}(t, X^{\omega}(t)) = 1$ for all even $t > 0$ and $U_{2,1}^{\omega}(t, X^{\omega}(t)) = 1$ for all odd t

Proof. a) By the assumption about the discounting functions, both firms get finite payoffs. Therefore the proof of a) is identical as that of Proposition 21.4.

b) We formulate simplified strategies of parents as in the proof of Proposition 21.4.

First let us note that if at least one of the discounting functions, without loss of generality the discounting function of firm 1, is such that $\sum_{t=0}^{N} \Xi^1(t)$ does not converge to a finite number, then each strategy with payoff equal to $+\infty$, including the strategy we examine, is an equilibrium strategy.

Therefore we only have to check the opposite case.

We cannot apply backward induction since we have infinite time horizon. However, we can try to estimate the payoff in the infinite horizon game by the payoffs in finite horizon games. We check firm 1 after assuming that parents choose the strategies as in the proof of Proposition 21.4, while firm 2 as formulated in this proposition.

Let us take any finite N. We know that there is a best response to strategies of the remaining players in the game with time horizon N such that $A_{1,1}(t) = A_{1,2}(t) = 0$, $Q_{1,1}(t) = D(t-1)$ and $Q_{1,2}(t) = 0$ for even $t > 0$; $A_{1,1}(t) = \varepsilon$, $A_{1,2}(t) = 0$, $Q_{1,1}(t) = 0$ and $Q_{1,2}(t) = D(t)$ for odd t. The payoff for this strategy in the infinite game is equal to $\sum_{t=1}^{\infty} \bar{D} \cdot \Xi^1(t) - \sum_{t=0}^{\infty} \varepsilon \cdot \Xi^1(2t+1)$, while in the game with time horizon N it is equal to $\sum_{t=1}^{N} \bar{D} \cdot \Xi^1(t) - \sum_{t=0}^{\left[\frac{N-1}{2}\right]} \varepsilon \cdot \Xi^1(2t+1)$. The optimal payoff in the infinite horizon game is finite, since it is constrained from above by e.g. $\sum_{t=0}^{\infty} 2\bar{D} \cdot \Xi^1(t)$. Let us take any small $\delta > 0$. Since $\sum_{t=N}^{\infty} 2\bar{D} \cdot \Xi^1(t) \to 0$, there exists N such that the sum of discounted payoffs for the strategy optimal in the infinite game from time $N+1$ to $+\infty$ is less than δ. The maximal sum of discounted payoffs from time 0 to time N is equal to $\sum_{t=1}^{N} \bar{D} \cdot \Xi^1(t) - \sum_{t=0}^{\left[\frac{N-1}{2}\right]} \varepsilon \cdot \Xi^1(2t+1)$, which converges to $\sum_{t=1}^{\infty} \bar{D} \cdot \Xi^1(t) - \sum_{t=0}^{\infty} \varepsilon \cdot \Xi^1(2t+1)$ - the payoff for the examined strategy in the infinite game, therefore our strategy is the best response also in the infinite horizon game. □

Remark 21.2. In the case of the cost function linear with switching-on cost $c(q_1, q_2) = \begin{cases} 0 & \text{if } q_1 = q_2 = 0, \\ 2\bar{c} + b \cdot (q_1 + q_2) & \text{if } q_1, q_2 > 0, \\ \bar{c} + b \cdot (q_1 + q_2) & \text{otherwise.} \end{cases}$ the complicated assumption can be reduced to a simple one: $(p - b) \cdot \bar{D} < \varepsilon$.

21.4 Conclusions

The compound model of an oligopolistic toy market studied in the paper led us to apparently strange results, among which was that it is reasonable for a firm to advertise a good it is not producing at the current stage. This is a result of exploiting the human need for consequence, according to which parents try to keep promises. Although this seems strange, such strategic behaviour of firms can be observed at toy markets as a way of increasing sales in January, which confirms the validity of the results proven.

Another interesting result is the existence of equilibria with alternating advertising efforts of the firms, which makes advertising more efficient. Such a result can also be observed at real world duopolistic markets (not necessarily toy markets).

Bibliography

Aumann, R. J. (1964). Markets with a Continuum of Traders, *Econometrica* **32**, pp. 39–50.

Aumann, R. J. (1966). Existence of Competitive Equilibrium in Markets with Continuum of Traders, *Econometrica* **34**, pp. 1–17.

Balder, E. (1995). A Unifying Approach to Existence of Nash Equilibria, *International Journal of Game Theory* **24**, pp. 79–94.

Cellini, R. Lambertini, L. (2003). Advertising with Spillover Effects in a Differential Oligopoly Game with Differentiated Goods, *Central European Journal of Operations Research* **11**, pp. 409-423.

Cellini, R., Lambertini, L. (2004). Dynamic Oligopoly with Sticky Prices: Closed-Loop, Feedback and Open-Loop Solutions, *Journal of Dynamical and Control Systems* **10**, pp. 303–314.

Cellini, R. Lambertini, L. (2007). A Differential Oligopoly Game with Differentiated Goods and Sticky Prices, *European Journal of Operational Research* **176**, pp. 1131–1144.

De Cesare, L. Di Liddo A. (2001). A Stackelberg Game of Innovation Diffusion: Pricing, Advertising and Subsidy Strategies, *International Game Theory Review*, **3**, pp. 325–339.

Dockner, E. Feichtinger, G. (1986). *Dynamic Advertising and Pricing in an Oligopoly: A Nash Equilibrium Approach*, 238-251, in T. Basar (ed.), 1986, *Dynamic Games and Applications in Economics*, Lecture Notes in Economics and Mathematical Systems **265**, (Springer).

E. Dockner, G. Feichtinger, G. Sorger (1985). *Interaction of Price and Advertising under Dynamic Conditions*, Working Paper, University of Economics and Technical University, (Vienna).

Dorfman, Steiner (1954). Optimal Advertising and Optimal Quality, *American Economic Review* **4**, pp. 826–836.

Feichtinger, G., Hartl, R. F., Sethi, S. P. (1994). Dynamic Optimal Control Models in Advertising: Recent Developments, *Management Science* **40**, pp. 195–226.

Feichtinger, G., Jørgensen S. J(1983). Differential Game Models in Management Science, *European Journal of Operations Research* **14**, pp. 137–155.

Fersthman, C., Kamien, M. I. (1987). Dynamic Duopolistic Competition with Sticky Prices, *Econometrica* **55**,pp. 1151-1164.

Fruchter, G. A. (2001), Advertising in a Competitive Product Line, *International Game Theory Review* **3**, pp. 301–314.

Jørgensen, S. (1982). A Survey of Some Differential Games in Advertising, *Journal of Economic Dynamics and Control* **4**, pp. 341–369.

Jørgensen, S. (1986). Optimal Dynamic Pricing in an Oligopolistic Market: A Survey, 179-237, in T. Basar (ed.), 1986, *Dynamic Games and Applications in Economics, Lecture Notes in Economics and Mathematical Systems* **265**, (Springer).

Karatzas,I., Shubik,M., Sudderth, W. D. (1994). Construction of Stationary Markov Equilibria in a Strategic Market Game, *Mathematics of Operations Research* **19**, pp. 975–1006.

Mas-Colell, A. (1984). On the Theorem of Schmeidler, *Journal of Mathematical Economics* **13**, pp. 201–206.

Schmalensee, R. (1972). *The Economics of Advertising*, (North-Holland).

Schmalensee, R. (1976). A Model of Promotional Competition in Oligopoly, *Review of Economic Studies* **43**, pp. 493–508.

Schmeidler, D. (1973). Equilibrium Points of Nonatomic Games, *Journal of Statistical Physics* **17**, pp. 295–300.

Sethi, S. P. (1977). A Dynamic Optimal Control Models in Advertising: A Survey, *SIAM Review* **19**, pp. 685–725.

Vind (1964). Edgeworth-Allocations is an Exchange Economy with Many Traders, *International Economic Review* **5**, pp. 165–177.

Wieczorek, A. (2004). Large Games with Only Small Players and Finite Strategy Sets, *Applicationes Mathematicae* **31**, 79-96.

Wieczorek, A. (2005). Large Games with Only Small Players and Strategy Sets in Euclidean Spaces, *Applicationes Mathematicae* **32**, pp. 183–193.

A. Wieczorek, A. Wiszniewska (Wiszniewska-Matyszkiel) (1999). A Game-Theoretic Model of Social Adaptation in an Infinite Population, *Applicationes Mathematicae* **25**, pp. 417–430.

Wiszniewska-Matyszkiel, A. (2000a). *Dynamic Game with Continuum of Players Modelling "the Tragedy of the Commons", Game Theory and Applications* **5** (Petrosjan, Mazalov eds.), pp. 162–187.

Wiszniewska-Matyszkiel, A. (2000b), Existence of Pure Equilibria in Games with Continuum of Players, *Topological Methods in Nonlinear Analysis*, **16**, pp. 339–349.

Wiszniewska-Matyszkiel, A. (2001).*"The Tragedy of the Commons" Modelled by Large Games, Annals of the International Society of Dynamic Games* **6**, (E. Altman, O. Pourtallier eds.), Birkhauser, pp. 323–345.

Wiszniewska-Matyszkiel, A. (2002). Discrete Time Dynamic Games with Continuum of Players I: Decomposable Games, *International Game Theory Review* **4**, pp. 331–342.

Wiszniewska-Matyszkiel, A. (2002). Static and Dynamic Equilibria in Games with Continuum of Players, *Positivity* **6**, pp. 433-453.

Wiszniewska-Matyszkiel, A. (2003a). Static and Dynamic Equilibria in Stochastic Games with Continuum of Players, *Control and Cybernetics* **32**, pp. 103–126.

Wiszniewska-Matyszkiel, A. (2003b). Discrete Time Dynamic Games with Continuum of Players II: Semi-Decomposable Games, *International Game Theory Review* **5**, 27-40.

Wiszniewska-Matyszkiel, A. (2005). A Dynamic Game with Continuum of Players and its Counterpart with Finitely Many Players, *Annals of the International Society of Dynamic Games* **7** (A. Nowak, K. Szajowski eds.), 445-469, (Birkhauser).

Wiszniewska-Matyszkiel, A. (2006). Modelling Stock Exchange by Games with Continuum of Players, in Polish, *Opere et Studio pro Oeconomia* **1**, 5-38.

Wiszniewska-Matyszkiel, A. (2007). Common Resources, Optimality and Taxes in Dynamic Games with Increasing Number of Players, to appear in *Journal of Mathematical Analysis and Applications*, in press – doi: 10.1016/j.jmaa.2007.03.033.

Wiszniewska-Matyszkiel, A. (2005). *Stock Exchange as a Game with Continuum of Players*, preprint 153/2005 Institute of Applied Mathematics and Mechanics, Warsaw University; available at http://www.mimuw.edu.pl/ english/research/ reports/imsm/; submitted.

Chapter 22

On Some Classes of Balanced Games

R. B. Bapat
Indian Statistical Institute
7, S. J. S. Sansanwal Marg
New Delhi, 110016, India
e-mail: rbb@isid.ac.in

Abstract

We introduce two classes of games and show that they are balanced. In *regression games,* the observations in a regression model are controlled by players, and the worth of a coalition is inversely proportional to the variance of the estimate of the regression parameter. In *connectivity games* the players control the edges of a graph and the worth of a coalition is directly proportional to the degree of connectivity of the subgraph formed by the corresponding edges.

Key Words: Balanced games, regression games, connectivity games

22.1 Introduction

A cooperative game (N, v) consists of a set N of players and a characteristic function (or worth function) $v : 2^N \longrightarrow (0, \infty)$. We assume that $|N| = n$ and set $N = \{1, 2, \ldots, n\}$. We also assume $v(\phi) = 0$. A fundamental problem in cooperative game theory is to prescribe a procedure to distribute $v(N)$ among the n players in a manner which is justified by some natural principles. Such a procedure is known as a solution concept. Particularly notable solution concepts are the von Neumann-Morgenstern stable set, the Shapley value, the nucleolus and the core. In this paper we shall be interested in the core, which we now proceed to define (see, for example, [Owen (1982)], [Tijs (2003)]).

A vector $x = (x_1, \ldots, x_n)'$ is called an *imputation* if $x_i \geq 0, i = 1, 2, \ldots, n$ and $\sum_{i=1}^{n} x_i = v(N)$. The *core* of the game (N, v) consists of all imputations x which satisfy $\sum_{i \in S} x_i \geq v(S)$ for every $S \subset N$.

We now introduce some notation. For $S \subset N$, let $\mathbf{1}_S$ denote the $n \times 1$ incidence vector of S. Thus the i-th coordinate of $\mathbf{1}_S$ is 1 if $i \in S$ and 0 otherwise. Note that $\mathbf{1}_N$ is the $n \times 1$ vector of all ones and we denote it simply by $\mathbf{1}$.

A game (N, v) is called *balanced* if for any nonnegative numbers $\lambda_S, S \subset N$, satisfying

$$\sum_S \lambda_S \mathbf{1}_S = \mathbf{1},$$

it is true that

$$\sum_S \lambda_S v(S) \leq v(N).$$

The following result, proved independently by [Bondareva (1963)] and [Shapley (1967)] is well-known and can be proved using the duality theorem of linear programming.

Theorem 22.1. *[Bondareva-Shapley] The game (N, v) has a nonempty core if and only if it is balanced.*

For $T \subset N$ we define the induced subgame (T, v_T) by setting $v_T(S) = v(S)$ for every $S \subset T$. A game is called *totally balanced* if all its induced subgames are balanced, i.e., if all its induced subgames have nonempty core.

22.2 Regression Games

Consider the linear regression model

$$y_i = u_i \beta_1 + v_i \beta_2 + \epsilon_i, i = 1, 2, \ldots, n; \tag{22.1}$$

where u_i, v_i are known, y_i are the observations, β_1, β_2 are unknown parameters and ϵ_i are uncorrelated errors with the common, unknown variance σ^2.

We assume that β_1 is the parameter of interest. We assume that β_1 is estimable, that is, there exists a linear function $c_1 y_1 + \cdots + c_n y_n$ with expectation β_1. The BLUEs (best linear unbiased estimates) of β_1 and β_2 are obtained by minimizing the sum of squared errors,

$$\sum_{i=1}^{n} \epsilon_i^2 = \sum_{i=1}^{n} (y_i - u_i \beta_1 - v_i \beta_2)^2.$$

The resulting least-squares estimate and its variance are given in the following well-known result, the first part of which is usually referred to as the Gauss-Markov Theorem.

Theorem 22.2. *With reference to the linear model (22.1), the BLUE $\hat{\beta}_1$ of β_1 is given by*

$$\hat{\beta}_1 = \frac{(\sum_{i=1}^n v_i^2)(\sum_{i=1}^n u_i y_i) - (\sum_{i=1}^n u_i v_i)(\sum_{i=1}^n v_i y_i)}{(\sum_{i=1}^n u_i^2)(\sum_{i=1}^n v_i^2) - (\sum_{i=1}^n u_i v_i)^2}.$$

Furthermore,

$$\frac{\sigma^2}{var(\hat{\beta}_1)} = \sum_{i=1}^n u_i^2 - \frac{(\sum_{i=1}^n u_i v_i)^2}{\sum_{i=1}^n v_i^2}.$$

We now introduce a cooperative game based on the model (22.1). Consider n players, $\{1, 2, \ldots, n\}$ and suppose the i-th player controls (that is, able to provide) the observation y_i. We may imagine a situation where each observation comes from a household and thus it is available only if the head of the family cooperates. If $S \subset N = \{1, 2, \ldots, n\}$, then we set $v(S)$ to be σ^2 times the reciprocal of the variance of the BLUE of β_1 in the linear model obtained by using the observations $y_i, i \in S$, provided β_1 is estimable in that model. Otherwise we set $v(S) = 0$. Also, as usual, we set $v(\phi) = 0$.

The interpretation of the worth function should be clear. A set of players stand to gain more if they are able to get a more precise estimate of the parameter. The precision is measured by the reciprocal of the variance of the BLUE. We remark that the choice of β_1 as the parameter of interest is merely a matter of convenience. We could take β_2, or, in fact, any linear combination of β_1 and β_2, as the parameter of interest. Typically, in a regression problem, $v_i = 1$ for $i = 1, 2, \ldots, n$, and β_1, the slope of the regression line, is the parameter of interest.

We must introduce some technical assumptions at this point. These assumptions are necessary to take care of the estimability of the parameter β_1. The assumptions are as follows:

(A0) For $i = 1, 2, \ldots, n$, $v_i \neq 0$.

(A1) for $i, j \in \{1, 2, \ldots, n\}$, $i \neq j$, the vectors (u_i, v_i) and (u_j, v_j) are linearly independent.

Assumption (A0) ensures that β_1 is not estimable from any single observation. Thus $v(S) = 0$ if $|S| = 1$. Assumption (A1) guarantees that β_1 is estimable in any model containing two or more observations. Therefore $v(S) > 0$ if $|S| \geq 2$.

We refer to this game as the regression game associated with the model (22.1).

Theorem 22.3. *In the presence of assumptions (A0) and (A1), the regression game associated with the model (22.1) is totally balanced.*

Proof. We first show that the regression game is balanced. Note that if $S \subset N = \{1, 2, \ldots, n\}$, and $|S| \geq 2$, then β_1 is estimable in the model obtained by taking the observations $y_i, i \in S$, and

$$v(S) = \sum_{i \in S} u_i^2 - \frac{(\sum_{i \in S} u_i v_i)^2}{\sum_{i \in S} v_i^2}.$$

Furthermore, $v(S) = 0$ if $|S| \leq 1$.

As before, for $S \subset N = \{1, 2, \ldots, n\}$, let $\mathbf{1}_S$ be the $n \times 1$ incidence vector of S. Suppose $\lambda_S \geq 0$ satisfy

$$\sum_S \lambda_S \mathbf{1}_S = \mathbf{1}. \tag{22.2}$$

Then we must show

$$\sum_S \lambda_S v(S) \leq v(N), \tag{22.3}$$

which is the same as

$$\sum_S \lambda_S \{\sum_{i \in S} u_i^2 - \frac{(\sum_{i \in S} u_i v_i)^2}{\sum_{i \in S} v_i^2}\} \leq \sum_{i=1}^n u_i^2 - \frac{(\sum_{i=1}^n u_i v_i)^2}{\sum_{i=1}^n v_i^2}. \tag{22.4}$$

Since, in view of (22.2),

$$\sum_S \lambda_S \sum_{i \in S} u_i^2 = \sum_{i=1}^n u_i^2,$$

(22.4) will be proved once we show

$$\frac{(\sum_{i=1}^n u_i v_i)^2}{\sum_{i=1}^n v_i^2} \leq \sum_S \lambda_S \frac{(\sum_{i \in S} u_i v_i)^2}{\sum_{i \in S} v_i^2}. \tag{22.5}$$

We have

$$\frac{(\sum_{i=1}^n u_i v_i)^2}{\sum_{i=1}^n v_i^2} = \frac{(\sum_S \lambda_S \sum_{i\in S} u_i v_i)^2}{\sum_{i=1}^n v_i^2}$$
$$= (\sum_S \frac{\lambda_S \sum_{i\in S} v_i^2}{\sum_{i=1}^n v_i^2} \frac{\sum_{i\in S} u_i v_i}{\sum_{i\in S} v_i^2})^2 \sum_{i=1}^n v_i^2$$
$$\leq \sum_S (\frac{\lambda_S \sum_{i\in S} v_i^2}{\sum_{i=1}^n v_i^2})(\frac{\sum_{i\in S} u_i v_i}{\sum_{i\in S} v_i^2})^2 \sum_{i=1}^n v_i^2$$

by Cauchy-Schwarz inequality

$$= \sum_S \lambda_S \frac{(\sum_{i\in S} u_i v_i)^2}{\sum_{i\in S} v_i^2},$$

and (22.5) is proved. It follows by Theorem 22.1 that the regression game is balanced.

Any induced subgame of the regression game associated with the model (22.1) is again a regression game based on the model obtained by taking a subset of the observations. Therefore using a similar proof we can show that any induced subgame is also balanced and therefore the regression game is totally balanced. □

Let us consider the simple linear regression model, which is a special case of the model (22.1), with $v_i = 1$ for $i = 1, 2, \ldots, n$ (after making a minor change in notation):

$$y_i = \beta_0 + u_i\beta_1 + \epsilon_i, i = 1, 2, \ldots, n. \tag{22.6}$$

The corresponding regression game (N, v) has worth function given by

$$v(S) = \sum_{i\in S}(u_i - \bar{u}_S)^2$$

for any nonempty $S \subset N$, and $v(\phi) = 0$. Here $\bar{u}_S$ is the mean of $u_i, i \in S$.

For this game we can give a core element explicitly: Set $x_i = (u_i - \bar{u})^2, i = 1, 2, \ldots, n$; where $\bar{u} = \bar{u}_N$ is the mean of $u_1, \ldots, u_n$. Then $x = (x_1, \ldots, x_n)'$ is in the core. This follows since for any $S \subset N$,

$$\sum_{i\in S}(u_i - \bar{u})^2 \geq \sum_{i\in S}(u_i - \bar{u}_S)^2,$$

in view of the well-known fact that the sum of squared deviations is minimized when the deviations are about the mean.

A similar argument shows that the following game is also totally balanced. Let $u_1, \ldots, u_n$ be real numbers and consider the game (N, w), with $w(S) = \sum_{i\in S} |u_i - \tilde{u}_S|$, for any nonempty $S \subset N$, where $\tilde{u}_S$ is the median of $u_i, i \in S$.

22.3 Connectivity Games

Consider a graph with m vertices and n edges. Suppose there are n players, $\{1, 2, \ldots, n\}$ and the i-th player controls (is able to provide) the edge $e_i, i = 1, 2, \ldots, n$. In many practical problems, particularly in transport and telecommunications, it is preferable to have a network with a high "degree of connectivty". This motivates the following cooperative game. For a coalition S, $v(S)$ is directly proportional to the degree of connectivity of the subgraph induced by the edges $\{e_i, i \in S\}$.

We begin by recalling some concepts from graph theory which will be required. We consider graphs which have no loops or parallel edges. Thus a *graph* $G = (V(G), E(G))$ consists of a finite set of *vertices,* $V(G)$, and a set of *edges,* $E(G)$, each of whose elements is a pair of distinct vertices. We will assume familiarity with basic graph-theoretic notions, see, for example, [Bondy and Murty (1976)],[West (2001)].

Given a graph, one associates a variety of matrices with the graph. Some of the important ones will be defined now. Let G be a graph with $V(G) = \{1, \ldots, m\}, E(G) = \{e_1, \ldots, e_n\}$.

The *adjacency matrix* $A(G)$ of G is an $m \times m$ matrix with its rows and columns indexed by $V(G)$ and with the (i, j)-entry equal to 1 if vertices i, j are adjacent (i.e., joined by an edge) and 0 otherwise. Thus $A(G)$ is a symmetric matrix with its i-th row (or column) sum equal to $d(i)$, which by definition is the degree of the vertex $i, i = 1, 2, \ldots, m$. Let $D(G)$ denote the $n \times n$ diagonal matrix, whose i-th diagonal entry is $d(i), i = 1, 2, \ldots, m$.

The *Laplacian matrix* of G, denoted by $L(G)$, is simply the matrix $D(G) - A(G)$.

There is another way to view the Laplacian matrix. First we introduce yet another important matrix associated with G. Suppose each edge of G is assigned an orientation, which is arbitrary but fixed. The (vertex-edge) *incidence matrix* of G, denoted by $Q(G)$, is the $m \times n$ matrix defined as follows. The rows and the columns of $Q(G)$ are indexed by $V(G), E(G)$ respectively. The (i, j)-entry of $Q(G)$ is 0 if vertex i and edge e_j are not incident and otherwise it is 1 or -1 according as e_j originates or terminates at i respectively.

A simple verification reveals that the Laplacian matrix $L(G)$ equals $Q(G)Q(G)'$. Observe that although we introduced an orientation for each edge while defining $Q(G)$, the matrix $L(G)$ does not depend upon the particular orientation.

Example: Let G be the graph with vertex set $\{1,2,3,4,5\}$ and edge set $\{12,23,13,24,34,45\}$. Then

$$A(G) = \begin{bmatrix} 0\,1\,1\,0\,0 \\ 1\,0\,1\,1\,0 \\ 1\,1\,0\,1\,0 \\ 0\,1\,1\,0\,1 \\ 0\,0\,0\,1\,0 \end{bmatrix}, Q(G) = \begin{bmatrix} 1 & 0 & 1 & 0 & 0 & 0 \\ -1 & 1 & 0 & 1 & 0 & 0 \\ 0 & -1 & -1 & 0 & 1 & 0 \\ 0 & 0 & 0 & -1 & -1 & 1 \\ 0 & 0 & 0 & 0 & 0 & -1 \end{bmatrix},$$

and

$$L(G) = Q(G)Q(G)' = \begin{bmatrix} 2 & -1 & -1 & 0 & 0 \\ -1 & 3 & -1 & -1 & 0 \\ -1 & -1 & 3 & -1 & 0 \\ 0 & -1 & -1 & 3 & -1 \\ 0 & 0 & 0 & -1 & 1 \end{bmatrix}.$$

Let G be a graph with $V(G) = \{1,\ldots,m\}, E(G) = \{e_1,\ldots,e_n\}$. Some basic properties of the Laplacian matrix are summarized below (see [Bapat (1996)],[Merris (1994)]).

(i) $L(G)$ is a symmetric, positive semidefinite matrix.

(ii) The off-diagonal entries of $L(G)$ are nonpositive (in fact, they are either 0 or -1).

(iii) The diagonal entries of $L(G)$ are the vertex degrees and the row sums and the column sums are all zero.

(iv) The quadratic form afforded by $L(G)$ has a rather simple description:

$$\langle L(G)x, x\rangle = \sum_{(i,j)\in E(G)} (x_i - x_j)^2.$$

(v) The rank of $L(G)$ is $n-k$, where k is the number of connected components of G. In particular, if G is connected, then the rank of $L(G)$ is $n-1$.

The Laplacian matrix is also known by several other names in the literature such as the Kirchhoff matrix or the Information matrix.

Let $\mathcal{L}$ be the set of $n \times n$ symmetric, positive semidefinite matrices with row sums zero and let f be a nonnegative real valued function f on $\mathcal{L}$. Let G be graph with n edges. We define a cooperative game as follows. The player set is $N = \{1,2,\ldots,n\}$. If $S \subset N$, let L_S be the Laplacian of the subgraph of G formed by the vertex set $V(G)$ and the edges in S. Let the worth function $v(S)$ be given by $v_f(S) = f(L_S)$. We now prove the following.

Theorem 22.4. *Suppose f satisfies the following conditions:*
(i) $f(\alpha L) = \alpha f(L)$ for any $\alpha \geq 0$ and for any $L \in \mathcal{L}$.
(ii) $f(\sum_{i=1}^k \alpha_i L_i) \geq \sum_{i=1}^k \alpha_i f(L_i)$ for any $\alpha_i \geq 0, i = 1, 2, \ldots, k$ satisfying $\sum_{i=1}^k \alpha_i = 1$ and for any $L_i \in \mathcal{L}, i = 1, 2, \ldots, k$.
Then the game (N, v_f) is balanced.

Proof. As before, for $S \subset N = \{1, 2, \ldots, n\}$, let $\mathbf{1}_S$ be the $n \times 1$ incidence vector of S. Suppose $\lambda_S \geq 0$ satisfy

$$\sum_S \lambda_S \mathbf{1}_S = \mathbf{1}. \tag{22.7}$$

Then, in view of Theorem 22.2, we must show

$$\sum_S \lambda_S v_f(S) \leq v_f(N), \tag{22.8}$$

Note that $\sum_S \lambda_S L_S = L_N$, the Laplacian of G. Hence, in view of the properties (i) and (ii) of f,

$$\begin{aligned} v_f(N) &= f(L_N) \\ &= f(\sum_S \lambda_S L_S) \\ &= \sum_S \lambda_S f(\frac{\sum_S \lambda_S L_S}{\sum_S \lambda_S}) \\ &\geq \sum_S \lambda_S \sum_S \lambda_S f(L_S) \\ &= \sum_S \lambda_S v_f(S), \end{aligned}$$

and the proof is complete. □

Some examples of functions f which satisfy (i) and (ii) of Theorem 22.4 are as follows (see, for example, [Ghosh and Boyd (2006)]). Let $0 = \mu_0 \leq \mu_1 \leq \cdots \leq \mu_{n-1}$ be the eigenvalues of the Laplacian L.

(1) $f(L) = \mu_1$, the algebraic connectivity, introduced by [Fiedler (1973)].
(2) $f(L) = \sum_{i=1}^k \mu_i$ for any $k = 2, \ldots, n-1$
(3) $f(L) = -\sum_{i=1}^k \mu_{n-i}$ for any $k = 2, \ldots, n-1$
(4) $f(L) = (\prod_{i=2}^{n-1} \mu_i)^{\frac{1}{n-1}}$

To conclude, we have introduced two classes of balanced games: regression games and connectivity games. The class of regression games is also shown to be totally balanced. These classes supplement other well-known classes of totally balanced games such as assignment games [Shapley and Shubik (1972)] and permutation games [Tijs et al. (1984)].

Bibliography

Bapat, R. B. (1996). The Laplacian matrix of a graph, *The Mathematics Student,* **65** pp. 214–223.

Bondareva, O. N. (1963). Some applications of linear programming methods to the theory of cooperative games (in Russian), *Problemy Kibernet,* **10**, pp. 119–139.

Bondy J. A. and Murty, U. S. R. (1976). *Graph Theory with Applications,* (Macmillan).

Fiedler, M. (1973). Algebraic connectivity of graphs, *Czech. Math. J.*, **23**, pp. 298–305.

Ghosh, A. and Boyd, S. (2006). Upper bounds on algebraic connectivity via convex optimization, *Linear Algebra and Its Applications,* **418**, pp. 693–707.

Owen, G. (1982). *Game Theory,* 2nd Ed. (Academic Press, New York).

Shapley, L. S. (1967). On balanced sets and cores, *Naval Research Logistics Quarterly,* **14**, pp. 453–460.

Shapley L. S. and Shubik, M. (1972). On assignment games, *International Journal of Game Theory,* **1**, pp. 111–130.

Tijs, S. H., Parthasarathy, T., Potters J. and Rajendra Prasad, V. (1984). Permutation games: another class of totally balanced games, *Operations Research Spektrum,* **6**, pp. 119–123.

Tijs, S. H. (2003). *Introduction to Game Theory,* (Hindustan Book Agency, New Delhi).

Merris, R.(1994). Laplacian matrices of graphs: a survey, *Linear Algebra Appl.,* **197,198**, pp. 143–176.

West, D. B. (2001). *Introduction to Graph Theory,* (2nd Edition, Prentice-Hall Inc.).

Chapter 23

Market Equilibrium for Combinatorial Auctions and the Matching Core of Nonnegative TU Games

Somdeb Lahiri[1]
Institute for Financial Management and Research
24, Kothari Road, Nungambakkam
Chennai- 600 034, Tamil Nadu, India
e-mail: lahiri@ifmr.ac.in, Or somdeb.lahiri@yahoo.co.in

Abstract

We introduce the concept of the induced combinatorial auction of a nonnegative TU game and show that the existence of market equilibrium of the induced combinatorial auction implies the existence of a possibly different market equilibrium as well, which corresponds very naturally to an outcome in the matching core of the TU game. Consequently we show that the matching core of the nonnegative TU game is non-empty if and only if the induced combinatorial auction has a market equilibrium.

Key Words: Combinatorial auctions, market equilibrium, constrained equilibrium, nonnegative TU game, matching, matching core.

23.1 Introduction

The formation of productive partnerships and sharing the yield that accrues from it has been an important concern of both economics and game theory. The theory of matching that originated with the seminal paper of Gale and Shapley (1962) is largely concerned with the formation of stable two agent partnerships. The theory of two-sided matching as discussed in Roth and Sotomayor (1990) that grew out of the work of Gale and Shapley,

[1]This is a revised version of an earlier paper "Market Equilibrium for Bundle Auctions and the Matching Core of Nonnegative TU Games"

emphasizes the formation of partnerships between pairs of agents, where each pair consists of agents on two distinct sides of a market. Similarly, the three-sided matching model of Alkan (1988) is concerned with the formation of possible triplets, each triplet comprising agents from three different sets. A related model due to Shapley and Scarf (1974) called the housing market, considers a private ownership economy, where each individual owns exactly one object and what is sought is the existence of an allocation in the core of the economy.

The model discussed here is based on the one proposed by Kaneko and Wooders (1982). They considered a feasible set of coalitions on which a worth function was defined. They investigated the existence of non-empty cores for such problems. Kaneko and Wooders (1982) called their model a TU partitioning game. Since all coalitions are feasible in our model and the worth of each coalition is non-negative, we call our model a nonnegative TU game. In a final section of our paper, we discuss how our model can be used to deal with situations represented by TU partitioning games.

Motivated by the work of Eriksson and Karlander (2001) and the literature on coalition formation, we investigate conditions under which there exists an outcome for non-negative TU games satisfying the following conditions:

(a) The realized coalitions form a partition of the set of agents.
(b) The worth of each coalition in the partition is distributed among the agents in the coalition.
(c) No feasible coalition is worth more than the sum of what its members receive.

We call the set of such outcomes, the matching core of the nonnegative TU game. The partition component of an outcome in the matching core is called a matching.

The matching core of a nonnegative TU game is different from the core of a TU game. Necessary and sufficient conditions for the non-emptiness of the core of a TU game were first obtained by Bondareva (1963) and Scarf (1967). Our results reported here, are of a different nature.

The result we seek is a pursuit similar to that of Proposition 2.4 of Eriksson and Karlander (2001). However, our result is different. In effect, we provide a necessary and sufficient condition for the existence of a non-empty matching core for such problems when pay-offs are transferable among the agents. In our definition of the matching core, we do not require that the sum of pay-offs to all players be equal to the worth of the grand coalition.

We require instead, that the sum of pay-offs to players in each coalition that is *realized*, is equal to the worth of that coalition. The possibility of the grand coalition not being realized is admissible in our framework, and hence "budget balancedness" being superimposed on our solution concept, appears irrelevant. Thus, our matching core is weaker than the core in Eriksson and Karlander (2001). Our investigation is very likely yet another representation of similar concerns raised in Kaneko and Wooders (1982).

Moulin (1995) contains a discussion of restrictions in the pattern of coalition formation for games where utilities are transferable. Games of this sort that in addition admit non-empty cores are called universally stable.

Our approach in this paper is based on results that may be derived for multi-unit auctions. Suppose each agent is represented as an indivisible item that, and each possible coalition is represented by a buyer of such items. A coalition realizes it worth as pay-off if and only if it is able to secure the items that initially belong to its members. Otherwise, the coalition gets zero pay-off. Such a combinatorial auction is said to be **induced** by the nonnegative TU game. We show that the matching core of the nonnegative TU game is non-empty if and only if the induced combinatorial auction has a market equilibrium. In fact, after deriving the induced combinatorial auction of the TU game, we show that the existence of a market equilibrium implies the existence of a possibly different market equilibrium as well, where each coalition either consumes the items initially owned by its members or nothing at all, and the price vector is such that the profit of each coalition is zero. Such a market equilibrium is then shown to correspond very naturally to an outcome in the matching core of the TU game, from which our main result follows. The interesting thing to note in this context, is that a nonnegative TU game may have a empty matching core, as Example 23.1 aptly illustrates.

The results obtained here for nonnegative TU games can be easily applied to situations where certain coalitions are prohibited from being realized, as with TU partitioning games. This can be achieved by setting the worth of a prohibited coalition to be zero. One of the most well-known examples of such games is the one due to Shapley and Shubik (1972), concerning assignment games. The game considered in Shapley and Shubik (1972) is itself derived from the framework of assignment problem modeled in the seminal paper by Koopmans and Beckmann (1957). The non-emptiness of the core of the related assignment game follows analogously as in the related results obtained by Koopmans and Beckmann (1957) for assignment problems. Thus, the implications of our analysis for partitioning games are as valid as they are for nonnegative TU games.

23.2 Combinatorial Auctions

The model in this section is adopted from Lahiri (2006). Let $Z = \mathcal{X} \cup \{0\}$, where $\mathcal{X}$ denotes the set of natural numbers. Let there be $H > 0$ agents and $L+1 > 1$ commodities. The first L commodities are used as inputs to produce the $L+1^{th}$ commodity, which is a numeraire consumption good.

Let e denote the vector in $\mathcal{R}^L$ all whose coordinates are equal to one and for $j = 1, \ldots, L$, let e^j denote the vector in $\mathcal{R}^L$ whose j^{th} coordinate is equal to one and all other coordinates are equal to zero.

The economy is initially endowed with exactly one unit of each of the L indivisible inputs. Thus the initial endowment of the economy is the vector e in Z^L.

A function $f : Z^L \rightarrow \mathcal{R}_+$ (: the set of non-negative real numbers) is said to be a discrete function.

Each agent i has preferences defined over Z^L which is represented by a discrete production function f^i, such that or all $i = 1, \ldots, H, f^i$ is non-decreasing (i.e. for all $x, y \in Z^L : [x \geq y]$ implies $[f^i(x) \geq f^i(y)]$).

A combinatorial auction is an H-tuple $[f^i/i = 1, \ldots, H]$.

Note : If x belongs to Z^L and $x \leq e$ then the set $S = \{j/x_j = 1\}$ is a subset of $\{1, \ldots, L\}$. Conversely if S is a subset of $\{1, \ldots, L\}$, then we can associate to S the point x in Z^L such that $x_j = 1$ if and only if $j \in S$. If $[f^i/i = 1, \ldots, H]$ is a combinatorial auction then $f^i(x)$ can be interpreted as agent i's bid (or valuation) for the bundle of items $\{j/x_j = 1\}$.

An input consumption vector of agent i is denoted by a vector $X^i \in Z^L$.

A price vector p is an element of $\mathcal{R}^L_+/\{0\}$, where for $j = 1, \ldots, L, p_j$ denotes the price of input j.

At a price vector p, the objective of agent i is to maximize profits subject to availability of the inputs:

Maximize $[f^i(X^i) - p^T X^i]$

Subject to $X^i \leq e$.

An allocation is an array $X = < X^i/i = 1, \ldots, H >$ such that $X^i \in Z^L$ for all $i = 1, \ldots, H$.

An allocation X is said to be feasible if $\sum_{i=1}^{H} X^i = e$.

A (constrained) market equilibrium is a pair $\langle p^*, X^* \rangle$ where p^* is a price

vector, X^* is a feasible allocation such that for all $i = 1, \ldots, H, X^{*i}$ solves:
Maximize $[f^i(X^i) - p^T X^i]$
Subject to $X^i \leq e$.

In what follows and in view of the definition of (constrained) market equilibrium no agent can consume more than one unit of any commodity. Thus, for our purpose it is enough to consider the restriction of each f^i to the unit cube. Hence, an alternative way of representing the combinatorial auction would be the following: Instead of representing a consumption bundle as a vector with either zero or one as its coordinates, we may view it as the set of items, which comprise the bundle. Thus if X^i is the input bundle consumed by i and $S = \{j/(e^j)^T X^i = 1\}$ then instead of $f^i(X^i)$, we may write $f^i(S)$ to represent the utility (output) that agent i derives (produces) as a consequence of utilizing the input bundle X^i.

In such a representation f^i would cease to be defined on L-tuples of integer valued vectors in the unit cube. Instead f^i would be a non-negative real valued function defined on subsets of $\{1, \ldots, L\}$.

In fact the representation using sets of items is more common in the auction theory literature. Pekec and Rothkopf (2003) contain a lucid survey of some of the major issues in combinatorial auctions. Our choice of representation was largely dictated by considerations that permit the use of simple algebra and thus lead to "smoother" proofs of the results that we obtain in this paper. However conceptualization of a combinatorial auction may be facilitated if we adopt the alternative representation.

23.3 Games with Transferable Utilities

Given a positive integer $n \geq 3$, and a set of agents $N = \{1, \ldots, n\}$, let $\prod$ be the set of all non-empty subsets of N.

A non-negative *TU* game (or game with transferable utilities) is a function $v : \prod \to \mathcal{R}_+$, such that: (i) For all $i \in N : v(\{i\}) = 0$; (ii) $v(S) > 0$ for at least one $S \in \prod$. $S \in \prod$ is said to be a coalition, and $v(S)$ is said to be the "worth" of the coalition S. The reason why we use the prefix "non-negative" in the above definition is because in general a *TU* game need not be non-negative. Further, a *TU* game normally specifies the worth of the empty set to be zero. We define our game on non-empty sets. By itself, requiring the worth of a singleton to be equal to zero, is a harmless normalization.

Requiring the worth of at least one coalition to be positive makes the game non-trivial, as will be observed shortly.

A matching is a partition A of N, i.e. A is a non-empty collection of mutually disjoint sets in $\prod$ whose union is N.

A pay-off vector is an element of $\mathcal{R}_{+}^{N}$.

An outcome of is a pair (A, x), where x is a pay-off vector and A is a matching.

An outcome (A, x) is said to belong to the matching core of the nonnegative TU game v if:

(1) for all $S \in A : \sum_{i \in S} x(i) = v(S)$;

(2) for all $S \in \prod : \sum_{i \in S} x(i) \geq v(S)$.

Let $C(v)$ denote the set of outcomes that belong to the matching core of v.

An outcome (A, x) in $C(v)$ is said to belong to the core of v if and only if $A = \{N\}$, i.e. the only realizable coalition in the partition A is the grand coalition.

Thus the concept of the matching core of a non-negative TU game is weaker than the core.

Since $v(S) > 0$ for at least one coalition $S, (A, x) \in C(v)$ implies $x \neq 0$.

If we had allowed the worth of every coalition to be zero, then for such a game $v, C(v) = \{(A, 0)/A$ is a partition of $N\}$. Assuming that the worth of at least coalition is positive, rules out such trivial possibilities.

The following example due to Ahmet Alkan shows that the matching core of a nonnegative TU game v may be empty.

Example 23.1. (due to Ahmet Alkan): Let $N = \{1, 2, 3, 4, 5\}$. Let $v(S) =$ 30 if S has exactly three agents and zero otherwise. Towards a contradiction suppose (A, x) belongs to $C(v)$. If S is a three agent set belonging to A, then at x, at least one member of S, say j, gets at most 10. Since the agents in $N \setminus S$ get zero, the total amount obtained by j and agents in $N \setminus S$ is less than 30, although they form a three-member set.

Thus, every agent in N gets zero at x. Clearly, (A, x) does not belong to $C(v)$.

23.4 Games with Transferable Utilities as Combinatorial Auctions

Let $L = n$ and $H = 2^n - 1$.

For $x \in Z^L$, let $e(x)$ be the L-vector whose i^{th} coordinate is $\min\{x_i, 1\}$. Let $\prod = \{S_1, \ldots, S_H\}$. For $i \in \{1, \ldots, H\}$, let $v^i : Z^L \to \mathcal{R}_+$ be defined as follows:

$$v^i(x) = v^i(e(x)) = v(S_i) \text{ if } S_i \subset \{j/x_j > 0\},$$
$$= 0, \text{ otherwise.}$$

$[v^i/i = 1, \ldots, H]$ is said to be the combinatorial auction induced by v.

The restriction of v^i to $\{x \in Z^L/x = e(x)\}$ corresponds to $v(S_i)$ times the S_i-unanimity game, if we consider the set $\{j/x_j = 1\}$ instead of x itself. For any $i \in \{1, \ldots, H\}$ and price vector p, if X^i solves

Maximize $[v^i(X^i) - p^T X^i]$

Subject to $X^i \leq e$

then it also solves

Maximize $[v^i(X^i) - p^T X^i]$.

If S_i is a singleton, then $v^i(x) = 0$ for all $x \in Z^L$.

For $S \in \prod$, let e^S be defined to be equal to $\sum_{j \in S} e^j$.

Proposition 23.1. *Let $\langle p^*, X^* \rangle$ be a market equilibrium for $[v^i/i = 1, \ldots, H]$. Then, there exists a market equilibrium $\langle p^*, X^{\&} \rangle$ such that for all $k = 1, \ldots, H : X^{\&k} \in \{0, e^{S_k}\}$.*

Proof. Let $S = S_i \in \prod$ and suppose $X^{*i} = e^Q \notin \{0, e^S\}$.

Case 1: S is a proper subset of Q.

Since $v^i(e^Q) - p^{*T} e^Q \geq v^i(e^S) - p^{*T} e^S = v^i(e^Q) - p^{*T} e^S$.

Thus, $0 \geq p^{*T}(e^Q - e^S) = p^{*T} e^{Q \setminus S} = \sum_{j \in Q \setminus S} p_j^*$.

Since $p^* \geq 0$, $p_j^* = 0$ for all $j \in Q \setminus S$.

Let $X^{\#i} = e^S, X^{\#k} = X^{*k} + e^j$ if $S_k = \{j\}$ and $j \in Q \setminus S, X^{\#k} = X^{*k}$ otherwise.

It is easy to verify that $\langle p^*, X^{\#} \rangle$ is a market equilibrium:

$v^i(X^{\#i}) - p^{*T} X^{\#i} = v(S) - \sum_{j \in S} p_j^* = v(S) - \sum_{j \in Q} p_j^* = v^i(X^{*i}) - p^{*T} X^{*i}$;

$v^k(X^{\#k}) - p^{*T} X^{\#k} = v^k(X^{*k} + e^j) - p^{*T} X^{*k} - p_j^* = v^k(X^{*k} + e^j) - p^{*T} X^{*k}$ if $S_k = \{j\}$ and $j \in Q \setminus S$.

The feasibility of $X^{\#}$ follows from the feasibility of X^* and the fact that $X^{\#}$ is obtained from X^* by transferring items from S to one or more coalitions outside S.
For $j \in Q \setminus S$ and $S_k = \{j\}, X^{*k} + e^j$ does not belong to $\{0, e^j\}$ if and only if X^{*k} is not equal to zero. Since X^{*k} cannot be equal to e^j, it follows that $|\{i/X^{*k} \notin \{0, e^{S_k}\}| > |\{i/X^{\#k} \notin \{0, e^{S_k}\}|$.
Case 2: $S \setminus Q$ is non-empty and $Q \setminus S$ is non-empty.
Thus, $v^i(e^Q) = 0$.
Since $-p^{*T}e^Q = v^i(e^Q) - p^{*T}e^Q \geq v^i(0) - p^{*T}0 = 0$ and since $p^* \geq 0$, we get $p_j^* = 0$ for all $j \in Q$.
Let $X^{\#i} = 0, X^{\#k} = X^{*k} + e^j$ if $S_k = \{j\}$ and $j \in Q, X^{\#k} = X^{*k}$ otherwise.
It is easy to verify that $\langle p^*, X^{\#}\rangle$ is a market equilibrium:
$v^i(X^{\#i}) - p^{*T}X^{\#i} = 0 = v^i(X^{*i}) - p^{*T}X^{*i}$;
$v^k(X^{\#k}) - p^{*T}X^{\#k} = v^k(X^{*k} + e^j) - p^{*T}X^{*k} - p_j^* = v^k(X^{*k} + e^j) - p^{*T}X^{*k}$ if $S_k = \{j\}$ and $j \in Q$.
The feasibility of $X^{\#}$ follows from the feasibility of X^* and the fact that $X^{\#}$ is obtained from X^* by transferring items belonging to coalition S to the members of Q.
For $j \in Q$ and $S_k = \{j\}, X^{*k} + e^j$ does not belong to $\{0, e^j\}$ if and only if X^{*k} is not equal to zero. Since X^{*k} cannot be equal to e^j, it follows that $|\{i/X^{*k} \notin \{0, e^{S_k}\}| > |\{i/X^{\#k} \notin \{0, e^{S_k}\}|$.
Case 3: Q is a non-empty proper subset of S.
Thus, $v^i(e^Q) = 0$.
Since $-p^{*T}e^Q = v^i(e^Q) - p^{*T}e^Q \geq v^i(0) - p^{*T}0 = 0$ and since $p^* \geq 0$, we get $p_j^* = 0$ for all $j \in Q$.
Let $X^{\#i} = 0, X^{\#k} = X^{*k} + e^j$ if $S_k = \{j\}$ and $j \in Q, X^{\#k} = X^{*k}$ otherwise.
It is easy to verify that $\langle p^*, X^{\#}\rangle$ is a market equilibrium:
$v^i(X^{\#i}) - p^{*T}X^{\#i} = 0 = v^i(X^{*i}) - p^{*T}X^{*i}$;
$v^k(X^{\#k}) - p^{*T}X^{\#k} = v^k(X^{*k} + e^j) - p^{*T}X^{*k} - p_j^* = v^k(X^{*k} + e^j) - p^{*T}X^{*k}$ if $S_k = \{j\}$ and $j \in Q$.
The feasibility of $X^{\#}$ follows from the feasibility of X^* and the fact that $X^{\#}$ is obtained from X^* by transferring items belonging to S to the members of Q.
For $j \in Q$ and $S_k = \{j\}$, $X^{*k} + e^j$ does not belong to $\{0, e^j\}$ if and only if X^{*k} is not equal to zero. Since X^{*k} cannot be equal to e^j, it follows that $|\{i/X^{*k} \notin \{0, e^{S_k}\}| > |\{i/X^{\#k} \notin \{0, e^{S_k}\}|$.
Thus, in each case we obtain a market equilibrium $\langle p^*, X^{\#}\rangle$ such that $|\{i/X^{*k} \notin \{0, e^{S_k}\}| > |\{i/X^{\#k} \notin \{0, e^{S_e}\}|$.

Repeating the process at most finitely many times, we arrive at a market equilibrium $\langle p^*, X^{\&}\rangle$ such that for all $k = 1, \ldots, H : X^{\&k} \in \{0, e^{S_k}\}$. □

Proposition 23.2. *Let $\langle p^*, X^*\rangle$ be a market equilibrium for $[v^i/i = 1, \ldots, H]$ such that for all $i = 1, \ldots, H : X^{*i} \in \{0, e^{S_i}\}$. Then there exists a market equilibrium $\langle q^*, X^*\rangle$ such that $v^i(X^{*i}) - q^{*T}X^{*i} = 0$ for all $i = 1, \ldots, H$.*

Proof. Since $\langle p^*, X^*\rangle$ is a equilibrium for the induced combinatorial auction $[v^i/i = 1, \ldots, H]$ and $X^{*^i} \in \{0, e^{s_i}\}$ for all $i = 1, \ldots, H$, it must be the case that $\{S_i/X^{*i} \neq 0\}$ is a partition of N. Further, $v^i(X^{*i}) - p^{*T}X^{*i} \geq 0$ for all $i = 1, \ldots, H$.
For $i \in \{1, \ldots, H\}$ and $X^{*i} = e^{s_i}$, let $q_j^* = p_j^* + \left(\frac{v(S_i)-p^{*T}X^{*i}}{|S_i|}\right)$ for all $j \in S_i$.
Hence $q^* \in \mathcal{R}_+^{\mathcal{L}}\{\prime\}$ and $q^* \geq p^*$.
Thus for $i \in \{1, \ldots, H\}$ and $X^{*i} = e^{s_i}$:

$$q^{*T}X^{*i} = q^{*T}e^{s_i} = \sum_{j\in S_i} q_j^* = \sum_{j\in S_i} p_j^* + v(S_i) - p^{*T}X^{*i}$$

$$= \sum_{j\in S_i} p_j^* + v(S_i) - \sum_{j\in S_i} p_j^* = v(S_i) = v_i(X^{*i}).$$

If $X^{*i} = 0$, then $v^i(X^{*i}) - q^{*T}X^{*i} = 0$.
Let $x \in Z^L$.
Case 1: $X^{*i} = e^{S_i}$. If $x \geq X^{*i}$, then $v^i(x) = v^i(X^{*i}) = v(S_i)$ and $q^{*T}x \geq q^{*T}X^{*i}$.
Thus, $v^i(x) - q^{*T}x \leq v^i(X^{*i}) - q^{*T}X^{*i}$.
If $\neg(x \geq X^{*i})$ then $v^i(x) = 0$ and $q^{*T}x \geq 0$.
Thus, $v^i(x) - q^{*T}x \leq 0 = v^i(X^{*i}) - q^{*T}X^{*i}$.

Case 2: $X^{*i} = 0$.
Thus, $v^i(X^{*i}) - q^{*T}X^{*i} = v^i(X^{*i}) - p^{*T}X^{*i} = 0 \geq v^i(x) - p^{*T}x \geq v^i(x) - q^{*T}x$, since $q^* \geq p*$.
Thus, $\langle q^*, X^*\rangle$ is a market equilibrium. □

In view of Proposition 23.2, we say that a market equilibrium $\langle p^*, X^*\rangle$ is a **zero-profit market equilibrium** if $v^i(X^{*i}) - q^{*T}X^{*i} = 0$ for all $i = 1, \ldots, H$.

Note: *If* the induced combinatorial auction *has* a market equilibrium (say) $\langle p, X\rangle$, *then* a combination of Propositions 23.1 and 23.2, provides a simple procedure by which a zero-profit market equilibrium can be constructed from $\langle p, X\rangle$. If $X^i \geq e^{S_i}$ then let $X^{*i} = e^{S_i}$. The singleton coalitions $\{\{j\}/j \in \{X^i(j) = 1, j \notin S_i\}\}$ are "disengaged" and each such coalition $\{j\}$ is assigned the agent (object) 'j' only at X^*. If $(X^i \geq e^{S_i})$, then the singleton coalitions $\{\{j\}/j \in \{X^i(j) = 1\}\}$ are all "disengaged" and each such coalition $\{j\}$ is assigned the agent (object) 'j' only at X^* once again. All other coalitions are assigned nothing at X^*. This is precisely the construction outlined in the proof of Proposition 23.1. Using the formula indicated in the proof of Proposition 23.2, we can now obtain a new price vector p^*, such that $\langle p^*, X^*\rangle$ is a zero-profit market equilibrium.

23.5 Market Equilibrium and Matching Cores

In this section we establish the main consequences of our present investigation.

Theorem 23.1. *(A, x) belongs to $C(v)$ if and only if $\langle x, X^*\rangle$ is a zero-profit market equilibrium, where for $i = 1, \ldots, H$: $[X^{*i} = e^{S_i}$ if and only if $S_i \in A; X^{*i} = 0$ if and only if $S_i \in \prod \backslash A]$.*

Proof. Let (A, x) belong to the matching core of v and $\langle x, X^*\rangle$ be as defined in the statement of this theorem. Thus, $x \geq 0$ and $x \neq 0$. This implies that x is a price vector.
Suppose $S_i \in A$.
Thus, $X^{*i} = e^{S_i}$ and $v^i(X^{*i}) - x^T X^{*i} = v(S_i) - \sum_{j \in S_i} x(j) = 0$. Further,

$$v^i(e^S) - x^T e^S = v(S_i) - \sum_{j \in S} x(j) \leq v(S_i) - \sum_{j \in S_i} x(j) = 0 \text{ if } S_i \subset S.$$

If $S_i \not\subset S$, then $v^i(e^S) - x^T e^S = 0 - \sum_{j \in S} x(j) \leq 0$ and $v^i(0) - x^T 0 = 0$.

Since $\sum_{i=1}^{H} X^{*i} = \sum_{S_i \in A} e^{S_i} = e$, $\langle x, X^*\rangle$ is a zero profit market equilibrium.
Now suppose $\langle x, X^*\rangle$ is a zero profit market equilibrium and let $A = \{S_i \in \prod / X^{*i} = e^{S_i}\}$.
If $v(S_i) - \sum_{j \in S_i} x(j) = v^i(e^{s_i}) - \sum_{j \in S_i} x(j) = 0$ if $S_i \in A$.
If $S_i \notin A$, then $v(S_i) - \sum_{j \in S_i} x(j) = v^i(e^{s_i}) - \sum_{j \in S_i} x(j) \leq v^i(0) - x^T 0 = 0$.

Thus, $(A, x) \in C(v)$. □

The following result follows directly from Propositions 23.1, 23.2 and Theorem 23.1, and is the main consequence of the analysis reported above.

Theorem 23.2. *Let v be a non-negative TU game and $[v^i/i = 1, \ldots, H]$ be the combinatorial auction induced by v. Then $C(v)$ is non-empty if and only if there exists a market equilibrium for $[v^i/i = 1, \ldots, H]$.*

It follows as a consequence of Theorem 23.2 that the induced combinatorial auction of the TU game discussed in Example 23.1 has no market equilibrium. We now verify it independently, without using Theorem 23.2.

Let $[v^i/i = 1, \ldots, H]$ be the induced combinatorial auction where $H = 2^5 - 1$ and $L = 5$.

Suppose towards a contradiction (x, X^*) is a market equilibrium for the induced combinatorial auction. By Proposition 23.1, we may assume that for all $k = 1, \ldots, H : X^{*k} \in \{0, e(S_k)\}$.

By Proposition 23.2, we may assume that (x, X^*) is a zero-profit equilibrium. If more than three goods have positive price, then there exists at least one three agent coalition represented by a buyer, who can make positive profits instead of the zero profit that it receives, contradicting that (x, X^*) is a market equilibrium. On the other hand, if every good has zero price, then every three agent coalition represented by a buyer can make positive profits, contradicting that (x, X^*) is a zero-profit equilibrium. Thus, at most three goods may have positive price and at least one good definitely has positive price at x.

Any agent who receives a good with a positive price will make higher profits if a good with zero price which is initially not allocated to it, but is subsequently allocated to it, instead of the one with a positive price. Hence one agent receives e at X^* and the rest get nothing.

If the sum of the prices of any three goods is less than thirty, then any $S \in \prod$ with $|S| = 3$ who did not get anything makes higher profits consuming these three goods, instead of the ones it consumes at the zero profit equilibrium (x, X^*).

Hence the sum of prices of any three goods must be equal to thirty. But then every good must have a positive price, leading to a contradiction.

23.6 Discussion

The concept of a matching core that we consider here is very similar to the various stability concepts that have been invoked in the literature for the study of assignment and matching problems.

Let x be any pay-off vector. In order that x be "viable" it is necessary that there exists a matching A, such that for every coalition S in A, the sum of its pay-offs does not exceed $v(S)$. If (A, x) does not belong to the matching core, then for all such matchings (i.e. those which make x viable), there should exist at least one coalition S whose sum of pay-offs at x is less than $v(S)$. Thus at any outcome in the matching core of v, members of S are clearly better off than at x. In a sense this along with the requirement that S itself be a member of a matching in the matching core, is precisely the content of external stability for a von Neumann Morgenstern stable set. It is worth verifying what an analogous interpretation of internal stability for stable sets would imply in our context. It may often be the case that the problem being considered prohibits the formation of certain coalitions. Thus for instance in a two-sided matching model, two or more agents on the same side of the market cannot form a coalition. Our statement of a nonnegative TU game is general enough to accommodate such possibilities.

If S is a coalition which is prohibited then we set its worth $v(S)$ to be equal to zero. Let (A, x) belong to $C(v)$. If S does not belong to A, then there is clearly no problem to be addressed. What if S belongs to A?

If S belongs to A, then $x(k)$ must be equal to zero for all k in S. If instead of (A, x) we considered the outcome (A^*, x) where $A^* = (A \backslash \{S\}) \cup \{\{k\}/k \in S\}$, then it is easily verified that this new outcome belongs to $C(v)$ as well. The difference between A and A^* is that all prohibited coalitions in A are replaced by their members. Thus, our model is general enough to cope with the exigencies that arise in matching problems, particularly in the context of markets.

Acknowledgements

The problem concerning matching cores of TU games was indirectly suggested to me by Ahmet Alkan. I would like to thank him for his comments on an ancient version of this paper, and in particular for Example 23.1. I would also like to put on record a very deep acknowledgment to Andrew McLennan for having commented on the stylistic requirements of the paper. An earlier version of this paper was presented at EVROPAEVM (5th Economic Workshop Game Theory and Applications in Problems of International Economics) held at University of Helsinki on 27-28 October 2006. I would like to thank all the participants at the workshop and in particular Andrey Garnaev, Klaus Kultti, Lina Mallozi, Tapio Palokangas, Juha Tervalla and Hannu Vartiainen for their thoughtful comments and observations. I would also like to put on record my gratitude to the participants of ISMPDM 07 and in particular Vijaya Krishna and S. K. Neogy for their observations and comments on my paper. Finally I happily put on record my sincere thanks to the anonymous referee of the present volume for suggestions towards over-all improvement of this paper.

Bibliography

Alkan A. (1988). Non-Existence of Stable Threesome Matchings, *Mathematical Social Sciences*, **16**, pp. 207–209.

Bondareva O. N. (1963). Some Applications of Linear Programming Methods to the Theory of Cooperative Games (in Russian), *Problemy Kibernetiki*, **10**, pp. 119–139.

Eriksson K. and Karlander J. (2001). Stable outcomes of the roommate game with transferable utility, *Int. J Game Theory*, **29**, pp. 555–569.

Gale D. and Shapley L. (1962). College Admissions and the stability of Marriage, *American Mathematical Monthly*, **69**, pp. 9–15.

Kaneko M. and Wooders M. (1982). Cores of partitioning games, *Mathematical Social Sciences*, **3**, pp. 313–327.

Koopmans T. C. and Beckmann M. (1957). Assignment Problems and the Location of Economic Activities, *Econometrica*, **25**, pp. 53–76.

Lahiri S. (2006). Existence of Market Equilibrium for Multi-unit Auctions.

Pekec A. and Rothkopf M. H. (2003). Combinatorial Auction Design, *Management Science*, **49**, pp. 1485–1503.

Moulin H.(1995). Cooperative Microeconomics: A Game-Theoretic Introduction, (Prentice Hall Harvester Wheatsheaf).

Roth A. and Sotomayor M. (1990). Two-Sided Matching, *Econometric Society Monograph* **18**, (Cambridge University Press).

Scarf H. (1967). The Core of a N Person Game, *Econometrica*, **35**, pp. 50–69.

Shapley L. and Scarf H. (1974). On Cores and Indivisibility, *Journal of Mathematical Economics*, **1**, pp. 23–28.

Shapley L. and Shubik M. (1972). The assignment game I: the core, *Int. J. Game Theory*, **1**, pp. 111–130.

Chapter 24

Continuity, Manifolds, and Arrow's Social Choice Problem

Kari Saukkonen[1]
Institutions and Social Mechanisms
FI-20014 University of Turku
FINLAND

Abstract

In the 1950s Arrow formulated an important conceptual framework enabling one to discuss various collective decision making problems in an axiomatic fashion. There is, nevertheless, no topological structure given in Arrow's social choice framework to make it possible to discuss continuity of social welfare functions. In the turn of 1980s Chichilnisky had a systematic framework to discuss continuity of certain type of social welfare functions. In this paper, it is explained what continuity of a social welfare function is for Chichilnisky. It is then pointed out that there are difficulties, if this viewpoint is extended to cover continuity of Arrovian social welfare function, because of too specific assumption about the topological structure and dimension of the state sets. The discussion suggests that Chichilnisky's framework is not much help in formulating appropriate topological foundations for the Arrovian social choice theory conceptualizing, for example, the workings of capitalistic democracy.

Key Words: Continuous Arrovian social choice processes, topological manifolds, forcing over states

24.1 Introduction

Bergson (1938) stated Abba Lerner's resource allocation problem in terms of maximizing differentiable, and thus continuous economic welfare func-

[1]I am grateful to an anonymous referee, Hannu Nurmi, and Hannu Salonen for comments. I also thank the Yrjö Jahnsson Foundation for financial support.

tion; this calculus based approach to economic welfare was summarized and distributed by Samuelson (1983, xxi–xxiv, 203–253). The broader, social decision aspect of this map retained as a kind of fixed, ceteris paribus part.

Arrow (1950, 1951, 1952, 1963, 1967) reformulated Bergson's approach and extended explicitly the social decision part of the problem, covering such issues as voting and legislation, instead of just market decisions. The new formulation of social welfare function contained some new principles and was in terms of set theory. As Bergson's calculus based reasoning became obsolete in this setting, continuity of a social welfare function was not discussed. Arrow's formulation set the broad standards for how social choice problems were since discussed.

Unlike Bergson, who assumed full divisibility of all variables and implicitly the Euclidean topology, Arrow made no fixed, specific assumption about the algebraic or topological structure of the set social states, including the question of cardinality. This was to attain full generality of the social choice problem. The tacit assumption was that these kind of structures are specified only if the particular social choice problem investigated requires it; to fix them beforehand would be unduly restrictive. Schofield (1977), who build on Kramer (1973), was among the first to discuss in a precise way the use of topological structures in social choice theory; in particular, Schofield considered topological manifold structures on the set of social states in spatial voting context; still, continuity properties of the social choice rule was not in the forefront.

In the turn of 1980s Chichilnisky (1979, 1980), and also Chichilnisky and Heal (1983), who referred to Arrow, Black, Condorcet, and the utility tradition of and arising from Antonelli, Debreu, and others, took up the question about continuity of social choice rule seriously. Saposnik (1975) had already defined continuity for a particular class of social choice rules, in case the set of social states is a compact connected subset of the Euclidean space. According to Chichilnisky (1979, 1980), one should be able to talk about continuity and discontinuity of social welfare functions. In Chichilnisky's approach, a specific topological manifold structure is first defined over the set of social states. The notion of preference is formulated as a certain type of generalized vector field over the space of states. Sets of preferences are given, for example, the sup norm topologies. Finally, continuity of a social welfare function—which is a mapping from the set of preference profiles into the set of (social) preferences—is defined in terms of these topologies, necessitating also some product considerations.

The amount of literature studying and using continuity properties of social choice rules is small as compared to the whole field of social choice. Nevertheless, continuity of a function is basic to much of mathematics and sciences, and its uses by analogy should be recalled. Chichilnisky's approach insists on continuity of certain type of social welfare functions, and it brings the use of topological methods, especially differential topology, in forefront. According to Lauwers (2000, 2), it can be considered as a "breakthrough".

To solve the problem of continuous transformations in general social choice theory adequately, Chichilnisky's (1980) definition of continuous social welfare function should be able to deal the arising social choice problems in an appealing way. However, it does not quite do that. This can be indicated indirectly. Before discussing Chichilnisky's approach, Gaertner (2006, 167) notes that it is "a step beyond a core of social choice" witnessing also involved discussions over the last 15–20 years. The whole idea of continuity is sometimes questioned. Baigent (1997, 176) notes in discussing Chichilnisky's paper that "[T]he fact that Arrow chose a non-topological framework which went unchallenged in social choice theory for some three decades suggests that continuity was not viewed as greatly compelling by many."

The controversy about continuity of social choice rules tends to run around in circles, and I wish to ask why is it, more precisely, that continuity of Chichilnisky's social welfare function is not viewed simply as an applicable "extension" to define continuity of the usual Arrovian social welfare function. Is the controversy really about the condition of continuity or is it about something else?

One can visualize the general problem situation as a two branching set of reasons—one substantial and one formal—which would need to amalgamate or "diamond" together in the end, so that the topological and the traditional social choice developments would be compatible and could be fitted into a common further development.

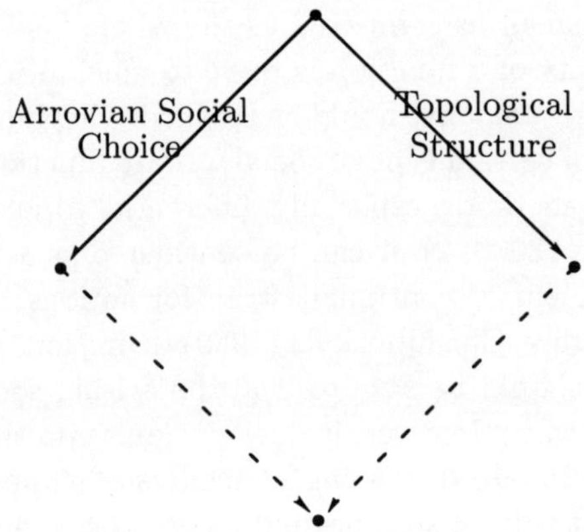

Fig. 1 The Diamond Property

In other words, first, any general definition of social welfare function should have a social choice footing covering the material that comes from the "core", mentioned by Gaertner. From here onwards, we assume that this set of problems comes from Arrow's (1963) definition of the social choice problem. Second, the definition of topological structure should not limit the problems to be investigated so that the basic social choice problems from the core become unmanageable. (The diamond property here is an informal analogue to the Church-Rosser diamond property used in logic.)

The purpose of this paper is to point out that there is a series of problems, that is, the diamond property for Arrovian social choice and topology will fail, if the definition of Chichilnisky continuous social welfare function is extended to cover the continuity of Arrovian social welfare function. This is because the topological structure assumed on the set of social states in Chichilnisky's framework precludes the sufficiently general approach on social choice, limiting out, for example, certain classes of indivisibilities, basic voting problems with candidates, problems of legislature involving judicial statements, and in general, because the assumed fixed class of topologies cannot adopt to all generic classes of Arrovian social choice problems.

24.2 Continuity via Topological Manifolds

24.2.1 *Background*

As Arrow (1951; 1963) reformulated Bergson's (1938) notion of continuous social welfare function in the turn of 1950s, a new notion, supported with axiomatic style of analysis, emerged which set the standards for discussing social choice problems ever since. But the social welfare function was now formulated without any idea of how to define, investigate, and make use of continuity properties of these kind of maps, and there was not much discussion or systematic effort either before the end of 1970s. In the turn of 1980s Chichilnisky had a systematic framework to discuss continuity of certain type of social welfare functions. This construction started from an inside view of manifold theory; it did not try to apply some more general frame of topological underpinnings, relating abstract set theory, algebra and logic, suggested by Arrow's social choice problem itself. Chichilnisky (1980) formulated the notion of continuous social welfare function by (1) restricting the allowed cardinality of the set of feasible social states by using a particular, fixed topology defined over the set of states, and by (2) using a different, non-Arrovian notion of preference, given by a certain mapping between manifolds, instead of using the usual notion of two-placed relation defined over the set of states.

Thus the concept of social welfare function was once again reformulated, as Arrow had reformulated the Bergson's welfare function, but with the difference that Arrow's social welfare function was, and still is, viewed as a kind of canonical standard of frame, accompanied by a large number of close variants. The class of topologies postulated for the set of states together with the dimension assumption used made the construction rather particular from the point of view of general topology and also atypical from the point of view of received social choice theory. Nevertheless, collective choice principles could now be explicitly posed for the map, as was also Arrow's original way of dealing social choice problems, but this time continuity was definable for the map; and new type of results were obtained, again mainly negative in character.

It is fundamental to Chichilnisky's (1980) formulation of continuous social welfare function that the set of social states on which preferences are to be defined is taken to be a topological manifold X for which $\dim(X) \geq 2$. Manifolds with boundary are also allowed. Informally, this kind of topological space can be viewed as a generalization of the usual Euclidean space having at least 2 dimensions, in the sense that locally (i.e. in a neighborhood of any of its points) it looks like the Euclidean space $\mathbb{R}^n$ for fixed $n \geq 2$; and in case the manifold has a boundary, we switch to the Euclidean closed half-space $\mathbb{H}^n$ for $n \geq 2$.

If this manifold is assumed to carry a differential structure, which is an extra pile top on the purely topological structure, basic notions familiar from elementary calculus, like "linear approximation"and "smoothness", can be extended to this manifold. The notion of preference, as used by Chichilnisky, becomes then definable over the manifold of states. In addition, the conceptual machine of general differential theory—differential topology, differential geometry, and differential equations—can be then applied in working with this type of state-preference structure.

According to Chichilnisky (1982c), Arrow's (1963) collective decision making problem and the vast literature following that formulation has focused on finite state social choice problems. This mode of analysis is described as "combinatorial" or "algebraic" by Chichilnisky (1982c, 337; 1980, 168). It is then asserted that the methods of analysis used in this approach may not provide proper intuitive geometric understanding of the social choice problem. Similar viewpoint is noted, for example, by Baryshnikov (1993, 404; 1997, 208) and Heal (1997, 158). It is then suggested that the problem should be formulated for certain generalizations of Euclidean state sets.

It is also noted (Chichilnisky 1982c, 337, 346) that in this context, continuity of a social welfare function becomes definable; and it is implied by her discussion, that in this context, viewed as being pervasive for calculus-based economic theory, it is also natural. This latter point is made also in Chichilnisky (1991, 315–316).

According to Chichilnisky (1982c, 337, 338), continuity of a social welfare function is not only a technical but also a desirable property, because (1) it makes mistakes in identifying preferences less crucial, and (2) it permits one to approximate social preferences—that is, images of profiles of individual preferences under the social welfare function—on the basis of a sample of individual preferences.

24.2.2 *Chickhilnisky's Smooth Social Welfare Function*

Let X be the set of collective alternatives over which preferences are to be defined. The set X is equipped with a topological structure having the following additional property (called locally Euclidean of dimension n): if $x \in X$, then there is an open set $U \subset X$ such that $x \in U$ and $U \cong \mathbb{R}^n$, where "$\cong$" denotes topological equivalence. It is assumed that n is constant and $n \geq 2$. (A sufficient condition for a locally Euclidean space to have a constant dimension is that it is connected, see e.g. Conlon (2001, 3; Corollary 1.1.12).) In other words, it is assumed that X is a topological manifold of dimension at least 2. A manifold is usually assumed to be also Hausdorff or second countable, or both. For example, $X = \mathbb{R}^2$ is a 2-dimensional manifold, and any open subset X' of X is a 2-dimensional manifold.

It is basic that a topological space X has a positive, finite dimension at most n, if and only if, X equals a union of its $(n+1)$ subspaces of dimension zero. For example, the 2-dimensional manifold $\mathbb{R}^2$ is the union of three subspaces $\mathbb{Q}^2$, $\mathbb{P}^2$, and $\mathbb{R}^2 - (\mathbb{Q}^2 \cup \mathbb{P}^2)$, all for which $\dim(\mathbb{Q}^2) = \dim(\mathbb{P}^2) = \dim[\mathbb{R}^2 - (\mathbb{Q}^2 \cup \mathbb{P}^2)] = 0$; here, $\mathbb{Q}$ denotes the set of rational numbers, and $\mathbb{P}$ denotes the set of irrational numbers.

Closed n-ball $\bar{B}^n = \{y \in \mathbb{R}^n : \|y\| \leq 1\}$ is not a manifold in the above sense, since a point on the boundary $\partial\bar{B}^n = S^{n-1}$ does not have a neighborhood $U \cong U'$, where U' is an open subset of $\mathbb{R}^n$. But it does have a neighborhood $U \cong U''$, such that U'' is open in a closed Euclidean half-space $\mathbb{H}^n = \{(x_1, \ldots, x_n) \in \mathbb{R}^n : x_n \geq 0\}$. If this is the exemplar case, the set X of collective alternatives is equipped with a topological structure having the following property: if $x \in X$, then there is an open neighborhood U_x of x in X such that $U_x \cong W_x$, and W_x is an open subset, containing x, of the Euclidean half-space $\mathbb{H}^n$ for $n \geq 2$. In words, X is a manifold of dimension at least 2 with boundary. Again, if needed, it may be assumed that X is also Hausdorff, or has a countable basis, or both. The closed unit 2-ball $\bar{B}^2$ is manifold of dimension 2 with boundary $\partial\bar{B}^2 = S^{2-1} = S^1$. The interior is the open ball $B^2 = \{y \in \mathbb{R}^2 : \|y\| < 1\}$.

Manifolds with boundary are not, in technical terms, manifolds, but their generalization. Nevertheless, to avoid clumsy wording in our discussion, the term "manifold" refers to both manifolds and manifolds with boundary, unless otherwise stated.

In Chichilnisky (1980) the set X of feasible alternatives is equipped with a manifold topology having dimension at least 2. The usual case investigated in many papers, like Chichilnisky (1979, 1982a, 1982b, 1983, 1986), is the case of manifold with boundary having dimension at least 2. Other, non-topological assumptions, like sufficient smooth structure on the manifold, is assumed whenever needed.

Arrow (1950, 1951, 1952, 1959, 1963) defined preference as a two-placed relation on the set of all social states. The structure of states was held arbitrary up to specification for a particular purpose of application. In contrast, Chichilnisky (1979, 348; 1980, 168–169; 1982a, 224; 1982b, 209–210; 1986, 132) and Chichilnisky and Heal (1983, 71–72) represented the notion of preference as a certain type of mapping, defined on a topological manifold X of feasible states. The manifold X is assumed to carry a tangent space at each point $x \in X$. The range of the map is then taken to be the manifold's tangent bundle TX, which is simply a collection of vector spaces, one glued to each point of the manifold. As only ordinal preferences are considered (no intensities of preference), vectors in each space are normalized to same length, say to length one, leaving only the direction of the preference at each point of the manifold. But there is more. Chichilnisky insists that a preference p on X is represented as, at least, a continuous map. So we need a topology on the set TX.

The tangent bundle TX comes with a natural topology. It is basic (see, for example, Lee (2006, 81–82)) that if X is (differentiable and) Hausdorff, second countable, and locally Euclidean topological space of dim n, then TX is (differentiable and) Hausdorff, second countable, and locally Euclidean topological space of dim $2n$.

Let X be such a differentiable manifold of dimension at least 2. A Chichilnisky preference on X is a continuous—and at least once continuously differentiable and locally integrable—mapping $p : X \longrightarrow TX$, written $p \longmapsto p(x)$, with the property that $s \circ p = 1_X$. Here, $\circ$ is a product of maps read from right to left, s is a continuous map $s : TX \longrightarrow X$, and 1_X is the mapping $1_X(z) = z$, for all $z \in X$, on X. Furthermore, $p(x)$ will be assumed to be normalized to length one.

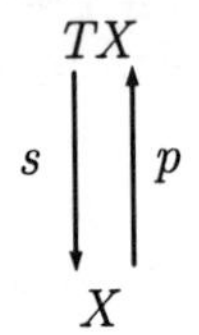

Fig. 2 A preference p on a manifold X of states

Informally, a preference p on X is an "arrow" of length one attached to each point of X, chosen to be tangent to X and vary continuously—and smoothly and locally integrable fashion—from point to point. In mathematical terms, it is simply a (locally integrable, C^1, normalized) vector field on a manifold X.

Let P denote the set of all preferences on X. In Chichilnisky (1980, 1986), the set P of preferences on X is topologized either with (1) the sup norm or with (2) a Sobolev norm, depending on the nature of the manifold. Sup norm is used, in case the manifold, over which preferences are defined, is a manifold with boundary, and the Sobolev norm is used, if the manifold does not have a boundary.

The k-fold product $P \times P \times \cdots \times P$ forms the domain of the social welfare function F; it is the set of preference profiles, k being the number of individuals in the society V. In Lauwers (2000, 5), the set P^V of profiles is equipped with the product topology.

We now have a definition of continuous social welfare function in the Chichilnisky sense. A Chichilnisky social welfare function $F : P^V \longrightarrow P$ is said to be continuous, if $F^{-1}(U)$ is open, whenever $U \subset P$ is open.

24.2.3 *Illustrations of the Framework*

Considers the following two postulates for the Chichilnisky continuous social welfare function F:

> *Unanimity*: $F(p, \ldots, p) = p$ for any $p \in P$.
> *Anonymity*: $F(p_1, \ldots p_k) = F(p_{h(1)}, \ldots, p_{h(k)})$ for any bijection h from $\{1, .., k\}$ to itself.

Chichilnisky (1980, 170–174) shows that, if the set X of social states is equipped with a manifold topology such that $\dim(X) \geq 2$ and preferences of finite number of individuals are represented as vector fields on X, as described above, then there exists no Chichilnisky continuous social welfare function satisfying unanimity and anonymity.

A simple version of this theorem is given in Chichilnisky (1979, 348–351). There, it is shown that for a two individual society, with two perfectly divisible Euclidean commodities, a Chichilnisky social welfare function $F : P \times P \longrightarrow P$, which is continuous, anonymous, and respects unanimity exists if, and only if, there exists a continuous mapping from the closed unit disk of $\mathbb{R}^2$ into itself without fixed point. Recall that a closed unit disk of $\mathbb{R}^2$ is compact convex set, and the Brouwer (1912) fixed point theorem, or rather its corollary, states that a continuous function from such a set into itself has a fixed point. Hence, a social choice "paradox", nonexistence of such a map F.

Baigent (1984, 1985) reformulates Chichilnisky's continuity condition by using the quotient topology (arising from the equivalence classes induced by anonymity) on the set of profiles, instead of product topology. Furthermore, he takes preference to be a connected, reflexive, and transitive relation on a set of topological space of alternatives, and he notes that the set of preferences can be endowed with various other topologies than

Chichilnisky's (1980), like Debreu (1969), Kannai (1970, 1972), Hildenbrand (1974), and others. Lauwers (2002, 265) weakens the condition of anonymity in Chichilnisky's theorem to one-individual anonymity, that is, if precisely one individual has a deviating preference on a particular profile, and the other individuals have identical preference, then the identity of this deviating individual does not matter in determining the collective preference.

Number of results have been published using Chichilnisky's notion of continuous social welfare function, or very similar construction, for example, Chichilnisky and Heal (1983), and Weinberger (2004). We mention here two further results by Chichilnisky. First is related to the majority decision making. Consider the following conditions for the Chichilnisky social welfare function F.

> *Pareto condition*: If all individuals in society V prefer social state $x \in X$ to social state $y \in X$, then according to the image of profile of preferences under F, x is socially preferred to y.
>
> *Decisive majority condition*: Let V be the set of individuals. If there is a subset $V' \subset V$ of individuals whose preferences agree on all social states $x \in X$ and for individuals in V'^c preferences agree on all states but in an opposite direction as those in V', and $\mathrm{card}(V') \neq \mathrm{card}(V'^c)$, then the image of this preference profile under F agrees with the majority of individuals.

Chichilnisky (1982b, 217–220) shows that, if the set of individuals is finite and the preferences of individuals are defined over a topological manifold X such that $\dim(X) \geq 2$, then there is no continuous Chichilnisky social welfare function $F : P^V \longrightarrow P$ satisfying both the Pareto condition and the decisive majority condition.

Yet another result by Chichilnisky is related to the relationship between continuous Pareto function and dictatorship. Consider the following collective choice principle for a Chichilnisky social welfare function F:

> *Weak Positive Association*: If $F(f) = p_k$ for some individual k and some profile $f \in P^V$, then $F(\bar{f}_{-k}, p_k) \neq p_k$ with $f = (-p_k, -p_k, \cdots, -p_k)$.

Chichilnisky (1982a, 228–233) shows the following result. Suppose that there are at least two individuals and preferences are defined over a topological manifold X such that $\dim(X) \geq 3$, and suppose that $F : P^V \longrightarrow P$ is a continuous Chichilnisky social welfare function satisfying the conditions of Pareto and Weak positive association. Then, there exists a homotopy h

defined on $P^V \times [0,1]$, from the function F to a dictatorial social welfare function F'.

According to Saari (1997, 221–224), this result is not to be interpreted as merely another dictatorial result, but for example, in the following way, which illustrates also number of nondictatorial Chichilnisky social welfare functions. Suppose a continuous Paretian function F is first given. Start deforming it continuously towards the dictatorial function F'; there exists a homotopy between such a maps according to the theorem. This deformation gives abundance of nondictatorial functions where one individual plays a dominant role, but does not exclusively determine the social preference. However, when this continuous deformation starts to approach the dictatorial function F', the influence of other, non-dominant individuals decrease, until F', the genuinely dictatorial function is reached. It is also said in this context that the continuous Pareto function F is topologically homotopic (continuously deformable) to dictatorial function F'.

According to Baryshnikov (2000, 124–124), the deformation process could be erratic, giving "wild" Chichilnisky social welfare functions. Thus, one might consider a more refined class of continuous functions, the Pareto-isotopic functions, which satisfy some given collective decision principles all the way from start to end. Baryshnikov (2000, 130 –131) shows then a kind of possibility result, stating that, if there are four individuals and preferences are defined over a 3-dimensional space, then there are continuous Pareto welfare functions that are not isotopic to dictatorial rules.

24.2.4 *Implications of the Manifold View*

The importance of Chichilnisky's view on social welfare function is that it re-inserts continuity. Although this is done for very particular class of social welfare functions it is still sufficiently general so that various axiomatic collective choice principles could be tested with a representative function, and so, certain type of social choice problems can be investigated axiomatically, like in ordinal Arrovian social choice theory, but now in a continuous framework.

Chichilnisky's assumption of topological manifold states implies that the set of states is equipped with not only a topology, but a very particular topology. Properties like local path connectedness, local compactness, second countability, and requirement that the space has at most countably many components, need to be assumed on the set of states. But more importantly, the assumption of manifold dimension at least 2 implies that the cardinality of the set of states must be at least $\mathfrak{c}$. This is easily seen, for example, as follows. Suppose that X is the manifold of states for which $\dim(X) \geq 2$. Let x be a social state in X. There must be an open set $U \subset X$, which contains the state x such that $U \cong U'$, and U' is some open set in $\mathbb{R}^n$ for $n \geq 2$. Every nonempty open set of $\mathbb{R}^n$, where n is a positive integer, has cardinality $\mathfrak{c}$. Since the open set U of states is bijected to such U', the set U must have cardinality $\mathfrak{c}$. But since U is a subset of X, the set X must have cardinality at least $\mathfrak{c}$.

This poses rather severe restrictions for purposes of general choice theory. To re-iterate: It is tacitly assumed that the set of social states must be infinite, thus all finite state social choice problems, having topology or not, are excluded. It is also assumed that the set of social states must be uncountable, and so all countably many infinite state social choice problems are excluded.

Finally, it is assumed that the set of uncountably infinite states must have a topological structure, and the structure cannot be freely fixed (consistent with the cardinality assumption), but it must be of quite special class—namely, a manifold topology X of $\dim(X) \geq 2$. This precludes vast amount of infinite state social choice problems, having topology over states or not.

As we have seen, the manifold topology could be in principle applied to finite and countably infinite set of states having the required topological structure, but it is the dimension assumption that precludes this possibility.

So there will be problems, if Chichilnisky's continuity construction should be applied to cover Arrovian social choice problems. We will discuss these more specifically for the rest of the paper, but in brief, they are as follows. To define continuity of the social welfare function it is necessary to define topology on the domain (the set of preference profiles) and on the range (the set of social preferences) of the function. Chichilnisky's approach provides a direction to do that in a certain way. However, putting aside the different definitions of preference in Chichilnisky and Arrow, the approach can hardly be regarded even as an adequate, not to say desirable, definition for continuity for Arrovian social choice problems. To do this, it would be necessary that (1) the definition works with both finite and infinite state social choice problems, (2) it works with highly disconnected state sets, (3) it allows generic structures on state sets, and (4) in general, it allows state sets, where there is no topology defined at all. It is readily seen that the Chichilnisky's definition, taken as such, cannot cover these.

As for example, Baigent (1997, 176–177) has demanded justification for continuity and particular topologies used, it is perhaps not so much that the continuity itself needs to be justificated or, for that matter, the topologies used, in so far as they can be freely fixed; but what is important, is that the definition of continuity should work in Arrovian social choice contexts. It should have a social choice footing.

24.3 Manifolds and Arrovian Social Choice Systems

24.3.1 *Simple Arrow-type Social Choice System*

One implication of Arrow's (1951; 1963) parsimonious treatment of collective alternatives is that the domain of choice and the possible structure it might have are open-ended subject to possible interpretation and parametrization of the social choice problem at hand. So, to evaluate any topological formulation for Arrovian social choice process, that proceeds by fixing a class of topological structures over the set of collective alternatives to define continuity of the preference transformations, it is necessary to identify some basic examples of what kind of sets and subsets are involved in the social choice process, when the final social state is arrived at. For this purpose, we consider a simple Arrow-type social choice system consisting of various exemplar parts—economic, political, and legal—related to the workings of capitalistic democracy. This is an important example of social choice problem in Arrow (1963). It is pointed out next that a general formulation of continuity for an Arrovian social welfare function, capturing the workings of capitalistic democracy, should not preclude simple abstract sets (e.g. countable sets of alternatives), topological spaces in general (e.g. zero-dimensional spaces), or first-order structures (e.g. Boolean algebras).

Consider first the case of various political decision making methods. If the continuity formulation assumes a category of objects representing the realm of collective alternatives that does not allow finite sets, then it cannot be assured that the social choice process covers properly various important voting methods. One basic example is the simple majority decision making method, investigated e.g. by Arrow (1963), by May (1952), and for countable society by Fey (2004). For a finite society, this method simply states that an alternative x is weakly preferred to alternative y if, and only if, the number of individuals that weakly prefers x to y is at least as great as the number of individuals that weakly prefers y to x. The method is already applicable to sets of alternatives that contain at least two distinct elements. Examples include two competing government officials $\{v, u\}$, two competing public projects $\{a, b\}$, and two competing legislative proposals $\{\mathbf{L}, \mathbf{L}'\}$. For precisely two alternatives, the simple majority decision making method satisfies Arrow's (1963, 46–48) conditions on positive association, independence, nonimposition, and nondictatorship. A social welfare function like this will yield, obviously, an ordering of two alternatives for every set of individual orderings over such a set. Arrow (1963, 48) notes that this Possibility Theorem for Two Alternatives is "the logical foundation of the Anglo-American two-party system."

If it cannot be assured that the topological (continuous, or for that matter, discontinuous) social choice process covers various important political decision making methods, including the simple majority decision making method, then this kind of formulation cannot be taken to represent properly continuous Arrovian social choice process, capturing the workings of capitalistic democracy. This is because the purpose of Arrow's (1963) original notion of social welfare function was to offer a conceptualization of the principal social choice methods used in the social organization, characterized by some axiomatization suited for that specific purpose, and for capitalistic democracy the map should, according to Arrow (1963), cover the various voting methods used to make political decisions. These will include the simple majority decision making method. As this method is already definable with precisely two alternatives, any formulation that precludes finite sets at the outset is unsatisfactory for our purposes.

Consider then the case of economic (private asset) decision making methods. If a formulation of general continuous social choice process assumes a formal category of objects representing sets of collective alternatives that does not allow at most countable sets, then it cannot be assured that the process covers the broad variety of economic exchange systems, including those addressing the problem of economic indivisibilities. Take for example the economic exchange system with integer number indivisibilities, proposed by Dierker (1971), for which there are finite number n of private, input-output good types, and finite set V of economic agents v. The purely individualistic consumption set $X_v \subset \mathbb{Z}^n$, that can include negative consumption, has cardinality at most $\aleph_0$ for each agent $v \in V$; and the set of all logically possible allocations with fixed private consumption sets, $\hom(V, \bigcup X_v) = \{x \mid x : V \longrightarrow \bigcup X_v\}$, has cardinality at most $\aleph_0$.

If it cannot be assured that the continuity formulation covers the broad variety of economic exchange systems, including those addressing the problem of economic indivisibilities, the formulation cannot be taken to represent properly continuous Arrovian social choice process conceptualizing e.g.

the workings of capitalistic democracy. For Arrow (1963), another part of the social welfare function, dealing capitalistic democracy, was purported to amalgamate price-mediated exchange systems. With this respect he makes it clear that the process should be amenable to economic indivisibilities. According to Arrow (1963, 17), "[I]n order to handle such problems as indivisibilities, which have been productive of so much controversy in the field of welfare economics, it is necessary to assume that some of the components of the social state are discrete variables." He (1978, 188) also points out, "[I]n some deep sense there are increasing returns to scale. The true basis for division of labor is the value to specialization, not merely in the economy but in society as a whole." So, if a continuity formulation does not allow finite or denumerable sets of alternatives, it cannot be properly extended to cover continuity of Arrovian social welfare functions.

Full divisibility of goods was not perhaps the true desideratum of the most general, final model of general economic equilibrium either. For example, in 1950s Debreu (1959, 30) notes in his classic *Theory of Value* before he proceeds to the formal analysis: "[I]t will be assumed instead that this quantity [of integer number of goods such as trucks] can be any real number. This assumption of perfect divisibility is imposed by the present stage of development of economics." Then a list of other examples followed: machine tools, linotypes, cranes, Bessemer converters, houses, refrigerators, trees, sheep, shoes, turbines, and so on. As we have seen above, Arrow shared this desideratum too in a more general social choice contexts.

From time to time, the following three persistent assertions about Arrovian social choice framework are made in the literature: (1) Arrovian social welfare functions are definable (only) with finite alternative sets (e.g. Baigent 1987, 161), (2) Arrovian social welfare functions are definable (only) with "discrete" sets of alternatives (e.g. Lauwers 2000, 1), and (3) continuity cannot be defined for finite sets of alternatives (e.g. Gaertner 2006, 168). For general Arrovian social choice framework, as a formal conceptualization, there is no need to assume (1) or (2). The statement (3) is also not true: take for example the trivial map $\emptyset \longrightarrow \emptyset$, where $\emptyset$ has all subsets open. This vacuous construction is as finite as the "finite" can possibly get, yet the mapping is continuous. (Although the space $\emptyset$ is also a topological manifold, the $\dim(\emptyset)$ can be any integer n, even negative.)

So far we have discussed only, in principle, about abstract sets (i.e. the abstract elements are simply parametrizable into particular alternatives under consideration), which do not necessarily have any formal external structure, besides the topological property of the cardinality of a set. This is the Cantorian freedom to keep in general social choice theory. For illustrative purposes we used also some intuitions arising from integer number systems. If topologies are really needed for sets of alternatives, the choice of topological properties may not be completely arbitrary. Take, for example, the set of positive integers $\mathbb{Z}_+$; let $\mathbb{Z}_+$ have all subsets open. The space $\mathbb{Z}_+$ is locally compact, locally path connected, second countable, and Hausdorff, but it is zero-dimensional and hence not of positive dimension. Some of these properties will change under infinite products. The space of all logically possible allotment of goods $\hom(V, \mathbb{Z}_+^n)$, under Tychonoff products and for V countable infinite, is Hausdorff, zero-dimensional, and totally disconnected; and although it is second countable, it is not anymore locally compact or locally path connected. Furthermore, if V is uncountable, the space $\hom(V, \mathbb{Z}_+^n)$ is not even second countable. The space $\mathbb{Z}_+$ is not compact but it has a Stone-Čech compactification. The compactified space is compact, zero-dimensional, and Hausdorff; and although it is also locally compact, it is not locally path connected or second countable.— This is also a good place to note that if a set is topologically connected and contains at least two distinct elements, it must have cardinality at least $\mathfrak{c}$.

Let us now get back to the main line of argument. Consider the case of legislation and the problem of designing a structure for the overall set of alternatives. Legislation is an important part of the original Arrow's collective decision making problem, although it seems that it is not much investigated by Arrow himself. The problem is mentioned by Arrow (1952, 46; 1963, 1), and it is briefly discussed again in Arrow (1997b) in view of the U.S. Supreme Court Decisions. The latter problem was studied by Easterbrook (1982), Stearns (1994), and later also by Stearns (2002). The issues there relate, for example, to "consistency" of legal decisions made at different time periods.

According to Knight (1942, 252) institutions and law first came into being from causal process and became a social problem in connection with the enforcement of conformity against "recalcitrant" individuals. Knight (1942, 254) notes that this must have happened long time before "deliberate change in law", that is, legislation, was thought by anyone. According to Arrow (1997a, xvi–xvii), only with the Age of Enlightment does there appear a systematic formal approach to methods of voting in legislatures and structure of law, with the work of Jean-Charles de Borda, Marquis de Condorcet, and others.

The existing complex capitalistic democracies rely substantively on contracting between agents or groups of agents. One economic reason why agents acting under complex exchange systems would wish to have an operational legal system is to have a formal contractual assurance in large-scale investments (McMillan 2002, 60). For instance, suppose a group of agents take a large investment project for which the resulting income can be collected only after a long period of time. Here, trust and honesty can be of suspect, and the gain from "taking the money and running" may exceed any costs to recalcitrant agent's reputation. Nevertheless, the law, backed up (e.g.) by government actions, can promote the investment.

In a more fundamental sense the adopted law, formally codified or not, is related to agents freedom to dispose private assets and their autonomy to design transactions with others, for example by contracting. To specify the legal and hence the enforced contractual alternatives, and also their structure, and the structure they induce on the overall set of alternatives, we need to specify a sufficiently common "language" for which the relevant parts of economic, political, and legal reality can be "named" and "operated" upon. This is in fact a deep assumption about the inner properties of the agents themselves, that is, that they belong to the same life form and could in principle apprehend to play "the same game", even if they at the moment do not seem to do so. Hence, the notion of "language", as used here, is not related to problems arising from any potential misunderstanding of the "meaning" of non-logical symbols of the "language", in so far as these misunderstandings can be, at least in principle, clarified by the agents.

Furthermore, we are not necessarily talking about written or spoken "languages" at all here, but about some sufficiently shared conceptions about economic reality. The notion of "language" is thus used in the same fashion as in abstract model theoretical logic. Also, it is basic that the logical theory of models and topology are closely related. For example, the notion of deductive closure on our state descriptions X can be defined as a generalization of topological structure by weakening Kuratowski's closure axioms on closure operation $\mathrm{Cl} : 2^X \longrightarrow 2^X$.

A general formulation of continuous social choice process should not preclude general topological state spaces, including, for example, zero-dimensional spaces; also, assuming (e.g.) that the actions of the agents can themselves affect the generic structure of the overall set of alternatives, it should not preclude any reasonable first-order structure over states. Otherwise, it cannot be assured that the purely formal features of legal statements and the agreements they enforce can be satisfactorily, if at all, captured; also, there is then no assurance that the formulation is compatible with the structure of the overall set of alternatives, conditioned, for example, by the content of the adopted legal design.

A propositional legal statement $\mathbf{L}$ that is "true" (i.e. has a value $\mathbf{1}$) under any given set of circumstances $\delta : U \longrightarrow \{\mathbf{0}, \mathbf{1}\}$ is a clopen set $\Delta(\mathbf{L})$ in a zero-dimensional topological space $\{\mathbf{0}, \mathbf{1}\}^U$, where U contains the atomic components of the used legal language. The legal statement $\mathbf{L}$ determines under logical equivalence $\sim$, a bundle $[\mathbf{L}]$ of logically equivalent legal statements. Letting A be the set of all law bundles, we have a Lindenbaum algebra $\mathbf{A} = (A, \oplus, \otimes, {}^c, \mathbf{0}, \mathbf{1})$ of law bundles, which is a formal theory of first-order language.

The preferences of agents are defined, for example, over $X' \subset A$. A social choice, say $[\mathbf{L}] \in X'$, is then made. Nevertheless, whether X' is taken separately from overall collective alternatives X or subsumed in X, the content of the adopted legal design $[\mathbf{L}]$ conditions also the structural features of X by (e.g.) enforcing certain agreements and contractual relationships but not others. A specific contracting problem, which "makes explicit the language used," is investigated in Battigalli and Maggi (2002).

Furthermore, a legal system is compiled in pieces and may take long time to emerge. Although law is common good to all agents, it can condition allocations and income distribution and hence, some groups of agents would wish to develop it in a certain direction, while another groups would wish to take it (perhaps) completely opposite direction, so there will be all kind of interferences, when the eventual law builds up.

In a simple formal description, we make use of an abstract game in which there are two players (representing e.g. groups of agents, which need not be fixed or finite) who build the structure on X. Agents $\exists$ wish to play certain direction, agents $\forall$ (perhaps) another direction. Players move, for example, in turns. Players operate on the restriction that they can write only finite amount of information at a time, which affects the structural design of X, and the players write these conditions so that they are compatible with what has been written so far. At each stage all previous moves done in the game are assumed to be known. For example, $\exists$ tries to choose moves so that the eventual structure on states X will have a property Φ (e.g. a design for increasing long term welfare of the agents), while $\forall$s moves may interfere with $\exists$s project. It is intuitive that, if Φ stays in effect for all subsequent stages of the social choice process, no matter what $\forall$ does, it is enforced by some conditions already written down.

Let x be a set of first-order legal statements conditioning the structure of X. We assume that each x is taken from a set C of sets of legal statements satisfying certain purely logical "consistency" features, and x is then called a condition. Also it is assumed that at most finitely many new logical constants occur in any x. $\bigcup \bar{x}$ is the union of a chain $\bar{x} = (x_i)_{i<\omega}$ of conditions. Note that $\bigcup \bar{x}$ determines a formal first-order structure on X, and the above referred Φ is a property which the set $\bigcup \bar{x}$ can have or fail to have.

Let E be a subset of ω. Players $\forall$ and $\exists$ play the game $\mathrm{G}(\Phi; E)$ by choosing a construction sequence $\bar{x}$ consisting of legal statement sets x_i. Player $\exists$ has the choice of x_i if $i \in E$, and otherwise $\forall$ chooses x_i. It can be assumed, for example, that $\forall$ moves first by choosing x_0 and both players have infinitely many moves (e.g. making choices in turn). And at each stage of the construction, the player who makes the choice knows all previous moves of the game. Player $\exists$ wins the game $\mathrm{G}(\Phi; E)$ if, and only if, at the end of the game $\bigcup \bar{x}$ has the property Φ.

If for any position $(x_0, \ldots, x_k)$ in a game $\mathrm{G}(\Phi; E)$ it holds that if a condition $y \subset x_k$, then the position is winning for player $\exists$, then it is said that y "forces" Φ; that is, as soon as $\exists$ has got y into $\bigcup \bar{x}$, the player $\exists$ can be sure of winning. The compiled first-order generic structure on social states X is a model of $\bigcup \bar{x}$.

This kind of game for enforcing properties on the set of states is an application of forcing, also called construction by games, which is used in logic for building models in general, see e.g. Hodges (2006). For our purposes, it should be clear that if it cannot be assured that the continuity formulation of social choice process is compatible with various "language" features, arising from legal statements and the induced contractual environments, and ideally also with various first-order structures on X, that can have quite peculiar properties, then this kind of formulation cannot be taken to represent properly continuous Arrovian social choice process, conceptualizing e.g. the workings of capitalistic democracy.

24.3.2 *Market Mechanism and Welfare*

Market mechanism and its economic welfare properties defines an important subclass of Arrow's collective decision making problem. Arrow (1963, 17) notes that in order to handle such issues as economic indivisibilities it is necessary to assume that some of the components of the collective alternative are "discrete" variables. It is pointed out here that in the presence of certain classes of indivisibilities, the Chichilnisky (1980) continuous social welfare function is not defined, and even if it is defined up to required topological homeomorphism of the state set structure, it may omit the economically important features of the collective alternatives amounting to dimension zero. Furthermore, to attain the fullest generality and compatibility with nonmarket contexts like voting, Arrow (1963, 11– 19) did not fix any apriori notions of satiation or nonsatiation on preference, whereas Chichilnisky works mainly with nonsatiated preferences and implies unnecessary strong assumptions with this respect. Third, the notion of "discrete" tends to embroil confusion in the literature, when the relationship between Arrow's and Chichilnisky's work is discussed, and the notion of discreteness is briefly commented at the end of this section.

In Chichilnisky's (1980) definition of continuous social welfare function preferences are defined over a topological manifold that has a dimension at least 2. This dimension specification implies that the set, over which preferences are defined, must have cardinality at least $\mathfrak{c}$. This is unnecessarily large cardinality assumption in view of general model of preference nonsatiation. The cardinality $\aleph_0$ suffices for this purpose.

There are, however, more important problems that arise, for example, from the commodity indivisibilities. In considering the context of market mechanism and its welfare properties, the only two possible social state models that Arrow (1963, 16–17; 1950, 25–26) wants to exclude right from the beginning are (1) the case where all commodities are perfectly divisible, on the grounds of potential indivisibilities, and (2) the case where there is only one commodity, on the grounds of practical non-relevance.

Thus, the simplest possible Arrow social choice problem, when there are commodities, is a two commodity world in which at least one commodity is indivisible. Let $X_v = X \times X'$ be the two commodity consumption set for the individual v in society V such that at least one of the sets describes indivisible commodity. A special case of Arrow's choice problem, which is assumed usually in welfare economics, is when individuals have preferences over only own consumption sets; these preferences are called taste-preferences by Arrow (1963). In this case, our simple example will be excluded from Chichilnisky's definition of continuous social welfare function. The Chichilnisky approach operates under the assumption of topological diffeomorphisms between applied state sets, when one set X'' of dimension at least 2 is given. However, here the dimension of X_v is at most 1, so there cannot be any continuous bijection, and hence no topological homeomorphism or any diffeomorphism, between X_v and a set X'' having dimension at least 2, although the sets may, or may not have the same cardinality. (Recall that cardinality of a set is a topological property preserved under continuous bijections, but the theorems of Hilbert, Brouwer, and Peano states that there is no continuous bijection between two continuums of different order.)

Consider then the following particular case. Let the consumption set for an individual v be $X_v = \mathbb{R}_+ \times \{0, 1, 2\}$, where the second commodity is indivisible. This one-dimensional problem is taken from Ellickson (1993, 108). Ellickson (1993) considers more generally various detailed examples using refined models of general equilibrium, like Aumann's (1964) infinite population framework, to handle the cases where at least one component is divisible, and other commodities may be indivisible. This reasoning presupposes, of course, the standard Euclidean topology on divisible factor $\mathbb{R}_+$. Ellickson (1993, 135) even goes on to say: "[A]rguably all commodities are indivisible."

Consider the following case investigated by Inoue (2006). Let the consumption set X_v for an individual v be a subset of the finite product $\mathbb{Z}^n$. Inoue (2006) reconsiders the technical assumption, using infinite population multimarket context, that at least one factor must be a divisible good, in the sense that he considers the case where all commodities are purely indivisible and can be consumed only in integer amounts. (For Inoue (2006), the consumption set $X_v \subset \mathbb{Z}^n_+$ is taken to be a subset of universal class χ of consumption sets, formed in a certain way, and χ is endowed with the topology of closed convergence. In this framework, a preference p can be taken to be a subset of $\mathbb{Z}^n \times \mathbb{Z}^n$, and the set P of preferences is endowed with the topology of closed convergence. But we do not follow necessarily this line of thinking here.)

For our purposes, a fixed $X_v \subset \mathbb{Z}^n$ can be taken to be a topological manifold, supposing that n is finite. Assume furthermore that $n \geq 2$. The proper topological dimension of X_v is 0 for any $v \in V$, and not n. Suppose that individual's preference can be defined not only over individual's own consumption set, but also over other individuals' consumption sets; for example, if preferences are what Arrow (1963) calls value-preferences. According to Chichilnisky (1980, 169), the dimension of collective consumption set X is calculated as follows: if it is possible that individual has preferences also over other individuals' choices, then the $\dim(X) = \dim(X) \cdot k$, k being the cardinality of finite society V.

Let κ be arbitrary, finite or infinite, cardinality of the set V. (If V is infinite, the set X of collective alternatives is not necessarily manifold, but it is still zero-dimensional.) Then if the dimension of the social choice problem X is calculated as in Chichilnisky (1980), the $\dim(X) = 0$, since by simple Cantor arithmetic $\dim(X) \cdot \kappa = \text{card}(\emptyset) \cdot \kappa = \text{card}(\emptyset \times V) = \text{card}(\emptyset) = 0$.

If there are, unlike in Inoue (2006), $\aleph_0$ or $\mathfrak{c}$ purely indivisible commodities, the topological dimension of X_v is still zero, and so the dimension of X is also zero. (Although X_v is then not necessarily manifold, it is zero-dimensional.) This is independent of the cardinality of the set of individuals, and whether individual's preferences are defined only over individual's own consumption set, or also over other individuals' consumption sets. Thus, in case of purely indivisible commodities, the collective alternative set X cannot be topologically equivalent to a manifold of dimension at least 2, as required in Chichilnisky (1980). Hence, the continuity of Chichilnisky's social welfare function is not defined in this case, and the definition is inapplicable for Arrow's social choice problem, which does not preclude the purely indivisible commodity world. This is so, whether there are finite or infinite kinds of such commodities, and whether there are finite or infinite number of individuals.

Inoue (2006) notes that most consumer commodities are available only on integer amounts, despite their physical appearance (e.g. wine is liquid, but it is usually sold in bottles). Some commodities could be thought as a kind of grained indivisibles (i.e. commodities that are like pebbles and for which the distinction may matter, but not so much as in the case of integer goods), or in rare cases, even more finely as a kind of liquid (e.g. gasoline for a car from the station). The grained indivisibilities work somewhat like rational numbers, which is a zero-dimensional space. The liquid case is usually thought as a set of real numbers, equipped with the usual Euclidean topology. It is, nevertheless, not necessary to conceptualize this situation as a Euclidean real number space; it can be viewed as a set of real numbers, or more generally any linearly ordered set, equipped with, for example, a non-Euclidean Sorgenfrey (1947) topology, which is again a zero-dimensional space. It is basic that any, finite or infinite, product of zero-dimensional spaces is zero-dimensional.

Taking the *discrete variable* literally in Arrow's description of social state, it means that the variable has countable number of possible realizations, and so the corresponding state factor set should have cardinality at most $\aleph_0$. This is not yet very fruitful for general theory. Suppose then that the factor set has the *discrete topology*. This does not by itself restrict the cardinality of the factor set to be $\aleph_0$; it can be countable or uncountable. Nonetheless, infinite products are not discrete spaces. Since the use of merely discrete topology for purposes of economic or other type of indivisibility excludes many if not all (nontrivial) topological spaces from considerations, it is appropriate to consider some generalization of topological discreteness, like zero-dimensionality. All discrete spaces are zero-dimensional, but the converse does not hold.

24.3.3 *Voting*

Collective decision problem, such as the selection of a person for office by a vote, is an important subclass of Arrow's social choice problem. It is pointed out next that there are problems, if Chichilnisky's (1980, 1982b) continuous social welfare function is extended to cover the continuity of Arrovian social welfare function, since it does not give sufficient possibility to fix the cardinality and dimension of the elementary political candidate sets. Furthermore, the state-factor structure of the candidate set becomes obscured and the tacitly assumed fixed cardinality is arbitrary for general nonmarket decision making framework.

The implied assumption of Chichilnisky's definition of continuous social welfare function is that the set of alternatives over which a preference is defined has a cardinality at least $\mathfrak{c}$. In view of modeling preference satiation, which is an important assumption in political decision science, this cardinality assumption is arbitrary. Preference satiation does not depend on the cardinality of the set of alternatives.

There are, however, more important problems. The simplest possible Arrow collective decision problem, which deals nonmarket decision making such as voting, is the case where there are precisely two collective alternatives. Let the preference field for an individual v be $X_v = X = \{a, b\}$. Suppose each individual v, in a finite society V of cardinality k, is associated a variable D_v as follows: $D_v = -1$, if v prefers a to b, $D_v = 0$, if v is indifferent between a and b, and $D_v = 1$, if v prefers b to a.

May's (1952) social welfare function $(D_1, \ldots, D_k) \longmapsto D$ maps the k-fold Cartesian product $\{-1, 0, 1\}^V$ surjectively to social preference $D \in \{-1, 0, 1\}$. May's social welfare function is a special case of Arrow's (1963) social welfare function, in case where there are precisely two social states.

May (1952) investigates the question what are the conditions for this function, so that it is the familiar method of making collective decisions by simple majority for two alternatives. Consider the following four principles for the May social welfare function F:

> *Universal domain and single-valuedness*: The mapping F is defined and single valued on $\{-1, 0, 1\}^V$.

Anonymity: $F(D_1, \ldots, D_k) = F(D_{h(1)}, \ldots, D_{h(k)})$, for any bijection h from $\{1, \ldots, k\}$ to itself.
Neutrality: $F(-D_1, \ldots, -D_k) = -F(D_1, \ldots, D_k)$, where "$-$" behaves like in case of integers.
Monotonicity: If $F(f) = 0$ or 1, and if for any $f, g \in \{-1, 0, 1\}^V$ it holds that $f = g$, except that for some vth projection $\mathrm{pr}_v(g) > \mathrm{pr}_v(f)$, then $F(g) = 1$.

May (1952, 682–683) shows that a social welfare function F is the method of simple majority if, and only if, it satisfies the conditions of universal domain and single-valuedness, anonymity, neutrality, and monotonicity. This is a basic theorem in classical social choice theory.

Arrow's (1963, 46–48) possibility theorem, for precisely two alternatives, states that the method of simple majority satisfies his well-known conditions of nondictatorship, independence of irrelevant alternatives, nonimposition and positive association (pp. 25–31), when applied to two alternatives. These same conditions amount to the famous Arrow's "paradox", once three or more alternatives are considered.

Challenging social choice problems emerge, in various disguises, once the number of alternatives is increased to three or more. We refrain commenting these here, except noting that there usually are not precisely two alternatives, but say three or four. Even if there are, preferences may still be thought to be defined over larger set, not necessarily infinite, from which the alternatives are drawn from.

Chichilnisky's (1980, 1982b) definition of continuous social welfare function is not applicable, if the set X of candidates consists of two elements, because of Chichilnisky's dimension requirement. It is not much so that the nature of states cannot exhibit cardinality at least $\mathfrak{c}$, but the fact that for Chichilnisky it cannot be otherwise; the cardinality of the set of states over which preferences are defined must be at least $\mathfrak{c}$. This effectively precludes, for example, all finite and countable infinite collective alternative problems.

However, suppose that the assumption $\dim(X) \geq 2$ is dropped. Then the set $X = \{a, b\}$ can be equipped with a zero-dimensional manifold structure. A Chichilnisky-type preference p on X is a map $p : X \longrightarrow TX$ such that $s \circ p = 1_X$, where $s : TX \longrightarrow X$. Then a topology on the set P need to be defined, after which continuity of maps $F : P^V \longrightarrow P$ can be investigated.

(Here, it may be useful to first observe that any zero-dimensional manifold admits trivially a unique differentiable C^∞ structure, and thus also C^2 structure: For each state $x \in X$ the only open set U containing x such that

$U \cong \mathbb{R}^0$, is $\{x\}$. Furthermore, there is precisely one pair (U, φ), such that U is open in X and $\varphi : U \longrightarrow U'$ is a homeomorphism from U to an open subset of $U' = \varphi(U) \subset \mathbb{R}^0$. There are only two such pairs, $(\{a\}, \phi)$ and $(\{b\}, \gamma)$, if X is manifold consisting of two elements, and the intersection $\{a\} \cap \{b\}$ of the domains of these maps is empty. So, the maps are what are called smoothly compatible, and the claimed structure follows. This is, of course, obvious. The finite alternative case is taken into consideration, for example, in Schofield (1984, 189).)

Manifolds occur in economics as indifference surfaces. Arrow (1963, 16–17), however, gave several reasons why it is better to represent the choice mechanism by abstract ordering relations, instead of indifference maps, in general social choice theory. It turns out that there is a simple way to define continuity of the social welfare function here, consistent with this line of thinking. Let $\{-1, 0, 1\}$ have all subsets open; it is then a zero-dimensional manifold. Take finite products of $\{-1, 0, 1\}$ and form the product topology on $\{-1, 0, 1\}^V$. Then all maps $F : \{-1, 0, 1\}^V \longrightarrow \{-1, 0, 1\}$ are trivially continuous. There is no need to define topology on the set of states itself, as is in Chichilnisky's approach. When May's conditions are applied, it is readily seen that there is a continuous function F satisfying these conditions, for two-alternatives and finite set of individuals.

Another point is that the form of the basic candidate decision problem does not naturally imply topological dimension of 2 or more, as is required by Chichilnisky's definition of continuity, or at least dimension 1, as is required by Saposnik (1975, 684), no matter what the cardinality of the (sub)set of individuals involved in the decision making. It is more likely that the proper topological dimension of this problem is simply 0. Of course, if one starts with a higher dimensional manifold X of states, the zero-dimensional submanifolds of X are precisely the discrete subsets of X.

Even if the set of collective alternatives is uncountable, it is dubious that it must have a topological structure at all in view of defining continuity of the social welfare function—the topology is essentially on the set of preferences—not to demand that this topological structure is to be many dimensional local Euclidean structure, instead of some disconnected structure, for example, the Sorgenfrey structure.

It may also be noted, that there is no state factor-structure at all in this simple candidate decision problem; the collective decision problem involves an abstract, nonstructured set. The continuity definition of Chichilnisky (1980, 1982b) tacitly assumes these classes of decision problems. However, Chichilnisky's (1982b) definition of continuity requires a collective alternative set which is diffeomorphically equivalent to unit cube, which is a Cartesian product of closed unit intervals. Topologically equivalent objects can be transformed to each other in a continuous fashion. But this does not remove the fact that in this case the factor-nature of the collective decision problem becomes blurred.

24.3.4 *Legislation*

Formation of an entire legislature for a community of individuals is an important, although undeveloped, class of Arrow's collective decision making problem. It is pointed out here that defining continuity of Arrovian social welfare function by using topological manifold structures, having dimension at least 2, over legal environments is a problem, since if there is a chosen class of topologies, it should abstract the essential sentential features of the judicial statements, but the manifold assumption together with the dimension assumption makes Chichilnisky's view inapplicable in this case.

Let $X = \{[\mathbf{L}], [\mathbf{L}']\}$ be the set of collective alternatives for society V. For example, let $[\mathbf{L}]$ be the pre-existing legislation, and let $[\mathbf{L}']$ be the proposed new legislation. The purpose is to choose collectively between them. This kind of example is given, but only for illustrative purposes, in Arrow (1967, 62). Chichilnisky (1980, 1982b) lacks the consideration of language that would be needed to describe this kind of problem properly, but Arrow (1963) too gives no specification of the formal structure from where these elements came from, or how to describe or talk about these elements. We briefly discuss these matters next.

For example, suppose that $[\mathbf{L}']$ describes the Code of Hammurabi. The system consists of 282 law statement forms, some of which are lost. These can be thought to be build from some finite number of atomic law statements—describing, for example, events, actions, and monetary transfers—with respect to some formal operations. For simplicity, we just take a the natural language law clauses, like "If a man put out the eye of another man, his eye shall be put out," to be translated into a formal language's specific atomic law statement, denoted by $\mathbf{L}_i$. Although the logical structure of these kind of statements can be analyzed further, that need not concern us here.

The overall idea is that preferences are first defined over the whole vague universe of law bundles, after which, when particular specification is considered, like X, the welfare function orders the bundles, so that a preferable collective choice can be made. The choice need not, of course, depend on preferences of every individual in V.

How the law bundles like $[\mathbf{L}]$ are formed? The first thought is that the law bundles are formed from the alphabet $H = \{\mathbf{L}_1, \mathbf{L}_2, \dots\} \cup \{\triangle\}$, where $\triangle$ is a binary operation symbol, as follows. Let $W(H)$ be the set of all finite sequences, that is, words on H. Let L be the smallest subset of $W(H)$ which includes $\{\mathbf{L}_1, \mathbf{L}_2, \dots\}$ and which has the property that whenever it contains words B and B' it also contains the word $(B \triangle B')$. That is, the set L is the smallest subset of $W(H)$ which includes $\{\mathbf{L}_1, \mathbf{L}_2, \dots\}$ and which is closed under the operation $(B, B') \longmapsto (B \triangle B')$.

In the most simpleminded model, one could take the law bundles, the terms of preference, to be represented by elements of this algebra $(L, \triangle)$, where the operation $\triangle$ corresponds to the intuitive use of "and". But what we probably want is that the law bundles are taken from some richer structure, so that they could be operated with the usual logical operations, somewhat like in Battigalli and Maggi (2002) in case of contracts. Let $[\mathbf{L}]$ stand for the set of all law statement forms that are logically equivalent to $\mathbf{L}$. Then we can use an algebra of equivalence classes $L/\sim$ of the law bundles $[\mathbf{L}]$, under the usual logical operations, giving us, for example, the structure $(L/\sim, \wedge, \not\Leftrightarrow)$.

One starts translating the elements of the codex universe into the parameters of the formal language:

> No. 122: "If any one give another silver, gold, or anything else to keep, he shall show everything to some witness, draw up a contract, and then hand it over for safe keeping" corresponds to $\mathbf{L}_1$.
>
> No. 242: "If any one hire oxen for a year, he shall pay four gur of corn for plow-oxen" corresponds to $\mathbf{L}_2$, and so on.

No matter how these are translated, if the translation of $(\mathbf{L}_1 \wedge \mathbf{L}_2)$ into the codex universe of laws is true, then $\mathbf{L}_1$ must be true. To take all possible translations to legal universe, including the Hammurabi statements and its variations, would be too much work for our purposes. However, one can use a shortcut. Map all elements $U = \{\mathbf{L}_1, \mathbf{L}_2, \dots\}$ to $\{\mathbf{0}, \mathbf{1}\}$, and consider elements $\lambda \in \{\mathbf{0}, \mathbf{1}\}^U$, and then consider extension $\bar{\lambda}$ preserving truth for law statement B built up from U. The set $\Lambda(B)$ of all these assignments is a clopen set in a zero-dimensional topological space $\{\mathbf{0}, \mathbf{1}\}^U$. Since one can establish a bijection between the topological space $\{\mathbf{0}, \mathbf{1}\}^U$ and the Stone space $S(L/\sim)$ of the algebra of equivalence classes $(L/\sim)$, of logically equivalent law statement forms, the terms of the preference—law bundles like $[\mathbf{L}]$ and $[\mathbf{L}']$—are drawn by correspondence from a topological space.

Preferences are defined over the alternative, competitive law bundles X'. Taking the k-fold product (k being the finite number of individuals in the society) of the set of preferences, we have the domain for the social welfare function. For a given a feasible subset of X', say $X = \{[\mathbf{L}], [\mathbf{L}']\}$, a particular collective choice problem is formed. Some desirable properties need to be fixed for the social welfare function. In case there are only two bundles $X = \{[\mathbf{L}], [\mathbf{L}']\}$, society might use, for example, the simple majority decision making method. In the original Babylonian choice problem, the official choice of the uniform codification might have been dictated by only few individuals, Hammurabi and some of his close associates, although the Code probably already existed in some form.

Social choice problems will become more challenging, in various guises, when the number of law bundles from which the choice is made is larger than two. Nevertheless, if the set U of atomic law statements is finite, and the set P of preferences is like in Arrow (1963), one can declare all subsets of P open, and then define the continuity of the Arrow's social welfare function in an obvious way. Clearly, one does not even need to define a topology on the set of law bundles to define the continuity of the social welfare function. The problems become more subtle when the set of alternatives or the set of individuals can be infinite. In any case, if topology is defined over the system of law statements, along the line sketched above, topological manifold structures X for which $\dim(X) \geq 2$ are not compatible with this view, because the most natural topological space in this case is zero-dimensional.

The usual formal models of legal reasoning developed in the 1990s and since, assume the property of logical nonmonotonicity, for example Sartor (1994). This means roughly that the following property fails: whenever B follows from a set S of law statements, then it follows also from every superset of S' of S. This is because, for example, the laws need to be interpreted in particular cases before they can be applied, or since new information may render new legal arguments possible. However, this idea presupposes a rather different problem than considered here, that is, applying the law rather than forming the whole system of law to be applied to begin with.

If the nonmonotonicity is, nevertheless, assumed there is no real change in case the set of atomic bundles is finite, since the topological space is compact. But one needs to be aware that in this case arbitrary law statement forms cannot be substituted for atomic laws (these are elements of U), whenever these occur in formulas, as in elementary logical constructions. In case of infinite atomic (components of) laws, one gets the nonmonotonicity by, for example, restricting the set of possible valuations $\lambda \in \{\mathbf{0}, \mathbf{1}\}^U$ in which case the compactness may fail. The topological space is still essentially disconnected space. For more about nonmonotonicity and restricting the set of valuations, see Makinson (2005).

24.3.5 *Arrow's Abstract Theory of Collective Decision Making*

Arrow's social choice theory views market mechanism, or core properties in general, and nonmarket collective decision making, such as voting and deliberate change of legislation, as a special cases of abstract social choice problems. To retain full generality, no algebraic, topological, measure theoretic, or other similar structure is fixed permanently for the set of states. In contrast, the Chichilnisky view assumes a specific class of topologies to be defined over set of states. It is pointed out here that this is a problem, if Chichilnisky's view about continuous social welfare function is extended to cover continuity of Arrovian social welfare function, since there may not be a single class of topologies, or for that matter, other formal structures, for the set social states that covers ubiquitously all Arrovian collective choice problems that may emerge.

As already discussed, according to Chichilnisky (1980) the set X of social states is equipped with a locally Euclidean topology $\dim(X) \geq 2$, which for applied purposes of differentiable manifolds, is usually taken to be second countable and Hausdorff. Letting X be a class C^n for $n \geq 2$. A preference p on X is then viewed as a locally integrable morphism $p : X \longrightarrow TX$ of class C^{n-1} such that $p(x)$ lies in the tangent space $T_x X$ for each $x \in X$. In contrast, the only thing Arrow (1963, 24, 103) assumes about the set X of social states (alternatives) is the following two conditions:

> (1) Among all the alternatives there is a set X' of three alternatives such that, for any set of individual orderings $r_1 \ldots r_k$ of the alternatives in X', there is an admissible set of individual orderings $p_1 \ldots p_k$ of all the alternatives such that, for each individual v, $\langle x, y\rangle \in p_v$ if and only if $\langle x, y\rangle \in r_v$ for x and y in X'.
>
> (2) Among all the triples of alternatives satisfying Condition 1, there is at least one on which no individual is a dictator.

There is no topological or other readily fixed structure on the realm from where the discussed subset of alternatives is drawn from, and there is, in particular, no cardinality assumption on that realm other than that it contains at least three alternatives. For Arrow (1963, 13), a preference is a relation $p \in X \times X$ which satisfies the following two properties: (1) for all x and y, either $\langle x, y\rangle \in p$ or $\langle y, x\rangle \in p$, and (2) for all x, y, and z, if $\langle x, y\rangle \in p$ and $\langle y, z\rangle \in p$, then $\langle x, z\rangle \in p$. Chichilnisky's view on the structure of states X is quite particular, out of those possible structures that might arise in Arrow's social choice problems; also, the notion of preference p on X is given as a certain differentiable relation between X and TX, instead of just as a simple ordering relation on X.

As discussed in the previous sections, the class of topological manifolds of dimension at least 2 does not cover properly the topological features of the original basic examples of Arrow's collective decision making problem. These include, for example, (1) certain classes of indivisibilities, like purely indivisible commodity sets, (2) certain political alternatives, such as basic candidates in nonspatial voting, and (3) alternatives in legal contractual environments. There could have been choices of topological structures entirely different, but at least equally natural, to topologize, for example the set of states. Instead of abstracting the calculus tradition and using differential topology and differential geometry, more general and simple topological approaches could have been used, for example using some class of disconnected spaces, like zero-dimensional compact spaces.

As markets, voting, and legislation were only particular example classes of Arrow's general social choice framework, there may be no single, fixed class of topologies for social states that covers ubiquitously all cases that may come up in particular applications. Thus, from the point of view of Arrow's framework, Chichilnisky's use of fixed, narrow class of topologies—together with the implied differential structures, which are not topological by their nature—takes away the inherent freedom incorporated in the Arrow's abstract social choice framework, that is, freedom to assume and freedom to construct various structures for the set of states. Closely related, is the overall freedom to construct and investigate continuity of various other types of social choice rules than Arrovian social welfare functions, for which different assumptions about the nature of binary relations may make an important difference.

The freedom to assume and freedom to construct algebraic and other structures for the states can be seen in two ways. First, a social choice theorist could get the structure from the "storehouse of abstract forms—the mathematical structures", as described by Bourbaki (1950, 231), and then claim that social choice universe can fit itself to these forms, as through a "kind of preadaption." Although the storehouse is vast and elegant, a particular social choice problem may exhibit complex properties that the readily fixed constructions do not necessarily have.

But there is no necessity to restrict oneself to the kind of structural imperialism of pure mathematics. The other possibility is to build the structure by oneself by forcing, so that the structure is guaranteed to have the properties needed for the social choice problem at hand. (This technique is mainly used in set theory and model theoretical logic, see e.g. Hodges (2006) or Jech (2006).) In these constructions, one cannot necessarily say that the constructed structure is such-and-such, like in some fixed, readily available structure retrieved from the Bourbaki's mathematical storehouse of abstract forms, but the best thing one can say is that the constructed structure can be guaranteed to have such-and-such properties, called enforceable properties. As we have already noted, this method of building structures is essentially an abstract game of some fixed length, measured by some ordinal. Here, different tasks are assigned to separate abstract players or builders of the construction. Each builder can regard the other builders as rivals who keep interfering in his attempts to carry out his tasks, while only finite amount of information can be piled to the structure by a player at a time.

These games can be viewed topologically as follows. Let $E = \omega$, and consider the set $\{\mathbf{0},\mathbf{1}\}^E$, that is, the set of all maps $\lambda : E \longrightarrow \{\mathbf{0},\mathbf{1}\}$. A *condition* is a map $\lambda : E' \longrightarrow \{\mathbf{0},\mathbf{1}\}$ where E' is a finite subset of E. Let $M(\lambda)$ be the set of all those maps of $\{\mathbf{0},\mathbf{1}\}^E$ which extend λ . The set $\{\mathbf{0},\mathbf{1}\}^E$ is given a topology by taking as basis the sets $M(\lambda)$ for conditions λ . Given a nonempty closed set $\Lambda \subset \{\mathbf{0},\mathbf{1}\}^E$ and a set $\Lambda' \subset \Lambda$, abstract players, called Abelard ($\forall$) and Eloise ($\exists$), play the game $G(\Lambda, \Lambda')$ of length E as follows. The players choose an increasing sequence $\lambda_0 \subset \lambda_1 \subset \ldots$ of conditions so that $\Lambda \cap M(\lambda_i)$ is nonempty for each $i < E$. Player $\forall$ chooses λ_i if and only if i is even. Player $\exists$ wins if and only if $\Lambda' \cap \bigcap_{i<E} M(\lambda_i)$ is nonempty. This topological form is given in Hodges (2006, 26).

To sum up, the question then arises why the class of topologies to be used is required to be the ones discussed above, as there clearly would have been need for more general class of topologies, and more importantly, why it is required that there must be a specified topological structure to be explicitly defined on the set of social states to begin with. It seems that to attain the needed full generality in social choice problem, defining an explicit topology or other structure on the set of states is, if no reason happens to come up in a particular application, completely unnecessary in defining the continuity of social welfare function. But this cannot happen in the framework defined by Chichilnisky in which the topology is fixed beforehand. A different approach is needed to define the notion of continuity for Arrovian social welfare function, and its generalizations.

24.4 Summary

The general problem of social choice is stated by Arrow (1963, 103) as follows : "The social choice from any given environment is an aggregation of individual preferences." For Arrow (1963), one important example providing various "environments" was capitalistic democracy. To this, one might add that in modern capitalistic democracies there are millions of individuals. So, it is intuitive that if the social organization is even moderately "competitive" or "democratic", a change in only few individuals' tastes or values, should not alter the end result of the social choice process very much. This is essentially the idea of continuity. To investigate continuity properties of social choice processes, topological structures are needed.

The question then arises, how to define a general notion of continuity for this kind of social choice process. Instead of addressing this question directly, we have opted for a much more modest, indirect approach, and we asked: can this continuous process be defined satisfactorily by using the category of manifolds (over sets of states) and smooth (hence continuous) maps, as is done in so-called topological social choice theory or topological social choice model, which have been around now for nearly forty years.

Our first main observation was that, if a formulation of continuous social choice process assumes a formal category of objects representing sets of collective alternatives that does not allow, for example, finite and countably infinite sets (that can be also abstract), various disconnected topological structures (like zero-dimensional structures), or various first-order structures (that can be generic), then this kind of formulation cannot be taken to represent properly continuous Arrovian social choice process conceptualizing e.g. the workings of capitalistic democracy. The second main observation was that the manifold view of social choice, that assumes manifold topologies (with or without boundary) with positive dimension, to be defined over the set of collective alternatives in order to define continuity of certain type of social welfare functions, does not allow abstract alternative sets (including finite or countably infinite sets), topological spaces of alternatives having dimension 0, or generic first-order structures over alternatives.

Hence, we conclude that the manifold view of social choice that assumes manifold topologies with positive dimension over sets of collective alternatives, in order to define continuity of certain type of social welfare functions, cannot be taken to represent properly continuous Arrovian social choice process conceptualizing e.g. the workings of capitalistic democracy.

Overall, if continuity is defined for Arrovian social welfare function, it should be defined in such a way that it is not necessary to restrict the category of objects (representing sets of collective alternatives and their possible structures) to be, for example, positive dimensional topological manifolds. To see that this general approach is possible, simply observe that the set of states and the set of preferences over states are separate concepts, even though preferences are defined over states. Ideally one should have certain Cantorian freedom to construct and assume about sets of states—with or without topological structures—so that the relevant intuitions arising from economic, political, and legal reality can be formally specified, if they need to be specified. Only then can continuous Arrovian social choice process have the general social choice footing, referred in the introduction. In this way we can also see that it is not so much the notion of *continuity* of the social welfare function that should turn out to be awkward to "classical" (nontopologized) social choice theory, but what one assumes about sets of states and preferences.

Bibliography

Arrow, K. J. (1984 [1950]). A difficulty in the concept of social welfare, in K. J. Arrow, *Collected Papers of Kenneth J. Arrow, Volume 1: Social Choice and Justice*, pp. 1–29, (Belknap Press).

Arrow, K. J. (1951). *Social Choice and Individual Values*, (Wiley).

Arrow, K. J. (1984 [1952]). The Principle of Rationality in Collective Decisions, in K. J. Arrow, *Collected Papers of Kenneth J. Arrow, Volume 1: Social Choice and Justice*, pp. 45–58, (Belknap Press).

Arrow, K. J (1959). Rational choice functions and orderings, *Economica*, **26**, pp. 121–127.

Arrow, K. J. (1963). *Social Choice and Individual Values*, (Yale University Press, 2nd ed.).

Arrow, K. J. (1984 [1967]). Values in collective decision making, in K. J. Arrow, *Collected Papers of Kenneth J. Arrow, Volume 1: Social Choice and Justice*, pp. 59–77. (Belknap Press).

Arrow, K. J. (1997a). Introduction, in *Social Choice Re-examined, Vol. 1*, eds: K.J. Arrow, A. Sen, and K. Suzumura, pp. xvi–xvii, (MacMillan).

Arrow, K. J. (1984 [1978]). Nozick's Entitlement Theory of Justice, in K. J. Arrow, *Collected Papers of Kenneth J. Arrow, Volume 1: Social Choice and Justice*, pp. 175–189, (Belknap Press).

Arrow, K. J. (1997b). The functions of social choice theory, in *Social Choice Re-examined, Vol. 1*, eds: K. J. Arrow, A. Sen, and K. Suzumura, pp. 3–9, (MacMillan).

Aumann, R. J. (1964). Markets with a continuum of traders, *Econometrica*, **32**, pp. 39–50.

Baigent, N. (1984). A reformulation of Chichilnisky's impossibility theorem, *Economics Letters*, **16**, pp. 23–25.

Baigent, N. (1985). Anonymity and continuous social choice, *Journal of Mathematical Economics*, **14**, pp. 1–4.

Baigent, N. (1987). Preference proximity and anonymous social choice, *The Quarterly Journal of Economics*, **102**, pp. 161–169.

Baigent, N. (1997). Discussion of Chichilnisky's paper, in *Social Choice Re-examined Vol. 1*, eds,: K. J. Arrow, A. Sen and K. Suzumura, pp. 175–178, (Macmillan Press).

Baryshnikov, Y. M. (1993). Unifying impossibility theorems: a topological approach, *Advances in Applied Mathematics*, **14**, pp. 404–415.

Baryshnikov, Y. M. (1997). Topological and discrete social choice: in a search of a theory, *Social Choice and Welfare*, **14**, pp. 199–209.

Baryshnikov, Y. M. (2000). On isotopic dictators and homological manipulators, *Journal of Mathematical Economics*, **33**, pp. 123–134.

Battigalli P. and Maggi G. (2002). Rigidity, discretion, and the costs of writing contracts, *The American Economic Review*, **92**, pp. 798–817.

Bergson, A. (1938). A reformulation of certain aspects of welfare economics, *The Quarterly Journal of Economics*, **52**, pp. 310–334.

Bourbaki, N. (1950). The architecture of mathematics, *American Mathematical Monthly*, **57**, pp. 221–232.

Brouwer, L. E. J. (1912). Über abbildung von mannigfaltikeiten, *Mathematische Annalen*, **71**, pp. 97–115.

Chichilnisky, G. (1979). On fixed point theorems and social choice paradoxes, *Economics Letters*, **3**, pp. 347–351.

Chichilnisky, G. (1980). Social choice and the topology of space of preferences, *Advances of Mathematics*, **37**, pp. 165–176.

Chichilnisky, G. (1982a). The topological equivalence of the Pareto condition and the existence of a dictator, *Journal of Mathematical Economics*, **9**, pp. 223–233.

Chichilnisky, G. (1982b). Structural instability of decisive majority rules, *Journal of Mathematical Economics*, **9**, pp. 207–221.

Chichilnisky, G. (1982c). Social aggregation rules and continuity, *The Quarterly Journal of Economics*, **97**, pp. 337–352.

Chichilnisky, G. (1986). Topological complexity of manifolds of preferences, in *Contributions to Mathematical Economics: In Honor of Gérard Debreu, pp. 131–141*, (North-Holland).

Chichilnisky, G. (1991). Social choice and the closed convergence topology, *Social Choice and Welfare*, **8**, pp. 307–317.

Chichilnisky, G. and Heal, G. (1983). Necessary and sufficient conditions for a resolution of the social choice paradox, *Journal of Economic Theory*, **31**, pp. 68–87.

Conlon, L. (2001). *Differentiable Manifolds*, (Birkhäuser, 2nd ed.).

Debreu, G. (1959). *Theory of Value: An Axiomatic Analysis of Economic Equilibrium*, (Yale University Press).

Debreu, G. (1969). Neighbouring economic agents, *La Décision* (Editions du Centre National de la Recherche Scientifique, Paris), pp. 85–90.

Dierker, E. (1971). Equilibrium analysis of exchange economies with indivisible commodities, *Econometrica*, **39**, pp. 997–1008.

Easterbrook, F. H. (1982). Ways of criticizing the court, *Harward Law Review*, **95**, pp. 802–832.

Ellickson, B. (1993). *Competitive Equilibrium: Theory and Applications*, (Cambridge University Press).

Fey, M. (2004). May's Theorem with an infinite population, *Social Choice and Welfare*, **23**, pp. 275–293.

Gaertner, W. (2006). *A Primer in Social Choice Theory*, (Oxford University Press).

Heal, G. (1997). Social choice and resource allocation: a topological perspective, *Social Choice and Welfare*, **14**, pp. 147–160.

Hildenbrand, W. (1974). *Core and Equilibria of a Large Economy*, (Princeton University Press).

Hodges, W. (2006). *Building Models by Games*, (Dover).

Inoue, T. (2006). Erratum to "Do pure indivisibilities prevent core equivalence? Core equivalence theorem in an atomless economy with purely indivisible commodities only", *Journal of Mathematical Economics*, **42**, pp. 228–254.

Jech, T. (2006). *Set Theory: The Third Millennium Edition, Revised and Expanded*, (Springer, corrected 4th printing).

Kannai, Y. (1970). Continuity properties of the core of a market, *Econometrica*, **38**, pp. 791–815.

Kannai, Y. (1972). Continuity properties of the core of a market: a correction, *Econometrica*, **40**, pp. 995–958.

Knight, F. H. (1982 [1942]). Science, philosophy, and social procedure, in F. H. Knight, *Freedom and Reform: Essays in Economics and Social Philosophy*, pp. 244–267, (Liberty Fund).

Kramer, G. H. (1973). On a class of equilibrium conditions for majority rule, **41**, pp. 285–297.

Lauwers, L. (2000). Topological social choice, *Mathematical Social Sciences*, **40**, pp. 1–39.

Lauwers, L. (2002). A note on Chichilnisky's social choice paradox, *Theory and Decision*, **52**, pp. 261–2666.

Lee, J. M. (2006). *Introduction to Smooth Manifolds*, (Springer).

May K. O. (1952). A set of independent necessary and sufficient conditions for simple majority decision, *Econometrica*, **20**, pp. 680–684.

Makinson, D. (2005). How to go nonmonotonic, in *Handbook of Philosophical Logic*, 2nd ed., eds: D. M. Gabbay and F. Guenthner, pp. 175–278, (Springer).

McMillan, J. (2002). *Reinventing the Bazaar: A Natural History of Markets*, (Norton).

Saari, D. (1997). Informational geometry of social choice, *Social Choice and Welfare*, **14**, pp. 211–232.

Samuelson, P. (1983 [1947]). *Foundations of Economic Analysis: Enlarged Edition*, (Harvard University Press).

Saposnik, R. (1975). Social choice with continuous expression of individual preferences, *Econometrica*, **43**, pp. 683–690.

Sartor, G. (1994). A formal model of legal argumentation, *Ratio Juris*, **7**, pp. 212–226.

Schofield, N. (1977). Transitivity of preferences on a smooth manifold of alternatives, *Journal of Economic Theory*, **14**, pp. 149–171.

Schofield, N. (1984). Classification theorem for smooth social choice on a manifold, *Social Choice and Welfare*, **1**, pp. 187–210.

Sorgenfrey, R. H. (1947). On the topological product of paracompact spaces, *Bulletin of the American Mathematical Society*, **53**, pp. 631–632.

Stearns, M. L. (1994). The misguided renaissance of social choice, *The Yale Law Journal*, 103, pp. 1219–1293.

Stearns, M. L. (2002). *Constitutional Process: A Social Choice Analysis of Supreme Court Decision Making*, (University of Michigan Press).

Weinberger, S. (2004). On the topological social choice model, *Journal of Economic Theory*, pp. 377–384.

Chapter 25

On a Mixture Class of Stochastic Game with Ordered Field Property

S. K. Neogy
Indian Statistical Institute, New Delhi-110016
e-mail: skn@isid.ac.in.
A. K. Das
Indian Statistical Institute, Kolkata-700108
e-mail: akdas@isical.ac.in
S. Sinha
Jadavpur University, Kolkata-700 032
e-mail: sagnik62@yahoo.co.in
A. Gupta
Indian Statistical Institute, Kolkata-700108
e-mail: agupta@isical.ac.in

Dedicated to the memory of Professor S. R. Mohan

Abstract

In this paper, we consider a mixture class of zero-sum stochastic game in which the set of states are partitioned into sets S_1, S_2 and S_3 so that the law of motion is controlled by Player I alone when the game is played in S_1, Player II alone when the game is played in S_2 and in S_3 the reward and transition probabilities are additive. We prove that the game with SC/AR-AT mixture has the ordered field property. This gives an alternative proof of the ordered field property that holds for such a mixture type of game. Finally we discuss about computation of value vector and optimal stationary strategies for SC/AR-AT mixture class of stochastic game.

Key Words: Structured stochastic game, switching control property, AR-AT property, vertical linear complementarity, pivotal algorithm

25.1 Introduction

The minimax value associated with a matrix game will lie in the same ordered field as that of entries of the payoff matrix. This is called *ordered field property* for matrix games and it is observed by [Weyl (1950)]. [Shapley (1953)] introduced stochastic game and established the existence of value and optimal stationary strategies for discounted stochastic games. [Gillette (1957)] studied the undiscounted case or limiting average payoff case. [Shapley (1953)] also noted that if the data comes from rational field, the solution may not lie in the rational field. Throughout this paper by field we mean *field of real numbers* only. In general, it is difficult to find the value vector and optimal strategies for the stochastic games. We expect to obtain finite step algorithms for this special class of stochastic games which possess ordered field property. In fact it is not known in general whether a finite step algorithm exists if a stochastic game possess ordered field property. The aim of this paper is to study a mixture class of stochastic game for which the ordered field property holds. We also look at undiscounted case of this special class of stochastic games in which there is reasonable hope for obtaining a computable solution using a finite step algorithm.

A stochastic game with a finite state space and action space is defined below.

A two-player finite state/action space zero-sum stochastic game is defined by the following objects.

(1) A state space $S = \{1, 2, \ldots, N\}$.
(2) For each $s \in S$, finite action sets $A(s) = \{1, 2, \ldots, m_s\}$ for Player I and $B(s) = \{1, 2, \ldots, n_s\}$ for Player II.
(3) A reward law $R(s)$ for $s \in S$ where $R(s) = [r(s, i, j)]$ is an $m_s \times n_s$ matrix whose $(i, j)^{th}$ entry denotes the payoff from Player II to Player I corresponding to the choices of action $i \in A(s)$, $j \in B(s)$ by Player I and Player II respectively.
(4) A transition law $q = (q_{ij}(s, s') : (s, s') \in S \times S, i \in A(s), j \in B(s))$, where $q_{ij}(s, s')$ denotes the probability of a transition from state s to state s' given that Player I and Player II choose actions $i \in A(s), j \in B(s)$ respectively.

The game is played in stages $t = 0, 1, 2, \ldots$ At some stage t, the players find themselves in a state $s \in S$ and independently choose actions $i \in A(s), j \in B(s)$. Player II pays Player I an amount $r(s, i, j)$ and at stage $(t + 1)$, the new state is s' with probability $q_{ij}(s, s')$. Play continues at this new state.

The players guide the game via strategies and in general, strategies can depend on complete histories of the game until the current stage. We are however concerned with the simpler class of *stationary strategies* which depend only on the current state s and not on stages. So for Player I, a stationary strategy

$$f \in F_s = \{f_i(s) \mid s \in S,\ i \in A(s),\ f_i(s) \geq 0,\ \sum_{i \in A(s)} f_i(s) = 1\}$$

indicates that the action $i \in A(s)$ should be chosen by Player I with probability $f_i(s)$ when the game is in state s.

Similarly for Player II, a stationary strategy

$$g \in G_s = \{g_j(s) \mid s \in S,\ j \in B(s),\ g_j(s) \geq 0,\ \sum_{j \in B(s)} g_j(s) = 1\}$$

indicates that the action $j \in B(s)$ should be chosen by Player II with probability $g_j(s)$ when the game is in state s.

Here F_s and G_s will denote the set of all stationary strategies for Player I and Player II, respectively. Let $f(s)$ and $g(s)$ are the m_s and n_s dimensional column vector, respectively.

Fixed stationary strategies f and g induce a Markov chain on S with transition matrix $P(f, g)$ whose $(s, s')^{th}$ entry is given by

$$P_{ss'}(f, g) = \sum_{i \in A(s)} \sum_{j \in B(s)} q_{ij}(s, s') f_i(s) g_j(s)$$

and the expected current reward vector $r(f, g)$ has entries defined by

$$r_s(f, g) = \sum_{i \in A(s)} \sum_{j \in B(s)} r(s, i, j) f_i(s) g_j(s) = f(s) R(s) g(s).$$

With fixed general strategies f, g and an initial state s, the stream of expected payoff to Player I at stage t, denoted by $v_s^t(f, g)$, $t = 0, 1, 2, \ldots$ is well defined and the resulting discounted and undiscounted payoffs are

$$\phi_s^\beta(f, g) = \sum_{t=0}^{\infty} \beta^t v_s^t(f, g) \text{ for a } \beta \in (0, 1)$$

and

$$\phi_s(f, g) = \liminf_{T \uparrow \infty} \frac{1}{T+1} \sum_{t=0}^{T} v_s^t(f, g).$$

A pair of strategies (f^*, g^*) is optimal for Player I and Player II in the undiscounted game if for all $s \in S$

$$\phi_s(f, g^*) \leq \phi_s(f^*, g^*) = v_s^* \leq \phi_s(f^*, g),$$

for any strategies f and g of Player I and Player II respectively. The number v_s^* is called the *value of the game* starting in state s and $v^* = (v_1^*, v_2^*, \ldots, v_N^*)$ is called the *value vector.* The definition for discounted case is similar.

We will first describe some known classes of games which possess ordered field property. As already mentioned earlier, in general, it is difficult to find a pair of equilibrium (optimal strategies) strategies. Of course one can approximate it in the discounted case as Shapley has done it in his seminal paper on stochastic games but it is not an efficient procedure. See also the excellent survey paper by [Raghavan and Filar (1991)] and [Mohan and Parthasarathy (1994)].

- **Stochastic games with perfect information:** These are stochastic games in which in every state the action space of one of the players is singleton.
- **Single controller stochastic games :** In the case where player II is *single controller* this means $q(s' \mid s, i, j) = q(s' \mid s, j) \; \forall \, i, j, s, s'$.
- **Switching controlled games :** In a switching control stochastic game the law of motion is controlled by Player I alone when the game is played in a certain subset of states and Player II alone when the game is played in other states. In other words, a switching control game is a stochastic game in which the set of states are partitioned into sets S_1 and S_2 where the transition function is given by

$$q_{i,j}(s, s') = \begin{cases} q_i(s, s'), \text{ for } s' \in S, s \in S_1, i \in A(s) \text{ and } \forall j \in B(s) \\ q_j(s, s'), \text{ for } s' \in S, s \in S_2, j \in B(s) \text{ and } \forall i \in A(s) \end{cases}$$

- **Ser-Sit games :** In this case rewards are assumed to be separable, namely $r(s, i, j) = c(s) + \rho(i, j)$ and the transitions are state independent, that is $q(t \mid s, i, j) = q(t \mid i, j)$ for all (s, i, j).
- **AR-AT games :** A stochastic game is said to be an *Additive Reward-Additive Transition game (AR-AT game)* if

  the reward (i) $r(s, i, j) = r_i^1(s) + r_j^2(s)$ for $i \in A(s)$, $j \in B(s)$, $s \in S$

  and the transition probabilities

  (ii) $q_{i,j}(s, s') = q_i^1(s, s') + q_j^2(s, s')$ for $i \in A(s), j \in B(s)$, $(s, s') \in S \times S$.

Remark 25.1. One major line of research that has evolved is focused on identifying those classes of zero-sum stochastic games for which there is a possibility of obtaining a finite step algorithm to compute a solution. We will refer to these class of zero-sum stochastic games as *structured stochastic games.* It is known that for all the cases identified above finite step algorithm exists. For more details see the survey paper by [Raghavan and Filar (1991)].

Theorem 25.1. *(Ordered field property) If a stochastic game belongs to any one of the five category described above, then in the zero-sum case, the stochastic game possesses ordered field property.*

Open Problems:

Now we mention a few *open problems* from [Raghavan and Filar (1991)].

Problem I: Characterize those stochastic games which possess the ordered field property (Recall that we consider ordered field from the set of reals).

Problem II: : Suppose it is known that a certain class of stochastic games possess the ordered field property. Is it possible to give finite step algorithm to solve such games?

This problem is known to have an affirmative answer in the classes of games discussed above in this section. The above theorem gives only sufficient conditions.

The class of switching control (SC) stochastic games is introduced by [Filar (1981)]. While the above transition structure is a natural generalization of the single control game from the algorithmic point of view this class of games appear to be more difficult. The game structure was used to develop a finite step algorithm in [Vrieze (1983)] but that algorithm requires solving a large number of single control stochastic games. [Mohan, Neogy and Parthasarathy (1997a,b)] formulated a single control game as solving a single linear complementarity problem and proved that Lemke's algorithm can solve such an LCP. [Mohan and Raghavan (1987)] proposed an algorithm for discounted switching control games which is based on two linear programs. [Schultz (1992)] formulated discounted switching control game as a linear complementarity problem.

AR-AT games have been studied in the literature earlier by [Raghavan, Tijs and Vrieze (1985)]. Both the discounted and the limiting average criterion of evaluation of strategies have been considered. It is known, for example, that for a β-discounted zero-sum AR-AT game, the value exists and both players have stationary optimal strategies, which may also be taken as pure strategies.

[Sinha (1989, 2000)] consider the mixture of the above five structured classes and studies the ordered field property. To be more specific, one such case is the mixture of AR-AT and the switching controller stochastic games whose data satisfy the AR-AT conditions in some state and the switching control conditions in the remaining state.

In this paper we consider only the following generalization of the two classes of stochastic games in which the state space S is the union of 3 disjoint subsets S_1, S_2 and S_3 such that the law of transition is controlled by Player-I in S_1 and player -II in S_2 and all the state in S_3 of the game has AR-AT state. More specifically, a zero-sum stochastic game is in SC/AR-AT *mixture class* if
(i). $S = S_1 \cup S_2 \cup S_3,\ S_i \cap S_j = \emptyset\ \forall\, i \neq j$
(ii). $q_{i,j}(s, s') = q_i(s, s')$, for $s' \in S, s \in S_1, i \in A(s)$ and $\forall j \in B(s)$.
(iii). $q_{i,j}(s, s') = q_j(s, s')$, for $s' \in S, s \in S_2, j \in B(s)$ and $\forall i \in A(s)$
(iv). the reward $r(s,i,j) = r_i^1(s) + r_j^2(s)$ for $i \in A(s)$, $j \in B(s)$, $s \in S_3$ and the transition probabilities $q_{i,j}(s,s') = q_i^1(s,s') + q_j^2(s,s')$ for $i \in A(s), j \in B(s)$, $(s,s') \in S_3 \times S$.

[Sinha (2000)] gives a nonconstructive proof to show that that the above SC/AR-AT mixture class of game has ordered field property and raises the question that whether a finite step algorithm can be developed in SC/AR-AT mixtures. In Section 25.2, we present the definitions and results required for discussions in subsequent sections. In Section 25.3, we formulate the problem of computing the value vector v_s^β and optimal stationary strategies $f^\beta(s)$ for Player I and $g^\beta(s)$ for Player II for the class of discounted stochastic game with SC/AR-AT mixture as a linear complementarity problem. The class of undiscounted stochastic game with SC/AR-AT mixture is presented as a vertical linear complementarity problem in Section 25.4. This complementarity formulation gives an alternative proof of the ordered field property. Finally we discuss the possibility of obtaining a finite algorithm for computation of value vector and optimal stationary strategies in Section 25.5.

25.2 Preliminaries

Given a real square matrix A of order n and a vector $q \in R^n$, the *linear complementarity problem* (LCP(q, A)) is to find $w \in R^n$ and $z \in R^n$ such that $w - Az = q$, $w \geq 0$, $z \geq 0$ and $w^t z = 0$. It is well studied in the literature on Mathematical Programming and arises in a number of applications in Operations Research, Mathematical Economics, Engineering and Stochastic Games. For recent books on this problem, see [Cottle, Pang, and Stone (1992)], [Murthy (1988)] and a survey on application of complementarity in stochastic games see [Mohan, Neogy and Parthasarathy (2001)].

[Cottle and Dantzig (1970)] extended the problem considered above to a problem in which the matrix A is not a square matrix. The generalization of the linear complementarity problem introduced by them is given below:

We say that an $m \times k$ matrix A with the partitioned form $A = \begin{bmatrix} A^1 \\ \vdots \\ A^k \end{bmatrix}$ is a vertical block matrix of type $(m_1, m_2, \ldots, m_k)$ if A^j is of order $m_j \times k$, $1 \leq j \leq k$ and $\sum_{j=1}^{k} m_j = m$.

Given a vertical block matrix $A \in R^{m \times k}$, $(m \geq k)$ of type $(m_1, \ldots, m_k)$ and $q \in R^m$ where $m = \sum_{j=1}^{k} m_j$, the generalized linear complementarity problem is to find $w \in R^m$ and $z \in R^k$ such that

$$w - Az = q, \quad w \geq 0, \; z \geq 0 \tag{25.1}$$

$$z_j \prod_{i=1}^{m_j} w_i^j = 0, \; j = 1, 2, \ldots, k \tag{25.2}$$

This generalization is also known as *vertical generalization of the linear complementarity problem* [Cottle and Dantzig (1970)] and it is denoted by VLCP(q, A). If $m_j = 1$ then VLCP reduces to well known complementarity problem.

[Lemke (1970)] anticipated many meaningful applications for the VLCP introduced by [Cottle and Dantzig (1970)]. Mohan, Neogy and Parthsarathy made a number of applications of vertical linear complementarity problem in stochastic games. See [Mohan, Neogy and Parthasarathy (2001)] and the references cited therein.

We require the following result from [Schultz (1992)] to prove our main result for discounted case in the next section.

Theorem 25.2. *([Schultz (1992)])[Theorem 1.1] A β-discounted zero-sum stochastic game has values v_s^β and optimal stationary strategies f^β for Player I and g^β for Player II if and only if there exists a solution $(v^\beta, f^\beta, g^\beta)$ that solves the following nonlinear system SYS1.*
SYS1: *Find $(v^\beta, f^\beta, g^\beta)$ such that*

$$v_s^\beta - \beta \sum_{s' \in S} v_{s'}^\beta \sum_{j=1}^{n_s} q_{ij}(s,s') g_j^\beta(s) - [R(s)g^\beta(s)]_i \geq 0, \; i \in A(s), s \in S \tag{25.3}$$

$$-v_s^\beta + \beta \sum_{s' \in S} v_{s'}^\beta \sum_{i=1}^{m_s} q_{ij}(s,s') f_i^\beta(s) + [f^\beta(s)R(s)]_j \geq 0, \; j \in B(s), s \in S \tag{25.4}$$

Corollary 25.1. *If $(v^\beta, f^\beta, g^\beta)$ satisfies (25.3) and (25.4) then*

$$v_s^\beta = \beta[P(f^\beta, g^\beta)v^\beta]_s + r_s(f^\beta, g^\beta) \tag{25.5}$$

We require the following definition and results established by [Filar and Schultz (1987)] to prove our subsequent results for undiscounted case.

Definition 25.1. A pair of optimal stationary strategies (f^*, g^*) for an undiscounted stochastic game is *asymptotically stable* if there exist a $\beta_0 \in (0,1)$ and stationary strategy pair (f^β, g^β) optimal in the β-discounted stochastic game for each $\beta \in (\beta_0, 1)$ such that

(i) $\lim_{\beta \uparrow 1} f^\beta = f^*, \lim_{\beta \uparrow 1} g^\beta = g^*$

(ii) for all $\beta \in (\beta_0, 1)$, $r(f^\beta, g^\beta) = r(f^*, g^*)$, $P(f, g^\beta) = P(f, g^*)$ for $f \in F_s$ and $P(f^\beta, g) = P(f^*, g)$ for $g \in G_s$ where $P(f,g)$ is the transition matrix and $r(f,g)$ is the current expected reward vector which are defined earlier.

Theorem 25.3. *([Filar and Schultz (1987)][Theorem 2.1]) An undiscounted stochastic game possesses value vector v^* and optimal stationary strategies f^* for Player I and g^* for Player II if and only if there exists a solution $(v^*, t^*, u^*, f^*, g^*)$ with t^*, $u^* \in R^{|S|}$ to the following nonlinear system SYS2a.*

SYS2a: *Find* (v, t, u, f, g) *where* $v, t, u \in R^{|S|}$, $f \in F_S$ *and* $g \in G_S$ *such that*

$$v_s - \sum_{s' \in S} v_{s'} \sum_{j=1}^{n_s} q_{ij}(s,s') g_j(s) \geq 0, \; i \in A(s), s \in S \tag{25.6}$$

$$v_s + t_s - \sum_{s' \in S} t_{s'} \sum_{j=1}^{n_s} q_{ij}(s,s') g_j(s) - [R(s)g(s)]_i \geq 0, \; i \in A(s), s \in S \tag{25.7}$$

$$-v_s + \sum_{s' \in S} v_{s'} \sum_{i=1}^{m_s} q_{ij}(s,s') f_i(s) \geq 0, \; j \in B(s), s \in S \tag{25.8}$$

$$-v_s - u_s + \sum_{s' \in S} u_{s'} \sum_{i=1}^{m_s} q_{ij}(s,s') f_i(s) + [f(s)R(s)]_j \geq 0, \; j \in B(s), s \in S \tag{25.9}$$

Theorem 25.4. *([Filar and Schultz (1987)][Theorem 2.2]) If a stochastic game possesses asymptotically stable stationary optimal strategies then feasibility of the nonlinear system (SYS2b) is both necessary and sufficient for existence of a stationary optimal solution.*
SYS2b: *Find* (v, t, f, g) *where* $v, t \in R^{|S|}$, $f \in F_S$ *and* $g \in G_S$ *such that (25.6),(25.7),(25.8) are satisfied and*

$$-v_s - t_s + \sum_{s' \in S} t_{s'} \sum_{i=1}^{m_s} q_{ij}(s,s') f_i(s) + [f(s)R(s)]_j \geq 0, \; j \in B(s), s \in S \tag{25.10}$$

25.3 Discounted Zero-sum SC/AR-AT Mixture Stochastic Game

Theorem 25.5. *A* β*-discounted zero-sum SC/AR-AT mixture stochastic game has values* v^β *where*

$$v_s^\beta = \begin{cases} v_s^\beta, & s \in S_1 \cup S_2 \\ \zeta_s^\beta + \eta_s^\beta, & s \in S_3 \end{cases}$$

and an optimal pair of stationary strategies $(f^\beta,\ g^\beta)$ *if and only if* v_s^β, $f^\beta(s)$ *and* $g^\beta(s)$ *are a part of a solution of SYS3.*

SYS3:

$$v_s^\beta - \beta \sum_{s' \in S_1 \cup S_2} v_{s'}^\beta q_i(s, s') - \beta \sum_{s' \in S_3} (\zeta_{s'}^\beta + \eta_{s'}^\beta) q_i(s, s')$$

$$-[R(s) g^\beta(s)]_i \geq 0, \; i \in A(s), s \in S_1 \tag{25.11}$$

$$v_s^\beta - \theta_s^\beta - [R(s) g^\beta(s)]_i \geq 0, \; i \in A(s) \; s \in S_2 \tag{25.12}$$

$$-v_s^\beta + \theta_s^\beta + [f^\beta(s) R(s)]_j \geq 0, \; j \in B(s) \; s \in S_1 \tag{25.13}$$

$$-v_s^\beta + \beta \sum_{s' \in S_1 \cup S_2} v_{s'}^\beta q_j(s, s') + \beta \sum_{s' \in S_3} (\zeta_{s'}^\beta + \eta_{s'}^\beta) q_j(s, s')$$

$$+[f^\beta(s) R(s)]_j \geq 0, \; j \in B(s), s \in S_2 \tag{25.14}$$

$$-\zeta_s^\beta + \beta \sum_{s' \in S_1 \cup S_2} v_{s'}^\beta q_j^2(s, s') + \beta \sum_{s' \in S_3} (\zeta_{s'}^\beta + \eta_{s'}^\beta) q_j^2(s, s')$$

$$+r_j^2(s) \geq 0, \; j \in B(s), s \in S_3 \tag{25.15}$$

$$\eta_s^\beta - \beta \sum_{s' \in S_1 \cup S_2} v_{s'}^\beta q_i^1(s, s') - \beta \sum_{s' \in S_3} (\zeta_{s'}^\beta + \eta_{s'}^\beta) q_i^1(s, s')$$

$$-r_i^1(s) \geq 0, \; i \in A(s), s \in S_3 \tag{25.16}$$

$$f_i^\beta(s)[v_s^\beta - \beta \sum_{s' \in S_1 \cup S_2} v_{s'}^\beta q_i(s, s') - \beta \sum_{s' \in S_3} (\zeta_{s'}^\beta + \eta_{s'}^\beta) q_i(s, s')$$

$$-[R(s) g^\beta(s)]_i] = 0, \; i \in A(s), s \in S_1 \tag{25.17}$$

$$f_i^\beta(s)[v_s^\beta - \theta_s^\beta - [R(s) g^\beta(s)]_i] = 0, \; i \in A(s) \; s \in S_2 \tag{25.18}$$

$$g_j^\beta(s)[-v_s^\beta + \theta_s^\beta + [f^\beta(s) R(s)]_j] = 0, \; j \in B(s) \; s \in S_1 \tag{25.19}$$

$$g_j^\beta(s)[-v_s^\beta + \beta \sum_{s' \in S_1 \cup S_2} v_{s'}^\beta q_j(s,s') + \beta \sum_{s' \in S_3} (\zeta_{s'}^\beta + \eta_{s'}^\beta) q_j(s,s')$$

$$+ [f^\beta(s) R(s)]_j] = 0,\ j \in B(s), s \in S_2 \tag{25.20}$$

$$g_j^\beta(s)[-\zeta_s^\beta + \beta \sum_{s' \in S_1 \cup S_2} v_{s'}^\beta q_j^2(s,s') + \beta \sum_{s' \in S_3} (\zeta_{s'}^\beta + \eta_{s'}^\beta) q_j^2(s,s')$$

$$+ r_j^2(s)] = 0,\ j \in B(s), s \in S_3 \tag{25.21}$$

$$f_i^\beta(s)[\eta_s^\beta - \beta \sum_{s' \in S_1 \cup S_2} v_{s'}^\beta q_i^1(s,s') - \beta \sum_{s' \in S_3} (\zeta_{s'}^\beta + \eta_{s'}^\beta) q_i^1(s,s')$$

$$- r_i^1(s)] = 0,\ i \in A(s), s \in S_3 \tag{25.22}$$

Proof. We prove this theorem by showing that a feasible solution to SYS3 is a solution of SYS1 and by Theorem 25.2, this solution solves the stochastic game with SC/AR-AT structure. Conversely, we show that any solution of SYS1 can be used to construct a solution of SYS3. For $s \in S_1 \cup S_2$, we follow the similar argument of the proof given in [Schultz (1992)][Theorem 2.1]. However, we provide the details for the sake completeness.

From (25.17) and (25.19) we get

$$v_s^\beta - \beta \sum_{s' \in S} \sum_{i \in A(s)} v_{s'}^\beta q_i(s,s') f_i^\beta(s) - f^\beta(s) R(s) g^\beta(s) = 0,\ s \in S_1 \tag{25.23}$$

$$-v_s^\beta + \theta_s^\beta + f^\beta(s) R(s) g^\beta(s) = 0,\ s \in S_1 \tag{25.24}$$

Now (25.23) and (25.24) together imply

$$\theta_s^\beta = \beta \sum_{s' \in S} \sum_{i \in A(s)} v_{s'}^\beta q_i(s,s') f_i^\beta(s),\ s \in S_1 \tag{25.25}$$

Similarly, from (25.18) and (25.20) we get

$$\theta_s^\beta = \beta \sum_{s' \in S} \sum_{j \in B(s)} v_{s'}^\beta q_j(s,s') g_j^\beta(s),\ s \in S_2 \tag{25.26}$$

From (25.26), (25.11), (25.12) and (25.25), (25.13), (25.14) we get (25.3) and (25.4) respectively for $s \in S_1 \cup S_2$.

For $s \in S_3$, using (25.21), (25.22) and noting that $v_s^\beta = \zeta_s^\beta + \eta_s^\beta$ we obtain

$$\zeta_s^\beta - \beta \sum_{s' \in S} \sum_{j=1}^{n_s} v_{s'}^\beta q_j^2(s,s') g_j^\beta(s) - \sum_{j \in B(s)} r_j^2(s) g_j^\beta(s) = 0 \qquad (25.27)$$

$$\eta_s^\beta - \beta \sum_{s' \in S} \sum_{i=1}^{m_s} v_{s'}^\beta q_i^1(s,s') f_i^\beta(s) - \sum_{i \in A(s)} r_i^1(s) f_i^\beta(s) = 0 \qquad (25.28)$$

Adding (25.16) and (25.27) we get the inequality (25.3) for $s \in S_3$.

$$v_s^\beta - \beta \sum_{s' \in S} \sum_{j=1}^{n_s} v_{s'}^\beta [q_i^1(s,s') + q_j^2(s,s')] g_j^\beta(s) - \sum_{j \in B(s)} [r_i^1(s) + r_j^2(s)] g_j^\beta(s) \geq 0$$

$\Rightarrow$

$$v_s^\beta - \beta \sum_{s' \in S} \sum_{j=1}^{n_s} v_{s'}^\beta q_{ij}(s,s') g_j^\beta(s) - \sum_{j \in B(s)} r(s,i,j) g_j^\beta(s) \geq 0, \; s \in S_3, \; i \in A(s)$$

$$(25.29)$$

Similarly, adding (25.15) and (25.28) we obtain inequality (25.4) for $j \in B(s), s \in S_3$ of SYS1. Therefore, by Theorem 25.2, $v_s^\beta, f^\beta(s), g^\beta(s)$ is optimal strategy for SYS3.

Conversely, from any solution $(v_s^\beta, \theta_s^\beta, f^\beta(s), g^\beta(s))$ for $s \in S_1$ and $s \in S_2$ of SYS1 we define θ_s^β as in (25.25), (25.26). Rewriting SYS1 using the switching control assumption, we get the inequalities (25.11) through (25.14) and (25.17) through (25.20). Similarly, from any solution $(v_s^\beta, \theta_s^\beta, f^\beta(s), g^\beta(s))$ of SYS1 for $s \in S_3$, we write $v_s^\beta = \zeta_s^\beta + \eta_s^\beta$ and define $\zeta_s^\beta, \eta_s^\beta$ as in (25.27) and (25.28). Using the AR-AT structure, we rewrite SYS1 to get the inequalities (25.15), (25.16) (25.21)and (25.22) of SYS3. Therefore any solution $v_s^\beta, f^\beta(s), g^\beta(s)$ of SYS1 can be used to construct a solution of SYS3. □

Lemma 25.1. *For a β-discounted zero-sum SC/AR-AT mixture stochastic game has values v_s^β for $s \in S$ and optimal stationary strategies $f(s)$ and $g(s)$ for $s \in S$ if and only if v_s^β, $f^\beta(s)$ and $g^\beta(s)$ are a part of a solution of a LCP given by (25.30) through (25.49) where*

$$v_s^\beta = \bar{v}_s^\beta - \hat{v}_s^\beta, \; s \in S_1 \cup S_2$$

$$\theta_s^\beta = \bar{\theta}_s - \hat{\theta}_s, \; s \in S_1 \cup S_2$$

$$\zeta_s^\beta = \bar{\zeta}_s^\beta - \hat{\zeta}_s^\beta, \;\; \eta_s^\beta = \bar{\eta}_s^\beta - \hat{\eta}_s^\beta, \; s \in S_3.$$

Proof. It is enough to show that SYS3 in Theorem 25.5 can be written as a LCP.

First we consider the inequalities (25.11),(25.12) and (25.16). Let

$$w_1^f(s,i) = v_s^\beta - \beta \sum_{s' \in S} v_{s'}^\beta q_i(s,s') - [R(s)g^\beta(s)]_i \geq 0,\ i \in A(s),\ s \in S_1 \tag{25.30}$$

$$w_2^f(s,i) = v_s^\beta - \theta_s^\beta - [R(s)g^\beta(s)]_i \geq 0,\ i \in A(s)\ s \in S_2 \tag{25.31}$$

$$w_3^f(s,i) = \eta_s^\beta - \beta \sum_{s' \in S_1 \cup S_2} v_{s'}^\beta q_i^1(s,s') - \beta \sum_{s' \in S_3} (\zeta_{s'}^\beta + \eta_{s'}^\beta) q_i^1(s,s')$$

$$-r_i^1(s) \geq 0,\ i \in A(s), s \in S_3 \tag{25.32}$$

Then we consider the inequalities (25.13),(25.14) and (25.15). Let

$$w_1^g(s,j) = -v_s^\beta + \theta_s^\beta + [f^\beta(s)R(s)]_j \geq 0,\ j \in B(s)\ s \in S_1 \tag{25.33}$$

$$w_2^g(s,j) = -v_s^\beta + \beta \sum_{s' \in S} v_{s'}^\beta q_j(s,s') + [f^\beta(s)R(s)]_j \geq 0,\ j \in B(s), s \in S_2 \tag{25.34}$$

$$w_3^g(s,j) = -\zeta_s^\beta + \beta \sum_{s' \in S_1 \cup S_2} v_{s'}^\beta q_j^2(s,s') + \beta \sum_{s' \in S_3} (\zeta_{s'}^\beta + \eta_{s'}^\beta) q_j^2(s,s')$$

$$+r_j^2(s) \geq 0,\ j \in B(s), s \in S_3 \tag{25.35}$$

Now we express the variables ζ_s^β, η_s^β and θ_s^β as difference of nonnegative variables as a standard method of representing unbounded variables, i.e.,

$$v_s^\beta = \bar{v}_s^\beta - \hat{v}_s^\beta,\ s \in S_1 \cup S_2$$

$$\theta_s^\beta = \bar{\theta}_s - \hat{\theta}_s,\ s \in S_1 \cup S_2$$

$$\zeta_s^\beta = \bar{\zeta}_s^\beta - \hat{\zeta}_s^\beta,\ \eta_s^\beta = \bar{\eta}_s^\beta - \hat{\eta}_s^\beta,\ s \in S_3$$

Now we write down the constraints pertaining to probability vector $f(s)$ and $g(s)$ as follows.

$$\bar{w}^v(s) = -1 + \sum_{i \in A(s)} f_i^\beta(s) \geq 0, s \in S_1 \cup S_2 \tag{25.36}$$

$$\hat{w}^v(s) = 1 - \sum_{i \in A(s)} f_i^\beta(s) \geq 0, s \in S_1 \cup S_2 \tag{25.37}$$

$$\bar{w}^{\theta}(s) = -1 + \sum_{j \in B(s)} g_j^{\beta}(s) \geq 0, s \in S_1 \cup S_2 \tag{25.38}$$

$$\hat{w}^{\theta}(s) = 1 - \sum_{j \in B(s)} g_j^{\beta}(s) \geq 0, s \in S_1 \cup S_2 \tag{25.39}$$

$$\bar{w}^{\zeta}(s) = -1 + \sum_{i \in A(s)} f_i^{\beta}(s) \geq 0, s \in S_3 \tag{25.40}$$

$$\hat{w}^{\zeta}(s) = 1 - \sum_{i \in A(s)} f_i^{\beta}(s) \geq 0, s \in S_3 \tag{25.41}$$

$$\bar{w}^{\eta}(s) = -1 + \sum_{j \in B(s)} g_j^{\beta}(s) \geq 0, s \in S_3 \tag{25.42}$$

$$\hat{w}^{\eta}(s) = 1 - \sum_{j \in B(s)} g_j^{\beta}(s) \geq 0, s \in S_3 \tag{25.43}$$

We write the complementarity condition as

$$\left.\begin{array}{l} f_i^{\beta}(s) w_1^f(s,i) = 0, \ i \in A(s), s \in S_1 \\ f_i^{\beta}(s) w_2^f(s,i) = 0, \ i \in A(s), s \in S_2 \\ f_i^{\beta}(s) w_3^f(s,i) = 0, \ i \in A(s), s \in S_3 \end{array}\right\} \tag{25.44}$$

$$\left.\begin{array}{l} g_j^{\beta}(s) w_1^g(s,j) = 0, \ j \in B(s), s \in S_1 \\ g_j^{\beta}(s) w_2^g(s,j) = 0, \ j \in B(s), s \in S_2 \\ g_j^{\beta}(s) w_3^j(s,j) = 0, \ j \in A(s), s \in S_3 \end{array}\right\} \tag{25.45}$$

$$\left.\begin{array}{l} \bar{v}_s^{\beta} \bar{w}^{\zeta}(s) = 0, s \in S_1 \cup S_2 \\ \hat{v}_s^{\beta} \hat{w}^{\zeta}(s) = 0, s \in S_1 \cup S_2 \end{array}\right\} \tag{25.46}$$

$$\left.\begin{array}{l} \bar{\theta}_s^{\beta} \bar{w}^{\theta}(s) = 0, s \in S_1 \cup S_2 \\ \hat{\theta}_s^{\beta} \hat{w}^{\theta}(s) = 0, s \in S_1 \cup S_2 \end{array}\right\} \tag{25.47}$$

$$\left.\begin{array}{l} \bar{\eta}_s^{\beta} \bar{w}^{\eta}(s) = 0, s \in S_3 \\ \hat{\eta}_s^{\beta} \hat{w}^{\eta}(s) = 0, s \in S_3 \end{array}\right\} \tag{25.48}$$

$$\left.\begin{array}{l} \bar{\zeta}_s^{\beta} \bar{w}^{\zeta}(s) = 0, s \in S_3 \\ \hat{\zeta}_s^{\beta} \hat{w}^{\zeta}(s) = 0, s \in S_3 \end{array}\right\} \tag{25.49}$$

The LCP is given by (25.30) through (25.49). □

25.4 Undiscounted Zero-sum SC/AR-AT Mixture Stochastic Game

We require the following lemma which was proved by [Filar and Schultz (1987)].

Lemma 25.2. *([Filar and Schultz (1987)][Lemma 2.4])*
(i) If $(v^*, t^*, u^*, f^*, g^*)$ satisfy SYS1a, then for all $s \in S$

$$v_s^* = [P(f^*, g^*)v^*]_s$$

(ii) If $(v^*, t^*, u^*, f^*, g^*)$ solves SYS1b, then for all $s \in S$

$$v_s^* + t_s^* = [P(f^*, g^*)t^* + r(f^*, g^*)]_s$$

Theorem 25.6. *For an undiscounted zero-sum SC/AR-AT mixture stochastic game, the value vector and an optimal pair of stationary strategies can be derived from any solution to the following system of linear and nonlinear inequalities (SYS4). Conversely, for such a game, a solution of the SYS4 can be derived from any pair of asymptotically stable stationary strategies.*

SYS4: *Find $(v, t, \rho^1, \rho^2, \theta, \eta, \phi, \gamma, f, g)$ where $v, t, \in R^{|S|}$, $\rho^1, \rho^2 \in R^{|S_1 \cup S_2|}$, $\theta, \eta, \phi, \gamma \in R^{|S_3|}$, $f \in F_S$ and $g \in G_S$ such that*

$$v_s - \sum_{s' \in S} v_{s'} q_i(s, s') \geq 0, \; i \in A(s), s \in S_1 \tag{25.50}$$

$$-v_s + \rho_s^1 \geq 0, \; s \in S_1 \tag{25.51}$$

$$v_s + t_s - \sum_{s' \in S} t_{s'} q_i(s, s') - [R(s)g(s)]_i \geq 0, \; i \in A(s), s \in S_1 \tag{25.52}$$

$$-v_s - t_s + \rho_s^2 + [f(s)R(s)]_j \geq 0, \; j \in B(s), s \in S_1 \tag{25.53}$$

$$-v_s + \sum_{s' \in S} v_{s'} q_j(s, s') \geq 0, \; j \in B(s), s \in S_2 \tag{25.54}$$

$$v_s - \rho_s^1 \geq 0, \; s \in S_2 \tag{25.55}$$

$$v_s + t_s - \rho_s^2 - [R(s)g(s)]_i \geq 0, \; i \in A(s), s \in S_2 \tag{25.56}$$

$$-v_s - t_s + \sum_{s' \in S} t_{s'} q_j(s, s') + [f(s)R(s)]_j \geq 0, \; j \in B(s), s \in S_2 \tag{25.57}$$

$$\phi_s - \sum_{s'=1}^{N}(\theta_{s'} + \phi_{s'})q_i^1(s,s') \geq 0,\ i \in A(s), s \in S_3 \tag{25.58}$$

$$\gamma_s - \sum_{s'=1}^{N}(\eta_{s'} + \gamma_{s'} - \theta_{s'} - \phi_{s'})q_i^1(s,s') - r_i^1(s) \geq 0,\ i \in A(s), s \in S_3 \tag{25.59}$$

$$-\theta_s + \sum_{s'=1}^{N}(\theta_{s'} + \phi_{s'})q_j^2(s,s') \geq 0,\ j \in B(s), s \in S_3 \tag{25.60}$$

$$-\eta_s + \sum_{s'=1}^{N}(\eta_{s'} + \gamma_{s'} - \theta_{s'} - \phi_{s'})q_j^2(s,s') + r_j^2(s) \geq 0,\ j \in B(s), s \in S_3 \tag{25.61}$$

$$f_i(s)[v_s - \sum_{s' \in S} v_{s'} q_i(s,s')] = 0,\ i \in A(s), s \in S_1 \tag{25.62}$$

$$f_i(s)[-v_s + \rho_s^1] = 0,\ s \in S_1,\ i \in A(s) \tag{25.63}$$

$$f_i(s)[v_s + t_s - \sum_{s' \in S} t_{s'} q_i(s,s') - [R(s)g(s)]_i] = 0,\ i \in A(s), s \in S_1 \tag{25.64}$$

$$g_j(s)[-v_s - t_s + \rho_s^2 + [f(s)R(s)]_j] = 0,\ j \in B(s), s \in S_1 \tag{25.65}$$

$$g_j(s)[v_s - \rho_s^1] = 0,\ s \in S_2, j \in B(s) \tag{25.66}$$

$$g_j(s)[-v_s + \sum_{s' \in S} v_{s'} q_j(s,s')] = 0,\ j \in B(s), s \in S_2 \tag{25.67}$$

$$f_i(s)[v_s + t_s - \rho_s^2 - [R(s)g(s)]_i] = 0,\ i \in A(s), s \in S_2 \tag{25.68}$$

$$g_j(s)[-v_s - t_s + \sum_{s' \in S} t_{s'} q_j(s,s') + [f(s)R(s)]_j] = 0,\ j \in B(s), s \in S_2 \tag{25.69}$$

$$f_i(s)[\phi_s - \sum_{s'=1}^{N}(\theta_{s'} + \phi_{s'})q_i^1(s,s')] = 0,\ i \in A(s), s \in S_3 \tag{25.70}$$

$$f_i(s)[\gamma_s - \sum_{s'=1}^{N}(\eta_{s'} + \gamma_{s'} - \theta_{s'} - \phi_{s'})q_i^1(s,s') - r_i^1(s)] = 0,\ i \in A(s), s \in S_3 \tag{25.71}$$

$$g_j(s)[-\theta_s + \sum_{s'=1}^{N} (\theta_{s'} + \phi_{s'}) q_j^2(s, s')] = 0, \; j \in B(s), s \in S_3 \tag{25.72}$$

$$g_j(s)[-\eta_s + \sum_{s'=1}^{N} (\eta_{s'} + \gamma_{s'} - \theta_{s'} - \phi_{s'}) q_j^2(s, s') + r_j^2(s)] = 0, \; j \in B(s), s \in S_3 \tag{25.73}$$

$$f \in F_s, \; g \in G_s \tag{25.74}$$

Proof. We prove this theorem by showing that a feasible solution to SYS4 can be used to derive a solution of SYS2b and by Theorem 25.4, it follows that this solution solves the undiscounted SC/AR-AT mixture stochastic game. Conversely, we show that any solution of SYS2b can be used to construct a solution of SYS4. For $s \in S_1 \cup S_2$, we follow a similar argument of the proof given in [Filar and Schultz (1987)][Theorem 3.1, 4.1]. Let $z^* = (v^*, t^*, \rho^{1*}, \rho^{2*}, \theta^*, \eta^*, \phi^*, \gamma^*, f^*, g^*)$ be a feasible solution of the SYS4. From (25.62) through (25.69) we get

$$\rho_s^{1*} = \begin{cases} \sum_{s' \in S} \sum_{i=1}^{m_s} v_{s'}^* q_i(s, s') f_i^*(s), \; s \in S_1 \\ \sum_{s' \in S} \sum_{j=1}^{n_s} v_{s'}^* q_j(s, s') g_j^*(s), \; s \in S_2 \end{cases} \tag{25.75}$$

$$\rho_s^{2*} = \begin{cases} \sum_{s' \in S} \sum_{i=1}^{m_s} t_{s'}^* q_i(s, s') f_i^*(s), \; s \in S_1 \\ \sum_{s' \in S} \sum_{j=1}^{n_s} t_{s'}^* q_j(s, s') g_j^*(s), \; s \in S_2 \end{cases} \tag{25.76}$$

Now substituting the value of ρ_s^{1*} and ρ_s^{2*} in the system of inequalities (25.50) through (25.57) we get the system of inequalities in SYS2b. Note that the inequalities (25.50) and (25.55) yield after substitution

$$v_s^* - \sum_{s' \in S} v_{s'}^* q_i(s, s') [\sum_{j=1}^{n_s} g_j^*(s)] \geq 0, \; i \in A(s), s \in S_1$$

$$\text{i.e.,} v_s^* - \sum_{s' \in S} v_{s'}^* \sum_{j=1}^{n_s} q_i(s, s') g_j^*(s) \geq 0, \; i \in A(s), s \in S_1$$

since $\sum_{j=1}^{n_s} g_j^*(s) = 1$. Substituting ρ_s^1 in (25.55) and combining with the above using the definition of a switching control game we get

$$\text{i.e.,} v_s^* - \sum_{s' \in S} v_{s'}^* \sum_{j=1}^{n_s} q_{i,j}(s,s') g_j^*(s) \geq 0, \; i \in A(s), s \in S_1 \cup S_2$$

which is same as (25.6). Similarly inequalities (25.7), (25.8) and (25.10) can be obtained.

We define

$$v_s^* = \theta_s^* + \phi_s^* \;\text{ for } s \in S_3 \tag{25.77}$$

$$t_s^* = \eta_s^* + \gamma_s^* - \theta_s^* - \phi_s^* \;\text{ for } s \in S_3 \tag{25.78}$$

From (25.77) and (25.78) we get

$$\eta_s^* + \gamma_s^* = v_s^* + t_s^* \;\text{ for } s \in S_3$$

Substituting v_s^* for $(\theta_s^* + \phi_s^*)$ and $(v_s^* + t_s^*)$ for $(\eta_s^* + \gamma_s^*)$ in (25.58) through (25.61) and (25.70) through (25.73) we get

$$\phi_s^* - \sum_{s'=1}^{N} v_{s'}^* q_i^1(s,s') \geq 0, \; i \in A(s), s \in S_3 \tag{25.79}$$

$$\gamma_s^* - \sum_{s'=1}^{N} t_{s'}^* q_i^1(s,s') - r_i^1(s) \geq 0, \; i \in A(s), s \in S_3 \tag{25.80}$$

$$-\theta_s^* + \sum_{s'=1}^{N} v_{s'}^* q_j^2(s,s') \geq 0, \; j \in B(s), s \in S_3 \tag{25.81}$$

$$-\eta_s^* + \sum_{s'=1}^{N} t_{s'}^* q_j^2(s,s') + r_j^2(s) \geq 0, \; j \in B(s), s \in S_3 \tag{25.82}$$

$$\phi_s^* = \sum_{s'=1}^{N} \sum_{i=1}^{m_s} v_{s'}^* q_i^1(s,s') f_i^*(s), \; s \in S_3 \tag{25.83}$$

$$\gamma_s^* = \sum_{s'=1}^{N} \sum_{j=1}^{n_s} t_{s'}^* q_i^1(s,s') f_i^*(s) + \sum_{i=1}^{m_s} r_i^1(s) f_i^*(s), \; s \in S_3 \tag{25.84}$$

$$\theta_s^* = \sum_{s'=1}^{N} \sum_{j=1}^{n_s} v_{s'}^* q_j^2(s,s') g_j^*(s), \;\; s \in S_3 \tag{25.85}$$

$$\eta_s^* = \sum_{s'=1}^{N} \sum_{j=1}^{n_s} t_{s'}^* q_j^2(s,s') g_j^*(s) + \sum_{j=1}^{n_s} r_j^2(s) g_j^*(s), \;\; s \in S_3 \tag{25.86}$$

Adding (25.79) and (25.85) we get

$$\theta_s^* + \phi_s^* - \sum_{s'=1}^{N} \sum_{j=1}^{n_s} v_{s'}^* q_j^2(s,s') g_j^*(s) - \sum_{s'=1}^{N} v_{s'}^* q_i^1(s,s') \geq 0, \; i \in A(s), \;\; s \in S_3 \tag{25.87}$$

Therefore

$$\theta_s^* + \phi_s^* - \sum_{s'=1}^{N} v_{s'}^* \sum_{j=1}^{n_s} [q_j^2(s,s') g_j^*(s) + q_i^1(s,s') g_j^*(s)] \geq 0, \; i \in A(s), \;\; s \in S_3 \tag{25.88}$$

Substituting v_s^* for $(\theta_s^* + \phi_s^*)$ we get (25.6).

$$v_s^* - \sum_{s'=1}^{N} \sum_{j=1}^{n_s} v_{s'}^* q_{ij}(s,s') g_j^*(s) \geq 0, \; i \in A(s), \;\; s \in S_3 \tag{25.89}$$

Adding (25.80) and (25.86) we get (25.7).

$$\eta_s^* + \gamma_s^* - \sum_{s'=1}^{N} t_{s'}^* [\sum_{j=1}^{n_s} q_j^2(s,s') + q_i^1(s,s')] g_j^*(s) - \sum_{j=1}^{n_s} [r_j^2(s) + r_i^1(s)] g_j^*(s) \geq 0,$$
$$i \in A(s), s \in S \tag{25.90}$$

This implies

$$v_s^* + t_s^* - \sum_{s'=1}^{N} t_{s'}^* \sum_{j=1}^{n_s} q_{ij}(s,s') g_j^*(s) - [R(s)g(s)]_i \geq 0, \; i \in A(s), s \in S \tag{25.91}$$

Subtracting (25.83) from (25.81) and subtracting (25.84) from (25.82) we get (25.8) and (25.10) respectively. Since $f \in F_s$ and $g \in G_s$ the variables satisfy SYS2b and by Theorem 25.4, this yields an optimal solution to undiscounted SC/AR-AT mixture stochastic game.

To prove the converse, we show that any solution to SYS2b which always exists for these games, since they possess asymptotically stable optimal stationary strategies can be used to derive a feasible solution for SYS4. Assume that (v^*, t^*, f^*, g^*) be a feasible solution of the SYS2b. We define ρ_s^1, ρ_s^2 as in (25.75), (25.76). Rewriting SYS2b using the switching control assumption and using (25.75), (25.76) we get (25.50) through (25.57). Using (25.75), (25.76) and (25.50) through (25.57) we get (25.62) through (25.69).

From (25.6), (25.7), (25.8) and (25.10) and using the definition of AR-AT game we get

$$v_s^* - \sum_{s'=1}^{N}\sum_{j=1}^{n_s} v_{s'}^* q_j^2(s,s') g_j^*(s) - \sum_{s'=1}^{N} v_{s'}^* q_i^1(s,s') \geq 0,\ i \in A(s),\ s \in S_3 \tag{25.92}$$

$$v_s^* + t_s^* - \sum_{s'=1}^{N}\sum_{j=1}^{n_s} t_{s'}^* q_j^2(s,s') g_j^*(s) - \sum_{s'=1}^{N} t_{s'}^* q_i^1(s,s') - \sum_{j=1}^{n_s} r_j^2(s) g_j^*(s)$$

$$-r_i^1(s) \geq 0,\ i \in A(s),\ s \in S_3 \tag{25.93}$$

$$-v_s^* + \sum_{s'=1}^{N}\sum_{i=1}^{m_s} v_{s'}^* q_i^1(s,s') f_i^*(s) + \sum_{s'=1}^{N} v_{s'}^* q_j^2(s,s') \geq 0,\ j \in B(s),\ s \in S_3 \tag{25.94}$$

$$-v_s^* - t_s^* + \sum_{s'=1}^{N}\sum_{i=1}^{m_s} t_{s'}^* q_i^1(s,s') f_i^*(s) + \sum_{s'=1}^{N} t_{s'}^* q_j^2(s,s') + \sum_{i=1}^{m_s} r_i^1(s) f_i^*(s)$$

$$+r_j^2(s) \geq 0,\ j \in B(s),\ s \in S_3 \tag{25.95}$$

Take $\theta_s^*, \eta_s^*, \phi_s^*$ and γ_s^* for $s \in S_3$ as in (25.83) through (25.86). Adding (25.83) and (25.85) we get

$$\theta_s^* + \phi_s^* = \sum_{s'=1}^{N} v_{s'}^* [\sum_{i=1}^{m_s} q_i^1(s,s') f_i^*(s) + \sum_{j=1}^{n_s} q_j^2(s,s') g_j^*(s)]$$

$$= [P(f^*, g^*) v^*]_s = v_s^* \tag{25.96}$$

by Lemma 25.2 (i). Similarly, using Lemma 25.2(ii) and from (25.84) and (25.86) we get

$$\eta_s^* + \gamma_s^* = [P(f^*, g^*) t^* + r(f^*, g^*)]_s = v_s^* + t_s^* \tag{25.97}$$

From (25.92), (25.96) and using the definition of θ_s^* in (25.85) we get (25.58).

$$\theta_s^* + \phi_s^* - \theta_s^* - \sum_{s'=1}^{N} (\theta_{s'}^* + \phi_{s'}^*) q_i^1(s,s') \geq 0,\ i \in A(s),\ s \in S_3 \tag{25.98}$$

From (25.93),(25.86),(25.96) and (25.97) we get (25.59) of SYS4. From (25.94), (25.96) and the definition of ϕ^* in (25.83) yields (25.60) of SYS4.

$$-\theta_s^* - \phi_s^* + \sum_{s'=1}^{N}(\theta_{s'}^* + \phi_{s'}^*)q_j^2(s,s') + \phi_s^* \geq 0,\ j \in B(s),\ s \in S_3 \qquad (25.99)$$

Similarly from (25.95), (25.96), (25.97) and (25.84) we get (25.61) of SYS4. From (25.83) through (25.86), (25.96) and (25.97), we get (25.70) through (25.73). Since, $f \in F_s$ and $g \in G_s$, we obtain a feasible solution of SYS4□

Lemma 25.3. *An undiscounted zero-sum SC/AR-AT mixture stochastic game has values v_s for $s \in S$ and optimal stationary strategies $f(s)$ and $g(s)$ for $s \in S$ if and only if v_s^β, $f^\beta(s)$ and $g^\beta(s)$ are a part of the solution of a VLCP given by (25.100) through (25.127).*

Proof. It is enough to show that SYS4 in Theorem 25.6 can be written as a VLCP.

First we consider the inequalities (25.50), (25.51), (25.52),(25.56), (25.58) and (25.59). Let

$$w_1^f(s,i) = v_s - \sum_{s' \in S} v_{s'} q_i(s,s') \geq 0,\ i \in A(s), s \in S_1 \qquad (25.100)$$

$$w_2^f(s,i) = -v_s + \rho_s^1 \geq 0,\ s \in S_1 \qquad (25.101)$$

$$w_3^f(s,i) = v_s + t_s - \sum_{s' \in S} t_{s'} q_i(s,s') - [R(s)g(s)]_i \geq 0,\ i \in A(s), s \in S_1 \qquad (25.102)$$

Let $w_{c1}^f = w_1^f(s,i) + w_2^f(s,i) + w_3^f(s,i)$. Therefore

$$w_{c1}^f = v_s + t_s + \rho_s^1 - \sum_{s' \in S} v_{s'} q_i(s,s') - \sum_{s' \in S} t_{s'} q_i(s,s')$$
$$-[R(s)g(s)]_i \geq 0,\ i \in A(s), s \in S_1 \qquad (25.103)$$

$$w_4^f(s,i) = v_s + t_s - \rho_s^2 - [R(s)g(s)]_i \geq 0,\ i \in A(s), s \in S_2 \qquad (25.104)$$

$$w_5^f(s,i) = \phi_s - \sum_{s'=1}^{N}(\theta_{s'} + \phi_{s'})q_i^1(s,s') \geq 0,\ i \in A(s), s \in S_3 \qquad (25.105)$$

$$w_6^f(s,i) = \gamma_s - \sum_{s'=1}^{N}(\eta_{s'} + \gamma_{s'} - \theta_{s'} - \phi_{s'})q_i^1(s,s') - r_i^1(s) \geq 0,$$
$$i \in A(s), s \in S_3 \qquad (25.106)$$

Let $w^f_{c2} = w^f_5(s,i) + w^f_6(s,i)$. Therefore

$$w^f_{c2} = \phi_s + \gamma_s - \sum_{s'=1}^{N}(\theta_{s'} + \phi_{s'})q^1_i(s,s') - \sum_{s'=1}^{N}(\eta_{s'} + \gamma_{s'} - \theta_{s'} - \phi_{s'})q^1_i(s,s')$$

$$-r^1_i(s) \geq 0,\ i \in A(s), s \in S_3 \tag{25.107}$$

Then we consider the inequalities (25.53), (25.54), (25.55), (25.57), (25.60) and (25.61). Let

$$w^g_1(s,j) = -v_s - t_s + \rho^2_s + [f(s)R(s)]_j \geq 0,\ j \in B(s), s \in S_1 \tag{25.108}$$

$$w^g_2(s,j) = -v_s + \sum_{s' \in S} v_{s'}q_j(s,s') \geq 0,\ j \in B(s), s \in S_2 \tag{25.109}$$

$$w^g_3(s,j) = v_s - \rho^1_s \geq 0,\ s \in S_2 \tag{25.110}$$

$$w^g_4(s,j) = -v_s - t_s + \sum_{s' \in S} t_{s'}q_j(s,s') + [f(s)R(s)]_j \geq 0,\ j \in B(s), s \in S_2 \tag{25.111}$$

Let $w^g_{c1}(s,j) = w^g_2(s,j) + w^g_3(s,j) + w^g_4(s,j)$. Therefore

$$w^g_{c1}(s,j) = -v_s - t_s - \rho^1_s + \sum_{s' \in S} v_{s'}q_j(s,s') + \sum_{s' \in S} t_{s'}q_j(s,s')$$

$$+[f(s)R(s)]_j \geq 0,\ j \in B(s), s \in S_2 \tag{25.112}$$

$$w^g_5(s,j) = -\theta_s + \sum_{s'=1}^{N}(\theta_{s'} + \phi_{s'})q^2_j(s,s') \geq 0,\ j \in B(s), s \in S_3 \tag{25.113}$$

$$w^g_6(s,j) = -\eta_s + \sum_{s'=1}^{N}(\eta_{s'} + \gamma_{s'} - \theta_{s'} - \phi_{s'})q^2_j(s,s') + r^2_j(s) \geq 0,$$
$$j \in B(s), s \in S_3 \tag{25.114}$$

Let $w^g_{c2}(s,j) = w^g_5(s,j) + w^g_6(s,j)$. Therefore

$$w^g_{c2}(s,j) = -\theta_s - \eta_s + \sum_{s'=1}^{N}(\theta_{s'} + \phi_{s'})q^2_j(s,s')$$

$$+\sum_{s'=1}^{N}(\eta_{s'} + \gamma_{s'} - \theta_{s'} - \phi_{s'})q^2_j(s,s') + r^2_j(s) \geq 0,\ j \in B(s), s \in S_3 \tag{25.115}$$

Now we express the variables $v_s, t_s, \rho_s^1, \rho_s^2, \theta_s, \eta_s, \phi_s, \gamma_s$ as difference of nonnegative variables as a standard method of representing unbounded variables, i.e.,

$$\left.\begin{aligned}
v_s &= \bar{v}_s - \hat{v}_s,\ s \in S_1 \cup S_2 \\
t_s &= \bar{t}_s - \hat{t}_s,\ s \in S_1 \cup S_2 \\
\rho_s^1 &= \bar{\rho^1}_s - \hat{\rho^1}_s,\ s \in S_1 \cup S_2 \\
\rho_s^2 &= \bar{\rho^2}_s - \hat{\rho^2}_s,\ s \in S_1 \cup S_2 \\
\theta_s &= \bar{\theta}_s - \hat{\theta}_s,\ s \in S_3 \\
\eta_s &= \bar{\eta}_s - \hat{\eta}_s,\ s \in S_3 \\
\phi_s &= \bar{\phi}_s - \hat{\phi}_s,\ s \in S_3 \\
\gamma_s &= \bar{\gamma}_s - \hat{\gamma}_s,\ s \in S_3
\end{aligned}\right\}$$

Now we write down the constraints pertaining to probability vector $f(s)$ and $g(s)$ as follows.

$$\left.\begin{aligned}
\bar{w}^v(s) &= -1 + \sum_{i \in A(s)} f_i(s) \geq 0, s \in S_1 \cup S_2 \\
\hat{w}^v(s) &= 1 - \sum_{i \in A(s)} f_i(s) \geq 0, s \in S_1 \cup S_2
\end{aligned}\right\} \tag{25.116}$$

$$\left.\begin{aligned}
\bar{w}^t(s) &= -1 + \sum_{j \in B(s)} g_j(s) \geq 0, s \in S_1 \cup S_2 \\
\hat{w}^t(s) &= 1 - \sum_{j \in B(s)} g_j(s) \geq 0, s \in S_1 \cup S_2
\end{aligned}\right\} \tag{25.117}$$

$$\left.\begin{aligned}
\bar{w}^{\rho_1}(s) &= -2 + \sum_{i \in A(s)} f_i(s) + \sum_{j \in B(s)} g_j(s) \geq 0,\ s \in S_1 \cup S_2 \\
\hat{w}^{\rho_1}(s) &= 2 - \sum_{i \in A(s)} f_i(s) - \sum_{j \in B(s)} g_j(s) \geq 0,\ s \in S_1 \cup S_2
\end{aligned}\right\} \tag{25.118}$$

$$\left.\begin{aligned}
\bar{w}^{\rho_2}(s) &= \sum_{i \in A(s)} f_i(s) - \sum_{j \in B(s)} g_j(s) \geq 0,\ s \in S_1 \cup S_2 \\
\hat{w}^{\rho_2}(s) &= -\sum_{i \in A(s)} f_i(s) + \sum_{j \in B(s)} g_j(s) \geq 0,\ s \in S_1 \cup S_2
\end{aligned}\right\} \tag{25.119}$$

$$\left.\begin{aligned}
\bar{w}^{\theta}(s) &= -1 + \sum_{i \in A(s)} f_i(s) \geq 0,\ s \in S_3 \\
\hat{w}^{\theta}(s) &= 1 - \sum_{i \in A(s)} f_i(s) \geq 0,\ s \in S_3
\end{aligned}\right\} \tag{25.120}$$

$$\left.\begin{aligned} \bar{w}^{\eta}(s) &= -1 + \sum_{j\in B(s)} g_j(s) \geq 0,\ s \in S_3 \\ \hat{w}^{\eta}(s) &= 1 - \sum_{j\in B(s)} g_j(s) \geq 0,\ s \in S_3 \end{aligned}\right\} \tag{25.121}$$

$$\left.\begin{aligned} \bar{w}^{\phi}(s) &= -2 + \sum_{i\in A(s)} f_i(s) + \sum_{j\in B(s)} g_j(s) \geq 0,\ s \in S_3 \\ \hat{w}^{\phi}(s) &= 2 - \sum_{i\in A(s)} f_i(s) - \sum_{j\in B(s)} g_j(s) \geq 0,\ s \in S_3 \end{aligned}\right\} \tag{25.122}$$

$$\left.\begin{aligned} \bar{w}^{\gamma}(s) &= \sum_{i\in A(s)} f_i(s) - \sum_{j\in B(s)} g_j(s) \geq 0,\ s \in S_3 \\ \hat{w}^{\gamma}(s) &= -\sum_{i\in A(s)} f_i(s) + \sum_{j\in B(s)} g_j(s) \geq 0,\ s \in S_3 \end{aligned}\right\} \tag{25.123}$$

The complementarity conditions involving the inequalities related to the probability vector constraints are

$$\left.\begin{array}{l} f_i(s)w_1^f(s,i)w_2^f(s,i)w_3^f(s,i)w_{c1}^f(s,i) = 0,\ i \in A(s), s \in S_1 \\ f_i(s)w_4^f(s,i) = 0,\ i \in A(s), s \in S_2 \\ f_i(s)w_5^f(s,i)w_6^f(s,i)w_{c2}^f(s,i) = 0,\ i \in A(s), s \in S_3. \end{array}\right\} \tag{25.124}$$

$$\left.\begin{array}{l} g_j(s)w_1^g(s,j) = 0,\ i \in A(s), s \in S_1 \\ g_j(s)w_2^g(s,j)(s)w_3^g(s,j)w_4^g(s,j)w_{c1}^g(s,j) = 0,\ j \in B(s), s \in S_2 \\ g_j(s)w_5^g(s,j)w_6^g(s,j)w_{c2}^g(s,j) = 0,\ j \in B(s), s \in S_3. \end{array}\right\} \tag{25.125}$$

The complementarity conditions related to other variables are

$$\left.\begin{array}{l} \bar{v}_s.\bar{w}^v(s) = 0,\ s \in S_1 \cup S_2 \\ \hat{v}_s.\hat{w}^v(s) = 0,\ s \in S_1 \cup S_2 \\ \bar{t}_s.\bar{w}^t(s) = 0,\ s \in S_1 \cup S_2 \\ \hat{t}_s.\hat{w}^t(s) = 0,\ s \in S_1 \cup S_2 \\ \bar{\rho}_s^1.\bar{w}^{\rho^1}(s) = 0,\ s \in S_1 \cup S_2 \\ \hat{\rho}_s^1.\hat{w}^{\rho^1}(s) = 0,\ s \in S_1 \cup S_2 \\ \bar{\rho}_s^2.\bar{w}^{\rho^2}(s) = 0,\ s \in S_1 \cup S_2 \\ \hat{\rho}_s^2.\hat{w}^{\rho^2}(s) = 0,\ s \in S_1 \cup S_2 \end{array}\right\} \tag{25.126}$$

$$\left.\begin{array}{l}\bar{\theta}_s.\bar{w}^{\theta}(s)=0,\ \ s\in S_3\\ \hat{\theta}_s.\hat{w}^{\theta}(s)=0,\ \ s\in S_3\\ \bar{\eta}_s.\bar{w}^{\eta}(s)=0,\ \ s\in S_3\\ \hat{\eta}_s.\hat{w}^{\eta}(s)=0,\ \ s\in S_3\\ \bar{\phi}_s.\bar{w}^{\phi}(s)=0,\ \ s\in S_3\\ \hat{\phi}_s.\hat{w}^{\phi}(s)=0,\ \ s\in S_3\\ \bar{\gamma}_s.\bar{w}^{\gamma}(s)=0,\ \ s\in S_3\\ \hat{\gamma}_s.\hat{w}^{\gamma}(s)=0,\ \ s\in S_3\end{array}\right\} \quad (25.127)$$

The VLCP is given by (25.100) through (25.127). □

25.5 Computation of Value Vector and Optimal Stationary Strategies for SC/AR-AT Mixture Class of Stochastic Game

[Sinha (1989, 2000)] raises the question that whether a finite step algorithm can be developed for SC/AR-AT mixture class. The main results proved in this paper is the computation of optimal strategies and the value vector for both discounted and undiscounted SC/AR-AT mixture games as a complementarity problem. This is essentially the first step for developing finite step algorithm. This also gives an alternative proof for ordered field property. Investigation concerning the applicability of [Cottle and Dantzig (1970)] algorithm for solving the complementarity formulation presented in earlier section should be explored. While implementing the available pivoting algorithms on these two formulations for discounted and undiscounted case, perhaps special initialization scheme may be necessary and use of suitable degeneracy resolving mechanism may be needed. Alternatively one may use the neural network approach presented in [Neogy, Das and Das (2007)]. The computational results using neural network approach presented in this paper seems to be very encouraging.

Bibliography

Cottle, R. W. and Dantzig, G. B. (1970). A generalization of the linear complementarity problem, *Journal of Combinatorial Theory*, **8**, pp. 79–90.

Cottle, R. W., Pang, J. S. and Stone, R. E. (1992). *The Linear Complementarity Problem*, (Academic Press, New York).

Filar, J. A. (1981). Orderfield property for stochastic games when the player who controls transitions changes from state to state, *JOTA*, **34**, pp. 503–515.

Filar, J. A. and Schultz, T. A. (1987). Bilinear programming and structured stochastic games, *JOTA*, **53**, pp. 85–104.

Filar, J. A. and Vrieze, O. J. (1997). *Competitive Markov Decision Processes*, (Springer, New York).

Fink, A. M. (1964). Equilibrium in a stochastic n-person game, *J. Sci., Hiroshima Univ., Ser. A.*, **28**, pp. 89–93.

Gillette, D. (1957). Stochastic game with zero step probabilities, in *Theory of Games,* eds. A. W. Tucker, M. Dresher and P. Wolfe, (Princeton University Press, Princeton, New Jersey).

Kaplansky, I. (1945). A contribution to von Neumann's theory of games, *Annals of Mathematics*, **46**, pp. 474–479.

Lemke, C. E. (1965). Bimatrix equilibrium points and mathematical programming, *Management Science*, **11**, pp. 681–689.

Lemke, C. E. and Howson, J. T. (1964). Equilibrium points of bimatrix games, *SIAM J. Appl. Math.,* **12**, pp. 413–423.

Lemke, C. E. (1970). Recent results on complementarity problems, in: J.B. Rosen, O.L. Mangasarian and K. Ritter, ed. *Nonlinear Programming*, (Academic Press, New York), pp. 349–384.

Mohan, S. R. and Neogy, S. K. (1996b). Generalized linear complementarity in a problem of n person games, *OR Spektrum* **18**, pp. 231–239

Mohan, S. R., Neogy, S. K. and Parthasarathy, T. (1997a). Linear complementarity and discounted polystochastic game when one player controls transitions, in *Complementarity and Variational Problems*, eds: M.C. Ferris and Jong-Shi Pang, SIAM, Philadelphia, pp. 284–294.

Mohan, S. R., Neogy, S. K. and Parthasarathy, T. (1997b). Linear complementarity and the irreducible polystochastic game with the average cost criterion when one player controls transitions, in *Game theoretical applications to Economics and Operations Research*, eds. T. Parthasarathy, B. Dutta, J. A. M. Potters, T. E. S. Raghavan, D. Ray and A. Sen, (Kluwer Academic Publishers, Dordrecht, The Netherlands), pp. 153–170.

Mohan, S. R. and Parthasarathy, T. (1994). *Ordered field property in Stochastic games*, unpublished article, Indian Statistical Institute, New Delhi.

Mohan, S. R. and Raghavan, T. E. S. (1987). An algorithm for discounted switching control games, *OR Spektrum*, **9**, pp. 41–45.

Mohan, S. R., Neogy, S. K. and Parthasarathy, T. (2001). Pivoting algorithms for some classes of stochastic games: A survey, *International Game Theory Review*, **3**, pp. 253–281.

Mohan, S. R., Neogy, S. K., Parthasarathy, T. and Sinha, S. (1999). Vertical linear complementarity and discounted zero-sum stochastic games with ARAT structure, *Mathematical Programming*, Series A pp. 637–648.

Mohan, S. R., Sridhar, R. and Parthasarathy, T. (1992). $\bar{N}$ matrices and the class Q, in B. Dutta et al. (Eds), *Lecture Notes in Economics and Mathematical Systems*, **389**, pp. 24–36, (Springer Verlag, Berlin).

Murty, K. G. (1988). *Linear Complementarity, Linear and Nonlinear Programming,* (Heldermann Verlag, West Berlin).

Nash, J. F.(1951). Noncooperative games, *Ann. of Math.*, **54**, pp. 286–295.

Neogy, S. K. and Das A. K., (2005). Linear complementarity and two classes of structured stochastic games, in *Operations Research with Economic and Industrial Applications: Emerging Trends*, eds: S. R. Mohan and S. K. Neogy, Anamaya Publishers, New Delhi, India pp. 156–180.

Neogy, S. K., Das A. K. and Das, P. (2007.) Block Matrices and its applications in Complementarity and game theory, Submitted to Platinum jubilee volume.

von Neumann, J. and Morgenstern, O. (1944). *Theory of Games and Economic Behaviour*, (Princeton University Press, Princeton, NJ).

Nowak, A. S. and Raghavan, T. E. S. (1993). A finite step algorithm via a bi-matrix game to a single controller non-zerosum stochastic game, *Math. Programming*, **59**, pp. 249-259.

Parthasarathy, T. and Raghavan, T. E. S. (1981). An orderfield property for stochastic games when one player controls transition probabilities, *JOTA*, **33**, pp. 375–392.

Raghavan T. E. S.and Filar, J. A. (1991). Algorithms for stochastic games, a survey, *Zietch. Oper. Res.*, **35**, pp. 437–472.

Raghavan, T. E. S., Tijs, S. H. and Vrieze, O. J. (1985). On stochastic games with additive reward and transition structure, *JOTA*, **47**, pp. 375–392.

Sinha, S. (1989). *A contribution to the theory of Stochastic Games*, Ph.D thesis, Indian Statistical Institute, Delhi Centre.

Sinha, S. (2000). *A new class of Stochastic Games having Ordered field property*, Unpublished Manuscript, Jadavpur University, Kolkata.

Schultz, T. A. (1992). Linear complementarity and discounted switching controller stochastic games, *JOTA*, **73**, pp. 89–99.

Shapley, L. S. (1953). Stochastic games, *Proc. Nat. Acad. Sci. USA.*, **39**, pp. 1095–1100.

Vrieze, O. J. (1981). Linear programming and undiscounted games in which one player controls transition, *OR Spektrum*, **3**, pp. 29–35.

Vrieze, O. J. (1983). A finite algorithm for the switching controller stochastic game, *OR Spektrum*, **5**, pp. 15–24.

Weyl, H. (1950). Elementary Proof of a Minimax Theorem due to von Neumann, in: H.W. KUHN & A.W. TUCKER (Eds) *Contributions to the Theory of Games Vol.I, Annals of Mathematics Studies* **24**, pp. 19–25, (Princeton University Press, Princeton, N.J.).